The Froehlich / Kent

ENCYCLOPEDIA OF TELECOMMUNICATIONS

VOLUME 6

The Froehlich / Kent
ENCYCLOPEDIA OF TELECOMMUNICATIONS

Editor-in-Chief

Fritz E. Froehlich, Ph.D.
Professor of Telecommunications
University of Pittsburgh
Pittsburgh, Pennsylvania

Co-Editor

Allen Kent
Distinguished Service Professor of Information Science
University of Pittsburgh
Pittsburgh, Pennsylvania

Administrative Editor

Carolyn M. Hall
Pittsburgh, Pennsylvania

VOLUME 6

DIGITAL MICROWAVE LINK DESIGN to ELECTRICAL FILTERS

CRC Press
Taylor & Francis Group
Boca Raton London New York

CRC Press is an imprint of the
Taylor & Francis Group, an **informa** business

CRC Press
Taylor & Francis Group
6000 Broken Sound Parkway NW, Suite 300
Boca Raton, FL 33487-2742

First issued in paperback 2019

ISBN-13: 978-0-8247-2904-2 (hbk)
ISBN-13: 978-0-367-40244-0 (pbk)

Library of Congress Cataloging-in-Publication Data

The Froehlich/Kent Encyclopedia of Telecommunications / editor-in-chief, Fritz E.
Froehlich ; co-editor, Allen Kent.
    p. cm.
    Includes bibliographical references and indexes.
    ISBN 0-8247-2902-1 (v. 1 : alk. paper)
    1. Telecommunication—Encyclopedias. I. Froehlich, Fritz E.,
Kent, Allen.
    TK5102.E646 1990   90-3966
    384′.03—dc20   CIP

Visit the Taylor & Francis Web site at
http://www.taylorandfrancis.com

and the CRC Press Web site at
http://www.crcpress.com

CONTENTS OF VOLUME 6

CONTRIBUTORS TO VOLUME 6

Fumio Amano Section Manager, Radio and Satellite Communication, Systems Laboratory, Communication and Space Division, Fujitsu Laboratories Ltd., Kawasaki, Japan: *Echo-Canceling Algorithms*

Frederick T. Andrews Consultant, Telecommunications Network Strategies, Colts Neck, New Jersey: *Divestiture of Bell Operating Companies: Technical Challenges and Achievements*

Joyce E. Bedi Hagley Fellow, Department of History, University of Delaware, Newark, Delaware: *Dolbear, Amos Emerson*

Pierre J. Catala Telecom Course Leader, Engineering Technology Department, Texas A&M University, College Station, Texas: *Digital Microwave Link Design*

Jeffrey E. Cohen, Ph.D. Associate Professor of Political Science, University of Kansas, Lawrence, Kansas: *Divestiture Impact on Local Telephone Rate Policy in the United States: Diffusion of Local Measured Service, 1977–1985*

Jose M. Costa, Ph.D. BNR, Ottawa, Ontario, Canada: *DS1 Services and Standards*

Kamilo Feher, Ph.D. Professor, Electrical Engineering and Computer Science Department, University of California, Davis, California, and Director, Consulting Group, DIGCOM, Inc.: *Digital Modulation Techniques*

Shozo Komaki, Ph.D. Associate Professor, Osaka University, Suita, Osaka, Japan: *Digital Microwave Radio*

Michele D. Mathys Program Manager and Senior Systems Engineer, ESL Inc., Sunnyvale, California: *Digital Signal Processing*

George S. Moschytz, Ph.D. Professor and Director, Signal and Information Processing Laboratory, Swiss Federal Institute of Technology (ETH-Zentrum), Zurich, Switzerland: *Electrical Filters*

Kazuo Murano, Ph.D. Deputy General Manager, Communication and Space Division, Fujitsu Laboratories Ltd., Kawasaki, Japan: *Echo-Canceling Algorithms*

Takehiro Murase, Dr. Eng. Executive Manager, NTT Radio Communication Systems Laboratories, Yokosuka, Japan: *Digital Radio Systems*

Eli M. Noam, Ph.D. Director of Columbia Institute for Tele-Information, Professor of Finance and Economics, Columbia University, New York, New York: *Economic Theories of Regulation in Telecommunications*

Lisa D. Noble Product Manager, Spectral Innovations, Inc., Santa Clara, California: *Digital Signal Processing*

Mahmood R. Noorchashm, Ph.D. Member of Technical Staff, AT&T Bell Laboratories, Holmdel, New Jersey: *Digital Modulation and Coding*

M. Javad Peyrovian, Ph.D. Member of Technical Staff, AT&T Bell Laboratories, Holmdel, New Jersey: *Digital Modulation and Coding*

Lawrence R. Rabiner, Ph.D. Director, Information Principles Research Laboratory, AT&T Bell Laboratories, Murray Hill, New Jersey: *Digital Speech Processing*

John J. Rugo   Manager, Operations Systems, AT&T Network Systems, Holmdel, New Jersey: *Echo Suppressors*

John A. Stankovic, Ph.D.   Professor, Department of Computer Science, University of Massachusetts, Amherst, Massachusetts: *Distributed Computing*

Frank F. Taylor   Consultant, AT&T Bell Laboratories, Holmdel, New Jersey: *Digital Switching Systems*

John T. Wenders, Ph.D.   Professor, Department of Economics, University of Idaho, Moscow, Idaho: *Economic Theories of Regulation in Telecommunications*

Kuang-Tsan Wu, Ph.D.   Member of Scientific Staff, Bell Northern Research, Ottawa, Ontario, Canada: *Digital Modulation Techniques*

Digital Microwave Link Design

Introduction

Digital microwave links are becoming an integral part of the public networks of almost all countries. Furthermore, in the countries where telecommunications competition is allowed, many large corporations are using single-hop digital microwave links as a "bypass" connection to the "point of presence" of a long-distance telephone company.

This article describes the design of digital microwave links from a user's point of view as opposed to an equipment designer's point of view, and thus concentrates on how to calculate the "power budgets" for microwave links.

How Can Microwave Be Digital?

There is sometimes confusion as to what is meant by digital microwave. Since a radio wave cannot "propagate" direct current (DC) pulses directly, how can the radio be digital? In this case, however, *digital* refers to the baseband signal being transmitted, not to the radio frequency (RF) signal. The term also indicates that a specific "digital" modulation (phase-shift keying [PSK], quadrature amplitude modulation [QAM], etc.) is being used in the radio, as opposed to simply modulating existing frequency modulation (FM) transmitters.

The main difference between digital and analog radios is thus the type of modulation; the design of digital microwave links is very similar to that of their analog counterparts. The only difference is that the starting criterion for digital microwave links is the bit error rate (BER) instead of the signal-to-noise ratio (SNR). High-capacity digital radio links, in addition, will require special considerations because their broad RF spectrum and high-level modulations are very vulnerable to frequency selective multipath fading that creates intersymbol interference and thus increases the probability of errors.

Required Carrier Level

Analog link power budget calculations start from the maximum acceptable noise level in a voice channel (1). Equivalent digital link calculations start from the maximum BER that is acceptable in a digitized voice channel or digital data channel.

The BER is a function of the SNR at the input of the receiver's demodulator. The relationship between BER and SNR under Gaussian noise conditions can be calculated by using the complementary error function appropriate for the type and level of digital modulation being used. The result of these formulas usually is presented in the form of such probability of error curves as those shown in Fig. 1.

Since the starting point for calculations is the maximum acceptable BER, the carrier-to-noise (C/N) ratio would be obtained from Fig. 1 and then the required minimum carrier level would be calculated using

$$C = N + C/N = (-174 + 10 \log B + NF) + (C/N) \tag{1}$$

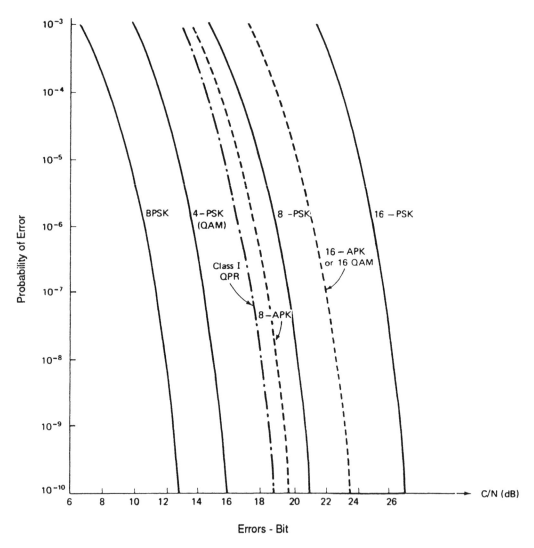

FIG. 1   Probability of error versus C/N for several digital modulations. The C/N is specified in the double-sided Nyquist bandwidth. (Adapted with permission from K. Feher, *Digital Communications: Microwave Applications*, Prentice-Hall, Englewood Cliffs, NJ, 1981, p. 71.)

where

$\quad C$ $\quad$ = the required carrier level in decibels relative to a milliwatt (dBm)

$\quad B$ $\quad$ = the intermediate frequency (IF) bandwidth in hertz (Hz)

$\quad NF$ = the receiver noise figure in decibels (dB)

$\quad C/N$ = the value, in dB, obtained from Fig. 1

Once the required received carrier level is known, the rest of the power budget calculations is the same as for an analog microwave link, that is, the needed antenna gains can be found from (1)

$$G_t + G_r = C - P_t + L_T + M \qquad (2)$$

where

$\quad G_t$ = the gain of the transmit antenna in dB

$\quad G_r$ = the gain of the receive antenna in dB

$\quad L_T$ = the sum of all losses (i.e., path, waveguides, etc.)

$\quad M$ = the required fade margin in dB

As a starting point, the gain is assumed to be the same at both ends; thus,

$$G_r = G_t = (G_t + G_r)/2$$

The diameter (in feet) of the parabolic antenna (dish) can be obtained from

$$D = (G/5.6F^2)^{1/2}$$

where

$\quad G$ = the numerical gain of the antenna (i.e., the inverse log of the value of G_r or G_t)

$\quad F$ = the frequency in gigahertz (GHz)

This equation assumes a 55% efficiency for the antenna, which is common for terrestrial microwave link antennas.

Example 1. A digital microwave link will use a 1-W (30-dBm), quaternary phase-shift keying (also known as 4PSK or QPSK), transmitter. The path loss is 140 dB, and the miscellaneous waveguide losses are 3.5 dB at each end. The performance requirements are a fade margin of 35 dB and a BER of 10^{-10}. What is the required gain of the antennas? Specifications of the receiver are $NF = 5$ dB, IF bandwidth $= 2.5$ megahertz (MHz).

Solution: From Fig. 1, a BER of 10^{-10} requires a C/N of 16 dB when QPSK modulation is used. From Eq. (1), the corresponding required carrier level is found:

$$C = -174 + 10 \log B + NF + C/N$$

$$= -174 + 64 + 5 + 16 = -89 \text{ dBm}$$

The overall antenna gain needed to produce the -89 dBm can be found from Eq. (2):

$$G_t + G_r = C - P_t + L_T + M$$
$$= -89 - 30 + (3.5 + 140 + 3.5) + 35 = 63 \text{ dB}$$

Equal gain can be assigned to each antenna thus:

$$G_t = G_r = 63/2 = 31.5 \text{ dB}$$

Energy per Bit and Noise Density

In digital systems, an E_b/N_0 ratio often is used rather than C/N. That ratio is based on the following parameters:

E_b is the energy per bit (i.e., the signal power at the input of the demodulator divided by the bit rate)

N_0 (pronounced "N zero") is the noise density (i.e., the noise power in a 1-Hz bandwidth). This is also the total noise power divided by the noise bandwidth of the receiver.

Therefore,

$$E_b/N_0 = (C/R)/N_0$$

or, in decibels,

$$E_b/N_0 = C - 10 \log R - N_0 \tag{3}$$

where
$\quad C \quad$ = the received carrier level in dBm
$\quad R \quad$ = the bit rate in bits per second (bps)
$\quad N_0 \quad$ = the noise density in dBm

BER curves such as the ones shown in Fig. 1 often have their horizontal axis labeled as E_b/N_0 rather than C/N.

The variable needed is usually C, so Eq. (3) is rewritten as

$$C = E_b/N_0 + 10 \log R + N_0$$

N_0 is related to the noise figure NF of the receiver by

$$N_0 = -174 \text{ dBm} + NF \text{ dB}$$

thus the required carrier level is given by

$$C = E_b/N_0 + 10 \log R - 174 + NF \tag{4}$$

and the value of C then can be entered in Eq. (2).

Practical Link Design

The BER curves shown in Fig. 1 represent the probability of error under a Gaussian noise condition but with otherwise ideal conditions: perfect (Nyquist) filters, no jitter, no intersymbol interference (ISI), and so on. In practical link design for real equipment, all the curves of Fig. 1 will be shifted to the right by an amount that depends on the actual radios being used. It therefore would be necessary to use BER curves given by the equipment manufacturer. Unfortunately, very few manufacturers actually provide those curves. Most equipment data sheets only show the required carrier level for one or two BER values (usually 10^{-3} and/or 10^{-6}). The level indicated then would be the value to be entered in Eq. (2).

The advantage, from the user's point of view, is that the data-sheet values eliminate the need for performing the calculations of Eqs. (1) or (4). If a different BER is required than the one shown in the data sheet, Fig. 1 can be used by entering the manufacturer's value of carrier level (for a given BER) on the curve of Fig. 1 that corresponds to the modulation used and "move" decibel for decibel on that curve to find the number of additional decibels needed to achieve the desired BER.

One must remember that the value of C given by the manufacturer is the "threshold" value of the received signal level (RSL), that is, its minimum value. It thus is necessary to add an appropriate fade margin as shown in Eq. (2).

Fade Margin and Outages

The required fade margin for digital microwave links often is calculated based on the percentage of the time that the link is allowed to exceed the desired maximum BER. In the United States, the AT&T objective has been a maximum yearly outage of 0.02% for a 400-kilometer (km) link, half of which is allocated to multipath propagation (2). The fade margin thus would be based on a 0.01% maximum outage.

If the link is less than 400 km, the allowed percentage of outage becomes $P = (0.01\ d)/400$, where d is the length of the link. If the link is long enough to require repeaters (multi-hop link), the percentage must be divided further by the number of hops. Everyone does not agree with this approach, but it is based on the assumption that deep fades are not likely to occur simultaneously on two or more hops. The simplification of just dividing the probability by the number of hops also assumes similar fading characteristics on all hops.

The fade margin calculation for the hop being designed now would be done in the same way as for an analog link by using the following rearranged form of Barnett's fading equation (3):

$$M = -10 \log (P) + 10 \log (2.5 \times 10^{-4})\, a \cdot b \cdot F \cdot D^3 \qquad (5)$$

where

M = the required fade margin in dB

P = the acceptable outage probability in percent

a = a terrain factor equal to 0.25 for mountainous terrains, 1 for average terrains, and 4 for very smooth terrains

b = a climatic factor equal to 0.125 for very dry climates or mountainous areas, 0.25 for temperate climates, and 0.5 for hot, humid climates

F = the frequency in GHz

D = the path length in miles. If D is in kilometers, the coefficient changes from 2.5×10^{-4} to 6×10^{-5}.

The required value of fade margin M then would be used in Eq. (2) to find the needed antenna gain. It should be noted that the required fade margin generally is based on "outage" objectives, thus this margin is the difference between the "nonfaded" carrier level and the level, calculated in Eq. (4), needed to produce the maximum acceptable BER. There is no general agreement on what this maximum acceptable BER or *outage threshold* should be. A value of 10^{-3} used to be common, based on the fact that this is the threshold on copper-pair T1 carrier systems transmitting pulse code modulation (PCM) voice channels. With more and more high-speed digital data transmission taking place, the threshold now is defined more often at 10^{-6}.

Errored Seconds

The increasing occurrence of data transmission also is leading to a change in quality criteria, which will affect the outage objectives on digital microwave links. The BER, which is an average value of bit errors over a given time, was a sufficient measure of quality for digitized voice transmission. Data, however, is transmitted in "packets" or "blocks" that need to be repeated whether all the bits in the packet are in error or just one. Since errors on digital links usually occur in bursts, an average error rate then is not very meaningful and new criteria such as "errored seconds" were needed. This need was reinforced by the appearance of Integrated Services Digital Network (ISDN) "channels."

The International Radio Consultative Committee (CCIR) has recommended

TABLE 1   CCIR Recommendations for Digital Microwave Links

Criterion	Objective
Degraded minutes (DM)	Fewer than 0.4% of 1-minute intervals having a BER worse than 10^{-6}
Severely errored seconds (SES)	Fewer than 0.054% of 1-second intervals having a BER worse than 10^{-3}
Errored seconds (ES)	Fewer than 0.32% of 1-second intervals having any errors
Unavailable seconds (US)	Fewer than 0.3% of consecutive 10-second intervals having a BER worse than 10^{-3}

the performance objectives shown in Table 1 (4). These objectives are likely to be adopted for digital microwave links in the United States as well. These objectives are for an ISDN "B" channel, that is, 64 kilobits per second (Kbps), on a 2500-km reference radio circuit. For links shorter than 2500 km, the above percentages are reduced proportionally (i.e., divided by 2500 and multiplied by the length of the link). On a per-hop basis, the link percentage is divided further by the number of hops in that link.

Note that the CCIR defines two BER thresholds, one at 10^{-3} for severely errored seconds (SES), the other at 10^{-6} for degraded minutes (DM). Two different calculations need to be made to see which one will require the highest fade margin. The errored second (ES) criterion defines a threshold that is bit-rate dependent (i.e., BER $= 1/R$) since even one bit in error during that second creates an "errored second." For example, the 64-Kbps channel will see errored seconds if the bit error rate is 1.56×10^{-5} or worse.

High-Capacity Links (Wideband Digital)

Flat Fading

Digital radio links occupying a spectrum of less than 15 MHz or so and using fairly simple modulation schemes such as frequency shift keying (FSK) and QPSK are designed in essentially the same way as analog radio links: the required threshold is selected and an appropriate fade margin is added based on the outage objectives. The margin is needed to compensate for the fades, caused by multipath propagation, that reduce the carrier-to-noise ratio (CNR) and therefore increase the BER. That fading is called *flat fading* or *thermal fading*. For higher-capacity digital links, another aspect of multipath fading becomes important, namely, *dispersive fading*, also called *selective fading* (5). In such cases, both the dispersive and the flat fading must be taken into account.

Dispersive Fading

When a signal with a wide spectrum is received, the multipath propagation causes different fade depths at different frequencies (dispersive fading) since the phase of arrival is frequency dependent and thus will create "notches" in the received signal spectrum. This distorted frequency spectrum, as well as differential delay distortion, will cause ISI, which increases the BER independently of the power level. That is, the BER cannot be compensated by increasing the power level and relatively small (frequency selective) fade depths can create high enough BER to cause an outage.

The dispersive fading impact, however, can be reduced substantially by incorporating one or more adaptive equalizers in the receiver. Frequency and/or space diversity also helps, in fact, it is interesting to note that using both adaptive equalization and diversity produces an improvement that is higher than the product of the two improvement factors. Thus, wideband digital radios make use of both mechanisms. It also is useful to know that frequency diversity is substantially more effective for digital links than for analog links, thus a smaller frequency separation is needed.

Even with the equalizers installed, different receivers will react differently to dispersive fading. Each receiver will have its own "response curve," called a *signature*, that describes its susceptibility to dispersive fading. The receiver's signature often is called an M *curve* (see Fig. 2).

The M curve shows the amount of fading at each frequency of the receiver IF bandwidth that will cause a given (threshold) BER. It is, in effect, the representation of the dispersive fade margin (DFM) of that receiver. Thus, the bigger the fade value shown, the more resistant the receiver is to selective fading at that frequency.

For power budget calculations, a single value of dispersive fade margin can be obtained by calculating the area under the M curve and inserting that number

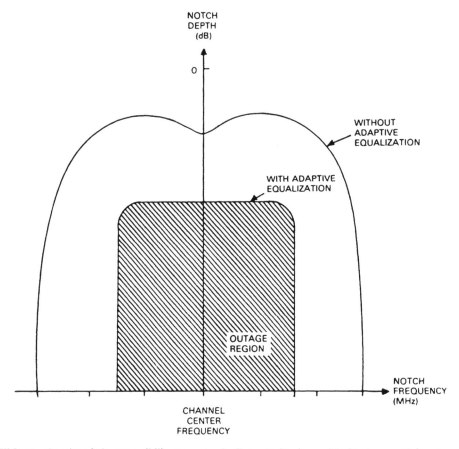

FIG. 2 Receiver fade susceptibility (*signature*). (From D. Taylor and P. Hartmann, Telecommunications by Microwave Digital Radio, *IEEE Commun.*, 24[8]:13 [1986] with permission of Institute of Electrical and Electronics Engineers.)

into an equation given in a Bellcore advisory (6). The resultant value of DFM, however, generally is available from the manufacturer of the receiver.

Composite Fade Margin

Because two different fading mechanisms take place, it is necessary to combine the two fade margins into a *composite fade margin* (also called *effective fade margin*), which can be calculated by using Eq. (6), derived by Greenstein (7):

$$CFM = -10 \log (10^{-FFM/10} + 10^{-DFM/10}) \qquad (6)$$

where
  CFM = the composite fade margin in dB
  FFM = the flat fade margin in dB from Eq. (5)
  DFM = the dispersive fade margin in dB

In fact, the impact of interference also can be taken into account by including an *interference fade margin* (IFM) component in Eq. (6) in the following manner:

$$CFM = -10 \log (10^{-FFM/10} + 10^{-DFM/10} + 10^{-IFM/10})$$

The value of CFM is the one to enter in the fade margin component M of Eq. (2).

The IFM is of special importance when "frequency reuse" schemes are employed to increase the bandwidth efficiency of the digital link. In such a case, the carrier frequency is reused by transmitting on an orthogonal polarization the second RF signal sent on that frequency. Because deep fades and high-density rain both can reduce the cross-polarization discrimination (XPD) between the two signals, an increased level of interference will take place and the IFM component of the equation becomes important. For the same reason, the high-capacity links must use antennas with very high values of XPD.

Conclusions

The power budget of digital microwave links with a capacity of 15 megabits per second (Mbps) or less can be designed in the same way as analog microwave links except that the starting point is the maximum acceptable BER instead of the maximum acceptable channel noise level. From the desired BER, the required minimum RF RSL is calculated. The final step is to add the fade margin necessary to achieve the maximum outage objective.

For high-capacity digital links, the same steps are followed but the fade margin added is a composite fade margin because dispersive fading must now be taken into account in addition to the flat fading considered so far. In addition, for links using cross-polarization frequency reuse, a further component, interference margin, is added to the composite fade margin.

Bibliography

Freeman, R., *Radio System Design for Telecommunications*, Wiley Interscience, New York, 1987.
Greenstein, L., and Shafi, M., *Microwave Digital Radio,* IEEE Press, New York, 1988.
IEEE Trans. Commun., special issue on Digital Radio, 27(12) (December 1979).

References

1. Catala, P., Analog Microwave Link Design. In: *The Froehlich/Kent Encyclopedia of Telecommunications*, Vol. 1 (F. E. Froehlich and A. Kent, eds.), Marcel Dekker, New York, 1990, pp. 233–247.
2. Kostal, H., Jeske, D., and Prabu, V., Advanced Engineering Methods for Digital Radio Route Design, *IEEE International Communications Conference*, 19B.6.1– 6.6 (1987).
3. Barnett, W., Multipath at 4, 6, and 11 GHz, *Bell Sys. Tech. J.,* 51(2):321–362 (February 1972).
4. International Radio Consultative Committee, CCIR Recommendations 557-1, 594-1, and 634, *CCIR Volume 9*, International Telecommunication Union, Geneva, 1989.
5. Stein, Seymour, Communication over Fading Radio Channels. In: *The Froehlich/ Kent Encyclopedia of Telecommunications*, Vol. 3 (F. E. Froehlich and A. Kent, eds.), Marcel Dekker, New York, 1991, pp. 261-303.
6. Bellcore, *Addendum to Requirements and Objectives for 4, 6, and 11 GHz Digital Radio Systems*, Bellcore Technical Advisory TA-TSY-000236, Issue 2, Bellcore, Morristown, NJ, July 1986.
7. Greenstein, L., and Shafi, M., Outage Calculation Methods for Microwave Digital Radio, *IEEE Commun.,* 25(2):30–39 (February 1987).

PIERRE J. CATALA

Digital Microwave Radio

Overview

The various kinds of radio systems for transmission of data are shown schematically in Fig. 1. The three mainstay technologies are terrestrial radio, satellite, and mobile communications. By capitalizing on the relative strengths of each, all three approaches will continue to play an important role in telecommunications network architectures in the years ahead.

Terrestrial microwave radio systems are not only economical but also offer enhanced reliability and flexibility. Given these advantages, these systems have been developed to carry the large traffic capacity between zonal centers in major cities, or to carry the medium or small traffic capacity between zonal centers and end offices. Microwave links also have been employed to connect end offices with customers. Here, we provide an overview of digital microwave radio and supporting technologies for application to trunk transmission routes.

Review of Digital Microwave Radio

Various Systems

The main, practical digital microwave systems that have been developed around the world are summarized in Table 1. The frequency bands most commonly used are 2, 4, 6, 11, and 18 gigahertz (GHz). Among these frequencies, the 4-, 6-, and 11-GHz bands have been found to yield the most value in terms of minimizing propagation loss while maximizing the amount of bandwidth used. For this reason, numerous systems have been developed centering around these three frequency bands (1–25).

In the late 1940s (the early days of microwave radio system development), such modulation methods as pulse position modulation (PPM) were studied (4–6). While efforts were made to compress the utilized spectrum as much as possible, the system still lagged far behind frequency modulation (FM) in spectrum utilization efficiency and, as a result, FM modulation continued in practical use for many years (7,8).

Later, against the backdrop of progress in semiconductor devices and digital processing technologies that developed along with computer technology, economical pulse code modulation (PCM) codecs and multiplexors became feasible. In 1969, a digital microwave system using 4PSK (quaternary phase-shift keying) modulation was developed and practically used in low-speed, short-haul routes (9,10).

However, the digital components available at that time were still too slow for application to long-haul routes with their twin requirements for speed and capacity. In addition, the 4PSK modulation scheme still could not equal the analog FM system for large-capacity transmission and high spectrum utilization efficiency.

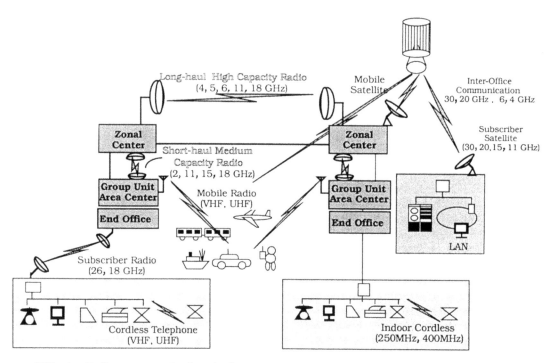

FIG. 1 Radio system applications in the networks (LAN = local-area network; VHF = very high frequency; UHF = ultra-high frequency).

From the end of the 1970s, with momentum building for an information society with the Integrated Services Digital Network (ISDN) as a key element, digitization of long-haul transmission routes suddenly emerged as an urgent requirement. This led to the development of 16QAM (quadrature amplitude modulation) and 9QPRS (quadrature partial response signaling) (11–13), the first amplitude modulation scheme to equal that of analog FM modulation in transmission capacity. The numbers signify the modulation levels, and the logarithm of the modulation numbers in base 2 denotes the transmission capacity. So, a 16QAM signal can carry 4 bits per symbol dulation. Dramatic progress also was seen in the development of digital LSI (large-scale integrated circuit), which soon was reflected in the falling cost of high-speed digital terminals. These combined developments contributed to the rapid digitalization of long-haul transmission routes. Research is continuing in laboratories around the world in an effort to push the number of modulation levels and the capacity even higher; today, practical 64QAM and 256QAM systems already have been developed (16–22).

Features of Microwave Radio

A variety of different transmission media is used to transmit data, and the sensible approach is to continue to employ all these different media in such a

TABLE 1 Typical Microwave Radios

Modulation Scheme	Bit Rate per RF Carrier (Mbps)	Voice Channels per RF Carrier	Carrier Separation (Copolar)	Carrier Separation (Cross polar)	RF Channel Arrangement	Spectrum Efficiency (bps/Hz)	Voice Circuits per MHz
Analog modulation schemes							
FM nonoverlapping sideband	—	1800	59.3	29.65	Interleaved	—	61
FM overlapping sideband	—	2700	59.3	29.65	Interleaved	—	91
SSB-AM	—	6000	59.3	29.65	Interleaved	—	202
	—	5400	59.3	29.65	Interleaved	—	182
Digital modulation schemes							
Nonsynchronous systems							
QPSK	45	672	59.3	29.65	Interleaved	1.52	23
16QAM nonoverlapping spectrum	90	1344	59.3	29.65	Interleaved	3.04	45
64QAM nonoverlapping spectrum	135	2016	59.3	29.65	Interleaved	4.55	68
64QAM overlapping sideband	135	2016	29.65	0	Co-channel	9.11	136
16QAM overlapping sideband	200	2880	80	0	Co-channel	5.00	72
256QAM overlapping sideband	400	5760	80	0	Co-channel	10.00	144
Synchronous systems							
64QAM hypothetical	155/2*155	2016/4096	40/80	0	Co-channel	7.75	101
256QAM hypothetical	155/2*155	2016/4096	29.65/60	0	Co-channel	10.33	134
256QAM hypothetical	4*155	8064	200/100	100/0	Interleaved/Co-channel	6.2/12.4	80/160

way as to exploit the advantage of each. Next, the special strengths of digital microwave radio are summarized (26,27).

Economical Transmission Cost

Areas particularly suited to the various transmission media (microwave radio, satellite communications, and optical fiber) in terms of traffic density and transmission distance are shown in Fig. 2. Fiber-optic systems are especially advantageous for very high capacity main trunks in which traffic is concentrated heavily. Satellite communication, on the other hand, is best suited to small and medium traffic capacity that is dispersed over the whole country. Microwave radio is especially appropriate when traffic is relatively dispersed and, on intermediate to large capacity routes, microwave often represents the most economic choice.

Enhanced Reliability and Resistance to Disasters

Since terrestrial radio-link systems are point to point, that is, the signal is transmitted between repeater stations as a beam through the air, the likelihood of such links being affected adversely by earthquakes and other natural disasters is substantially lower than other means of transmission over fixed facilities (see Fig. 3). Restoration of service after a disaster is also much faster for the same reason. The reliability of telecommunications networks is enhanced markedly by developing a mix of transmission media—including digital microwave—in such a way that they mutually reinforce each other.

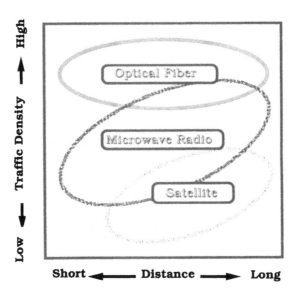

FIG. 2   Suitable application areas of various transmission media.

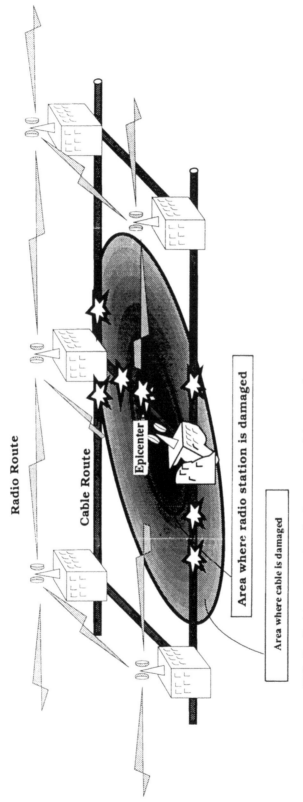

FIG. 3 Resistance to earthquake damage: one of the two radio routes can operate, but cable routes do not operate.

Ease and Speed of Route Construction

Microwave systems have been developed extensively and used around the world to carry analog voice circuits; therefore, a network of repeater stations is already in place throughout many countries. Since digital microwave routes can piggy-back on this existing infrastructure of repeater stations and towers, conversion to digital systems is relatively simple. The cost savings of this approach is obvious, and the ease and speed with which digital routes can be constructed represent significant advantages.

Even when new construction is necessary to accommodate new commercial subscribers or routes, digital radio is one of the easiest and least expensive options available; such systems require no right-of-way acquisition between towers and no excavation work to lay conduits. Radio links also have an advantage over buried facilities that require considerably more maintenance and support and often have to be repositioned, a task involving extensive excavation work.

State-of-the-Art Technologies

The driving force sustaining worldwide development of practical digital micro-wave systems is the migration to ever-higher modulation signal levels, going first from 16QAM to 64QAM, and more recently to 256QAM. Simply moving to a higher modulation level is not the whole story, however. This is because susceptibility to the effects of fading dramatically increases at the same time. The move to higher levels of modulation, therefore, also necessarily has in-volved considerable progress in devising techniques to compensate for the in-creased susceptibility to anomalous propagation impairments. (28–33).

Next, we describe a block diagram of repeater equipment for an introduction to multilevel modulation systems, modulation technology, the impairment from fading that can occur in these systems, and the adaptive countermeasures that have been devised to deal with these propagation impairments.

Repeater Equipment

A typical block diagram of a repeater station is shown in Fig. 4. In this section, the diagram is described on the basis of the signal flows.

The microwave signal received by the two antennas is divided into horizon-tally polarized and vertically polarized signals, and they are introduced to the radio signal branching filter circuit. Thereafter, these two signals are fed to the receiver equipment.

In the receiver, signals are preamplified at the radio frequency (RF) stage to compensate for signal losses that occur in the receiver circuits and then are converted into such intermediate frequency (IF) signals as the 140-megahertz (MHz) or 70-MHz band. They then are combined by the diversity reception combiner and the combined signal level fluctuation made by the

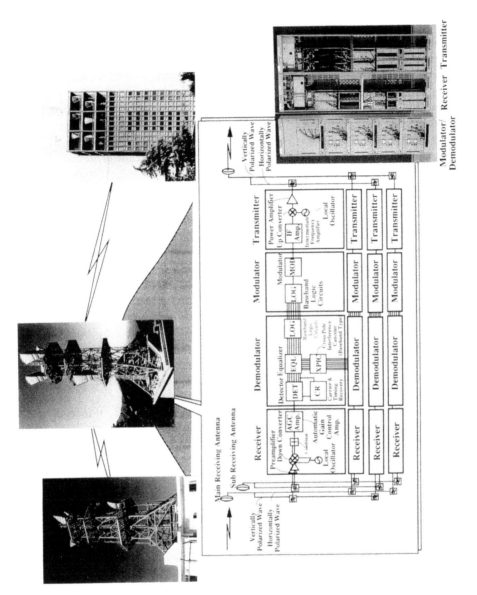

FIG. 4 Block diagram of repeater station equipment.

propagation path fading is reduced by the automatic gain controlled amplifier. After the signal level is fixed at the appropriate value, the signal is fed into the demodulator equipment that is connected serially to the receiver output port.

In the demodulator, the signal is detected and converted to baseband signals by mixing with the recovered carrier signal, and thereafter the baseband signals are retimed by the recovered baseband clock signals by using the QAM detector circuit. Then, the waveform distortion involved in the baseband signals made by the multipath fading in the propagation path is equalized by the digital transversal equalizer. At the same part of the circuit, the interference from the cross-polarized signal is canceled using the cross-polarized reference signal when the system utilizes the dual-polarized signal simultaneously. Then the waveform-reshaped baseband signals are regenerated to the baseband bits. These bits are fed into the baseband logic circuit in which the errors that occurred in the propagation path are corrected by using the forward-error-correction code. Finally, the information bits to be repeated are obtained and fed into the modulator equipment in the next stage.

In the modulator, the forward-error-correction bits are added and the output baseband signals are fed into the QAM modulation circuit to convert them to a QAM-modulated signal that has high spectrum and power efficiency features. The modulation circuit is constructed from two balanced amplitude modulators. The carrier input to an amplitude modulator is in phase with the local oscillator and the input to the other amplitude modulator is in quadrature with the local oscillator. The output signal from the modulators is fed into the transmitter equipment.

In the transmitter, the IF signal is amplified and then converted into the RF signal by using an up converter circuit, and then the RF signal is amplified by a power amplifier as an output signal from the transmitter. The output signal from the transmitter is fed into the antenna and feeder system in the final stage of the repeater station.

The output signals from the transmitters are multiplexed by the RF duplexing filters and then fed into an antenna through the polarization duplexer. And, finally, the signal is transmitted to the next radio repeater station.

In the actual repeater station, another set of repeater equipment (not shown in Fig. 4) is necessary to carry the voice or data circuits bound in the opposite direction of the circuits shown in Fig. 4. The diversity reception circuit shown in the figure is not required for a path with a hop length shorter than 30 kilometers (km) and the modulation levels are lower than 16QAM. In this diagram, the equalizer and cross-pole interference canceler are installed in the baseband stage. However, in the case of circuits being constructed in the IF stage, the equalizer and cross-pole interference canceler are installed between the AGC amplifier and the detector circuit.

Modulation Technology

In order to increase transmission capacity within a limited band of frequencies, multilevel modulation schemes usually have been adopted (29). The range of multilevel digital modulation schemes is shown in Fig. 5. In the figure, the left-hand illustrations denote the baseband signal waveform named eye diagram, and the right-hand circles and rectangles denote the signal phasor (amplitude in radial, phase in rotation angle). Modulation methods illustrated in the figure are categorized in phase-shift keying (PSK) and QAM. PSK signals (a) through (c) carry the information by changing their phase, for example, 2 states in BPSK (binary-phase PSK) as shown by the black-point in (a), 4 states in QPSK (quadri-phase PSK), and 8 states in 8PSK. These signals have no amplitude fluctuations because the signal phasors are located on the circle.

On the other hand, QAM carries the information by changing both amplitude and phase as shown by the signal phasor shown in (d) through (f); for example, 16QAM signal has 16 states that have different amplitude and/or phase.

It can be seen that, as the number of modulation levels increases, the distance between the nearest signal points shrinks, which increases the susceptibility to noise. It is more clearly shown by the baseband signal eye diagrams illustrated in the figure. As the baseband signal is sampled at time 0, $\pm T$, $\pm 2T$, . . . , and the information can be extracted from the sampled value, the difference between sampled signals shows the susceptibility to noise. Signal constellations that maximize both the number of signal points and distance between signal points offer the most desirable tradeoff. QAM modulation is more prevalent in today's digital radio systems than PSK modulation.

Recently, the transmission capacity within a limited band of frequencies, commonly referred to as the *spectrum utilization efficiency*, is measured by the information bits per second per unit frequency. It has reached 10 bits per second

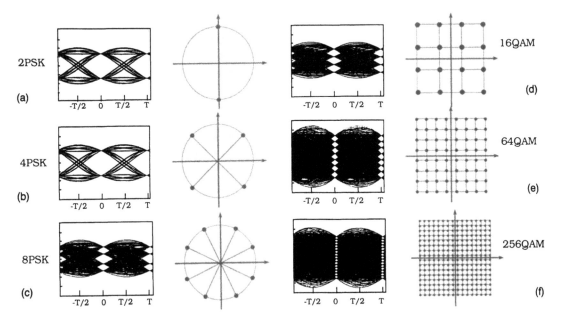

FIG. 5 Various modulation methods.

per hertz (bps/Hz) by using the 256QAM modulation scheme and dual-polarization operation (17).

To enhance the spectrum utilization efficiency, a narrower radio signal spectrum also is required. To realize the narrower spectrum shape, Nyquist's roll-off spectrum shaping usually is used. However, a 30% to 50% excess band extension from Nyquist's theoretical bandwidth limit is allowed to prevent fading impairment because waveform degradation caused by multipath fading becomes larger as spectrum shaping becomes narrower.

Multipath Fading

Resistance to fading on propagation routes decreases in roughly the same proportion as the number of modulation levels increases. Thus, while the allowable linear amplitude dispersion over the Nyquist bandwidth is about 10 decibels (dB) for 4PSK, the level falls to 5 db for 16QAM, and to a scant 2 dB for 256QAM. Moreover, as was observed in the preceding section, the tolerance value for noise becomes strained severely. This means that, as the number of levels increases, more sophisticated devices become necessary to compensate for the increased susceptibility. Some of the compensating techniques available to realize such devices are considered in the next section. But first, a simple explanation of fading is helpful.

Measured data collected under actual fading conditions are shown in Fig. 6-*a*. By comparing Point A with Point B in the figure, we can see that both received signal loss and in-band amplitude dispersion happen simultaneously at observation point A. Generally speaking, fading not only degrades the received signal level, but also results in the degradation of in-band frequency characteristics (30).

These properties are depicted in general by the three-ray interference model in Fig. 6-*b*. Even though the propagation-delay difference between the two rays is small compared to the symbol duration, the model can be approximated by a two-ray interference model with flat amplitude attenuation. In the figure, the direct and diffracted rays can be approximated by the simple sum of them, and the propagation time difference between them can be neglected, because the propagation time difference between these rays is shorter than the time difference between the direct ray and the reflected ray from the ground or sea surface. Therefore, the direct ray and the diffracted ray can be approximated by a single ray. In Fig. 6-*c*, the transfer function of the simplified two-ray model with flat attenuation is shown, and this transfer function approximates the data in Fig. 6-*a* well.

Fading Countermeasures

Bit errors as a result of fading do not occur frequently. Nevertheless, extremely stringent standards have been established to support essentially error-free transmission over large-scale public routes comprised of thousands of telephone

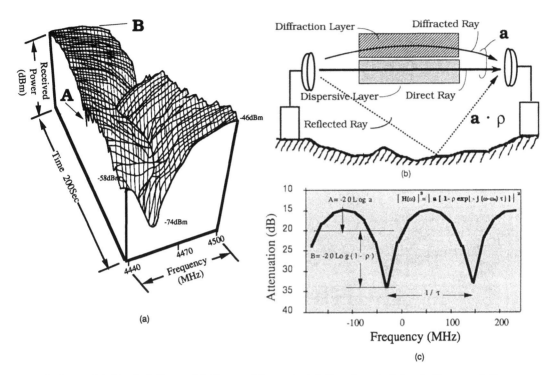

FIG. 6 Multipath fading and fading model: *a*, actual fading characteristics; *b*, three-ray fading model; *c*, amplitude of two-ray model with flat attenuation.

circuits. This is also the case to safeguard valuable data for digital transmission. The international standard, for example, stipulates that the 1-second average bit error rate cannot exceed 10^{-3} for more than 0.04% of the time during a month over a continuous 2500-km route. To satisfy such an exacting standard, antifading countermeasures have been developed (31). Next, we consider some of these countermeasures.

Equalizers

It has been observed that waveform distortion is one of the major causes of the bit errors occurring on transmission routes. Various kinds of automatic equalizers have been developed with the aim of compensating for this waveform distortion; they can be categorized broadly as being based on either frequency-domain equalization or time-domain equalization. In general, with the former type the circuit scale is small but equalization of the delay characteristics is relatively difficult. The latter type has a clear advantage because its characteristics are just the opposite. Performance results for different kinds of equalizers are summarized in Table 2.

TABLE 2   Characteristics of Various Equalizers

Type of Equalizer	Equalizer Performance		
	Minimum Phase-Shift Fading	Nonminimum Phase-Shift Fading	Complexity of Implementation
Frequency-domain equalizer			
Amplitude Slope	Fair	Fair	Simple
Amplitude Equalizer = Amplitude Slope + Parabolic Amplitude	Good	Good	Simple
Amplitude Equalizer + Delay Slope	Good	Good	Complex
Amplitude Equalizer + Delay Slope + Parabolic Delay	Excellent	Very Good	Complex
Single Tuned Circuit	Excellent	Good	Simple
Time-domain equalizer			
Linear transversal equalizer	Good	Very Good	Complex
Linear transversal + decision-feedback equalizer	Excellent	Very Good	Complex

Interference Canceler

Because susceptibility to noise and interference increases as the number of modulation levels increases, it is essential that the levels of interference are reduced through the use of a canceler. The main sources of interference are from adjacent routes and inter–cross polarization. Since both types of interference are exacerbated under fading conditions, it is important that fading is taken into consideration in the system design phase.

Until recently, the main methods available to reduce interference were improvement of antenna characteristics and the use of filters. Neither of these approaches, however, can be considered positive countermeasures. As our knowledge of interference-compensation techniques increasingly has become sophisticated, interference-reduction requirements have become more and more stringent. We currently are making good progress toward a positive compensation technology to deal with the most intractable source of interference — cross-polarization interference. In Fig. 7, a typical cross-polarization interference canceler is shown. In this case, it is better to use the co-channel frequency arrangement because the synchronous cancellation between cross-polarized signals is easy to obtain and interference-cancellation performance becomes high.

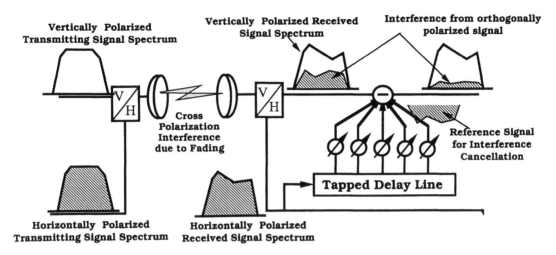

FIG. 7   Block diagram of cross-polarization interference canceler.

Diversity Reception

Diversity reception takes advantage of the fact that signals carrying the same information received at multiple antennas are unlikely to suffer impairment from fading at the same time. The technique prevents signal power deterioration and minimizes waveform distortion. Diversity techniques, especially in combination with automatic equalizers to take advantage of the synergistic effect, are being applied widely to digital microwave routes around the world. To obtain an unimpaired signal from multiple received signals, the combination method is used more commonly than the selection method because delay compensation between signals is unnecessary with the combination method.

Coding Technology

Error correction has been used in many telecommunications fields but generally has not been applied to digital microwave systems. This is because the additional bits required for coding purposes lower the spectrum utilization efficiency, and also because the effectiveness of error-correction techniques is rather low against the bursty, extremely high concentration of bit errors that result from multipath fading. With the development of such techniques as multicarrier transmission to compensate for waveform distortion, however, efforts are turning to interference noise, the main cause of bit errors. Multicarrier transmission also is starting to be applied actively to protect against degradations that come from implementing more complex devices. Increases in the checking rate on the order of about 10% are fairly commonplace. In recent years, the application of such coded modulation as trellis-coded modulation also has begun to be studied (33).

Performance Calculation

The following procedure is used to calculate transmission characteristics with the application of different fading-compensation methods (32).

1. Choose a performance threshold criterion
2. Specify a statistical channel model.
3. Find the parameter region R over which the performance threshold is exceeded.
4. Find the fraction of fade responses for which the model parameters lie within R.
5. If the conditional probability (given fading) of the performance threshold is exceeded, multiply by the total number of fading seconds.

Conclusion and Future Trends

The range of modulation schemes, the characteristics of digital microwave radio, and the current state of some key supporting technologies are surveyed briefly. Any progress toward more effective utilization of the limited bandwidth available that is the common property of all is a significant development, and much effort has been expended in the pursuit of this objective over the past several decades. As a result, spectrum utilization efficiency has been enhanced substantially so that, today, efficiency has reached 10 bps/Hz based on 256QAM modulation with dual-polarization operation. Certainly, in the years ahead we must continue to consolidate and build on the progress that already has been made in utilizing the finite bandwidth effectively.

Bibliography

Feher, K., *Digital Communications; Microwave Applications,* Prentice-Hall, Englewood Cliffs, NJ, 1981.
Greenstein, L. J., and Shafi, M., *Microwave Digital Radio*, IEEE Press, New York, 1988.
International Radio Consultative Committee, *Recommendations and Reports of the CCIR Study Group 9,* International Telecommunication Union, Geneva, 1986.

References

1. Taylor, D. P., and Hartmann, P. R., Telecommunications by Microwave Digital Radio, *IEEE Commun.*, 24(8):11–16 (1986).

2. Dewitt, R. G., Digital Microwave Radio—Another Building Block for the Integrated Digital Network, *IEEE 11th International Communications Conference*, 21.5–21.9 (1975).
3. Hayashi, Y., CCIR Work in Radio-Relay Systems, *IEEE 15th International Communications Conference*, 9.5.1–9.5.5 (1979).
4. Black, H. S., AN/TRC-6, A Microwave Relay System, *Bell Lab. Record*, 23(12) (1945).
5. Feldman, C. B., A 96-ch Pulse Code Modulation System, *Bell Lab. Record*, 26(9) (1948).
6. Someya, I., Fukami, T., and Morita, M., 4000Mc Time-Division 23 Channel Radio Relay System, *NTT Electrical Comm. Lab. Tech. J.*, 2(1):155–179 (1953).
7. Roetkin, A. A., Smith, K. D., and Fries, R. W., The TD-2 Microwave Radio Relay System, *Bell Sys. Tech. J.*, 30(4):1041–1077 (1951).
8. Hathaway, S. D., Hensel, W. G., Jordan, D. R., and Prime, R. C., The TD-3 Microwave Radio Relay System, *Bell Sys. Tech. J.*, 47(7):1143–1188 (1968).
9. Yoshida, K., Tachikawa, K., Tanaka, R., and Fukeda, H., 2GHz Microwave PCM System, *Japan Telecomm. Rev.*, 11(1):18–29 (1969).
10. Miyagawa, S., Yamamoto, H., Performance of 2GHz PCM System Repeating Equipment, *Review of the Electrical Communication Laboratory Japan*, 7(3–4): 241–259 (1969).
11. Godier, I., DRS-8 Digital Radio for Long Haul Transmission, *International Communications Conference*, 5.4.102–5.4.105 (June 1977).
12. Bates, C. P., DR6-30-135 System Design and Application, *International Communications Conference*, 539–546 (1984).
13. Yamamoto, H., and Morita, K., 4/5/6L-D1 Digital Microwave Radio, *Review of the Electrical Communication Laboratory Japan*, 30(5):836–845 (1982).
14. Steel, J., and Nowotny, H. G., Digital Radio and Fiber Optics in Australian Telecommunications, *Globecom '87*, 5.1.1–5.1.7 (1987).
15. Wilkinson, R. B., An Overview of Digital Transmission Systems in New Zealand, *Globecom '87*, 5.2.1–5.2.5 (1987).
16. Crossett, J. A., and Hartmann, P. R., 64QAM Digital Radio Transmission System Integration and Performance, *ICC Conf. Rec.*, 636–641 (1984).
17. Murase, T., 256QAM 400 Mbit/s Microwave Radio System, *J. Telecommunications Research*, 30(2):23–30 (1988).
18. Deluca, O., 560 Mbit/s Digital Radio System Using 256 QAM Modulation, *2d European Conference of Radio Relay*, 202–208 (1989).
19. Nossek, J. A., Steinkamp, J. A., Thaler, J., and Vogel, K., Impact of the Digital Hierarchy on the Design of a New Generation, *2d European Conference of Radio Relay*, 21–28 (1989).
20. Candeo, S., Zenobio, D., and Russo, E., Implications of Synchronous Digital Hierarchies on Radio Relay System, *2d European Conference of Radio Relay*, 29–34 (1989).
21. Kohiyama, K. et al., Outage-free Digital Microwave Radio, *NTT Review*, 3(1):19–27 (1991).
22. Langer, O. M., Radio-Relay Systems between Economy and Frequency Efficiency in the Telecommunication Network of the Deutche Bundespost, *2d European Conference of Radio Relay*, 43–50 (1989).
23. Nishino, K., Nakamura, Y., Hosoda, A., and Mukai, T., 20 GHz PCM Radio-Relay System, *Japan Telecomm. Rev.*, 18(1):23–36 (1976).
24. Yamamoto, H., and Kohiyama, K., Construction and Overall Performance of Experimental 20 GHz Digital Radio Repeater, *Rev. ECL*, 22(7–8):571–578 (1974).

25. Yamamoto, H. et al., Future Trend in Microwave Digital Radio, *IEEE Commun.*, 25(2):40–52 (1987).
26. Kirby, R. C., *Radio-Relay Systems in International Telecommunications*, ECRR, Munich, (November 1986).
27. Fuketa, H., Shinji, M., and Komaki, S., Present Status and Future Prospects of Digital Microwave Radio, *2d ECRR*, 35–42 (1989).
28. Ivanek, F., Microwave Technology in Terrestrial Digital Communications, *Microwave J.*, 28–38 (January 1988).
29. Noguchi, T., Daido, Y., and Nossek, J. A., Modulation Techniques for Microwave Digital Radio, *IEEE Commun.*, 24(10):21–30 (1986).
30. Rummler, W. D., Coutts, R. P., and Liniger, M., Multipath Fading Channel Models for Microwave Digital Radio, *IEEE Commun.*, 24(11):30–42 (1986).
31. Chamberlain, J. K., Clayton, F. M., Sari, H., and Vandamme, P., Receiver Techniques for Microwave Digital Radio, *IEEE Commun.*, 24(11):43–54 (1986).
32. Greenstein, L. J., and Shafi, M., Outage Calculation Methods for Microwave Digital Radio, *IEEE Commun.*, 25(2):30–39 (1987).
33. Ungerboeck, G., Trellis-Coded Modulation with Redundant Signal Sets, Part I: Introduction, *IEEE Commun.*, 25(2):5–11 (February 1987).

SHOZO KOMAKI

Digital Modulation and Coding

Digital Signals and Systems

Transmission of information end to end consists of processes that are transparent to the end users. Whether it is voice, data, or image, the underlying processes involve conversion of messages into signals suitable for transmission over the available media.

In so-called analog communications systems, the message signals preserve their continuous variations throughout the course of transmission processes and include various forms of modulation and demodulation. Digital communications systems, however, convert, transmit, and deliver the information signals in digital form in all or part of their transmission functions. The delivered digital information has to be converted back into a form recognizable by the end user, who actually is not concerned whether the transmission took place in an analog or a digital form.

For the communications engineer, though, the terms *analog* and *digital* are reminiscent of two different paradigms that are the subject of this and other articles in this encyclopedia (1,2). It is important to realize, however, that the underlying structure of all digital signals and systems remains analog. This means that the electrical and optical waveforms are physically analog; they are varying continuously no matter how discrete our view of them. Therefore, it can be stated that the digital paradigm is an alternate way of shaping and using the technology to make it more efficient, cost effective, and easier to understand and develop.

Evolution to the use of digital signals and systems from analog ones is driven by numerous facts regarding the advantages of digital systems. These advantages again are looked at from the point of view of the end users, the designers, and the operators of the communications systems. The criteria used for comparison are efficiency of conversion and transmission, reliability of transmission, and overall performance.

In these regards, digital systems are superior to their analog counterparts. One advantage is that digital communications systems are more immune to noise and distortion because regenerative repeaters reproduce the signals periodically along the communications path to prevent error accumulation due to noise. Moreover, through appropriate coding schemes, the rates of the remaining errors can be controlled to a desired level. Use of an appropriate coding method also will make the signal independent of the channel distortion.

From the hardware point of view, digital transmission requires increased transmission bandwidth and involves increased system complexity. However, digital equipment presents a high degree of reliability, flexibility, and higher modularity in design, which counteracts the increased complexity. Present-day digital circuits with various levels of integration make implementation of complex systems much easier compared to their analog counterparts. This is due partly to the fact that digital representation of signals provides a uniform format for such different kinds of baseband signals as video and voice as well as still image, graphics, and data. Processes underlying digital information transmission involve conversions from analog to digital and vice-versa. This includes sampling of the analog signal, subsequent quantization of the sample, and final coding of the quantized sample. These subjects are covered in the following sections.

Coded Pulse Modulation

In analog pulse modulation, that is, pulse amplitude modulation (PAM), pulse duration modulation (PDM), and pulse position modulation (PPM), only time is expressed in discrete form, whereas the respective modulation parameters (namely, pulse amplitude, duration, and position, respectively) are varied in a continuous manner in accordance with the variations of the message signal (1). On the other hand, in coded pulse modulation, the message signal is sampled and the amplitude of each sample is rounded off (quantized) to the nearest value of a finite set of allowable values so that both time and amplitude are in discrete form. The modulation system that performs these operations is known as pulse code modulation (PCM).

In the next section, PCM, which lies at the heart of digital communications systems, is described. Increased capacity of communications channels and higher capabilities of electronic devices today have made implementation of PCM possible.

Elements of Pulse Code Modulation

As mentioned above, PCM systems are more complex than analog pulse modulation systems in that the message signal is subject to a larger number of operations. In the transmitter, the essential operations are sampling, quantizing, and coding (see Fig. 1). The quantizing and encoding operations usually are done in the same circuit, the analog-to-digital converter. The essential operations in the receiver are regeneration of impaired signal, decoding, and demodulation of the train of quantized samples. Depending on the length of the transmission channel and impairments, regenerators may be needed at intermediate points along the transmission path. In time division multiplexing (TDM), it is necessary to synchronize the receiver to the transmitter for proper operation of the overall system. Sampling is described in this encyclopedia in the article, "Analog Modulation" (1). The following sections are devoted to earlier, essential PCM operations.

Quantizing

The samples of a continuous signal, such as voice, have a continuous amplitude range. In other words, an infinite number of amplitude levels may be found within the finite amplitude range of the signal. It is not necessary to transmit

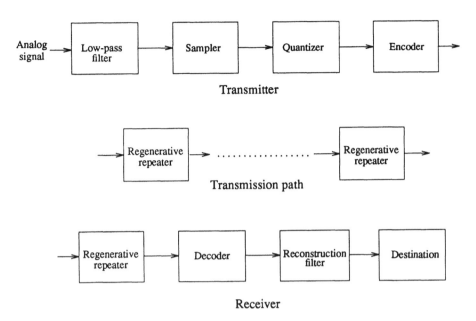

FIG. 1 Elements of a pulse code modulation system.

the exact amplitudes of the samples. The human senses (e.g., hearing) as the ultimate receivers only can detect finite intensity differences. Therefore, the original continuous signal may be approximated by a signal constructed of discrete amplitudes. The number of discrete amplitude levels is finite.

The conversion of analog samples of a signal into discrete amplitude levels is called *quantization*. This process has a staircase characteristic (Fig. 2). A

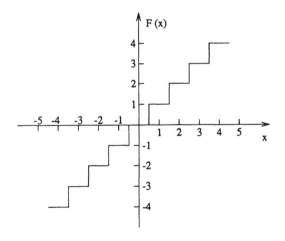

FIG. 2 Characteristic of uniform quantizing.

quantum is the difference between two adjacent discrete values. When a continuous signal is applied to a quantizer, it will be converted to a staircase signal. The *quantization error* is the difference between the input and output of the quantizer. The maximum instantaneous value of this error is half of one quantum step. In the quantization process of Fig. 3, a uniform separation between the quantizing values is used. In such an application as voice, it is preferred to use a variable separation between the quantizing levels for several reasons. First, voice signals have a wide dynamic range. A single talker may have a peak factor of 25 decibels (dB); in order to take the variation in volume from talker to talker into account, it becomes necessary to accommodate a range of signal amplitudes of at least 40 dB. Since quantizing distortion in a uniform quantizer is independent of signal amplitude, use of uniform quantization will result in a 40-dB difference in the signal-to-noise ratio for strong and weak talkers. Finally, the amplitude distortion of signals is not uniform; some amplitudes have higher probability of occurrence than others, and it is possible to optimize the average signal-to-distortion (S/D) power ratio performance by allowing more distortion for less probable amplitudes and less distortion for levels with higher probability. By using a nonuniform quantizer, the weak passages of a talk, which need more protection, are favored at the expense of the loud passages. The use of a nonuniform quantizer is equivalent to passing the baseband signal through a compressor and then applying the compressed signal to a uniform quantizer.

In order to restore the signal samples to their correct level, a device must be used in the receiver with a characteristic complementary to the compressor. Such a device is called an *expander*. The combination of a compressor and an expander is called a *compander*.

There are two commonly used methods of nonuniform quantization for voice transmission. North America and Japan have adopted a standard compression curve known as μ-law companding. Europe has adopted a different, but similar, standard known as A-law companding.

The μ-law is defined, for normalized range of $[-1, +1]$, by the following equation (3):

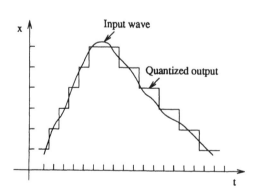

FIG. 3   Quantized signal wave.

$$F_{\mu}(x) = sgn(x) \frac{\ln(1 + \mu|x|)}{\ln(1 + \mu)} \qquad |x| \le 1 \qquad (1)$$

where x and $F_{\mu}(x)$ are normalized input and output voltages, respectively, μ is a positive constant, and sgn is the sign function. The A-law is defined by

$$F_A(x) = sgn(x) \frac{1 + \ln A|x|}{1 + \ln A} \qquad \frac{1}{A} \le |x| \le 1$$

$$F_A(x) = sgn(x) \frac{A|x|}{1 + \ln A} \qquad 0 \le |x| \le \frac{1}{A} \qquad (2)$$

The case of uniform quantizing corresponds to $\mu = 0$ for μ-law and $A = 1$ for A-law. The ratio of signal power to quantizing-distortion power (S/D) for A-law is somewhat flatter than μ-law over the range $1/A \le |x| \le 1$ at the expense of somewhat poorer S/D performance for low-level signals. The typical values for μ and A are 255 and 87.6, respectively.

Encoding

The process of translating the discrete set of sample values to a more appropriate signal form for transmission is called *encoding*. The signals for transmission are called *symbols*. A *code* is a plan for representing each discrete value as a particular arrangement of symbols. This particular arrangement is called a *codeword* or *character*.

In a binary code, each symbol may be either of two distinct values, such as presence or absence of a pulse, denoted as 1 or 0. A binary code with N binary digits (*bits*) can represent a total of 2^N distinct values. For example, a sample quantized into 1 of 8 levels may be represented by a 3-bit codeword. In Table 1, the ordinal number of the quantized level is expressed as a binary number.

The binary symbols 1 and 0 can be represented by electrical signals in several ways (Fig. 4):

TABLE 1 Binary Codewords for Eight-Level Quantization

Quantized Level	Binary Number
0	000
1	001
2	010
3	011
4	100
5	101
6	110
7	111

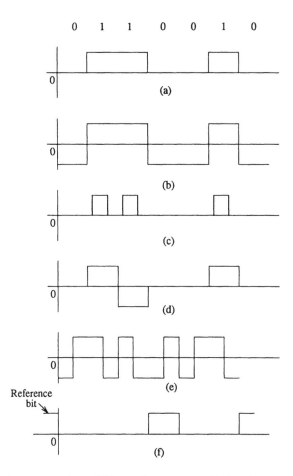

FIG. 4 Electrical representation of binary data: *a*, on–off signaling; *b*, polar signaling; *c*, return-to-zero signaling; *d*, bipolar signaling; *e*, Manchester code; *f*, differential coding.

On–off signal: Symbol 1 is represented by a rectangular pulse for a duration equal to that of the symbol; Symbol 0 is represented by switching off the pulse.

NRZ signal: Symbols 1 and 0 are represented by pulses of equal positive and negative amplitudes. This type of signal is called a *polar signal* or *non-return-to-zero* (NRZ) signal.

RZ signal: A rectangular pulse (half-symbol wide) is used for a 1 and no pulse for a 0. This type of signal is called a *return-to-zero* (RZ) signal.

Bipolar signal: Equal amplitude positive and negative pulses are used alternately for Symbol 1 and no pulse is used for Symbol 0. A useful property of this signal is that the power spectrum of the transmitted signal has no direct current (DC) component and relatively insignificant low-frequency components when the symbols 1 and 0 occur with equal probability.

Manchester code: Symbol 1 is represented by a positive pulse followed by a negative pulse with the same amplitude and half-symbol wide. For Symbol 0, the polarities of these symbols are reversed.

Differential encoding: The information is encoded in the transitions of a binary PCM wave. For example, a transition in a binary PCM wave may be used to designate symbol 0, while no transition is used to designate symbol 1 (see Fig. 4-*f*).

These possible binary coding methods are known as *line codes*. They are used to represent zeros and ones physically in the transmission process. However, the coding process is not confined to representation of quantized levels alone. Extra binary digits may be added to a codeword in the form of redundancy in order to build such various capabilities as error detection and/or forward error correction. This subject is discussed in the section, "Digital Coding," below.

Differential Pulse Code Modulation

When the voice or video signal is sampled at a rate slightly higher than the Nyquist rate, the adjacent samples will exhibit a high correlation. In other words, the signal does not change rapidly from one sample to the next and therefore the difference between adjacent samples has a variance that is smaller than the variance of the original signal. If highly correlated samples are encoded in a standard PCM, the resulting encoded signal contains redundant information. By removing this redundancy before encoding, we obtain a more efficient coded signal.

In a differential pulse code modulation (DPCM) system, the difference between a sample and its predicted value is encoded. From the past behavior of a signal up to a certain point in time, it is possible to make some inference about its future values; such a process commonly is called *prediction*. Suppose that a baseband signal $x(t)$ is sampled at a rate of $1/T_s$ to produce a sequence of samples denoted $x(nT_s)$. The prediction is made on the basis of previous samples. The symbol $\tilde{x}(nT_s)$ is used to denote the predicted value of $x(nT_s)$. Then the difference $e(nT_s) = x(nT_s) - \tilde{x}(nT_s)$ will be quantized and encoded.

Delta Modulation

Delta modulation (DM) is a simple technique for reducing the dynamic range of the numbers to be coded. Instead of sending each sample value independently, we send the difference between a sample and the previous one. If sampling is being performed at the Nyquist rate, this difference has a dynamic range twice as large as that of the original samples. At Nyquist rate, each sample is independent of the previous one. However, if we sample at a rate higher than the Nyquist rate, the samples are correlated and the dynamic range of the difference between two samples is less than that of the samples themselves. The net result of sampling at a faster rate but reducing the dynamic range will generate fewer binary digits for transmission.

Delta modulation uses only one bit of quantization. For example, a 1 is sent if the difference is positive and a 0 is sent if the difference is negative. Since there is only one bit of quantization, the differences are coded into only one of two levels, referred to as $+\Delta$ or $-\Delta$. At every sample point, the quantized waveform can only increase or decrease by Δ (Fig. 5). The transmitted train of bits for the example shown in Fig. 5 is

$$1\ 1\ 1\ 1\ 1\ 1\ 1\ 0\ 0\ 0\ 1\ 1\ 1\ 1\ 1\ 1\ 0\ 0\ \ldots$$

The key to effective use of DM is the intelligent choice of the two parameters, step size and sampling rate. These parameters must be chosen such that the staircase signal is a close approximation to the actual analog waveform. Figures 6-*a* and 6-*b* show the consequences of incorrect step size. When the steps are too small, the staircase cannot track rapid changes in the analog signal. When the steps are too large, considerable overshoot occurs during flat portions of the signal, which results in significant quantization noise.

Adaptive DM is a scheme that permits adjustment of the step size depending on the characteristics of the analog signal. In the song algorithm (Fig. 6-*c*), both in the transmitter and receiver, the transmitted bit is compared with the previous bit. If the two are the same, the step size is increased by a fixed amount Δ. If they are different, the step size is reduced by the fixed amount Δ. Note that a damped oscillation occurs following a rapid change in the signal. The space shuttle algorithm is a modification of the song algorithm that eliminates the damped oscillations (Fig. 6-*d*). When the present and previous bits are the same, it works the same way as song algorithm. However, when the bits disagree, the step size reverts immediately to its minimum size, Δ.

Digital Multiplexing

For multiple sources, TDM and PCM may be used to transmit the messages over a single transmission facility. TDM is the process of adding messages such

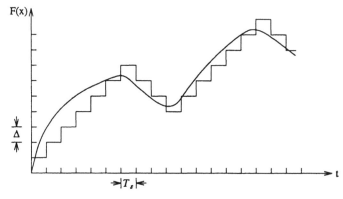

FIG. 5   Delta modulation.

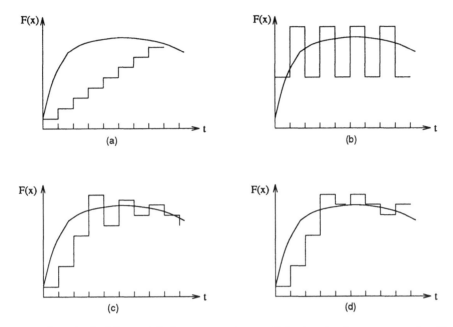

FIG. 6 Step size in delta modulation: *a*, steps too small; *b*, steps too large; *c*, song algorithm; *d*, space shuttle algorithm.

that they do not overlap in time. It is possible to separate the signals in time when they are represented by their samples. In a multiplexed PCM system, the pulses of each sample must be transmitted during the sampling period. The T1 carrier system designed and developed by Bell Laboratories in the early 1960s is an example of PCM multiplexing.

The T1 system has been adopted for use throughout the United States, Canada, and Japan. It has a bandwidth of 1.544 megabits per second (Mbps) that carries 24 voice signals (Fig. 7). Each voice signal is sampled first at 8000 samples/second. The samples then are quantized using 8-bit μ-law companding. The least significant bit is devoted periodically (typically once every 6 samples) to signaling information. The quantized and coded samples from 24 channels

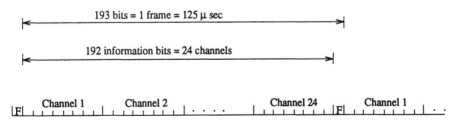

FIG. 7 T1 frame structure.

form a frame, and since sampling is done at an 8-kilohertz (kHz) rate, each frame occupies $1/8000 = 125$ microseconds (μs).

There are 192 information and signaling bits in each frame (24 bits times 8 bits), and a 193rd bit is added for frame synchronization. This framing signal has a fixed pattern of 100011011100 extended over 12 frames. The combination of 12 frames, including synchronization bits, is called a *superframe*. The product of 8000 frames per second with 193 bits per frame yields a transmission rate of 1.544 Mbps. The T1 structure for Europe is different and multiplexes 32 channels instead of 24 with a bit rate of 2.048 Mbps instead of 1.544 Mbps.

Synchronization

In a digital communications system, the receiver should be able to establish symbol timing before it can decide which of the various symbols are being received (e.g., zeros or ones in a binary system). Once the symbol timing is established, it is necessary to establish other timing relationships, including character/word and block/frame/message synchronization. It is not sufficient to decode a long string of symbols unless these symbols can be associated with the correct message. This is particularly important in TDM transmission in which the symbols must be decommutated.

Symbol Synchronization

The received signal is usually in the form of an electrical waveform extending over all time. It is important to chop the time axis into segments corresponding to the signal segment for each symbol. The problem of symbol synchronization is simplified if frequent transitions exist in the incoming waveform (e.g., Manchester code).

One problem associated with unipolar transmission is that the DC component of the waveform is a function of the density of ones. This causes problems if the electronics do not transmit DC (AC coupled devices). Bipolar signal overcomes this problem. There are two problems associated with an NRZ signal. First, when the data are static (e.g., long strings of ones or zeros), there is no transition in the transmitted waveform. This causes timing problems when we try to establish bit synchronization. The second problem occurs with data inversion. If the levels are reversed during transmission, the entire data train will be inverted and every bit will be in error. Due to these reasons, we often choose differential forms of encoding. In such techniques, the data are represented as changes in levels rather than by the particular signal level (differential encoding).

Let us now return to the timing problem. We mention three synchronization techniques: the maximum a posteriori, the early–late gate, and the digital data transition tracking loop (DTTL). In the maximum a posteriori technique, the received signal over a finite-length interval is observed and decoded by using a locally generated clock. The decoded signal then is correlated with a stored replica of the known symbol sequence, and the clock is adjusted to achieve maximum correlation. Perfect correlation is hard to attain due to bit errors, and this scheme requires that part of the transmitted signal is dedicated to this function. In the *early–late gate*, the error signal is defined as the difference between the square of the area under the *early gate* and the area under the *late gate*. This error goes to zero when the early and late gates coincide. In this case, the correct timing is at the midpoint of the gate periods. When the error is nonzero, the timing changes in a direction to cause the error to go to zero. In the digital DTTL, when the loop is in lock, one integrator integrates the input pulse over periods corresponding to the bit period, while the other integrator integrates over intervals spanning transitions. The output of the integrators are sampled, processed, and multiplied to determine the error. If the product is not zero, the timing pulse generator is adjusted in a direction to force the error toward zero (4).

Digital Coding

As mentioned above, coding in digital systems plays an important role in efficient information transmission and control of error performance. In the process of coding that succeeds sampling and quantization of analog signals, a sequence of digits called *codewords* is formed for each sample or letter of the source alphabet. In binary communications systems, the symbols used are zeros and ones.

The coding process can be designed with different objectives in mind. In one scheme, the efficiency of transmission is of primary concern. In such designs, which are called *entropy coding* or *minimum-length coding*, the objective is to transmit a maximum amount of information with a minimum number of bits. Huffman codes are examples of entropy coding systems. In these codes, the codewords are assigned according to the frequency of each source alphabet. The codewords are variable in length and more frequent source letters are assigned the shorter-length codes and vice-versa. Huffman coding is optimum in terms of average word length as it relates to the entropy of the source. Huffman coding is used for data, video, and text compression (5).

However, presence of noise, interference, distortion, and occasional artifacts in communications systems will not always allow the transmission efficiency to remain the primary concern. These disturbances will cause errors in the stream of bits so that some zeros and ones in the bit stream change to their complement values. To combat such situations, error-control coding is used. In error-control codes, extra digits that convey no information are added to the codewords. However, these additional bits help detect, correct, or even reconstruct errored or missing codewords. An alternative in this regard is to retransmit the garbled codewords upon detection of errors. Means have to be provided in the coding system, however, to detect presence of errors. Such coding systems are called *error-detecting* codes. But, in cases in which retransmission is not possible, such as simplex channels in which communication is one way or in the case of real-time two-way communications, means have to be provided to correct the errors on a real-time basis. In all such cases, we have to compromise among efficiency, reliability, and system complexity.

Error Detection

Detection of bit errors can be built into the design of the coding system. In this manner, enough redundant bits are added to the codewords in order to detect a corresponding number of errored bits in that codeword. A simple example of this kind of coding is addition of a parity-check bit to a codeword of arbitrary length in order to make the number of zeros/ones in that codeword even/odd. This kind of parity checking is quite popular from the point of view of simplicity and efficiency. Moreover, since for many practical purposes single errors in a codeword are the most frequent kind of errors, this method of detection is quite adequate. An example in which single parity-check bits can be used is transfer of American Standard Code for Information Interchange (ASCII) codes over telephone lines using voiceband data modems. However, it should be noted that this kind of coding only detects errored words with an odd number of errors without any indication of the number or position of the errors in the codeword. Errored codewords with an even number of errors remain undetected in this scheme. For sensitive data, therefore, more complicated error detection is required. However, this will require extra overhead bits and a corresponding bandwidth increase. *Coding efficiency R* is defined as the ratio of the number of information bits b over the word length w in bits including the parity bits. In mathematical form, this is

$$R = \frac{b}{w} \times 100\% \qquad (3)$$

For applications with stringent requirements, the coding system should be designed in such a way that the codewords are far enough apart in bit pattern that when errored bits occur, one codeword does not change completely into another valid codeword of the alphabet. This has a price in terms of coding efficiency and implementation. To put the argument on a quantitative footing, we define the *Hamming distance*.

Hamming distance between two codewords is the number of binary digits in which the two codewords differ from one another. For instance, the two codewords 10111001 and 11010101 differ in four bits and therefore have a Hamming distance of four. The Hamming distance of a set of codewords is the shortest Hamming distance between the codewords of that set and is denoted d here.

To detect e error bits in the transmission of the codewords, a code set must have a Hamming distance of at least $e + 1$. This is due to the fact that in such a set, e error bits would be insufficient to convert one transmitted codeword into another legitimate one. However, more than e bit errors may pass undetected. For instance, assuming that the shortest Hamming distance for a code set of 8-bit codewords occurs between words 10111001 and 11010101, four bit errors in transmission can change one codeword into another and therefore stay undetected. But three error bits in transmission always will be detected because none of the legitimate codewords of the system will be identified on receiving such an errored codeword.

The number of possible codewords for a given word length is limited by the required Hamming distance. This is discussed in the next section in regard to error-correcting codes.

Error Correction

Error-correcting codes are used to correct for transmission errors in the forward direction. Hence, they also are referred to as *forward error-correcting* (FEC) codes. This kind of coding system is used when acknowledgment of correct reception or request for retransmission, as in the case of automatic request for retransmission (ARQ) codes, is not possible.

In the system of FEC codes, the Hamming distance should be large enough so that if a codeword undergoes a certain number of bit errors, it can be decided which codeword was transmitted. It can be shown that in order to be able to correct for e bit errors in the codeword, a Hamming distance d of at least $2e + 1$ is required. Therefore, we must have

$$d \geq 2e + 1 \tag{4}$$

Note that up to $2e$ bit errors may be detected but they are not correctable. In error-correcting systems, received codewords are compared against the legitimate ones. Restoration to the codeword with shortest Hamming distance from the received one is carried out if the received codeword does not match any of the legitimate ones. More than e error bits will make the codeword closer to other codewords than the intended one and result in decoding error.

An example of error-correcting codes is the Hamming code. A possible Hamming code is given in Table 2. This is a (7,4) code in which 7-bit codewords contain 4 information bits and 3 redundancy bits added for single-error correction and double-error detection. The coding efficiency for this system is therefore 57%.

Note that each of the redundancy bits in the codewords of this example is the result of parity checking on a subset of the information and parity bits. For instance, R_1 is the even parity for positions 1, 3, 5, and 7. To find out which redundant bits check a given bit position in the codeword, we convert the corresponding position into binary code. The position of bits that are 1 indicates the bits that are involved in the parity calculation for that position. For instance, Bit Position 5 is equivalent to 101. Therefore, R_1 and R_3 are involved in the parity calculation for this position. Moreover, when the check bits are calculated at the receiver according to this rule, two cases can happen. If calculated parity bits are the same as the received ones, we conclude that no errors have happened. If not the same, we can calculate the position of error by finding the decimal equivalent of the $R_3R_2R_1$ sequence. For instance, if we calculate the parity bits and detect violations at R_1 and R_2, the sequence 011 is calculated for $R_3R_2R_1$ and therefore we conclude that a single error has occurred at the third position.

TABLE 2   Hamming Code

	Data			R_3	Data	R_2	R_1
Data	7	6	5	4	3	2	1
0	0	0	0	0	0	0	0
1	1	0	0	1	0	1	1
2	0	1	0	1	0	1	0
3	1	1	0	0	0	0	1
4	0	0	1	1	0	0	1
5	1	0	1	0	0	1	0
6	0	1	1	0	0	1	1
7	1	1	1	1	0	0	0
8	0	0	0	0	1	1	1
9	1	0	0	1	1	0	0
10	0	1	0	1	1	0	1
11	1	1	0	0	1	1	0
12	0	0	1	1	1	1	0
13	1	0	1	0	1	0	1
14	0	1	1	0	1	0	0
15	1	1	1	1	1	1	1

Block Codes

Block codes (N,b) are defined as the set of N-tuples of zeros and ones representing blocks of b message bits each. It can be shown that the necessary condition for an (N,b) block code to be capable of correcting e error bits is (6)

$$2^{N-b} \geq \sum_{i=0}^{e} \binom{N}{i} \tag{5}$$

This is called the *Hamming bound* and is a necessary and sufficient condition for single-error-correcting codes only. For other than the single-error-correcting codes, the Hamming bound is only a necessary condition and there is no guarantee that a coding system can be constructed corresponding to e, N, and b, satisfying the Hamming bound.

A *linear code* is defined as one that includes the N-tuple of all zeros $(0,0, \dots ,0)$ as a codeword and the set of codewords is closed under mod-2 addition. The *weight* of a codeword is defined as the number of nonzero elements of the N-tuple representing that codeword. It can be shown that, for linear block codes, the Hamming distance is the same as the weight of the codeword with the smallest nonzero weight. In so-called systematic block codes, the codewords consist of N-tuples with first b elements the same as the message bits, and the last $N - b$ elements used as parity bits. The N-tuples of codewords then would be of the form

$$C = [B \mid P] \tag{6}$$

where
 C = N-tuple of codewords
 B = b-tuple of message bits
 P = a p-tuple of parity bits corresponding to the message

For the systematic linear (N, b) block codes, the code vectors can be written in matrix form as

$$C = BG \tag{7}$$

where G is called the generator matrix and is given by

$$G = [I \mid Q] \tag{8}$$

Here, I is the $b \times b$ identity matrix and Q is a $b \times (N - b)$ matrix. The identity matrix produces the first b elements of the N-tuple of the codeword through the matrix multiplication

$$B = BI. \tag{9}$$

Q matrix generates the parity $(N - b)$-tuple through the matrix multiplication

$$P = BQ \tag{10}$$

Matrix multiplications use mod-2 addition instead of ordinary addition. This kind of representation for the coding system helps implementation of the encoding and decoding processes besides providing analytic simplicity. The task of finding appropriate G for a desired coding efficiency and error-correction scheme is the essential part of a coding system design. Hamming code presented in the section, "Error Correction" above, is an example of the linear block codes with the generator matrix G given by

$$G = \begin{bmatrix} 1 & 0 & 0 & 0 & | & 0 & 1 & 1 \\ 0 & 1 & 0 & 0 & | & 1 & 0 & 1 \\ 0 & 0 & 1 & 0 & | & 1 & 1 & 0 \\ 0 & 0 & 0 & 1 & | & 1 & 1 & 1 \end{bmatrix} \tag{11}$$

The parity bits corresponding to this generator matrix are calculated in the following manner:

$$P = B \begin{bmatrix} 0 & 1 & 1 \\ 1 & 0 & 1 \\ 1 & 1 & 0 \\ 1 & 1 & 1 \end{bmatrix} = [b_2 \oplus b_3 \oplus b_4 \quad b_1 \oplus b_3 \oplus b_4 \quad b_1 \oplus b_2 \oplus b_4] \tag{12}$$

Here, $\oplus$ denotes mod-2 addition or EXCLUSIVE-OR operation. From this process, we can calculate the codewords, including the parity or check bits, in the following manner:

$$
\begin{bmatrix}
0\ 0\ 0\ 0 \\
0\ 0\ 0\ 1 \\
0\ 0\ 1\ 0 \\
0\ 0\ 1\ 1 \\
0\ 1\ 0\ 0 \\
0\ 1\ 0\ 1 \\
0\ 1\ 1\ 0 \\
0\ 1\ 1\ 1 \\
1\ 0\ 0\ 0 \\
1\ 0\ 0\ 1 \\
1\ 0\ 1\ 0 \\
1\ 0\ 1\ 1 \\
1\ 1\ 0\ 0 \\
1\ 1\ 0\ 1 \\
1\ 1\ 1\ 0 \\
1\ 1\ 1\ 1
\end{bmatrix}
\left[
\begin{array}{cccc|ccc}
1 & 0 & 0 & 0 & 0 & 1 & 1 \\
0 & 1 & 0 & 0 & 1 & 0 & 1 \\
0 & 0 & 1 & 0 & 1 & 1 & 0 \\
0 & 0 & 0 & 1 & 1 & 1 & 1
\end{array}
\right]
=
\begin{bmatrix}
0\ 0\ 0\ 0\ 0\ 0\ 0 \\
0\ 0\ 0\ 1\ 1\ 1\ 1 \\
0\ 0\ 1\ 0\ 1\ 1\ 0 \\
0\ 0\ 1\ 1\ 0\ 0\ 1 \\
0\ 1\ 0\ 0\ 1\ 0\ 1 \\
0\ 1\ 0\ 1\ 0\ 1\ 0 \\
0\ 1\ 1\ 0\ 0\ 1\ 1 \\
0\ 1\ 1\ 1\ 1\ 0\ 0 \\
1\ 0\ 0\ 0\ 0\ 1\ 1 \\
1\ 0\ 0\ 1\ 1\ 0\ 0 \\
1\ 0\ 1\ 0\ 1\ 0\ 1 \\
1\ 0\ 1\ 1\ 0\ 1\ 0 \\
1\ 1\ 0\ 0\ 1\ 1\ 0 \\
1\ 1\ 0\ 1\ 0\ 0\ 1 \\
1\ 1\ 1\ 0\ 0\ 0\ 0 \\
1\ 1\ 1\ 1\ 1\ 1\ 1
\end{bmatrix}
\tag{13}
$$

The logic circuit used to implement this process is given in Fig. 8. Note that this circuit does not show the interleaving of the message and parity bits and additional circuitry is required for that purpose. However, it shows the calculation of parity bits.

In order to explain the function of the receiver, we define

$$
H = [Q'I] \tag{14}
$$

where prime denotes the transposition operator. This is called the parity-check matrix and has the property that

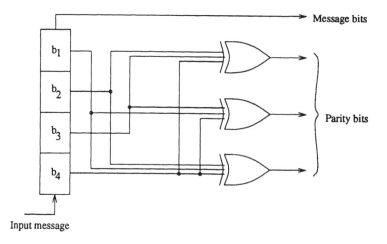

FIG. 8 Hamming encoder.

$$CH' = (0,0, \ldots ,0) \tag{15}$$

Now, suppose the received codeword C_r is the original codeword and an N-tuple of errors is added mod-2 to it. Therefore,

$$C_r = C \oplus E \tag{16}$$

The product of C_r and H' (modulo-2) will produce the so-called syndrome S.

$$S = C_r H' = (C \oplus E)H' = EH' \tag{17}$$

If the error N-tuple is $(0,0, \ldots ,0)$, S will be zero and the transmission is declared error free. However, less likely situations exist in which the syndrome is zero while errors still have occurred. Nonetheless, we choose the codeword having the smallest Hamming distance from the received codeword. This method is based on the *maximum-likelihood detection rule*. The systematic method for decoding involves calculation of the syndrome from the received codeword and a table lookup for the most likely error pattern. This estimated error pattern then is added modulo-2 to the received vector to calculate the most likely codeword. The schematic diagram for the decoder is given in Fig. 9.

Cyclic Codes

Cyclic codes are a subset of linear block codes. These codes are such that each codeword is a cyclically shifted version of a previous codeword. If we assume that a codeword is loaded into a shift register with the output of the last flip-flop connected to the input of the first, each shift of the register will generate another member of the code set. A codeword of length N can be represented as a polynomial of degree $N - 1$ as follows:

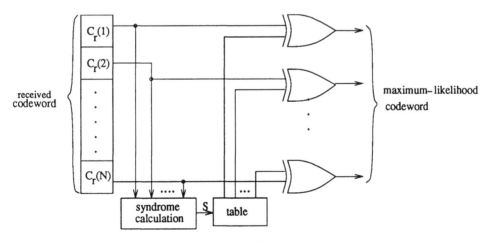

FIG. 9 Maximum-likelihood decoder.

$$C(x) = c_{N-1}x^{N-1} + c_{N-2}x^{N-2} + \ldots + c_0 \tag{18}$$

Polynomials representing codewords have binary coefficients and their addition and subtraction is carried out according to mod-2 addition.

For systematic codes, we can define the codeblock and parity polynomials as

$$B(x) = B_{b-1}x^{b-1} + B_{b-2}x^{b-2} + \ldots + B_0 \tag{19}$$

$$P(x) = P_{N-b-1}x^{N-b-1} + P_{N-b-2}x^{N-b-2} + \ldots + P_0 \tag{20}$$

and the codeword polynomials as

$$C(x) = x^{N-b}B(x) + P(x) \tag{21}$$

A cyclic code (N,b) is defined by a generator polynomial:

$$G(x) = x^{N-b} + g_{N-b-1}x^{N-b-1} + \ldots + g_1x + 1 \tag{22}$$

$G(x)$ divides each and every code polynomial in the set. Therefore, we have

$$C(x) = Q(x) \cdot G(x) \tag{23}$$

Equations (21) and (23) can be combined to give Eq. (24) with due regard to the fact that subtraction is the same as mod-2 addition.

$$\frac{x^{N-b}\,B(x)}{G(x)} = Q(x) + \frac{P(x)}{G(x)} \tag{24}$$

This equation signifies that $P(x)$ is the remainder of the division of $x^{N-k}B(x)$ by $G(x)$. To implement such an encoder, we calculate $P(x)$ and transmit it after the corresponding message. At the receiver, the syndrome polynomial is calculated in a similar manner.

$$S(x) = Rem\left[\frac{C_r(x)}{G(x)}\right] \tag{25}$$

Here, Rem denotes "remainder of the division." If $S(x)$ is zero, $C_r(x)$ is a valid code polynomial. Otherwise, an error is detected.

Convolutional Codes

In convolutional codes, the code bits are generated in such a way that they not only are dependent on the message bit in that time interval, but also on the block of message bits within the previous $M + 1$ time intervals ($M > 1$). The ith element of the codeword is a weighted modulo-2 sum of the present and previous message bits. Thus, c_i can be written mathematically as follows:

$$c_i = b_{i-M} \, g_M \oplus \cdots \oplus b_{i-1} \, g_1 \oplus b_i \, g_0 = \sum_{j=0}^{M} b_{i-j} \cdot g_j \qquad (\text{mod} - 2) \quad (26)$$

Equation (26) is analogous to the convolution integral in linear systems theory that can be written as

$$c(t) = \int b(t - \tau) g(\tau) d\tau \qquad (27)$$

It is obvious that each bit has an influence on M succeeding bits of encoded information. To generate the extra bits needed for error control, enough EXCLUSIVE-OR gates are connected to the shift register carrying the message bits. The output of these gates then are multiplexed through the use of a commutator (see Fig. 10). In this figure, $N = 2$, $b = 1$, and $M = 2$. Also, $g_2 = g_1 = g_0 = 1$.

The code and parity bits are thus

$$c_i = b_{i-2} \oplus b_{i-1} \oplus b_i$$
$$p_i = b_{i-2} \oplus b_i \qquad (28)$$

The values c_i and p_i are interleaved by the commutator A. The output bit stream will be

$$C = c_1 p_1 c_2 p_2 c_3 p_3 \cdots \qquad (29)$$

For more information on convolutional and error-correcting codes in general, the reader is referred to Refs. 7 and 8.

Digital Transmission

The digital baseband signals, because of significant low-frequency content, cannot be transmitted through channels with band-pass characteristics. There are

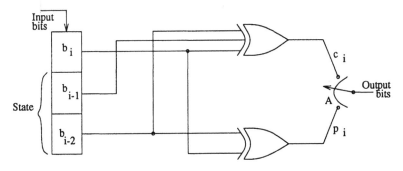

FIG. 10 Convolutional encoder.

three general methods of modifying the baseband signal so that it can be transmitted efficiently through band-pass channels: amplitude-shift keying (ASK), frequency-shift keying, (FSK), and phase-shift keying (PSK). These techniques are similar to the analog modulations amplitude modulation (AM), frequency modulation (FM), and pulse modulation (PM), respectively.

Amplitude-Shift Keying

In ASK, the binary digits 0 and 1 are transmitted through the channel by sinusoidal waveforms with the same frequency but different amplitudes (Fig. 11-*a*). A special case of ASK is on–off keying (OOK) in which a zero amplitude signal is transmitted for binary digit 0 (Fig. 11-*b*).

There are two approaches for generating ASK waveforms. One technique starts with the baseband signal and uses this to amplitude modulate a sinusoidal carrier. Another approach is to generate the AM wave by switching an oscillator on and off, depending on the binary digits. Similarly, for detection of ASK, two approaches can be used, coherent and incoherent detection (Fig. 12). In coherent detection, a synchronous local oscillator is used. In incoherent detection, the detector is a nonlinear device.

Frequency-Shift Keying

In binary FSK, the frequency of the carrier assumes only one of the two values depending on the binary digits (Fig. 11-*c*). In analog communication, FM is preferred over AM because of better performance in the presence of noise.

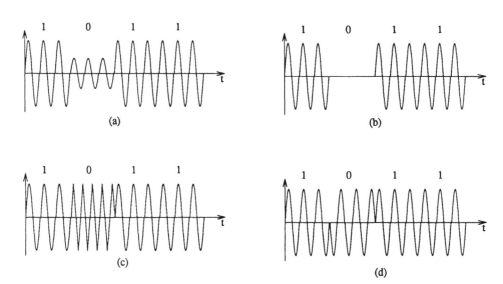

FIG. 11 Digital transmission waveforms: *a*, amplitude-shift keying; *b*, on–off keying; *c*, frequency-shift keying; *d*, phase-shift keying.

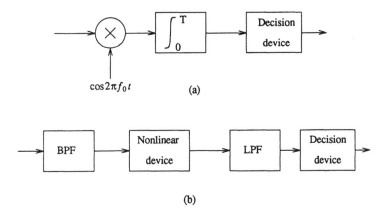

(a)

(b)

FIG. 12 Amplitude-shift keying detection: *a*, coherent; *b*, incoherent.

There are other reasons to use FM for data communications. Among these are some extremely simple implementations for the encoders and decoders.

The FSK waveform can be considered as the superposition of two ASK signals. One is the ASK signal that results from modulating a carrier with frequency f_1 from the baseband signal. The other results from modulating a carrier with frequency f_0 with the complement of the baseband signal. Figure 13 shows coherent and incoherent detection of FSK.

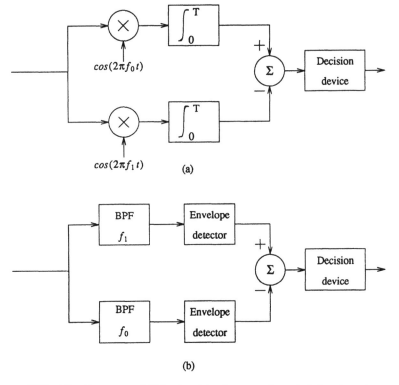

(a)

(b)

FIG. 13 Frequency-shift keying detection: *a*, coherent; *b*, incoherent.

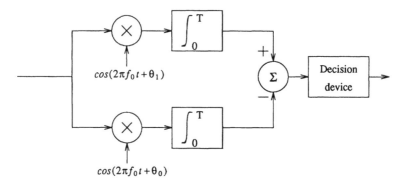

FIG. 14 Coherent detection for phase-shift keying.

Phase-Shift Keying

In binary PSK, the frequency of the carrier stays constant while the phase shift takes on one of the two constant values. The signals that transmit a 0 or 1 are expressed as

$$s_0(t) = A \cos(2\pi f_0 t + \theta_0)$$
$$s_1(t) = A \cos(2\pi f_0 t + \theta_1)$$

where θ_0 and θ_1 are constant phase shifts. For PSK detection, a matched filter (coherent detector) may be used (Fig. 14). Note that incoherent detectors could not be used for PSK. Incoherent detectors "throw away" all phase information.

In differential PSK, the information resides in the changes in the phase of the received signal and incoherent detection is possible (Fig. 15).

Summary

In this article, we discuss the advantages of digital communications systems as compared to their analog counterparts. The comparison criteria are noise

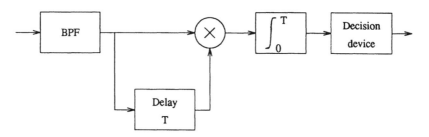

FIG. 15 Incoherent detection for differential phase-shift keying.

immunity, efficiency, and modularity in design. Coded pulse modulation is treated in some depth along with discussions of quantization and companding. Synchronization and timing recovery in digital communications systems also is covered in some detail.

Coding in digital communications systems is discussed in general from the point of view of efficiency and error performance. Error-control coding is discussed in detail and includes error detection and correction. Linear block codes, cyclic codes, and convolutional coding systems are discussed with various degrees of detail.

Finally, various digital modulation systems are described briefly. Due to the complexity of subject matter, especially in the coding section, relevant reference material is cited.

References

1. Noorchashm, M. R., Analog Modulation. In: *The Froehlich/Kent Encyclopedia of Telecommunications*, Vol. 1 (F. E. Froehlich and A. Kent, eds.), Marcel Dekker, New York, 1991, pp. 249–268.
2. Noorchashm, M. R., Conversion of Modulated Signals into Baseband: Analog Demodulation and Detection. In: *The Froehlich/Kent Encyclopedia of Telecommunications*, Vol. 4 (F. E. Froehlich and A. Kent, eds.), Marcel Dekker, New York, 1992, pp. 411–428.
3. Haykin, S., *Communication Systems*, John Wiley and Sons, New York, 1978.
4. Roden, M. S., *Analog and Digital Communication Systems*, Prentice-Hall, Englewood Cliffs, NJ, 1991.
5. Halapin, R. A., Compression of Data and Text. In: *The Froehlich/Kent Encyclopedia of Telecommunications*, Vol. 4 (F. E. Froehlich and A. Kent, eds.), Marcel Dekker, New York, 1992, pp. 165–177.
6. Lathi, B. P., *Digital and Analog Communication Systems*, 2d ed., Holt, Rinehart and Winston, Orlando, FL, 1989.
7. Carlson, A. B., *Communication Systems*, McGraw-Hill, New York, 1986.
8. Peterson, W. W., and Weldon, E. J., *Error Correcting Codes*, MIT Press, Cambridge, MA, 1990.

M. JAVAD PEYROVIAN
MAHMOOD R. NOORCHASHM

Digital Modulation Techniques

Introduction

Digital modulation techniques have played a crucial role in modern digital transmission. The overwhelming demand for digital communications toward integrated services digital network (ISDN) and toward digital mobile and digital cellular radio systems stimulates more and more research in this area (1,2). Digital modulator-demodulator (modem) principles, applications, and state-of-the-art developments are described in this article. The main thrust is on modem techniques that have been developed by the article co-authors and their colleagues (Dr. Feher's research teams at the University of California, Davis, and at the University of Ottawa, Canada). Modern techniques invented by other research engineers and teams are mentioned only briefly.

A digital modem can be regarded as a means to transfer a data source of a binary digit (bit) stream (i.e., a series of 1s and 0s) from one place to another through such various channels as cable, radio, satellite, and optical fibers. A few important terms are described first since they are used very often in the digital modulation world.

Spectral Efficiency.   Spectral efficiency refers to the number of bits transmitted per second in a 1-hertz (Hz) bandwidth (bps/Hz). If we transmit data at 135 megabits (mega meaning 10^6) per second (Mbps) in a 30-megahertz (MHz) bandwidth, the spectral efficiency is 135/30 = 4.5 bps/Hz, which can be achieved by a 64-state quadrature amplitude modulation (QAM) modem. High spectral efficiency (i.e., more bps/Hz) increasingly is desired in many applications, such as voiceband telephone channels and digital radios. State-of-the-art 256QAM modems achieving 6.67 bps/Hz have been developed and are in operation (1,3). Such higher spectral efficiency modems as 512QAM and beyond are likely to be developed in the 1990s throughout the world.

Bit Error Ratio.   The bit error ratio (BER), also known as bit error rate, is a performance measure of digital systems. A BER of 10^{-13} in a 135-Mbps system would mean 1 bit error occurs in a day. Such a low BER performance is required in many applications (e.g., long-haul digital radio).

Additive White Gaussian Noise.   Noise is encountered in practical systems and generally is assumed to have the following properties: (1) it is additive to the desired signal, (2) it has a flat spectrum in the frequency domain, and (3) the statistical distribution of noise voltage in the time domain is Gaussian (or normal). Of course, there are other noise sources besides additive white Gaussian noise (AWGN) that are impulsive in nature and require different treatments.

Signal-to-Noise Ratio and Carrier-to-Noise Ratio.   Generally speaking, BER is a function of the signal-to-noise ratio (SNR) or carrier-to-noise ratio (CNR). Plots of BER versus CNR (or SNR) are a performance measure of digital transmission systems. The plots resemble waterfalls so that "waterfall" curves also are used informally. Normally, SNR refers to a baseband system (i.e., without modulation) and CNR refers to modulated systems since a carrier is involved. In any case, SNR and CNR are equivalent.

Figure 1 summarizes spectral efficiency as a function of CNR for various modulation schemes without coding for probability of error (P_e) being equal to 10^{-8}. Also shown in Fig. 1 is the famous Shannon limit, which represents the best performance one can achieve. Note that, for uncoded systems, in terms of CNR the performance is about 9 decibels (dB) away from the Shannon limit. Thus, there is ample room for improvement by coding. As discussed in the section on trellis-coded modulation (TCM), TCM reduces the gap between the practical system performance and the Shannon limit and has been an exciting research area over the past decade.

A plot of BER versus CNR for various modulation formats is shown in Fig. 2. This serves as a guide for system designers. A number of digital modulation techniques and their abbreviations are summarized in Table 1.

In the next section, the performance criteria for digital transmission are introduced. The Nyquist criterion of zero intersymbol interference (ISI) in a band-limited channel and such important parameters as peak eye closure and root mean squared (rms) eye closure are detailed.

In the third section, digital modems with spectral efficiency less than 2 bps/Hz are described. The principles of quaternary phase-shift keying (QPSK), offset QPSK (OQPSK), and minimum-shift keying (MSK) are explained. Continuous phase modulation (CPM) is mentioned briefly.

In the fourth section, digital modems with spectral efficiency higher than 2 bps/Hz are introduced. State-of-the-art 1024QAM and 256QAM systems with spectral efficiency higher than 6 bps/Hz are described. Performance degradations of these highly spectrum-efficient QAM systems due to various distortions are presented. An interesting aspect of quadrature partial response signaling (QPRS) systems operated above the Nyquist rate is reported.

In the fifth section, pilot-aided QAM systems are described. The reason for pilot insertion is explained. Techniques recently developed for pilot insertion are highlighted.

In the sixth section, TCM, a combined coding and modulation technique, is described. This technique is important to the development of modern high-speed telephony modems and satellite modems. Both two-dimensional (2D) and multi-dimensional TCM schemes are highlighted.

In the seventh section, the need for digital modulation schemes without requiring carrier recovery is explained. Advanced differential detection techniques are presented. In the eighth section, such related work as operation in nonlinear channels is covered.

The concluding section of this article includes useful references for those interested in further exploration.

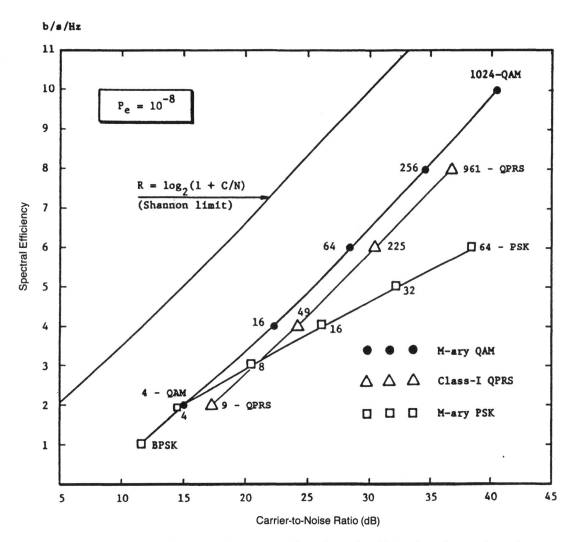

FIG. 1 Spectral efficiency or bits per second per hertz (bps/Hz) and carrier-to-noise ratio (CNR) requirement at $P_e = 10^{-8}$ of various modulation systems. The average CNR is specified in the double-sided Nyquist bandwidth that equals the symbol rate (QAM = quadrature amplitude modulation; QPRS = quadrature partial response signaling; PSK = phase-shift keying; BPSK = binary phase-shift keying).

Fundamentals of Pulse Shaping

Filtering is inevitable in almost all modems. Many modem system problems can be perceived by studying the simplest type of pulse amplitude modulation (PAM) systems. For example, QAM consists simply of two PAM signals modulated in quadrature.

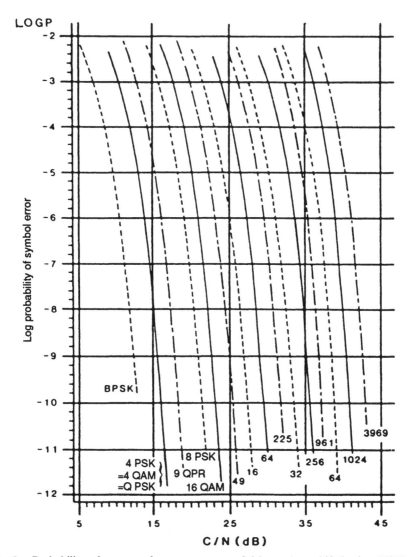

FIG. 2   Probability of error performance curves of *M*-ary phase-shift keying (PSK), *M*-ary quadrature amplitude modulation (QAM), and *M*-ary quadrature partial response (QPR) modulation systems versus the carrier-to-thermal noise ratio in decibels, white Gaussian noise channel only (QPSK = quaternary phase-shift keying). (From Ref. 3, printed with permission of the Institute of Electrical and Electronics Engineers, © 1987 IEEE.)

Pulse Amplitude Modulation Systems and the Nyquist Criterion

A PAM system block diagram is shown in Fig. 3 (4). For simplicity, the input symbols a_n (with a signaling interval T) are envisioned as a sequence of narrow impulses the shape $g(t - nT)$ of which at the channel input is determined by the transmit filter $G_T(\omega)$. For an L-level system, we assume that the transmitted

TABLE 1   Typical Digital Modulation Techniques

Abbreviation	Alternate Abbreviation	Descriptive Name of Modem Technique
DSBSCAM	DSB-SC	Double sideband suppressed carrier amplitude modulation
PSK	BPSK	Phase-shift keying; binary PSK
DPSK	DBPSK	Differential PSK; differential binary PSK (*no* carrier recovery)
DEPSK	DEBPSK	Differentially encoded binary PSK (*with* carrier recovery)
QPSK	CQPSK	Quadrature (quaternary) PSK; coherent QPSK
OQPSK	OKQPSK or SQPSK	Offset QPSK; offset-keyed QPSK; staggered QPSK
DQPSK		Differential QPSK (*no* carrier recovery)
DEQPSK		Differentially encoded QPSK (*with* carrier recovery)
MSK	FFSK	Minimum-shift keying; fast frequency shift keying
DMSK		Differential MSK
GMSK		Generalized or Gaussian MSK
TFM		Tamed frequency modulation
Multi-h FM	Correlative FM	Multi-index FM; correlative FM; duobinary FM
IJF-OQPSK	NLF-OQPSK	Intersymbol jitter-free OQPSK; Nonlinearly filtered OQPSK
SQAM		Superposed quadrature amplitude modulation
CPM		Continuous phase modulation
CPFSK		Continuous phase frequency shift keying
QAM	*M*-ary QAM	Quadrature amplitude modulation
APK		Amplitude phase keying
QPRS	QPR	Quadrature partial response signaling
SSB		Single-sideband amplitude modulation
TCM		Trellis-coded modulation

levels are $\pm d$, $\pm 3d$, . . . , $\pm(L - 1)d$ where $2d$ is the distance between adjacent levels. In what follows, we shall derive the Nyquist criterion for zero intersymbol interference (ISI) transmission.

Passing through the channel, the pulses $g(t)$ suffer time dispersion and added noise. The receive filter effects signal and noise shaping and is followed by a sampler and a slicer (threshold detector) that look at the received waveform at times $kT + t_0$ and determine in each case which of the L allowed symbols is most likely. (The reference time t_0 is inserted to account for the delay in transmission through the channel.) At the output of the receive filter, a series of these pulses $x(t)$ is determined by the overall system frequency response $X(\omega)$ = $G_T(\omega)C(\omega)G_R(\omega)$, that is,

$$x(t) = \frac{1}{2\pi} \int_{-\infty}^{\infty} G_T(\omega)C(\omega)G_R(\omega)e^{i\omega t}d\omega \qquad (1)$$

With the addition of noise, the output waveform $y(t)$ becomes

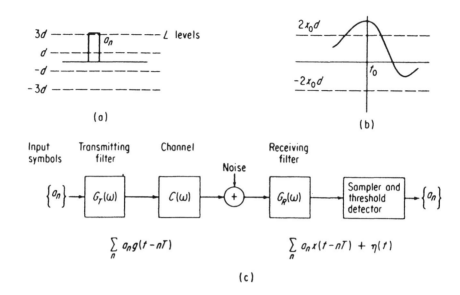

FIG. 3 A pulse amplitude modulation (PAM) system: a, allowed transmitter levels; b, decision thresholds (indicated by dashed lines); c, PAM system with simple threshold detection. (After Ref. 4, with permission from McGraw-Hill, © 1968.)

$$y(t) = \sum_n a_n x(t - nT) + \eta(t) \tag{2}$$

where $\eta(t)$ is the additive noise. At time $kT + t_0$, the desired output voltage is a_k; however, the actual value is

$$y_k = y(kT + t_0) = \sum_n a_n x(kT - nT + t_0) + \eta(kT + t_0)$$

$$= \sum_n a_n x_{k-n} + \eta_k \tag{3}$$

To isolate the desired amplitude a_k, Eq. (3) is rewritten in the form

$$y_k = x_0\left(a_k + \frac{1}{x_0} \sum_{\substack{n \\ n \neq k}} a_n x_{k-n} + \frac{\eta_k}{x_0}\right) \tag{4}$$

The second and third terms of Eq. (4) represent ISI and noise, respectively. The ISI arises from the overlapping tails of other pulses adding to the particular pulse $\mathbf{a}_k\, x(t - kT)$ that is examined at the kth sampling time. Notice from Eq. (4) that ISI can be eliminated only by making $x_n = 0$ for all $n \neq 0$. Examples of such a pulse can be seen in Fig. 4-a. It is clear that these pulses can be amplitude modulated and transmitted at T-second (s) intervals without overlap at the sampling instants. The reference time delay t_0 and the pulse amplitude x_0 are arbitrary. To be specific, we set $t_0 = 0$ and $x_0 = 1$. To change these values,

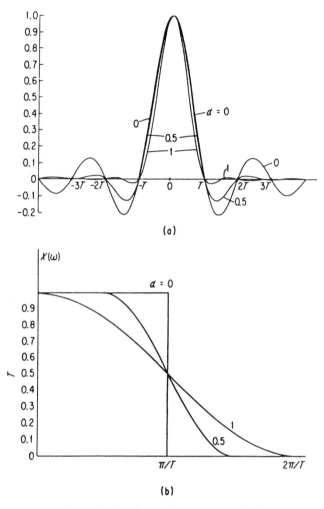

FIG. 4 Raised-cosine pulse shaping: a, time responses; b, frequency responses.

it is only necessary to add ωt_0 to the phase of $X(\omega)$ and to scale $X(\omega)$ by the desired gain x_0. These requirements that $X(\omega)$ must meet now can be written in the form

$$x_n = \frac{1}{2\pi} \int_{-\infty}^{\infty} X(\omega)e^{i\omega nT}d\omega = \delta_{n0} = \begin{cases} 0 & n \neq 0 \\ 1 & n = 0 \end{cases} \tag{5}$$

The integrals in Eq. (5) can be divided into segments corresponding to Nyquist bandwidths of the frequency range

$$x_n = \frac{1}{2\pi} \sum_k \int_{(2k-1)\pi/T}^{(2k+1)\pi/T} X(\omega)e^{i\omega nT}d\omega = \delta_{n0} \tag{6}$$

A change of variables gives

$$x_n = \frac{1}{2\pi} \sum_k \int_{-\pi/T}^{\pi/T} X\left(\omega + \frac{2k\pi}{T}\right) e^{i\omega T} d\omega = \delta_{n0} \qquad (7)$$

and, assuming the integration and summation may be interchanged, Eq. (7) may be written in terms of a single equivalent Nyquist bandwidth channel $X_{eq}(\omega)$,

$$x_n = \frac{1}{2\pi} \int_{-\pi/T}^{\pi/T} X_{eq}(\omega) e^{i\omega nT} d\omega = \delta_{n0} \qquad (8)$$

where

$$X_{eq}(\omega) = \begin{cases} \sum_k X\left(\omega + \frac{2k\pi}{T}\right) & |\omega| \leq \frac{\pi}{T} \\ 0 & |\omega| > \frac{\pi}{T} \end{cases} \qquad (9)$$

The only $X_{eq}(\omega)$ meeting Eq. (8) is the rectangular, zero phase characteristic

$$X_{eq}(\text{Nyquist}) = \begin{cases} T & |\omega| \leq \frac{\pi}{T} \\ 0 & \text{elsewhere} \end{cases} \qquad (10)$$

Thus, the equivalent Nyquist bandwidth channel must satisfy Eq. (10) if $X(\omega)$ is to be free from causing ISI. This way of stating the Nyquist criterion is very useful when a real channel performance is evaluated.

In nearly all practical cases, the actual bandwidth available does not exceed twice the minimum Nyquist bandwidth, which is equal to the symbol rate. If this restriction is made, that is, if

$$X(\omega) = 0 \quad |\omega| > \frac{2\pi}{T} \qquad (11)$$

then

$$X_{eq}(\omega) = X(\omega) + X^*\left(\frac{2\pi}{T} - \omega\right) \qquad (12)$$

where * denotes "complex conjugate."

Construction of $X_{eq}(\omega)$ is illustrated in Fig. 5, in which $X(\omega)$ is assumed to be real. One class of Nyquist characteristics is the so-called raised-cosine characteristic (4):

$$X(\omega) = \begin{cases} T & 0 \leq \omega \leq \frac{\pi}{T}(1 - \alpha) \\ \frac{T}{2}\left\{1 - \sin\left[\frac{T}{2\alpha}\left(\omega - \frac{\pi}{T}\right)\right]\right\} & \frac{\pi}{T}(1 - \alpha) \leq \omega \leq \frac{\pi}{T}(1 + \alpha) \end{cases} \qquad (13)$$

The impulse responses corresponding to these characteristics are

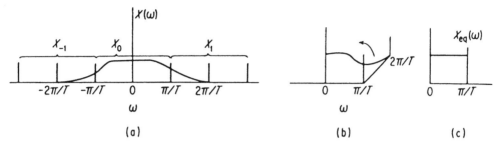

FIG. 5 Construction of $X_{eq}(\omega)$: a, equivalent Nyquist bandwidth channel is superposition of X_{-1}, X_0, X_1; b, folding of the portion of characteristic in excess of Nyquist bandwidth; c, equivalent channel must be perfect. (After Ref. 4, with permission from McGraw-Hill, © 1968.)

$$x(t) = \frac{\sin \pi t/T}{\pi t/T} \frac{\cos \alpha\pi t/T}{1 - 4\alpha^2 t^2/T^2} \tag{14}$$

Some of the raised-cosine characteristics and their impulse responses are shown in Fig. 4. The smaller the roll-off factor α is, the less bandwidth is required for a given rate of transmission, resulting in more efficient utilization of the spectrum. However, the problems of timing and channel equalization become more difficult. It can be observed from Fig. 4-a that the response $x(t)$ becomes more oscillatory as α decreases. State-of-the-art filter design can achieve small α in the range between 0.1 and 0.2 with reasonably small residual ISI.

Performance of Ideal Pulse Amplitude Modulation Systems with Additive Gaussian Noise

The probability of error P_e is probably the most important performance measure of digital systems. Here, we derive the P_e performance for ideal Nyquist-filtered PAM systems (i.e., no ISI) with AWGN that has a one-sided power spectral density (PSD) N_0. We assume that the pulse shaping $X(\omega)$ is raised cosine and is split equally between transmit and receive filters, each of which has the characteristic $X(\omega)^{1/2}$.

The receiver looks for the unattenuated transmitted levels and places slicing levels at 0, $\pm 2d$, . . . , $\pm(L - 2)d$. An error occurs when the noise at a sampling time exceeds in magnitude the decision distance d. However, the outside two levels can be in error only in one direction each; thus, a factor of $(1 - 1/L)$ is used to weight the probability of noise magnitude exceeding d. The P_e is

$$P_e = \left(1 - \frac{1}{L}\right)P[|\eta| > d] \tag{15}$$

The noise voltage at the receive filter output is still Gaussian with variance

$$\sigma^2 = \frac{N_0/2}{2\pi} \int_{-\infty}^{\infty} |X^{1/2}\omega)|^2 \, d\omega = N_0/2 \qquad (16)$$

and the probability of error is shown easily as

$$P_e = 2\left(1 - \frac{1}{L}\right)Q\left[\frac{d}{\sigma}\right] \qquad (17)$$

and $Q(x)$ is the Gaussian integral function (see Table 2). It is instructive to express P_e in terms of the transmitted power S ($S = (L^2 - 1)d^2/3T$) and the noise power measured in the minimum Nyquist bandwidth $[-\pi/T, \pi/T]$ before the receive filter, which is denoted by $N = (N_0/2T)$. After a straightforward manipulation of Eq. (17), the expression becomes (4)

$$P_e = 2\left(1 - \frac{1}{L}\right)Q\left[\left(\frac{3}{L^2 - 1}\frac{S}{N}\right)^{1/2}\right] \qquad (18)$$

A sequence of curves for P_e as a function of SNR for $L = 2, 4, 8,$ and 16 is shown in Fig. 6. As L becomes large, about 6 dB additional power is required to double the number of levels and retain the same error rate.

Very often, P_e is expressed in terms of energy per bit to noise-density ratio (E_b/N_0). For the binary ($L = 2$) case, $E_b = S \cdot T$, $N = N_0/2T$, and SNR $= 2E_b/N_0$. Equation (18) can be simplified as

$$P_e = Q\left(\sqrt{\frac{S}{N}}\right) = Q\left(\sqrt{\frac{2E_b}{N_0}}\right) = \frac{1}{2} \, erfc\left(\sqrt{\frac{E_b}{N_0}}\right). \qquad (19)$$

As discussed in the section, "Digital Modulation with Spectral Efficiency Less than 2 Bits per Second per Hertz," binary phase-shift keying (BPSK) and QPSK have exactly the same P_e expression in terms of E_b/N_0 since they are equivalent to a baseband binary system.

The Gaussian integral function $Q(x)$ and the complementary error function erfc(y) are very important for understanding the probability of error of digital transmission systems. Table 2 summarizes the relationships and numerical results of $Q(x)$ and erfc(y). Note that the numerical results in Table 2 are tabulated intentionally for $2Q(x)$ instead of $Q(x)$. This is because, in a multilevel PAM system, it is shown in Eq. (17) that P_e is proportional to $2Q(d/\sigma)$. Indeed, $2Q(d/\sigma)$ represents the probability that the noise voltage exceeds the decision distance. Thus, a plot of $2Q(x)$ such as that in Fig. 7 is very useful since it gives an approximation of symbol error rate versus decision-distance-to-noise-rms-voltage-ratio (ddnr). The plot resembles a waterfall, that is, the curve becomes steeper as ddnr increases. A probability of error distribution is illustrated in Fig. 8 for a four-level PAM system.

Eye-Closure Criteria

Due to hardware imperfection and channel dispersion, the received signal can never completely satisfy the Nyquist criterion. Therefore, it is useful to define

TABLE 2   The $Q(x)$ and erfc (y) Functions

Definition: $Q(x) = \int_x^\infty \frac{1}{\sqrt{2\pi}} e^{-z^2/2} \, dz$

1. $Q(0) = \frac{1}{2}; Q(-y) = 1 - Q(y),$ when $y \geq 0$

2. $Q(y) \approx \frac{1}{y\sqrt{2\pi}} e^{-y^2/2}$ when $y \gg 1$ (approximation may be used for $y > 4$)

3. $\text{erfc}^a (y) \triangleq \frac{2}{\sqrt{\pi}} \int_y^\infty e^{-z^2} \, dz = 2Q(\sqrt{2}y), y > 0$

4. $\text{erfc} (y) = 1 - \text{erf}(y)$

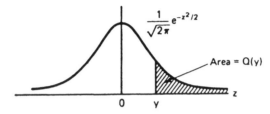

X(dB)	X(Linear)	$2Q(X)$	X(dB)	X(Linear)	$2Q(X)$
0.0	1.0000	$0.3173E + 00$	15.2	5.7544	$0.8695E - 08$
0.5	1.0593	$0.2895E + 00$	15.4	5.8884	$0.3899E - 08$
1.0	1.1220	$0.2619E + 00$	15.6	0.0256	$0.1685E - 08$
1.5	1.1885	$0.2346E + 00$	15.8	6.1659	$0.7006E - 09$
2.0	1.2589	$0.2081E + 00$	16.0	6.3096	$0.2798E - 09$
2.5	1.3335	$0.1824E + 00$	16.2	6.4565	$0.1071E - 09$
3.0	1.4125	$0.1578E + 00$	16.4	6.6069	$0.3924E - 10$
3.5	1.4962	$0.1346E + 00$	16.6	6.7608	$0.1372E - 10$
4.0	1.5849	$0.1130E + 00$	16.8	6.9183	$0.4571E - 11$
4.5	1.6788	$0.9319E - 01$	17.0	7.0794	$0.1447E - 11$
5.0	1.7783	$0.7536E - 01$	17.2	7.2444	$0.4345E - 12$
5.5	1.8836	$0.5961E - 01$	17.4	7.4131	$0.1234E - 12$
6.0	1.9953	$0.4601E - 01$	17.6	7.5858	$0.3305E - 13$
6.5	2.1135	$0.3456E - 01$	17.8	7.7625	$0.8330E - 14$
7.0	2.2387	$0.2517E - 01$	18.0	7.9433	$0.1969E - 14$
7.5	2.3714	$0.1772E - 01$	18.2	8.1283	$0.4353E - 15$
8.0	2.5119	$0.1201E - 01$	18.4	8.3176	$0.8974E - 16$
8.5	2.6607	$0.7797E - 02$	18.6	8.5114	$0.1719E - 16$
9.0	2.8184	$0.4827E - 02$	18.8	8.7096	$0.3049E - 17$
9.5	2.9854	$0.2832E - 02$	19.0	8.9125	$0.4990E - 18$
10.0	3.1623	$0.1565E - 02$	19.2	9.1201	$0.7505E - 19$
10.5	3.3497	$0.8091E - 03$	19.4	9.3325	$0.1034E - 19$
11.0	3.5481	$0.3880E - 03$	19.6	9.5499	$0.1298E - 20$
11.5	3.7584	$0.1710E - 03$	19.8	9.7724	$0.1480E - 21$
12.0	3.9811	$0.6861E - 04$	20.0	10.0000	$0.1524E - 22$
12.5	4.2170	$0.2476E - 04$	20.2	10.2329	$0.1412E - 23$
13.0	4.4668	$0.7938E - 05$	20.4	10.4713	$0.1171E - 24$
13.5	4.7315	$0.2229E - 05$	20.6	10.7152	$0.8640E - 26$

(continued)

TABLE 2 Continued

X(dB)	X(Linear)	$2Q(X)$	X(dB)	X(Linear)	$2Q(X)$
14.0	5.0119	$0.5390E - 06$	20.8	10.9648	$0.5645E - 27$
14.5	5.3088	$0.1103E - 06$	21.0	11.2202	$0.3247E - 28$
15.0	5.6234	$0.1872E - 07$	21.2	11.4815	$0.1634E - 29$
15.5	5.9566	$0.2575E - 08$	21.4	11.7490	$0.7149E - 31$
16.0	6.3096	$0.2798E - 09$	21.6	12.0226	$0.2703E - 32$
16.5	6.6834	$0.2334E - 10$	21.8	12.3027	$0.8763E - 34$
17.0	7.0795	$0.1447E - 11$	22.0	12.5892	$0.2421E - 35$
17.5	7.4989	$0.6433E - 13$	22.2	12.8825	$0.5648E - 37$
18.0	7.9433	$0.1969E - 14$	22.4	13.1825	$0.1106E - 38$
18.5	8.4140	$0.3964E - 16$	22.6	13.4896	$0.1801E - 40$
19.0	8.9125	$0.4989E - 18$	22.8	13.8038	$0.2417E - 42$
19.5	9.4406	$0.3706E - 20$	23.0	14.1253	$0.2651E - 44$
20.0	10.0000	$0.1524E - 22$	23.2	14.4544	$0.2352E - 46$
20.5	10.5925	$0.3227E - 25$	23.4	14.7911	$0.1673E - 48$
21.0	11.2202	$0.3247E - 28$	23.6	15.1356	$0.9433E - 51$
21.5	11.8850	$0.1416E - 31$	23.8	15.4881	$0.4173E - 53$
22.0	12.5892	$0.2420E - 35$	24.0	15.8489	$0.1431E - 55$
22.5	13.3352	$0.1444E - 39$	24.2	16.2181	$0.3758E - 58$
23.0	14.1254	$0.2650E - 44$	24.4	16.5958	$0.7469E - 61$
23.5	14.9623	$0.1294E - 49$	24.6	16.9824	$0.1108E - 63$
24.0	15.8489	$0.1430E - 55$	24.8	17.3780	$0.1212E - 66$
24.5	16.7880	$0.2986E - 62$	25.0	17.7828	$0.9612E - 70$
25.0	17.7828	$0.9607E - 70$	25.2	18.1970	$0.5453E - 73$

[a]In some references, the error function is defined somewhat differently.

some engineering parameters in order to judge system performance. The peak and rms eye-closure criteria are introduced in Ref. 4. Referring to Eq. (4), in the absence of noise the ISI for the sample y_k can be written as

$$\text{ISI} = y_k - x_0 a_k = \sum_{\substack{n \\ n \neq 0}} a_n x_{k-n} \tag{20}$$

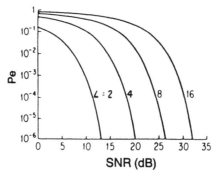

FIG. 6 Probability of error P_e for L-level pulse amplitude modulation where SNR is the signal-to-noise power ratio in the Nyquist bandwidth.

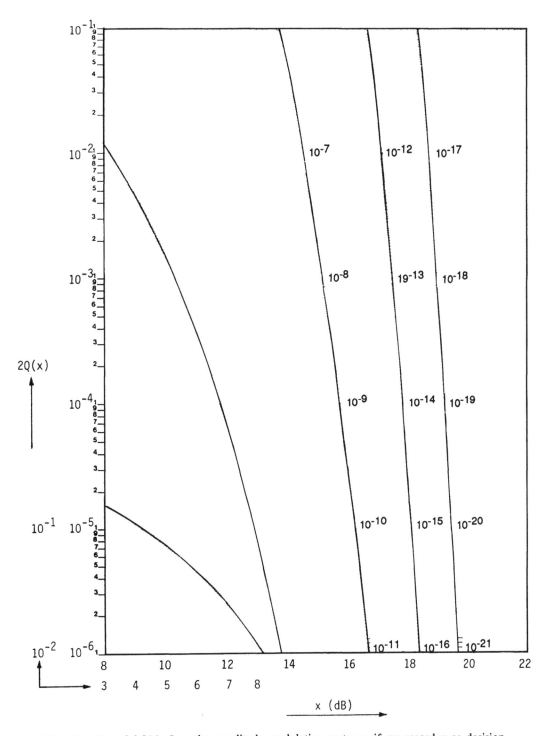

FIG. 7 Plot of $2Q(x)$. In pulse amplitude modulation systems, if we regard x as decision distance-to-noise-rms-voltage ratio (ddnr), then $2Q(x)$ represents the probability that the noise voltage exceeds the decision distance. In the plot, x is expressed in decibels (dB) (e.g., $x = 20$ dB if $x = 10$).

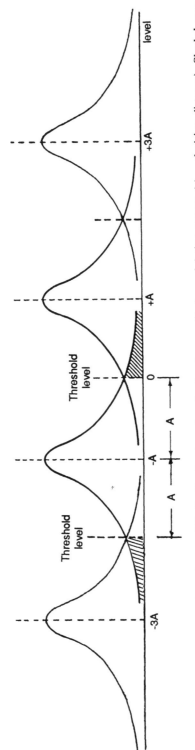

FIG. 8 Illustrative probability of error (P_e) distribution in a four-level pulse amplitude modulation system (A = decision distance). Shaded area represents P_e for a transmitted -1 state ($-A$ volts). In the absence of noise, the received signal sample would equal $-A$ volts. To make an error, the noise has to have a magnitude exceeding the decision distance of A volts.

The maximum perturbation of the received level is then

$$\text{Maximum perturbation} = (L - 1)d \sum_{\substack{n \\ n \neq 0}} |x_n| \tag{21}$$

Peak eye closure is defined as the maximum perturbation normalized to the decision distance $x_0 d$, that is,

$$\text{Peak eye closure} = (L - 1)D \tag{22}$$

where

$$D = \frac{1}{x_0} \sum_{\substack{n \\ n \neq 0}} |x_n| \tag{23}$$

is defined as the peak distortion of the system impulse response. In terms of the peak distortion D, the eye opening is

$$\text{Eye opening} = 1 - (L - 1)D \tag{24}$$

The eye opening may be used to bound the probability of error if the noise distribution is known. Gaussian noise commonly is assumed. For Gaussian noise of variance σ_N^2, the bound is (4)

$$\text{Max } P_e = P\{\eta > x_0 d[1 - (L - 1)D]\}$$
$$= Q\left\{\frac{x_0 d[1 - (L - 1)D]}{\sigma_N}\right\} \tag{25}$$

Thus, the problem of evaluating system performance using the eye-opening criterion reduces to calculation of the peak distortion D for the system impulse response.

Another useful parameter is the rms eye closure. The mean-squared ISI distortion in the system impulse response is defined as (4)

$$\mathcal{E}^2 = \frac{1}{x_0^2} \sum_{\substack{n \\ n \neq 0}} x_n^2 \tag{26}$$

$\mathcal{E}$ can be called *binary rms eye closure*. From Ref. 4, it can be shown that for an L-level PAM system the mean-squared eye closure $= (L^2 - 1)\mathcal{E}^2/3$. If we assume that noise and ISI are independent and Gaussian, the total mean-squared error is

$$\sigma^2 = \frac{d^2(L^2 - 1)}{3} \sum_{\substack{n \\ n \neq 0}} x_n^2 + \sigma_N^2 \tag{27}$$

and the probability of error would be

$$P_e = 2Q\left(\frac{x_0 d}{\sigma}\right) = 2Q\left\{\left[\frac{L^2 - 1}{3}\,\mathcal{E}^2 + \frac{\sigma_N^2}{x_0^2 d^2}\right]^{-\frac{1}{2}}\right\} \tag{28}$$

The assumption of Gaussian-distributed ISI is strictly not true. Thus, Eq. (28) should be used only as an approximation. However, the total rms error is still a very useful parameter for optimization of digital systems. Table 3 summarizes the definitions and formulas for the peak and rms eye-closure criteria.

Though the raised-cosine characteristic has been used widely, we should keep in mind that it is not the only one that satisfies the Nyquist criterion. What is important is that the "equivalent Nyquist bandwidth channel" should approximate, as much as possible, the rectangular shape with zero phase for minimum ISI.

In the beginning of this section, an ideal impulse train is assumed. In practice, a rectangular (so-called non-return-to-zero [NRZ]) pulse or any other pulse shapes can be used. Figure 9 illustrates a few commonly used pulse waveforms (also known as *line coding*) and the corresponding power spectra. If NRZ is used, all that is required is to equalize the NRZ spectrum to that of an ideal impulse. One can accomplish this by introducing an inversed spectrum of NRZ in the transmitter and/or receiver. For example, it often is seen in many applications that the transmit and receive filters are specified as the square root of the raised cosine plus x/sinx equalization. The x/sinx is to compensate for the NRZ-induced sinx/x spectrum.

Eye-Diagram Fundamentals

Channel imperfections frequently are evaluated by means of "eye diagrams" or "eye patterns" (5). These patterns can be observed on an oscilloscope if a signal such as $v_o(t)$ in Fig. 10-*a* is fed to the vertical input of an oscilloscope. The symbol-rate clock $c(t - nT_s)$ is fed to the external trigger of the oscilloscope. The

TABLE 3   Definitions for Peak and Root-Mean-Squared Distortion Criteria

Criterion	Formula	
	Peak	Rms
Impulse response, distortion	$D = \dfrac{1}{x_0} \displaystyle\sum_{\substack{n \\ n \neq 0}} \lvert x_n \rvert$	$\mathcal{E} = \dfrac{1}{x_0}\left(\displaystyle\sum_{\substack{n \\ n \neq 0}} x_n^2\right)^{\frac{1}{2}}$
L-level PAM, eye closure	$(L - 1)D$	$\left(\dfrac{L^2 - 1}{3}\right)^{\frac{1}{2}} \mathcal{E}$
Approximate probability of error for Gaussian noise, var σ_N^2	$Q\left\{\dfrac{x_0 d[1 - (L - 1)D]}{\sigma_N}\right\}$ (upper bound)	$2Q\left(\dfrac{L^2 - 1}{3}\,\mathcal{E}^2 + \dfrac{\sigma_N^2}{x_0^2\,d^2}\right)^{-\frac{1}{2}}$

Source: After Ref. 4, with permission from McGraw-Hill, © 1968.

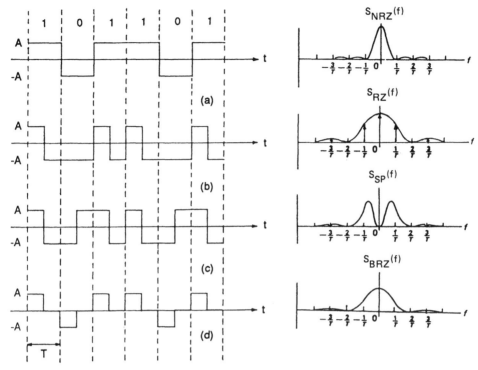

FIG. 9 Examples of digital data waveforms (also known as *line coding*) and corresponding power spectra: *a*, non-return to zero (NRZ); *b*, return to zero (RZ); *c*, split phase (SP or Manchester); *d*, bipolar return to zero (BRZ).

front trigger-delay adjustment, conveniently available on most oscilloscopes, assures that the displayed eye pattern is centered on the screen. The horizontal time base is set approximately equal to the symbol duration. The inherent persistence of the cathode-ray tube displays the superimposed segments of the $v_o(t)$ signal. The eye pattern of the pseudorandom binary sequence (PRBS) generator is displayed if the data output of this generator is connected directly to the vertical input of the oscilloscope, as shown in Fig. 10-*b*. The filtered (band-limited) PRBS signal would result in an eye pattern, illustrated in Fig. 10-*c*.

Digital Modulation with Spectral Efficiency Less than 2 Bits per Second per Hertz

BPSK and QPSK modulation schemes have been used widely in many such applications as satellite communications, early generation terrestrial microwave radios, and voiceband telephone channels. In this section, we present modulation schemes related to QPSK with spectral efficiency less than 2 bps/Hz.

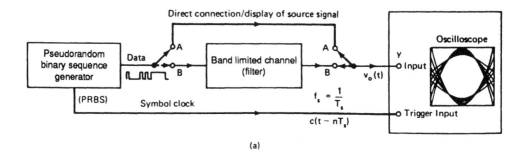

(a)

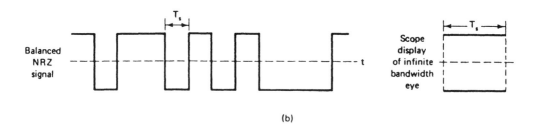

(b)

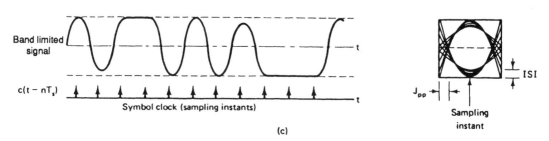

(c)

FIG. 10   Eye-diagram measurement setup and display: *a*, setup connections; *b*, illustrative eye diagram for an unfiltered nonreturn-to-zero signal; *c*, illustrative eye diagram for the corresponding filtered signal. (After Ref. 5, with permission from Prentice-Hall, © 1983.)

Quaternary Phase-Shift Keying, Offset Quaternary Phase-Shift Keying, and Minimum-Shift Keying Modulation Schemes

A block diagram of a conventional QPSK modem is shown in Fig. 11. The input NRZ datastream with a bit rate f_b is converted by a serial-to-parallel converter into two separate NRZ streams. One stream, $I(t)$, is in phase, and the other, $Q(t)$, is quadrature phase with a symbol rate $f_s = f_b/2$. The relationship between the input datastream and the I and Q streams is shown in Fig. 11-*b*. Both I and Q streams are applied separately to multipliers (also called *mixers*). The second input to the I multiplier is the carrier signal $\cos \omega_o t$, and the second input to the Q multiplier is the carrier signal shifted by 90° (i.e., $\sin \omega_o t$). The outputs of both multipliers are BPSK signals. The multiplier outputs then are

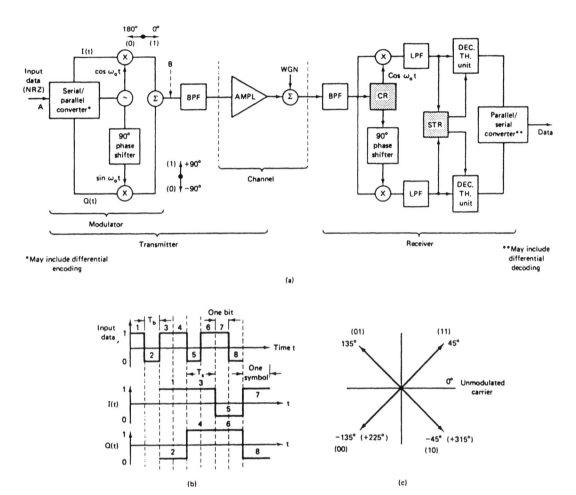

FIG. 11 Quaternary phase-shift keying system representation (Gray-coded vector presentation of signal states): *a*, block diagram (CR = carrier-recovery circuit; STR = symbol timing-recovery circuit; NRZ = nonreturn to zero; BPF = band-pass filter; LPF = low-pass filter; WGN = white Gaussian noise; DEC. TH. = decision threshold; AMPL = amplifier); *b*, modulator datastreams (data in bit number *n* is converted to *I/Q* baud number *n*; bit rate = $1/T_b$; baud or symbol rate = $1/T_s$; $T_b = T_s/2$); *c*, signal-space diagram. (Courtesy Ref. 77, © 1981.)

summed to give a four-phase signal. Thus, QPSK can be regarded as two BPSK systems operating in quadrature. We also can state that QPSK is 4QAM.

The four possible outputs of the modulator and their corresponding *IQ* digit combinations are shown in Fig. 11-*c*. Note that either 90° or 180° phase transitions are possible. For an unfiltered QPSK signal, phase transitions occur instantaneously and the signal has a constant-amplitude envelope. However, phase changes for filtered QPSK signals result in a varying envelope amplitude. In particular, a 180° phase change results in a momentary change to zero in envelope amplitude.

The QPSK signal at the modulator output normally is filtered to limit the radiated spectrum, amplified, then transmitted over the channel to the receiver input. Because the I and Q modulated signals are in quadrature (orthogonal), the receiver is able to demodulate and regenerate them independently, operating effectively as two BPSK receivers. The regenerated I and Q streams then are recombined in a parallel-to-serial converter to form the original datastream; however, this stream, of course, is subject to error because of the effect of noise.

A block diagram of an OQPSK system is shown in Fig. 12-*a*. A delay of T_b = $T_s/2$ is introduced in the Q stream. This results in the relationship between the I and Q streams and the input datastream shown in Fig. 12-*b*. The important result is that only 90° phase transitions occur in the modulator output signals. Like QPSK, an unfiltered OQPSK signal has a constant-amplitude envelope. However, for filtered OQPSK, the result is a maximum envelope fluctuation of 3 dB (70%), compared to 100% for the conventional QPSK signal. This lower envelope variation imparts certain advantages to OQPSK as compared to QPSK in both nonlinear satellite and line-of-sight microwave systems (6). The receiver shown in Fig. 12-*a* is identical to that shown in Fig. 11-*a* for QPSK, with the exception that the regenerated I datastream is delayed by T_b so that, when combined with the regenerated Q stream, the original datastream is recreated.

If sinusoidal pulse shaping is added to OQPSK, the 90° phase transitions can be avoided. The resultant modulation is MSK, which has a continuous phase and a constant envelope (Fig. 13-*a*). MSK is functionally equivalent to fast frequency shift keying (FFSK), as is seen in the following discussion.

The block diagram of an FFSK modulator is shown in Fig.13-*b*. The voltage-controlled oscillator (VCO) represents a possible implementation of the modulator. Logic state 1 corresponds to transmit frequency f_2, logic state 0 (-1 volt (V) data level) corresponds to f_1. The frequency deviation in FFSK is

$$\Delta f = \frac{f_2 - f_1}{2} = \frac{1}{4T_b} \qquad (29)$$

where T_b is the bit duration of the input datastream. Note that a coherent relation between the transmitted frequencies and the bit rate is required. A frequency shift keying (FSK) signal $s_{\text{FSK}}(t)$ can be considered as the transmission of a sinusoid, the frequency of which is shifted between the two frequencies

$$f_1 = f_c - \Delta f = f_c - \frac{1}{4T_b}$$

$$f_2 = f_c + \Delta f = f_c + \frac{1}{4T_b} \qquad (30)$$

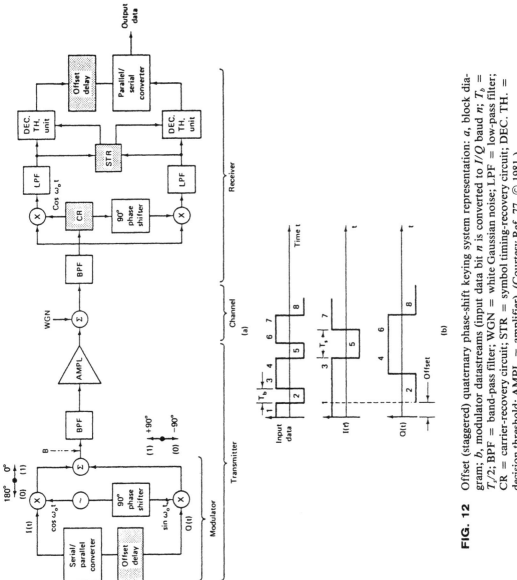

FIG. 12   Offset (staggered) quaternary phase-shift keying system representation: a, block diagram; b, modulator datastreams (input data bit n is converted to I/Q baud n; $T_b = T_s/2$; BPF = band-pass filter; WGN = white Gaussian noise; LPF = low-pass filter; CR = carrier-recovery circuit; STR = symbol timing-recovery circuit; DEC. TH. = decision threshold; AMPL = amplifier). (Courtesy Ref. 77, © 1981.)

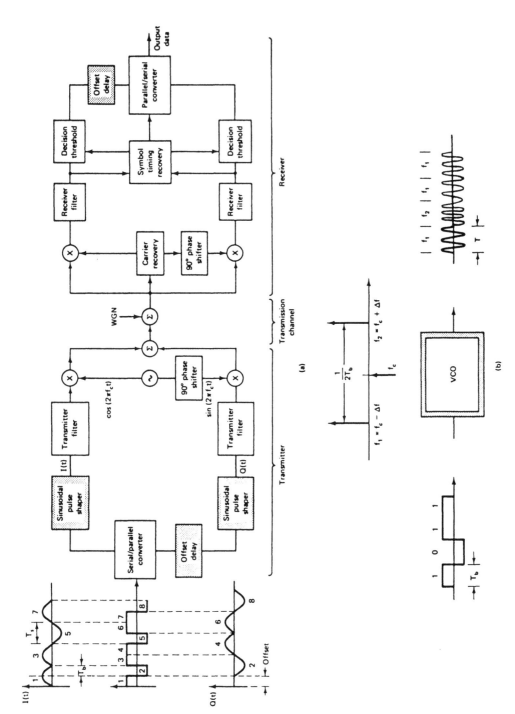

FIG. 13   Block diagram of minimum-shift keying (MSK) and fast frequency shift keying (FFSK) modems: *a*, MSK representation; *b*, equivalent FFSK representation (data in bit *n* is converted to *I/Q* baud *n*; $T_s = T_b$; VCO = voltage-controlled oscillator). (After Ref. 3, with permission from Prentice-Hall, © 1983.)

It is described by

$$s_{FSK}(t) = A \cos [2\pi(f_c \pm \Delta f)t]$$

$$= A \cos (\pm 2\pi \, \Delta f t) \cos (2\pi f_c t) - A \sin (\pm 2\pi \, \Delta f t) \sin (2\pi f_c t) \quad (31)$$

In the FFSK case, the frequency deviation is $\Delta f = 1/4T_b$; thus, the MSK signal $s_{MSK}(t)$ is

$$s_{MSK}(t = A \cos (\pm \pi t/2T_b) \cos (2\pi f_c t) - A \sin (\pm \pi t/2T_b) \sin 2\pi F_c t \quad (32)$$

Thus, we have proved that FFSK and MSK are equivalent. The sinusoidal pulse shaping means that the modulator output has either a positive or negative linear phase-change rate relative to the carrier, depending on the input data (7). The amplitude and phase of the output signals of the modulating multipliers is such that the sum of these signals (which is the unfiltered MSK modulated signal) has a constant-amplitude envelope. As a result, it can be transmitted through amplitude-limiting devices with minimal additional signal degradation in the same way that conventional frequency modulation (FM) modulated signals are transmitted.

The normalized power spectral densities of QPSK, OQPSK, and MSK as a function of frequency and normalized to the binary bit rate are shown in Fig. 14 (76). The QPSK main lobe has a width of $\pm 1/2T_b$; that of the MSK signal is wider ($\pm 3/4T_b$). For large values $(f - f_c)T_b$, the unfiltered MSK spectrum falls off at a rate proportional to f^{-4}, the unfiltered QPSK spectrum at a rate proportional to f^{-2}. The unfiltered infinite bandwidth spectral properties are of particular importance not only in earth station design, in which the high-power amplifier (HPA) usually is operated in a nonlinear (saturated) mode, but also in satellite system applications in which it is not economic to design Nyquist-shaped filters that follow the nonlinear output amplifiers.

It has been shown that the P_e performance of an MSK system is identical to that of QPSK and OQPSK (7), that is,

$$P_{e(MSK)} = \frac{1}{2} \, erfc \left(\sqrt{\frac{E_b}{N_0}} \right) = P_{e(QPSK)} \quad (33)$$

The design of carrier-recovery circuits is a difficult task, particularly if a fast modem synchronization (e.g., for burst modems) is required. To avoid the need for carrier recovery and to improve the synchronization speed, differential demodulation may be employed (4).

A typical differential demodulation of QPSK is shown in Fig. 15. Following the receive band-pass filter (BPF), the modulated carrier is split and sent to two differential phase-shift keying (DPSK) demodulators. Improved differential detection of QPSK is discussed in the section, "Advances in Differentially Coherent Modems." The theoretical BER curves of QPSK-related modulation schemes are summarized in Fig. 16. Detailed derivations can be found in Ref. 5.

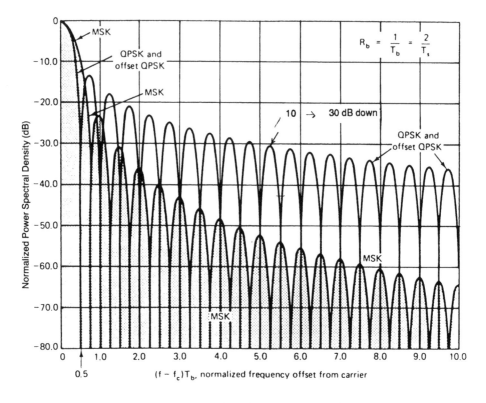

$$G_{MSK}(f) = \frac{8P_c T_b [1 + \cos 4\pi(f - f_c)T_b]}{\pi^2 [1 - 16T_b^2 (f - f_c)^2]^2}$$

$$G_{OKQPSK}(f) = 2P_c T \left[\frac{\sin 2\pi(f - f_c)T_b}{2\pi(f - f_c)T_b} \right]^2$$

FIG. 14 Normalized power spectral densities of unfiltered quaternary phase-shift keying (QPSK), offset QPSK, and minimum-shift keying systems. The modulated spectrum is symmetrical around the carrier frequency. For this reason, only the upper sideband is shown. (After Gronemeyer and McBride, 1976, with permission from the Institute of Electrical and Electronics Engineers, © 1976.)

Continuous Phase Modulation Schemes

As shown in the preceding section, the unfiltered MSK signal has a constant envelope and a continuous phase. However, it is noted that the MSK spectrum has a wider main lobe. Thus, a question arises as to whether further improvement can be achieved on MSK while maintaining a constant envelope. By improvement, we mean a narrower power spectrum, lower spectral side lobes, better error probability, or all of these. This leads to the development of CPM schemes (8). The modulated signal for a large class of CPM schemes can be described by

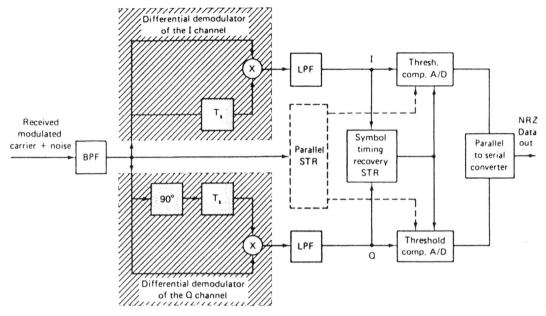

FIG. 15 Differential quaternary phase-shift keying demodulator block diagram (BPF = band-pass filter; LPF = low-pass filter; STR = symbol timing-recovery circuit; NRZ = nonreturn to zero). (After Ref. 5, with permission from Prentice-Hall, © 1983.)

$$s(t) = \sqrt{\frac{2E}{T}} \cos \left[2\pi f_0 t + \phi(t,\underline{\alpha})\right] \qquad (34)$$

where the transmitted information is contained in the phase

$$\phi(t,\underline{\alpha}) = 2\pi h \sum_{i=-\infty}^{\infty} \alpha_i q(t - iT) \qquad (35)$$

with $q(t) = \int_{-\infty}^{t} g(\tau)d\tau$. Normally, the function $g(t)$ is a smooth pulse shape over a finite time interval $0 \le t \le LT$ and 0 outside. Thus, LT is the length of the pulse and T is the symbol time. E is the energy per symbol, f_0 is the carrier frequency, and h is the modulation index. The M-ary data symbols α_i take values of $\pm 1, \pm 3, \ldots, \pm(M - 1)$. M is normally a power of 2. Values most commonly used are $M = 2$ for binary systems, $M = 4$ for quaternary systems, and $M = 8$ for octal systems.

From the definition of the class of constant-amplitude modulation schemes, we observe that the pulse $g(t)$ is defined in instantaneous frequency and its integral $q(t)$ is the phase response. The shape of $g(t)$ determines the smoothness of the transmitted information-carrying phase. The rate of change of the phase (or instantaneous frequency) is proportional to the parameter h, which normally is referred to as the *modulation index*. We have normalized the pulse shape $g(t)$

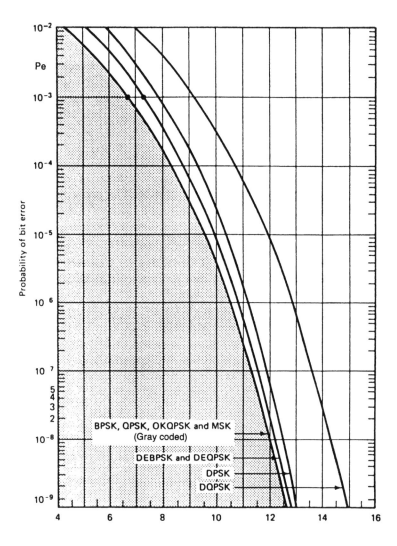

FIG. 16 Theoretical $P_e = f(E_b/N_o)$ performance of coherent binary phase-shift keying, differentially encoded phase-shift keying, coherent quaternary phase-shift keying (QPSK), and differential QPSK modems (Gray encoded) (additive white Gaussian noise and intersymbol-interference-free model). (After Ref. 5, with permission from Prentice-Hall, © 1983.)

in such a way that $\int_{-\infty}^{\infty} g(t)dt$ is ½. This means that, for schemes with positive pulses of finite length, the maximum phase change over any symbol interval is $(M - 1)h\pi$.

Thus, by choosing different pulses $g(t)$ and varying the parameters h and M, a great variety of CPM schemes can be obtained, such as continuous phase frequency shift keying (CPFSK) (7,9), generalized tamed frequency modulation (GTFM) (10), Gaussian MSK (GMSK) (11), and so on.

Detailed descriptions of pulse shape, phase trellis, power spectra, and error probability of CPM schemes can be found in the references mentioned above. As a summary, we present an example of an average power spectrum in Fig. 17 and an example of error probability in Fig. 18. As seen from Fig. 17, GMSK has a very compact spectrum and thus has a potential application in digital mobile radio. The interesting topic of differential detection of GMSK is presented in a separate section.

A New Family of Modulation Schemes

A relatively new nonlinear filtering technique combined with OQPSK modulation has significant spectral spreading advantages. This modulator, known as *nonlinearly filtered OQPSK* (NLF-OQPSK), is shown in Fig. 19. Detailed spectral density derivations, applications, and implementation diagrams are presented in Refs. 5 and 12.

In Fig. 19, the serial-to-parallel converted data are fed to the I and Q baseband channels. Prior to modulation, the $f_s = f_b/2$ rate symbol streams in I and Q are processed individually through a nonlinear device named *nonlinearly switched filter* (NLSF), as shown in Fig. 19. The operation principle of these nonlinearly switched filters is illustrated in Fig. 20. For a -1 to $+1$ transition of the unfiltered NRZ input signal, the rising segment of a sinusoid is switched on (connected to the transmission medium). For a $+1$ to -1 transition, the falling segment of a sinusoid is switched on. The $0°$ and the $180°$ reference sinusoidal generators shown in Fig. 20-*b* provide the required sinusoidal waves. For a continuous sequence of 1s or -1s (more than one input bit without transition), a positive or negative direct current (DC) segment is switched on.

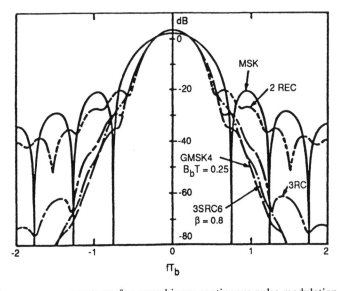

FIG. 17 Average power spectrum for some binary continuous pulse modulation schemes with $h = 1/2$ (MSK = minimum-shift keying; GMSK = generalized MSK). (After Ref. 8, with permission from the Institute of Electrical and Electronics Engineers, © 1986.)

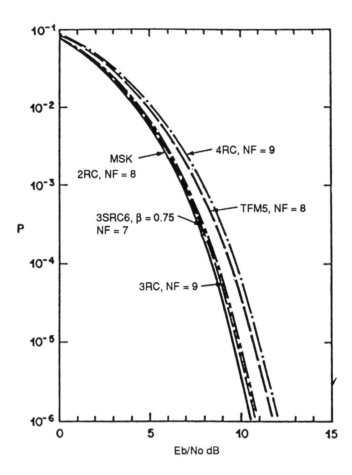

FIG. 18 The error probability for some binary $h = 1/2$ schemes. The minimum-shift keying type receivers use asymptotically optimum receiver filters for length N_F. (From Ref. 8, reprinted with permission of the Institute of Electrical and Electronics Engineers, © 1986 IEEE.)

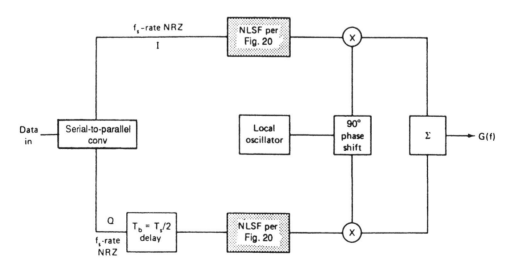

FIG. 19 Nonlinearly filtered offset-keyed quaternary phase-shift keying modulator.

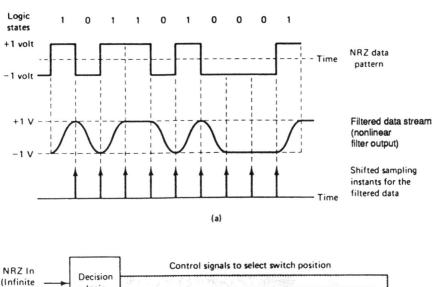

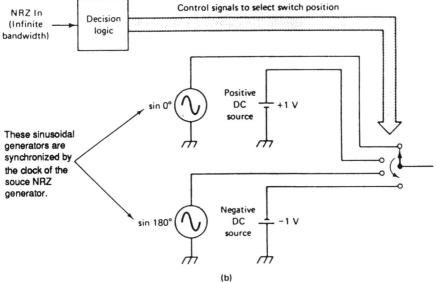

FIG. 20 Principle of operation of Feher's bandwidth-efficient, intersymbol-interference-free, and jitter-free nonlinear filter: *a*, generated wave shapes; *b*, block diagram (NRZ = nonreturn to zero). (After Ref. 12, with permission from Prentice-Hall, © 1981.)

The decision logic provides the switch position control signals. The measured eye diagram and PSD position control signals are shown in Fig. 21.

Relationship between Bit Energy–to–Noise Density Ratio and Carrier-to-Noise Ratio

The required bit energy (E_b) to noise density (N_0) ratio E_b/N_0 is a convenient quantity for system calculations and performance comparisons (5). However,

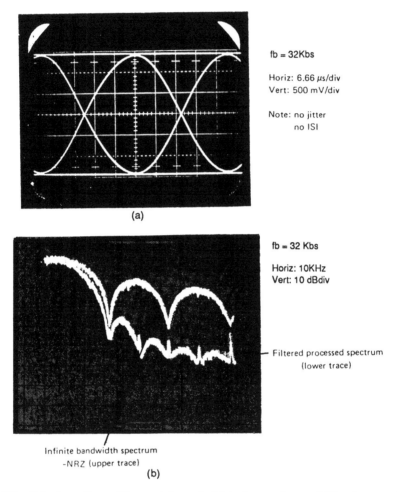

fb = 32Kbs

Horiz: 6.66 μs/div
Vert: 500 mV/div

Note: no jitter
      no ISI

(a)

fb = 32 Kbs

Horiz: 10KHz
Vert: 10 dBdiv

Filtered processed spectrum
(lower trace)

Infinite bandwidth spectrum
-NRZ (upper trace)

(b)

FIG. 21   Feher's nonlinear filter: *a*, intersymbol-interference-free and jitter-free eye diagram; *b*, power spectrum. (After Ref. 12, with permission from Prentice-Hall, © 1981.)

in practical measurements it is more convenient to measure the average carrier-to-noise power ratio (CNR). Most tests are performed using power and rms voltage meters, which are readily available. The following simple relations are useful for the E_b/N_0 to CNR transformations:

$$E_b = CT_b = C\left(\frac{1}{f_b}\right)$$

$$N_0 = \frac{N}{B_w}$$

$$\frac{E_b}{N_0} = \frac{CT_b}{N/B_w} = \frac{C/f_b}{N/B_w} = \frac{CB_w}{Nf_b}$$

$$\boxed{\frac{E_b}{N_0} = \frac{C}{N} \cdot \frac{B_w}{f_b}} \qquad (36)$$

The E_b/N_0 ratio equals the product of the CNR and the receiver noise bandwidth-to-bit-rate ratio (B_w/f_b). For example, suppose that a coherent BPSK modem operates at a rate of $f_b = 10$ Mbps. At $P_e = 10^{-4}$, the required E_b/N_0 ratio is 8.4 dB. Assume that Nyquist channel shaping is employed. Then, the corresponding CNR = $E_b/N_0 \cdot B_w/f_b$ = 8.4 dB. If QPSK is used, CNR = 11.4 dB.

Digital Modulation with Spectral Efficiency Higher than 2 Bits per Second per Hertz

In such band-limited channels as voiceband telephone channels and terrestrial microwave radio, modems with a spectral efficiency higher than 2 bps/Hz are demanded. Generally speaking, multilevel QAM and QPRS are used in these applications.

Principles of Quadrature Amplitude Modulation and Quadrature Partial Response Signaling Modulation Techniques

A block diagram for a QAM suppressed carrier modulator and its corresponding signal-state-space diagram is shown in Fig. 22 (5). The information is contained in both the amplitude and the phase of the modulated signal; thus, the term amplitude phase keying (APK) also is used in the literature.

The f_b rate binary source is commuted into two binary symbol streams, each having a rate of $f_b/2$. The 2-to-L-level converters convert these $f_b/2$-rate datastreams into L-level PAM signals having a symbol rate of

$$f_s = (f_b/2)/(\log_2 L) \qquad \text{symbols/s} \qquad (37)$$

For example, if the source bit rate is $f_b = 10$ Mbps, then the commuted binary streams have an $f_b/2 = 5$-Mbps rate. If an $M = 16$-ary QAM-modulated signal having a theoretical efficiency of 4 bps/Hz is desired, these commuted binary streams are converted into $L = 4$-level baseband streams. The resultant 4-level symbol streams of the I and Q channels are $5/\log_2 4$, or 2.5 M symbols/s. If premodulation low-pass filters (LPFs) are used, then the theoretical minimum bandwidth of these filters is 1.25 MHz. The minimum intermediate frequency (IF) bandwidth requirement is identical to the double-sided minimum baseband bandwidth (i.e., 2.5 MHz). This example illustrates that a 10-Mbps, 16QAM signal can be transmitted in a theoretical minimum bandwidth of 2.5 MHz, that is, with a theoretical spectral efficiency of 4 bps/Hz. Practical 16QAM systems have been achieving a bandwidth efficiency of approximately 3.7 bps/Hz. The

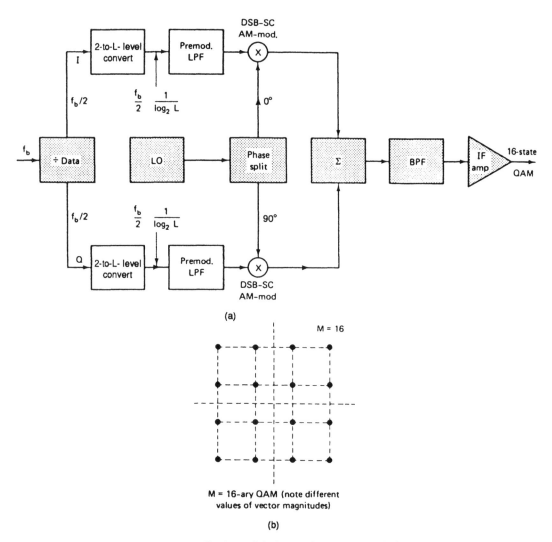

(a)

(b)

FIG. 22 *M*-ary quadrature amplitude modulation modulator: *a*, block diagram; *b*, corresponding signal-state-space diagram (LPF = low-pass filter; BPF = band-pass filter; DSBSCAM = double sideband suppressed carrier amplitude modulation; LO = local oscillator; IF = intermediate frequency). (After Ref. 5, with permission from Prentice-Hall, © 1983.)

block diagram of a coherent *M*-ary QAM demodulator is shown in Fig. 23. The operation of the *M*-ary QAM demodulator is self-explanatory from Fig. 23.

Another spectrally efficient modulation technique is QPRS. First, we have to understand the principle of the partial-response signaling (PRS) technique. Recall that the Nyquist theorem aims at no ISI at the sampling instants so that symbols are uncorrelated and can be recovered independently. Such systems often are called *zero-memory systems*. According to the Nyquist theorem, we know the maximum spectral efficiency of PAM systems is 2 symbols/s/Hz,

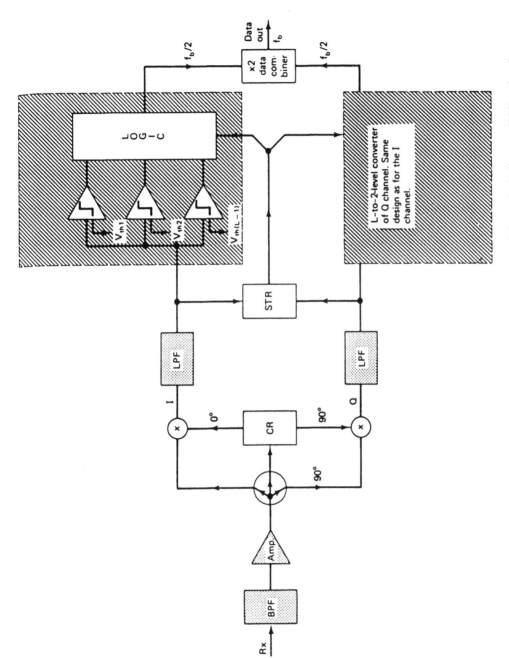

FIG. 23 *M*-ary quadrature amplitude modulation or phase-shift keying demodulator block diagram (BPF = band-pass filter; LPF = low-pass filter; STR = symbol timing-recovery circuit; CR = carrier-recovery circuit). (From Ref. 12, with permission from Prentice-Hall, © 1981.)

which sometimes is called the Nyquist rate and can be achieved only by $\alpha = 0$ filters. In practice, it is impossible to implement $\alpha = 0$ filters and thus impossible to achieve the Nyquist rate.

During the early 1960s, A. Lender discovered the correlative transmission method. This method, often referred to as *duobinary* or *partial response*, found numerous applications in terrestrial microwave, cable, and other data-transmission systems. In PRS systems, a controlled amount of ISI is introduced in order to simplify the filter design (particularly the phase-equalization problem) and to enable transmission at the Nyquist rate and even higher rates. An in-depth treatment of PRS may be found in Ref. 13. Here, we summarize the principle of duobinary (also called Class I) PRS.

In Fig. 24, the cascade of a simple one-tap transversal filter $H_1(f)$ with a very steep (ideally, brick-wall) filter $H_2(f)$ is shown. The impulse response of $H_1(f)$ is

$$h_1(t) = \delta(t) + \delta(t - T) \tag{38}$$

and the corresponding transfer function is

$$H_1(f) = F[h_1(t)] = 1 + e^{-j2\pi fT} \tag{39}$$

thus,

$$|H_1(f)| = 2 \cos \pi fT \tag{40}$$

The absolute value of $H_1(f)$ assures a continuous spectral rolloff between zero and the Nyquist frequency $f_N = 1/2T_s$. Close to this frequency, the attenuation of this phase-linear filter is very significant. The $H_2(f)$ brick-wall filter can be approximated in practical designs with a steep LPF (e.g., an eighth-order Chebychev filter). In almost all practical filter designs, it is very difficult to equalize the group delay at the edge of the passband and in the frequency region where the filter has significant attenuation, say more than 3 dB. In the design of a duobinary system such as that shown in Fig. 24, the phase-linear transversal filter $H_1(f)$ has a significant attenuation at the edge of the LPF (i.e., f_N); thus, the signal energy around this frequency is negligible. For this reason, in the duobinary system it is not required to have a carefully phase-equalized filter.

The transfer function of the cascaded duobinary filter is

$$|H(f)| = |H_1(f)H_2(f)| = \begin{cases} 2T \cos \pi fT & \text{for } f \leq \dfrac{1}{2T} \\ 0 & \text{elsewhere} \end{cases} \tag{41}$$

The corresponding impulse response is obtained from the inverse Fourier transformation of Eq. (41). It is

$$h(t) = \frac{\sin \pi t/T}{\pi t/T} + \frac{\sin \pi (t - T)/T}{\pi (t - T)/T} \tag{42}$$

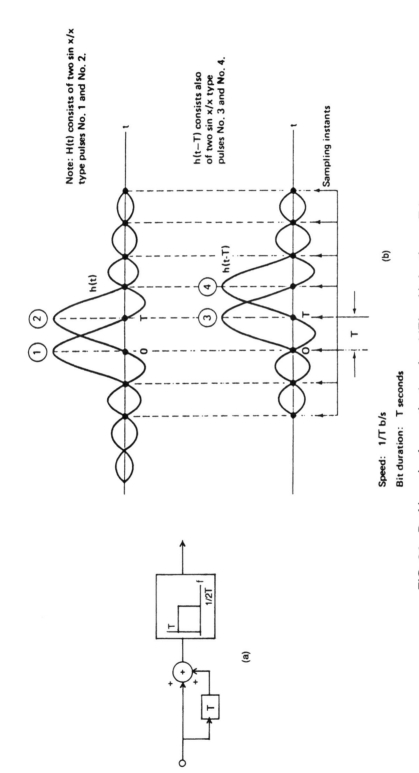

FIG. 24 Duobinary signal generation (speed $= 1/T$ bps; bit duration $= T$ s).

Thus, the response of the $H_1(f)H_2(f)$ filter to one input impulse $\delta(t)$ consists of two $\sin x/x$ terms. Since the signaling interval is T seconds, there will be an overlap of ISI between the symbols. For a synchronous input impulse sequence consisting of positive and negative impulses and having a signaling rate of $f_s = 2f_N = 1/T$, the sampled output has three distinct amplitude levels: $+2$, 0, or -2. In other words, a three-level correlated signal is obtained at the output of the duobinary encoder. This signal is not as sensitive to filter imperfections as conventional PAM, and it is possible to transmit at a rate about 25% higher than the Nyquist rate without significant degradation. However, as the signal contains three levels, for the same P_e performance the SNR requirement is about 2.1 dB higher than that of binary PAM (14,15). The 2.1-dB loss can be recovered by employing maximum-likelihood sequence estimation (MLSE) for the three-level PRS system (16). However, as the signal set size is increased, the performance of MLSE deteriorates. In a recent paper, it shows that 49QPRS with MLSE is still 1 dB worse than 16QAM (17).

A 9QPRS-modulated signal is obtained by quadrature modulating two three-level duobinary signals. Thus, the block diagram of a QPRS modulator is essentially the same as that of a QPSK modulator. However, as shown in Fig. 25, the QPRS modulator is equipped with duobinary premodulation LPFs (encoders) in the I and Q channels. The coherent QPRS demodulator is similar to the one shown in Fig. 23. It is noted that, among the many classes of PRS systems, Class I and Class IV (modified duobinary) are most useful in practice (14,18). 9QPRS radio systems having duobinary encoders have attained a practical spectral efficiency of 2.25 bps/Hz (14,19).

Modems for Microwave and Cable System Applications

For a number of applications, if it is required to carry data at a rate from 56 kilobits per second (Kbps) up to 8.448 Mbps, it is more economical to modify a frequency division multiplexed (FDM) analog system into a hybrid system than to install a completely new, dedicated digital system (12,20). During 1986, a number of 1.544-Mbps-rate (T1 rate) modems, requiring a single supergroup (SG) band of 240 KHz have been installed and are already operational over long-distance FDM-FM and single-sideband (SSB) microwave systems, as well as cable systems. These T1/SG modems can cover a single nonregenerative span of 500–1500 kilometers (km) (3,20). This new generation of hybrid data-in-voice (DIV) systems as well as the less spectral-efficient data-under-voice (DUV) and data-above-voice/video (DAV) systems of the 1970s and 1980s are suitable for the simultaneous transmission of time division multiplexed (TDM) data and of frequency division multiplexed (FDM) voice channels or video signals (12).

For standard CEPT (i.e., 2.048-Mbps digital transmission rate in a 240-kHz SG band, called CEPT-1/SG) a further significant increase in the spectral efficiency of candidate QAM modem techniques is required. To achieve such high spectral efficiency, such modulation techniques as 512QAM, 961QPRS, and 1024QAM are required.

Table 4 summarizes the spectral efficiency requirements for both of these DIV systems. We assume that both DIV modems have a 3.6% overhead for

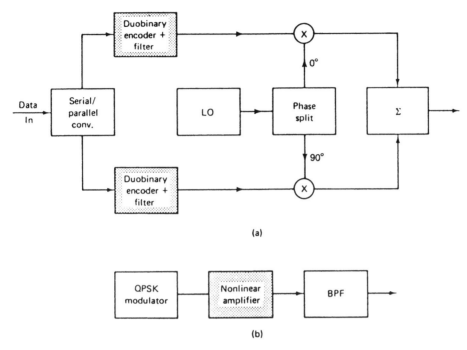

(a)

(b)

FIG. 25 Two methods to generate quadrature partial response signaling (QPRS) modulated signals: *a*, conventional duobinary QPRS modulator block diagram; *b*, QPRS modulation generated by bandlimiting a nonlinearly amplified quaternary phase-shift keying signal. The cascade of the transmit band-pass filter and the filters in the demodulator approach the duobinary conversion filter described in Fig. 24.

forward error correction (FEC), synchronization, and possible training sequence requirements of baseband adaptive-time-domain digital equalizers (BATE), which are essential for group delay and amplitude equalization of the numerous filters, translators, multiplexers, and radio or cable transmission system distortions.

From Table 4, we note that the spectral efficiencies are 8.84 and 6.66 bps/Hz for CEPT-1/SG and T1/SG, respectively. For the T1/SG application, a 256QAM modem is a logical choice. In this application, a roll-off factor between 0.1 and 0.2 is used and the out-of-band attenuation is in excess of 55 dB in order to avoid significant interference into the adjacent FDM bands (20). For the CEPT-1/SG application, in principle we could use 512QAM, 961QPRS, or 1024QAM.

*Modem Requirements for the Modulated Data Carrier Power
in the Supergroup Band*

This section discusses the required theoretical and practical modem CNR requirements, and the CNR available from cable and microwave transmission

TABLE 4   Spectral Efficiency Requirements for 1024QAM Systems for CEPT-1 Rate
Modems and T-1 Rate Systems Operated in a Supergroup Band

	CEPT-1/SG-FEC Modem	T1/SG-FEC Modem
Source rate	2.048 Mbps	1.544 Mbps
Redundancy for forward error correction and synchronization overhead bits	3.6%	3.6%
f_b = transmission rate	2.122 Mbps	1.6 Mbps
SG nominal bandwidth	240 kHz	240 kHz
f_b/SG bandwidth (spectral efficiency)	8.84 bps/Hz	6.66 bps/Hz

Source: Ref. 3, reprinted by permission of the Institute of Electrical and Electronics Engineers, ©
1987 IEEE.

systems. Specifically, we define C_D as the modulated data carrier power in the
SG band, and N as the noise in the double-sideband Nyquist bandwidth that
equals the symbol rate. The symbol rate for the 256QAM modem coded at the
T1 rate corresponds to 1.6 Mbps : 8 = 200 kHz, while for the 1024QAM modem coded at the CEPT rate it is 2.122 : 10 = 212 kHz. This receiver noise
bandwidth is practically the same (within 0.7 dB) as that of the standard SG
band.

From Table 5, we conclude that for $P_e = 10^{-8}$, a $C_D/N = 40.5$ dB is required for the CEPT-1 rate, theoretical, uncoded modem. All practical modems
have some design imperfections, such as less than perfect channel equalization,
not sufficient carrier-phase-noise cancellation, symbol jitter suppression, and
so on. For this reason, and to reduce the practical C_D/N requirement in general,

TABLE 5 C_D/N Requirements of Theoretical Uncoded and Practical FEC Coded
256QAM and 1024QAM Modems

Modem	BER	Theoretical Uncoded C_D/N (dB)	Practical* Uncoded Modem C_D/N (dB)	Practical* FEC Coded Modem C_D/N (dB) Requirement
2.048 Mbps CEPT-1 rate	10^{-8}	40.5	43	38
1024QAM	10^{-6}	39	41	37
1.544 Mbps	10^{-8}	34.5	37	33
T1 rate, 256QAM	10^{-6}	33	35	31

Source: Ref. 3, reprinted by permission of the Institute of Electrical and Electronics Engineers, ©
1987 IEEE.
Note: C_D = modulated data power in the standard SG band; N = noise power in the double-
sideband Nyquist band that equals the symbol rate of about 200 kHz.
*"Practical" implies modem with degradations due to transmission system and modem design
imperfections.

FEC coding may be used. In summary, based on Table 5, we assume that for $P_e = 10^{-6}$ threshold performance, the practically coded QAM modems require C_D/N of 37 dB and 31 dB for the CEPT-1 and T1 modems, respectively.

Available Modulated Data Carrier Power in the Supergroup Band

In this section, we summarize the available C_D/N ratio for an illustrative system, namely, a microwave SSB system. The detailed calculations of CNR are well documented in the readily available transmission literature. For this reason, only a brief summary of the principal transmission parameters in regard to the available CNR is given here. From Table 6, we note that the C_D/N provided by 5400-voice-channel capacity SSB systems is sufficiently high for the operation of the T1 modem and also for the CEPT-1 modem (20). The per-hop fade margin of the modem without FEC is about 26 dB. This fade margin is increased by 4 dB with FEC encoding. For an additional increase of the fade margin of the SSB system, the data load should be increased above the nominal C-load and/or, in the FDM-FM system application, a SG centered at a lower frequency of the radio baseband should be utilized (3).

In Table 6, we use a $C_D/N = 38$ dB requirement for the SSB microwave system application of 256QAM modems without FEC. As shown in Table 5, the C_D/N requirement of the FEC-coded 256QAM system is only 33 dB, thus an increase in the fade margin is obtained by FEC.

The fade margin of the FEC-encoded 1024QAM system having a CEPT-1 rate can be obtained from Tables 5 and 6. The required C_D/N is 38 dB for a BER $= 10^{-8}$ in the FEC-encoded 1024QAM case. Thus, the fade margin of the CEPT-1 rate, FEC-encoded modem is practically (within 1 dB) the same as that of the T1 modem without FEC listed in Table 6.

The performance of such highly spectral-efficient QAM as 256QAM and 1024QAM can be degraded significantly by residual errors of imperfect adaptive equalizers, channel filters, automatic gain control (AGC), and other system imperfections. In the next two sections, the performance of these systems under the effect of amplitude and group-delay distortions and residual AGC error is examined. The simulation results can be very useful in budgeting system degradations for system designers.

Performance Degradations of High-Level Quadrature Amplitude Modulation Systems

Performance degradations for high-level QAM systems are evaluated by computer simulations with the simulation model shown in Fig. 26. The equivalent baseband of the channel distortion transfer function can be expressed as

$$C(f) = |\, C(f)\, |\, e^{j\theta(f)} \tag{43}$$

$$A(f) = |\, C(f)\, | \tag{44}$$

TABLE 6 5400-Voice-Channel Single Sideband Radio System: Illustrative Link Budget for a T1/SG 1.6-Mbps 256QAM Modem (without FEC) (BER = 10^{-8})

Loading factor	−15.0 dBmO
Number of 4-kHz voice channels in a supergroup	60
Carrier (data) power	
$C_D = -15 + 10 \log 60$	2.8 dBmO
Bit rate	1.544 Mbps
Overhead bits	3.6 percent
Modulation type	256QAM
Baud rate	200.0 kBaud
Channel filter rolloff (alpha)	0.2
Noise bandwidth	200.0 kHz
Practical required − without FEC	
C_D/N (BER = 10^{-8})	38.0 dB
Modem span distance − no fading based on CCIR performance (3 pW per km)	807.0 km
Allowable fade margin per hop (approximation based on 1-hop fading)	26.9 dB
Circuit length (nondiversity)	962.0 miles
Circuit length (nondiversity)	1547.9 km
Circuit length (diversity)*	6500.0 miles
Circuit length (diversity)*	10458.5 km

Source: Ref. 3, reprinted by permission of the Institute of Electrical and Electronics Engineers, © 1987 IEEE.

Note: The FEC-coded 1024QAM modem has the same fade margin (within 1 dB) as the 256QAM modem without FEC.

*Circuit length is defined as the maximum distance at which the objective of 0.5% errored seconds is not exceeded.

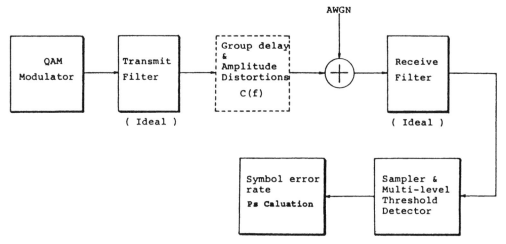

FIG. 26 Computer simulation model (QAM = quadrature amplitude modulation; AWGN = additive white Gaussian noise).

$$D(f) = \frac{-1}{2\pi} \frac{d\theta(f)}{df} \tag{45}$$

The distortionless channel is assumed to satisfy the Nyquist criterion.

Amplitude Distortions

To investigate the performance degradation due to residual amplitude distortions (i.e., after adaptive equalization), we assume that the phase distortions are equalized, that is, $D(f) =$ constant. The amplitude distortions are defined as follows:

$$A(f) = \begin{cases} L_A f & \text{for linear amplitude distortion} \\ P_A f^2 & \text{for parabolic amplitude distortion} \\ S_A \sin(2\pi K f/2f_{BW}) & \text{for sinusoidal amplitude distortion} \end{cases} \tag{46}$$

where K, chosen to be equal to 4, is the number of ripples of the sinusoidal distortion in the passband, f_{BW} is the system baseband bandwidth, and L_A, P_A, and S_A are parameters characterizing the amount of the amplitude distortion. Clearly, $f_{BW} = (1 + \alpha)f_N$ and $f_N = f_s/2$ with f_s the baud rate and α the roll-off factor.

The CNR degradations at $P_e = 10^{-4}$ for the three amplitude imperfection characteristics against the maximum amplitude distortion A_m, defined as

$$A_m|_{dB} = \begin{cases} L_A(2f_{BW}) & \text{for linear amplitude distortion} \\ P_A(f_{BW})^2 & \text{for parabolic amplitude distortion} \\ S_A & \text{for sinusoidal amplitude distortion} \end{cases} \tag{47}$$

and are given in Fig. 27 for 256QAM, 512QAM, and 1024QAM. The comparison here is based on equal symbol rate and equal filtering strategy. Furthermore, similar to the 256QAM modem study (21), for a given value of A_m the sinusoidal amplitude distortion causes the worst degradation, whereas the linear distortion causes the least.

Group-Delay Distortions

We now assume that the amplitude distortions are equalized that is, $A(f) = 1$. Again, three different characteristics for the group-delay distortions are defined:

$$D(f) = \begin{cases} L_D f & \text{for linear group delay} \\ P_D f^2 & \text{for parabolic group delay} \\ S_D \sin(2\pi f K/2f_{BW}) & \text{for sinusoidal group delay} \end{cases} \tag{48}$$

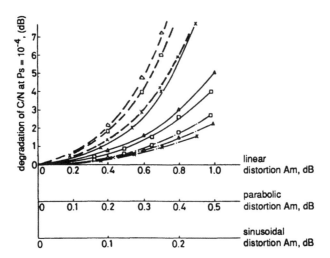

FIG. 27   Degradation of carrier-to-noise ratio as a function of maximal amplitude distortion
A_m for 256QAM, 512QAM, and 1024QAM for linear, parabolic, and sinusoidal amplitude distortions. Both QAM systems have the same baud rate of 15 megabauds
(Mbaud) per second and are raised-cosine filtered with a roll-off factor $\alpha = 0.2$:
$-\cdot-\cdot-$ = 256QAM; ———— = 512QAM; $- - -$ = 1024QAM; x = linear amplitude distortion; $\square$ = parabolic amplitude distortion; $\triangle$ = sinusoidal amplitude distortion.

where L_D, P_D, and S_D are parameters characterizing the amount of group delay.
We compute again the probability of a symbol error for the three characteristics
and for different values of the distortion. Define the maximum group delay τ_m
as

$$\tau_m \big|_{ns} = \begin{cases} L_D(2f_{BW}) & \text{for linear group delay} \\ P_D(f_{BW})^2 & \text{for parabolic group delay} \\ S_D & \text{for sinusoidal group delay} \end{cases} \qquad (49)$$

where ns = nanoseconds. We present the CNR degradation plotted against τ_m
in Fig. 28. We note that, for these systems for a given value of τ_m, the linear
group delay causes the most severe degradation, followed by sinusoidal and
parabolic group-delay distortions.

Impact of Residual Amplitude Fluctuation on High-Level Quadrature Amplitude Modulation Systems

Fading is known as one of the most destructive impairments of the performance
of digital radio systems. AGC amplifiers at the receiver may compensate for
the major part of amplitude fluctuation caused by fading. However, a certain
amount of amplitude fluctuation still would remain in the system. For a system
designer, it is important to know the impact of residual AGC error on the
performance of high-level QAM so that a degradation budget can be specified.
    As an analytically tractable model, let us characterize amplitude fluctuation

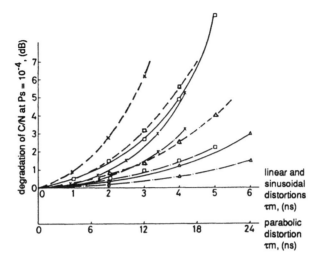

FIG. 28 Degradation of carrier-to-noise ratio as a function of maximal group delay τ_m for 256QAM, 512QAM, and 1024QAM for linear, parabolic, and sinusoidal group-delay distortions. Both systems have the same baud rate of 15 Mbaud/s and have raised-cosine filters with $\alpha = 0.2$: $-\cdot-\cdot-\cdot$ = 256QAM; $\underline{\quad\quad}$ = 512QAM; $-\ -\ -$ = 1024QAM; x = linear amplitude distortion; $\square$ = parabolic amplitude distortion; $\triangle$ = sinusoidal amplitude distortion.

by a relatively slowly varying sinusoidal waveform having a normalized (i.e., in terms of a signal envelope A_k) amplitude u and a frequency f_u. An equivalent baseband M-ary QAM signal is represented in a complex form as

$$X_k + jY_k = A_k e^{j\theta_k} \qquad (50)$$

where

$$\begin{aligned}
k &= 1,2,\ldots,M \\
\{X_k\}, \{Y_k\} &= \{\pm 1, \pm 3, \ldots, \pm(M-1)\} \\
A_k &= \text{the signal envelope of the } M\text{-ary QAM} \\
\theta_k &= \text{the phase of the } M\text{-ary QAM}
\end{aligned}$$

In the presence of residual amplitude fluctuation and AWGN, the received signal is represented as

$$r(t) = A_k[1 + u \sin(2\pi f_u t + \phi)]e^{j\theta_k} + n(t) \qquad (51)$$

where ϕ is a relative initial phase of the amplitude fluctuation uniformly distributed in $[0,2\pi]$, and $n(t)$ is an equivalent baseband thermal noise.

An analysis model of this system is represented as Fig. 29. A simulated state-space diagram of a 64QAM signal perturbed by residual amplitude fluctuations of $u = 0.04$ (or 0.34 dB) is shown in Fig. 30. Note that the amplitude fluctuation causes more ISI in the outer states. In the simulation, ideal square

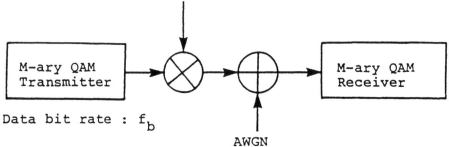

FIG. 29 Analysis model of *M*-ary quadrature amplitude modulation system in the presence of residual amplitude fluctuations and additive white Gaussian noise (u = normalized amplitude; ϕ = relative initial phase; f_u = frequency of the fluctuation; $f_u \ll f_b$). (Courtesy of Ref. 22.)

root of $\alpha = 0.2$ raised-cosine filters are used in the transmitter and receiver with $x/\sin x$ aperture equalizer in the transmitter.

The probability density function (pdf) of the in-phase component of the sum of the residual amplitude fluctuations and AWGN is obtained as (22)

$$p_k(n) = \frac{1}{\pi\sqrt{2\pi\sigma_G^2}} \int_{-\pi/2}^{\pi/2} \exp\left[\frac{-(n - a_k \sin \phi)^2}{2\sigma_G^2}\right] d\phi \qquad (52)$$

where a_k is the peak amplitude of the in-phase component fluctuation defined as $a_k = uA_k \cos \theta_k$. The pdf of the quadrature component of the total noise interference is obtained by replacing a_k with $b_k = uA_k \sin \theta_k$ in Eq. (52).

Error occurs when the total noise interference exceeds $d_{min}/2$, where d_{min} represents the minimum distance between two ideal neighboring signal states. Therefore, the symbol error probability is obtained as

$$P_{e_k} = 2 \int_{d_{min}/2}^{+\infty} p_k(n)dn + 2 \int_{d_{min}/2}^{+\infty} p_k(q)dq$$

$$= \frac{1}{\pi} \int_{-\pi/2}^{\pi/2} \left[\text{erfc}\left(\sqrt{\frac{\text{CNR}}{42}} - \sqrt{\frac{7\text{CNR}}{3}} \rho_k u \cos \theta_k \sin \phi\right) \right.$$

$$\left. + \text{erfc}\left(\sqrt{\frac{\text{CNR}}{42}} - \sqrt{\frac{7\text{CNR}}{3}} \rho_k u \sin \theta_k \sin \phi\right) \right] d\phi \qquad (53)$$

where $\rho_k = A_k/A_1$, $k = 1, 2, \ldots, 16$, and CNR is the mean carrier-to-noise power ratio of 64QAM given by

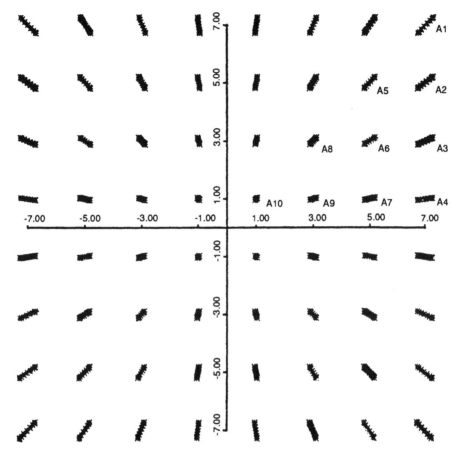

FIG. 30 State-space diagram of a 64 quadrature amplitude modulation signal perturbed by the residual fluctuation of $u = 0.04$ (0.34 dB). (Courtesy Ref. 22.)

$$\text{CNR} = \frac{1}{16} \sum_{k=1}^{16} \left(\frac{A_k}{\sqrt{2}\sigma_G} \right)^2 \tag{54}$$

The exact overall average symbol error probability for 64QAM is given by

$$P(e) = \frac{1}{32} [P_{e_1} + 3(P_{e_2} + P_{e_3} + P_{e_4}) + 2(P_{e_5} + P_{e_8} + P_{e_{10}})$$
$$+ 4(P_{e_6} + P_{e_7} + P_{e_9})] \tag{55}$$

Simulated state-space diagrams reveal that the outer signal states are more susceptible to residual amplitude fluctuation than the inner signal states. We analyze particular symbol error probabilities, including the best and the worst symbol error probabilities of 64QAM. Figure 31 shows typical symbol error probabilities and an average symbol error probability of 64QAM for $u = 0.1$

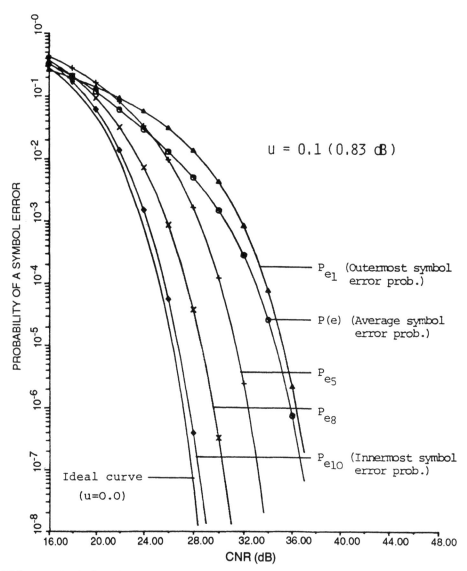

FIG. 31 Typical symbol error probabilities and an overall average symbol error probability $P(e)$ of 64QAM in the presence of additive white Gaussian noise and the residual amplitude fluctuation of $u = 0.1$ (0.83 dB). The symbol error probability P_{e_k} is calculated from Eq. (53) (P_{e_1} = error probability of the outermost symbol; $P_{e_{10}}$ = error probability of the innermost symbol). (Courtesy Ref. 22.)

(0.83 dB). The symbol error probabilities P_{e_1}, P_{e_5}, P_{e_8}, and $P_{e_{10}}$ correspond to those of illustrative symbols A_1, A_5, A_8, and A_{10} in Fig. 30, respectively, and are calculated from Eq. (53). The result in Fig. 31 shows that the outermost symbol (i.e., A_1) performs worst, whereas the innermost symbol A_{10} performs best, and the overall error performance is determined mostly by the worst symbol error probability. A loose upper bound (i.e., the worst) symbol error probability, thus, can be obtained from the outermost symbol error probability P_{e_1}.

Exact average symbol error probabilities of higher level QAM systems (e.g., 256QAM, 1024QAM, etc.) may be calculated following the same procedures done in the 64QAM analysis. However, calculations might be very tedious and require a fairly long computing time. As we notice from the typical symbol error probabilities of 64QAM in Fig. 31, those signal states on the axis connecting A_1 and A_{10} (i.e., A_1, A_5, A_8, and A_{10}) show typical order ranges of error probability, and the other signal states behave similarly to one of these four signal states. Thus, we can calculate tight-bound (TB) average symbol error probabilities of high-level QAM systems.

A TB average symbol error probability of 64QAM is obtained as (22)

$$P(e)_{TB} = \frac{1}{4} \left(\frac{1}{2} P_{e_1} + P_{e_5} + P_{e_8} + P_{e_{10}} \right) \qquad (56)$$

Figure 32 illustrates both the exact average symbol error probabilities and the TB symbol error probabilities of 64QAM in the presence of residual amplitude fluctuations and AWGN. Note that the TB symbol error probabilities fit the exact symbol error probabilities closely. Following the same approach, tight bounds on the symbol error probability of 256QAM and 1024QAM have been calculated, with results summarized in Figs. 33 and 34, respectively (22). An illustrative range of the amplitude fluctuations in terms of a normalized amplitude u is [0.0,0.035] or, equivalently, [0.0,0.30 dB]. Note that 1024QAM is extremely sensitive to the residual amplitude fluctuation of AGC amplifiers. For example, with only 0.09-dB residual amplitude fluctuation in the system, the CNR degradation at $P_e = 10^{-6}$ is about 2 dB. This reveals that for successful operation of very high level QAM systems, such as 1024QAM and beyond, the AGC circuit must be designed with extreme care.

Quadrature Partial Response Signaling Systems Operated above the Nyquist Rate

Recently, an improved-efficiency technique for Class I and Class IV PRS above the Nyquist rate with multilevel inputs has been proposed (23,24). In this section, we present the Class IV case. Class IV PRS has applications in such systems as transformer coupled circuits, SSB modems, and carrier systems with carrier-pilot tones, in which reduced low-frequency components in the spectrum are required.

First, let us review the speed tolerance of PRS systems. The conventional method for Class IV PRS above the Nyquist rate transmission is illustrated in

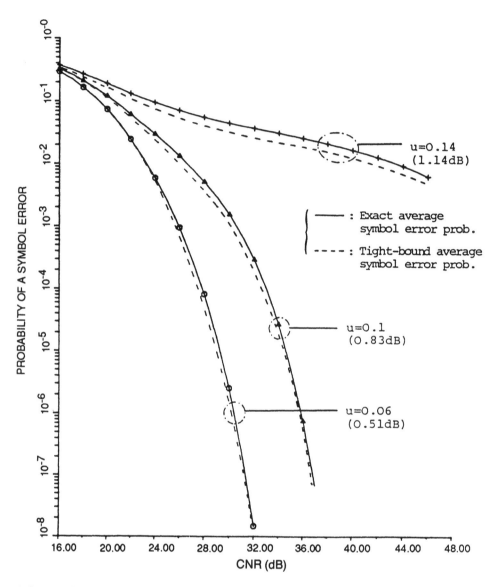

FIG. 32 Exact average symbol error probabilities (solid curves) and tight-bound average symbol error probabilities (dotted curves) of 64QAM in the presence of the residual amplitude fluctuations and additive white Gaussian noise (u = normalized amplitude of the residual amplitude fluctuation). (Courtesy of Ref. 22.)

Fig. 35 and is called *Model A*. In Fig. 35, the original system is designed to transmit $1/T$ symbols/s over a bandwidth of $1/2T$ Hz, resulting in a spectral efficiency of 2 symbols/s/Hz, that is, the so-called Nyquist rate. Transmission above the Nyquist rate is achieved by increasing the signaling rate to $(1 + a)/T$ symbols/s while keeping the transmission characteristic unchanged, as is the case of speed tolerance evaluation (14,18,25). Thus, the spectral efficiency of

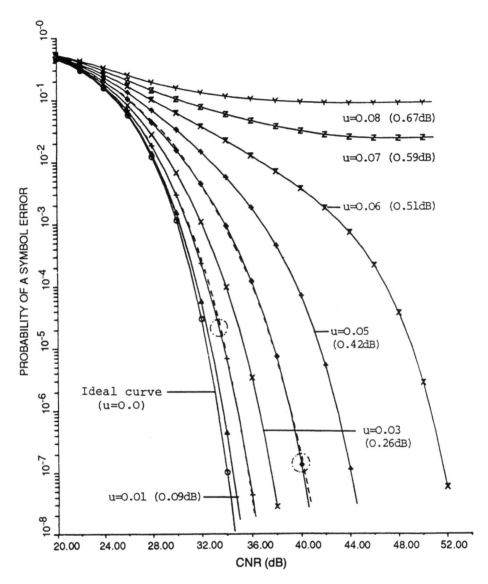

FIG. 33 Tight-bound average symbol error probabilities of 256QAM in the presence of the residual amplitude fluctuations and additive white Gaussian noise. Computer-simulation results are shown in dotted curves (u = normalized amplitude of the residual amplitude fluctuation). (Courtesy Ref. 22.)

Model A is $(1 + a)/T : 1/2T = 2(1 + a)$ symbols/s/Hz, where $100a$ represents the number (in percent) above the Nyquist rate. *Speed tolerance* for binary inputs is defined as the percentage increase over the Nyquist rate that will just cause overlap between adjacent levels (*eye closure*) (18). In the case of multilevel (*m*-ary) inputs, speed tolerance can be defined as the percentage increase over the Nyquist rate that will just cause at least one eye closure. Based on this

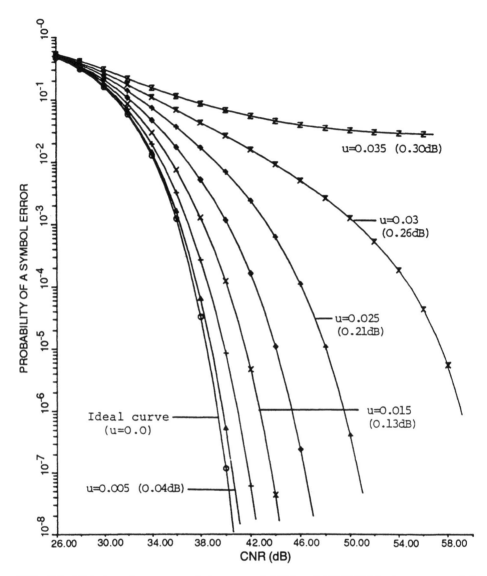

FIG. 34   Tight-bound average symbol error probabilities of 1024QAM in the presence of the residual amplitude fluctuations and additive white Gaussian noise (u = normalized amplitude of the residual amplitude fluctuation). (Courtesy Ref. 22.)

definition, speed tolerance of Class IV PRS with multilevel inputs was evaluated by computer simulation using "worst possible" data sequences. The results are summarized in Table 7.

We notice that speed tolerance of Class IV PRS has reduced dramatically to about 1% for eight-level inputs. This is due to inherently higher ISI in multilevel systems. Hence, by simple symbol-by-symbol receiver structures, it is nearly impossible to increase the spectral efficiency of Class IV PRS above the Nyquist rate by the conventional method.

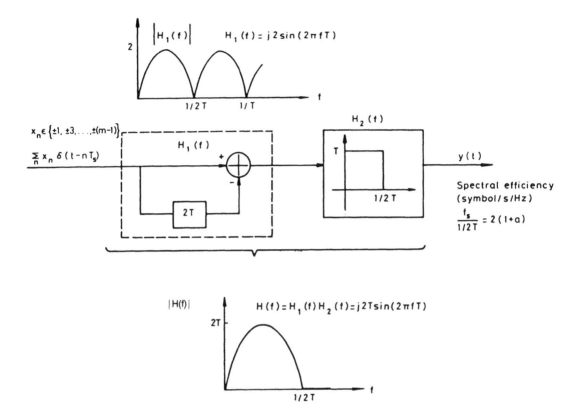

FIG. 35 Conventional system model, Model A, for Class IV partial response signaling (PRS) above the Nyquist rate and the cascaded transfer function. The spectral efficiency of this system is $2(1 + a)$ symbols/s/Hz (i.e., $100a\%$ higher than the so-called Nyquist rate of 2 symbols/s/Hz ($f_s = 1/T_s = (1 + a)/T$ = the actual signaling rate; when $a = 0$ [i.e., $T_s = T$], it is an ideal Class IV PRS system). (From Ref. 24, reprinted by permission of IEE, © 1988 IEE.)

TABLE 7 Speed Tolerance of Class IV Partial Response Signaling for Multilevel Inputs

Number of Input Levels	Speed Tolerance
2	16%
4	3%
8	1%

Source: Ref. 24, reprinted by permission of IEE, © 1988 IEE.

Fig. 36 shows a new system model (called *Model B*) using the spectral chopping technique for Class IV PRS above the Nyquist rate (24). It represents a scheme in which the original modified duobinary filter is followed by an ideal brick-wall filter with a cutoff frequency $f_N/(1 + a)$, where $f_N = 1/2T_s$ is the Nyquist frequency. The spectral efficiency of this new model is $1/T_s : f_N/(1 + a) = 2(1 + a)$ symbols/s/Hz, that is, the same as that of conventional Model A. The parameter a represents the fractional increase above the Nyquist rate (e.g., $a = 0.03$ means a 3% increase). Two computer-generated eye diagrams are presented in Fig. 37. The eye diagram of the new Model B with eight-level inputs and $a = 0.04$ is shown in Fig. 37-*b*. The quality of this eye pattern is much better than that of conventional Model A, shown in Fig. 37-*a* for the same efficiency of 6.24 bps/Hz.

An equivalent way of comparing PRS operation above the Nyquist rate for the two models is to keep the signaling rate f_s the same but reduce the required

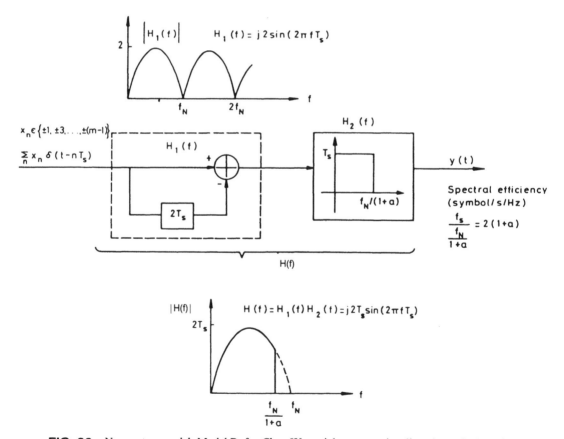

FIG. 36 New system model, Model B, for Class IV partial response signaling above the Nyquist rate with *m*-ary inputs and the corresponding transfer function. This also corresponds to the system model of the equivalent baseband in-phase or quadrature channel of a quadrature-modulated system ($f_s = 1/T_s = $ the actual signaling rate; $f_N = f_s/2 = $ the Nyquist frequency of the signal; when $a = 0$, it is an ideal Class IV PRS system). (From Ref. 24, reprinted by permission of IEE, © 1988 IEE.)

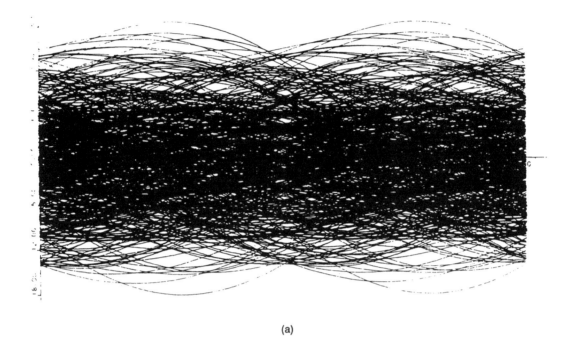

(a)

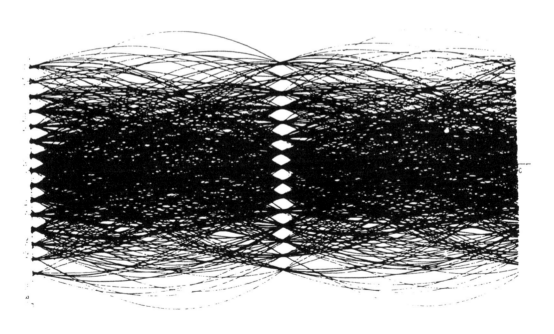

(b)

FIG. 37 Comparison of eye diagrams of Class IV, 15-level partial response signaling, 225 quadrature partial response signaling (QPRS) with 4% increase above the Nyquist rate (spectral efficiency = 6.24 bps/Hz): *a*, using conventional Model A; *b*, using the new Model B. (From Ref. 24, reprinted by permission of IEE, © 1988 IEE.)

bandwidth. Using this viewpoint, operation above the Nyquist rate in Model A is illustrated in Fig. 38-*a*, in which a scaled sine shape spectrum is retained. Thus, only the frequency scale is changed, resulting in all the frequency components suffering distortion. This distortion introduces very significant undesired ISI. On the other hand, in Model B, shown in Fig. 38-*b*, the old frequency response is simply "chopped off" beyond $f_N/(1 + a)$. Most of the original frequency components thus are retained and the only frequency components that suffer distortion are the ones that have been cut off. Note that the energy near f_N is very small and thus can be cut off without causing too much undesired ISI. This explains why Model B is better than the conventional Model A, especially when *m*-ary inputs are used.

We now present the computer-simulated performance of a more complicated system, Class IV 225QPRS using Model B. Note that 225QPRS is a band-pass system with 3 bits/symbol/axis, corresponding to $n = 8$ input levels/rail and $(2n - 1) = 15$ output levels/axis.

Essentially, a Class IV 225QPRS above the Nyquist rate can be generated by modulating in quadrature two baseband 15-level systems using Model B shown in Fig. 36. A bit rate of 1.6 Mbps is used in the simulation (i.e., with 800 Kbps in both in-phase and quadrature channels). The Class IV filtering is fully

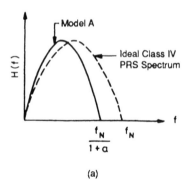

(a)

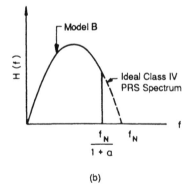

(b)

FIG. 38   A frequency-domain comparison of the two models for Class IV partial response signaling above the Nyquist rate: *a*, conventional Model A; *b*, new Model B.

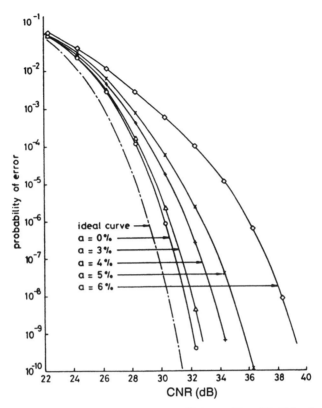

FIG. 39   Probability of error of the improved-efficiency Class IV 15-level partial response signaling, 225 quadrature partial response signaling above the Nyquist rate against carrier-to-noise ratio (the parameter a represents the fractional increase in spectral efficiency above the Nyquist rate).

at the transmitter, whereas the receiver filter merely bandlimits the channel noise. This commonly used filter shaping would result in about 1-dB degradation in CNR with respect to the case in which the filter shaping is divided optimally between the transmitter and the receiver. A perfect synchronization is assumed and symbol-by-symbol decision is used. Note that, under our simulation conditions, the 225QPRS system is equivalent to a baseband 15-level PRS system. Results are plotted in Fig. 39. Table 8 summarizes the required CNR at $P_e = 10^{-9}$ for different percentage increases above the Nyquist rate using the improved-efficiency Class IV PRS.

The advantage of the new technique is evident from Fig. 39 and Table 8. For example, for 225QPRS with a 3% increase above the Nyquist rate using our new Model B, only a 0.6-dB degradation is observed. Also listed in Table 8 is the corresponding result of the improved-efficiency Class I 225QPRS for comparison (23). We notice that, with new Model B for 225QPRS above the Nyquist rate for an increase of up to 5%, the performance of Class IV PRS is nearly as good as that of Class I PRS. For example, for the same 4% increase using Class IV PRS, there is only a 0.2-dB loss in CNR at $P_e = 10^{-9}$ with

TABLE 8   Required Carrier-to-Noise Ratio at $P_e = 10^{-9}$ for 225
Quadratic Partial Response Signaling above the Nyquist Rate

% above the Nyquist Rate	Spectral Efficiency bps/Hz	Class IV 225QPRS CNR (dB)	Class I 225QPRS* CNR (dB)
0	6	32.0	32.0
3	6.18	32.6	32.4
4	6.24	34.2	34.0
5	6.3	35.6	34.9
6	6.36	39.1	35.3

Source: Ref. 24, reprinted by permission of the IEE, © 1988 IEE.
*See Ref. 23.

respect to that of using Class I PRS. This is interesting since it might be tempting to say that for transmission above the Nyquist rate, Class I PRS is better than Class IV PRS. This is because the speed tolerance of Class I PRS is 43% for binary inputs whereas it is only 16% for Class IV PRS when the conventional method is used. However, it is not true if the spectral chopping technique is used because both Class I and Class IV PRS have a very small amount of energy near the Nyquist frequency so that part of it can be cut off, resulting in only a small performance degradation. In addition, Class IV PRS has some other advantages, described next.

An Efficient Method for Service Channel Transmission

Service channel transmission is illustrated in Fig. 40, in which the modified duobinary filter is followed by an ideal band-pass filter (BPF) with $f_L = bf_N$ and $f_H = f_N/(1 + a)$ representing the cutoff frequencies at the low-frequency end and at the high-frequency end, respectively. A service channel then can be inserted between DC and f_L (24). The effect of the location of f_L on the performance of a 15-level PRS/225QPRS above the Nyquist rate has been studied (24). The simulation result is shown in Fig. 41. It is seen that as long as $b < 0.03$, the degradation is in fact negligible. This indicates that, under the condition of a 4% increase above the Nyquist rate for Class IV 225QPRS, a bandwidth of $0.03f_N$ at the low-frequency end can be used for service channel transmission, which is not achievable by Class I PRS.

Higher Spectral Efficiency in Single-Sideband Systems

It should be noted that not only a service channel can be used at the low-frequency end of the Class IV PRS spectrum using the spectral chopping technique, but an increase in spectral efficiency also can be obtained in an FDM SSB system. From the results shown in Fig. 41, we observe that with a bandwidth of 3% of f_N cut at the low-frequency end and 4% at the high-frequency end, the

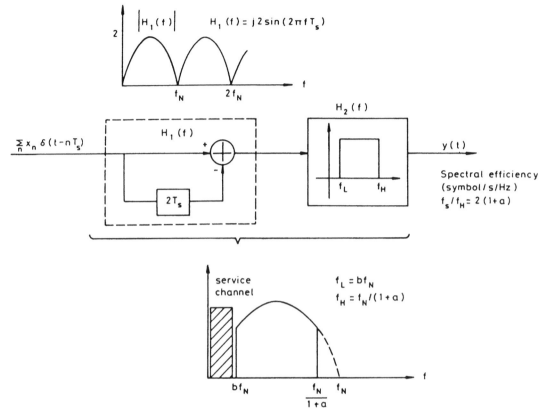

FIG. 40   System model of an improved-efficiency Class IV partial response signaling above the Nyquist rate for service channel transmission. (From Ref. 24, reprinted by permission of the IEE, © 1988 IEE.)

resultant performance degradation is reasonably small. Recall that in SSB systems only one side of the modulated spectrum is needed for transmission. Thus, it may be even better to chop the Class IV PRS spectrum symmetrically in an FDM SSB system application (24).

Constellation Considerations

Many digital modulation schemes are represented commonly and usefully by 2-dimensional (2D) constellations of all possible signal points (also called *states*). In this section, we discuss signal constellations. During the early 1960s, a fair amount of effort went into developing 2D signal constellations from different viewpoints (26). There are two important parameters associated with a constellation: rms (average) power and peak-to-rms power ratio. When choosing a constellation, it is desirable to have a lower average power and a lower peak-to-rms power ratio for the same distance separation between states.

A lower average power for the same distance separation between signal

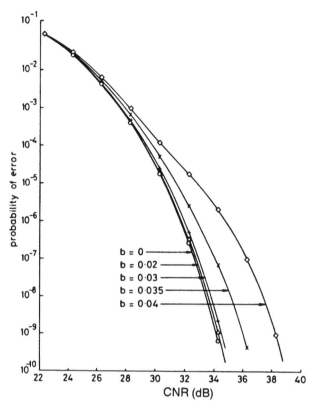

FIG. 41   Performance of the improved-efficiency Class IV 15-level partial response signaling, 225 quadrature partial response signaling with 4% above the Nyquist rate and various bandwidths carved out for a service channel. These results were obtained for the application in Fig. 40 for $f_H = f_N/(1 + a)$ with $a = 4\%$ and for $f_L = bf_N$ with $b = 0$, 2%, 3%, 3.5%, and 4%. (From Ref. 24, reprinted by permission of the IEE, © 1988 IEE.)

states directly means a lower CNR requirement. A lower peak-to-rms power ratio is preferred because, in any practical systems, the signal has to go through power amplifiers that are nonlinear, especially near saturation. As an example, for the transmission of 8 bits per signaling interval, a square 256QAM (Fig. 42-*a*) can be used, but this is not the best choice. By modifying the constellation using a circular boundary, a circular 256QAM (Fig. 42-*b*) is obtained, which is also known as stepped-square 256QAM (SS256QAM) (27,28). A straightforward calculation reveals that SS256QAM has the advantage of its average power and peak-to-rms power ratio being 0.19 and 1.38 dB, respectively, smaller than those of square 256QAM, yielding a total 1.57-dB advantage in system gain. The same principle can be applied to such other QAM schemes as 64QAM, 1024QAM, and so on. It is worth mentioning that, although hexagonal constellations seem to be most compact, the advantage over rectangular ones is insignificant or not at all (26).

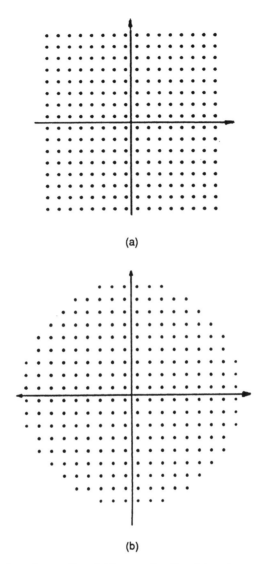

(a)

(b)

FIG. 42 Constellations for sending 8 bits per signaling interval: *a*, square 256 quadrature amplitude modulation; *b*, stepped-square 256QAM.

Pilot-Aided Digital Modulation Systems

Traditionally, QAM modems require such carrier-recovery techniques as Costasloops, quadruplers, or data-directed loops (5). An alternative approach is the pilot-aided transmission, which is emerging as a promising choice to combat multipath fading in some advanced digital communications systems. In a 64QAM high-capacity digital microwave radio system, a pilot inserted at the carrier frequency has been shown to be able to sustain a multipath notch fading depth of 45 dB at the carrier frequency (29). In a recent 256QAM digital transmission system application, pilots have been used for improved carrier acquisition (3). For mobile communications systems and other applications, more elaborate techniques have been proposed (30,31).

Whereas pilots are inserted either in band or out of band, the former is preferable. However, the in-band schemes require the suppression of the data signal PSD around the pilot frequencies to minimize interference from the data spectrum into the pilot and vice versa. Spectral suppression normally is achieved by applying certain balanced line-coding techniques (32,33). Some examples are bipolar code (34), feedback balanced code (35), duobinary code (13), and multilevel balanced codes (36,37).

In-Band Spectral-Suppression Techniques for Pulse Amplitude Modulation and Quadrature Amplitude Modulation Systems

Three types of basic coders for suppressing in-band spectrum are described: FJ, OF, and KF (38). Here, FJ, OF, and KF coders are due to Fujitsu researchers (29,35,37), Orton and Feher (41), and Kim and Feher (38), respectively. The input to a basic coder is an L-level PAM symbol stream, and the output symbols are also of L levels. For encoding, the input stream is sectioned into frames of length W symbols. Then, a block of one or more frames is mapped to a block of the same number of frames plus one flag symbol. In coder FJ, a block consists of a single frame so that W symbols are mapped to $W + 1$ symbols. The redundancy is

$$R_{\mathrm{FJ}} = \frac{1}{W} \tag{57}$$

In coders OF and KF, a block consists of N frames, where

$$N = \log_2 L \tag{58}$$

Figure 43 depicts one such output block for $L = 8$. Since NW symbols are mapped to $NW + 1$ symbols, the redundancy in this case, is

$$R_{\mathrm{OF}} = R_{\mathrm{KF}} = \frac{1}{NW}. \tag{59}$$

Each output frame is either a replica or an inversion of the corresponding input frame depending on the criteria described below. Whether a frame has been inverted or not is encoded into a bit. One such bit constitutes a flag symbol in FJ, whereas N such bits do the same in OF and KF. Comparing Eqs. (57) and (59), it is clear that OF and KF coders are more efficient than the FJ coder. In FJ and OF, the running digital sum (RDS) is controlled (32,39). In OF, for example, the RDS of the jth frame of the kth input block

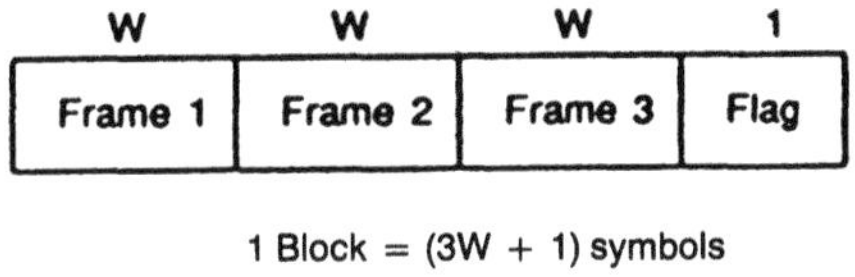

W	W	W	1
Frame 1	Frame 2	Frame 3	Flag

1 Block = (3W + 1) symbols

FIG. 43 Block structure of the basic coders OF and KF for $N = \log_2 8 = 3$ (8PAM), where one block = $(3W + 1)$ symbols.

$$d[j,k] = \sum_{i=1}^{W} x_{(k-1)NW+(j-1)W+1} \tag{60}$$

is compared with the RDS accumulated up to the previous frame,

$$D[j,k] = \sum_{i=1}^{(k-1)(NW+1)+(j-1)W} y_i. \tag{61}$$

In Eqs. (60) and (61), x_n is the nth input symbol, y_n is the nth output symbol, and the system is assumed to have started at $i = 0$ with $x_i = y_i = 0$ for $i < 0$. The coding rule is

$$y_{k(NW+1)+jW+i} = \begin{cases} x_{kNW+jW+i} & \text{if} \quad d[j,k]D[j,k] \le 0 \\ -x_{kNW+jW+i} & \text{if} \quad d[j,k]D[j,k] > 0 \end{cases}$$

$$\text{for} \quad i = 1, 2, 3, \ldots, W, j = 1, 2, 3, \ldots, N, k = 0, 1, 2, \ldots \tag{62}$$

That is, the current frame is inverted if the sign of its RDS is the same as that of the RDS accumulated up to the previous frame. The same holds true for FJ, except that $N = 1$ in Eqs. (60) to (62). The effect of controlling the RDS in this way is to have it wander about zero and so limit its peak-to-peak variation, called *digital sum variation* (DSV). It is well known that a code of a finite DSV has a PSD notch at $f = 0$ (32,39).

In coder KF, on the other hand, the running ISI sum (RIS) is controlled (40):

$$s[j,k] = \sum_{i=1}^{W} (-1)^{(k-1)(NW+1)+(j-1)W+i} x_{(k-1)NW+(j-1)W+i}. \tag{63}$$

$$S[j,k] = \sum_{i=1}^{(k-1)(NW+i)+(j-1)W} (-1)^i y_i.$$

The coding rule is

$$y_{k(NW+1)+jW+i} = \begin{cases} x_{kNW+jW+i} & \text{if} \quad s[j,k]S[j,k] \le 0 \\ -x_{kNW+jW+i} & \text{if} \quad s[j,k]S[j,k] > 0 \end{cases}$$

$$\text{for} \quad i = 1, 2, 3, \ldots, W, j = 1, 2, 3, \ldots, N, k = 0, 1, 2, \ldots \tag{64}$$

The current frame is inverted if the sign of its RIS is the same as that of the RIS accumulated up to the previous frame. The effect of this coding is to limit its peak-to-peak variation, called *ISI sum variation* (ISV) (40). It has been shown that a code of a finite ISV has a PSD notch at the Nyquist frequency $f = 1/2T$, where T is the symbol duration (32,40).

Separately from the basic coders, we identify two coder arrangements, Arrangement A and Arrangement B (Fig. 44). Arrangement A is self-explanatory. In Arrangement B, the input PAM stream is demultiplexed, each substream is processed by a separate basic coder of the same type, and then two coded outputs are multiplexed back again. The operation of Arrangement B usually is called *interleaving*.

Now, combinations of the basic coders and the coder arrangements yield resultant codes. Table 9 lists such combinations. Code FJ00 is known as feedback balanced code (FBC) (35). Use of FJ00 was reported in Ref. 37. Orton and Feher recently proposed the code OF01 (41). The codes OF00 and KF10 were introduced in Ref. 38. Regarding the notation of codes, the first 0 indicates the existence of a PSD notch at $f = 0$, whereas the second 0 indicates its existence at $f = 1/2T$. Hence, FJ01 and OF01 have only one notch at $f = 0$, whereas KF10 has only one notch at $f = 1/2T$. On the other hand, OF00 and FJ00 have notches at $f = 0$ and $f = 1/2T$. Next, we show the excellence of OF over FJ at $f = 0$ and/or at $f = 1/2T$ and the excellence of KF10 over all other codes at $f = 1/2T$.

It can be shown that, for FJ01 and OF01 (38),

$$\mathrm{DSV}_{01} = W. \tag{65}$$

FJ00 and OF00 also have finite but twice as large DSVs due to interleaving.

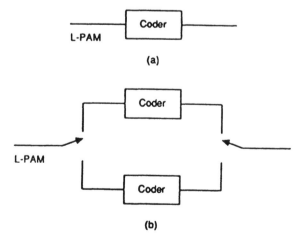

(a)

(b)

FIG. 44   Coder arrangements: *a*, Arrangement A; *b*, Arrangement B (PAM = pulse amplitude modulation).

TABLE 9 Codes Formed through Different Combinations of the Basic Coders and the Basic Arrangements

Coder	Arrangement A	Arrangement B
FJ	FJ01	FJ00
OF	OF01	OF00
KF	KF10	—

Source: Ref. 38, reprinted by permission of the Institute of Electrical and Electronics Engineers, © 1988 IEEE. The first 0 of each code designation indicates a power spectral density (PSD) notch at $f = 0$ while the second 0 indicates a PSD notch at the Nyquist frequency. A 1 indicates a full PSD value at the corresponding frequency.

$$\text{DSV}_{00} = 2W. \tag{66}$$

A smaller DSV is preferred since it means less spectral power at low frequencies (32,39). Similarly, at $f = 1/2T$, the ISV of the KF10 code is found to be (38)

$$\text{ISV}_{10} = W. \tag{67}$$

FJ00 and OF00 also have a notch at $f = 1/2T$. Their ISVs are

$$\text{ISV}_{00} = 2W. \tag{68}$$

Here, a smaller ISV also is preferred for a more pronounced notch at the Nyquist frequency (40).

Computer simulations with baseband M-ary QAM equivalents have been performed (38). Each channel is then an L-PAM system, where $L = \sqrt{M}$. A pair of coders chosen from Table 9 were applied independently to each channel. The symbol rate was 2 Mbaud, and the raised-cosine filter of a roll-off factor $\alpha = 0.2$ was used for pulse shaping.

Figure 45 compares KF10, OF00, and FJ00 at the Nyquist frequency for 256QAM with the redundancy of 1%. Here, the quality of the notches should be evaluated best in terms of their widths at a specified suppression depth. At -20 dB, for example, their relative widths are seen to be inversely proportional to the 1 : 2 : 8 ratio. If we vary the redundancy of the codings, the PSDs overlap (Fig. 46). Figure 47 shows that performance apparently varies with redundancy, whereas it hardly changes with the number of levels (see Fig. 48).

A disadvantage of employing power-suppression codes is an increase in bandwidth. However, out-of-band schemes like Simon's dual-pilot tone calibration technique (DPTCT) also require an extra bandwidth other than for the pure data signal (31). It is believed that a bandwidth increase of 1% to 2% due to coding will not be greater than that for out-of-band schemes. Besides, pilots of the in-band arrangement are usually more favorable.

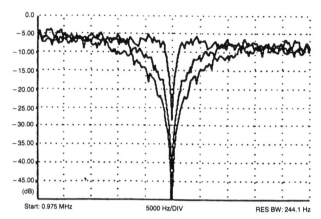

FIG. 45   Simulated power spectral densities of coded 256QAM signals at the Nyquist frequency, 2 Mbaud. Redundancy is 1% for all signals. KF10, OF00, and FJ00 codings are compared. (From Ref. 38, reprinted by permission of the Institute of Electrical and Electronics Engineers, © 1988 IEEE.)

Block-Inversion-Coded Quadrature Amplitude Modulation Systems

The block-inversion-coded (BIC) QAM systems employ a coding scheme very similar to the FJ coder (42). The BIC encoder transforms a block of m input data symbols $\{d_1, d_2, \ldots, d_m\}$ into an output codeword of $(m + 1)$ symbols. Figures 49-a and 49-b show the formats of a data block and its codeword, respectively. Each codeword consists of a binary status symbol I and m multi-level coded data symbols $\{c_1, c_2, \ldots, c_m\}$.

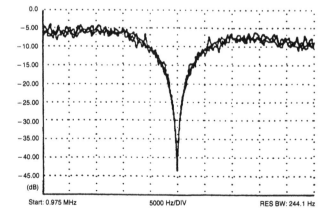

FIG. 46   Simulated power spectral densities of coded 256QAM signals at the Nyquist frequency, 2Mbaud. KF10 with 0.5%, OF00 with 1%, and FJ00 with 4% redundancy are used for coding. (From Ref. 38, reprinted by permission of the Institute of Electrical and Electronics Engineers, © 1988 IEEE.)

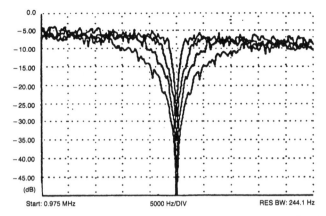

FIG. 47 Simulated power spectral densities of KF10-coded 256QAM signals at the Nyquist frequency, 2 Mbaud. Redundancies of 1%, 0.5%, 0.25%, and 0.125% are used. (From Ref. 38, reprinted by permission of the Institute of Electrical and Electronics Engineers, © 1988 IEEE.)

Encoding is performed by comparing the digital sum $DS_i = \Sigma_{j=1}^m d_j$ of the input data block with the digital sum of the previously transmitted codeword $DS_0 = I + \Sigma_{j=1}^m c_j$. If they are of the same polarity, all data symbols in the input block are inverted prior to transmission. Otherwise, no data inversion takes place. To identify the codeword with inverted data, a status symbol of -1 is used. Similarly, a status symbol of $+1$ identifies those codewords with uninverted data.

The net effect of this simple coding rule is to minimize the RDS of a coded signal, suppressing the low-frequency components of the signal. The normalized

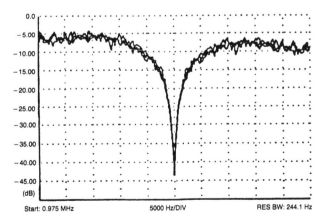

FIG. 48 Simulated power spectral densities of OF00-coded QAM signals at the Nyquist frequency, 2 Mbaud. Redundancy is 1% for all signals. 4QAM, 16QAM, 64QAM, and 256QAM signals are used. (From Ref. 38, reprinted by permission of the Institute of Electrical and Electronics Engineers, © 1988 IEEE.)

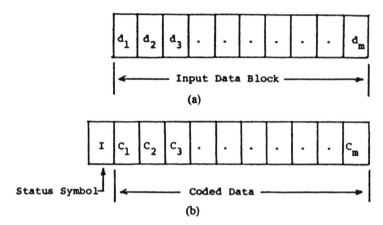

FIG. 49   Format of: *a*, an input data block; *b*, its BIC code block.

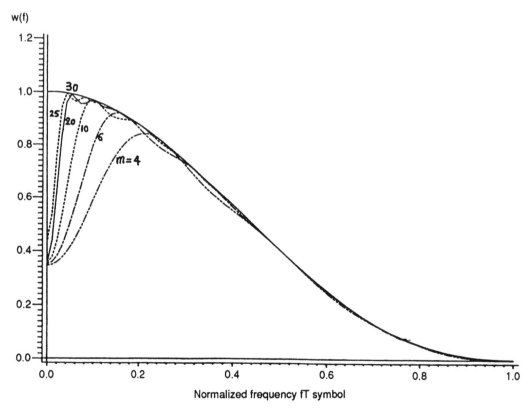

FIG. 50   Normalized power spectrum of 16-level BIC code. (From Ref. 42, reprinted by permission of the Institute of Electrical and Electronics Engineers, © 1988 IEEE.)

power spectra of the BIC code with 16-level signaling are shown in Fig. 50. The spectral control is observed to be excellent for codes with small data-block sizes. As the data-block size increases, the code's spectral control capability deteriorates.

The performance of a 256QAM modem using BIC and high-pass filters (HPFs) to facilitate carrier-pilot insertion also is reported in Ref. 42. A simplified block diagram is shown in Fig. 51. In the transmitter, a HPF and a BIC coder are used to accommodate carrier-pilot insertion. In the receiver, the carrier pilot is recovered by a phase-locked loop. A HPF identical to that in the transmitter is used to block out the DC component caused by the demodulated

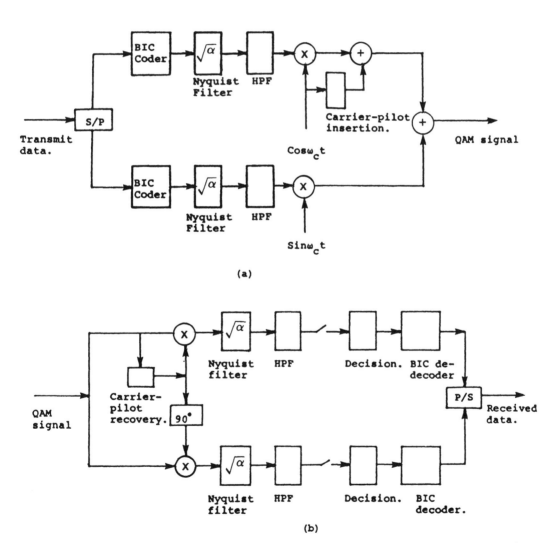

(a)

(b)

FIG. 51 Block diagram of a 256 quadrature amplitude modulation modem with BIC coder and carrier-pilot insertion: a, transmitter; b, receiver (HPF = high-pass filter).

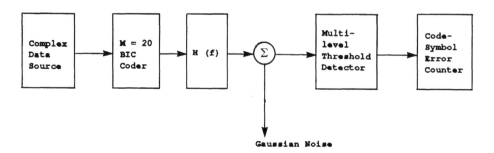

FIG. 52   Computer-simulation model of the 256QAM system baseband equivalent.

carrier pilot. This modem also incorporates a raised-cosine Nyquist filter with a roll-off of $\alpha = 0.2$.

The choice of the BIC code's data-block size m is an engineering compromise. A code with small m has more effective spectral control on the signal's low-frequency components. However, small m also implies a higher degree of signal redundancy and, hence, a wider additional transmission bandwidth requirement. In this study, m is chosen to be 20. While this code has adequate spectral control, it requires 1/20 or 5% additional transmission bandwidth.

The performance of this modem is investigated using the Monte Carlo method with a computer-simulation model (see Fig. 52). The filter $H(f)$ is described by a transfer function (Fig. 53) that consists of two components:

$$H(f) = H_1(f) - H_2(f). \qquad (69)$$

$H_1(f)$ represents the modem's Nyquist filter with its Nyquist frequency f_N normalized to 1 and roll-off parameter $\alpha = 0.2$. $H_2(f)$, on the other hand, represents the effect of the modem's HPFs. HPFs are known to cause ISI over many symbol periods. In this study, $H_2(f)$ is chosen to have full raised-cosine characteristics with cutoff frequency $f_c < 1$ to minimize the extent of ISI.

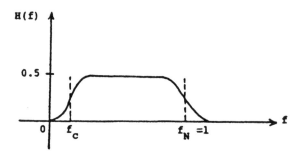

FIG. 53   Characteristics of the $H(f)$ used in the simulation model in Fig. 52.

$$H_2(f) = \begin{cases} 0.25\left\{1 - \sin\left(\dfrac{\pi}{2f_c}(f - f_c)\right)\right\} \\ \qquad\qquad\qquad 0 \le f \le 2f_c \\ 0 \qquad\qquad\qquad f > 2f_c \end{cases} \tag{70}$$

The result of the simulation study is shown in Fig. 54. Curves 1 and 2 show the modem's code-symbol error rate when the cutoff frequency f_c of $H_2(f)$ is 0.05 and 0.1% f_N, respectively. For comparison, a 256QAM modem with HPFs but without BIC coding to accommodate carrier-pilot insertion also is simulated. Its performance with the corresponding $f_c = 0.05$ and 0.1% f_N is shown as Curves 4 and 5. Finally, as a reference, the performance of a conventional 256QAM modem is plotted as Curve 3. Comparison of these curves shows that the HPF has a severe detrimental effect on the performance of a QAM modem without coding. A filter with the cutoff frequency $f_c = 0.05\%$ f_N causes a performance degradation of 4.2 dB at a symbol error rate of 10^{-5} (see Curves 3 and 4). If the filter's cutoff frequency is increased to $f_c = 0.1\%$ f_N, the modem's performance curve displays a flareout, and the system appears to have an error floor of 2×10^{-4} at approximately a high signal-to-noise ratio (see Curve 5). In general, excessive degradation is not acceptable.

With the introduction of $m = 20$ BIC coding, however, the modem's performance improves. For example, for $f_c = 0.1\%$ f_N, the coded modem's performance curve (Curve 2) no longer displays a tail flareout over the P_e range of interest in this study. But, it still suffers a performance degradation of 2.1 dB at $P_e = 10^{-5}$ (see Curves 2 and 3). However, if f_c is reduced further to 0.05% f_N, the coded modem's performance degradation becomes 0.9 dB at $P_e = 10^{-5}$ (see Curves 1 and 3).

Trellis-Coded Modulation

Such conventional coding schemes as block codes and convolutional codes improve modem performance by adding redundancy in the transmitted bit stream, resulting in either increasing bandwidth or sacrificing the information rate (43). Increased bandwidth also means an increased noise in the receive filter. Note that in the conventional coding approach modulation and coding are separated. Trellis-coded modulation (TCM) has evolved over the past decade as a combined coding and modulation technique for digital transmission over bandlimited channels. Its main attraction comes from the fact that it achieves significant coding gains over conventional uncoded multilevel modulation without compromising bandwidth efficiency. The first TCM schemes were proposed in 1976 (44). The basic principles of TCM were published by Ungerboeck in 1982 (45). According to Ungerboeck, the motivation for developing TCM initially came from his work on multilevel systems that employ the Viterbi algorithm to improve signal detection in the presence of ISI (46). This work provided him with ample evidence of the importance of Euclidean distance between signal

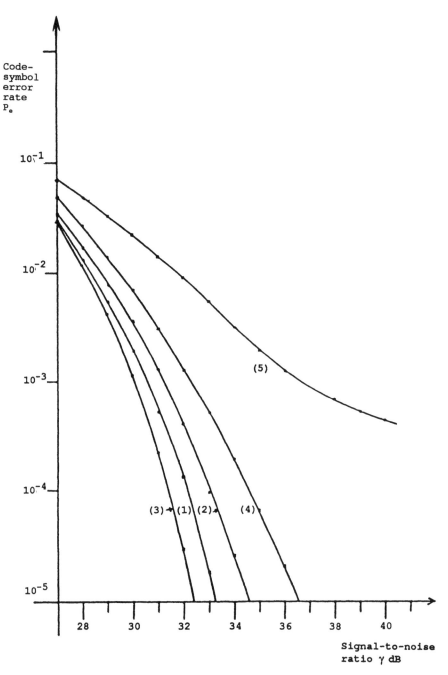

FIG. 54 Performance of the 256 quadrature amplitude modulation modem with carrier-pilot insertion. Curve 1 is BIC coded ($m = 20$) with $f_c = 0.05\%\ f_{NQ}$. Curve 2 is BIC coded ($m = 20$) with $f_c = 0.1\%\ f_{NQ}$. Curve 3 represents a conventional modem (uncoded and without a high-pass filter (HPF)). Curve 4 is uncoded with $f_c = 0.05\%\ f_{NQ}$. Curve 5 is uncoded with $f_c = 0.1\%\ f_{NQ}$. (From Ref. 42, reprinted by permission of the Institute of Electrical and Electronics Engineers, © 1988 IEEE.)

sequences. Since improvements over the established technique of adaptive equalization to eliminate ISI and then making independent signal decisions in most cases did not turn out to be very significant (47), he turned his attention to using coding to improve performance. In this connection, it was clear to him that codes should be designed for maximum free Euclidean distance rather than Hamming distance, and that the redundancy necessary for coding would have to come from expanding the signal set to avoid bandwidth expansion.

To understand the potential improvements to be expected by this approach, Ungerboeck computed the channel capacity of channels with additive Gaussian noise for the case of discrete multilevel modulation at the channel input and unquantized signal observation at the channel output. One-dimensional (1D) and 2D modulation constellations are considered, as illustrated in Fig. 55. The results of these calculations (Fig. 56) allowed two observations: first, in principle, coding gains of about 7–8 dB over conventional uncoded multilevel modulation should be achievable and, second, most of the achievable coding gain could be obtained by expanding the signal sets used for uncoded modulation only by the factor of two. Thus, the approach is to find trellis-based signaling schemes that use signal sets of size 2^{m+1} for transmission of m bits per modulation interval. The term *trellis* is used because these schemes can be described by a state-transition (trellis) diagram similar to the trellis diagrams of binary convolutional codes. The difference is that in TCM schemes the trellis branches are labeled with redundant nonbinary modulation signals rather than with binary code symbols.

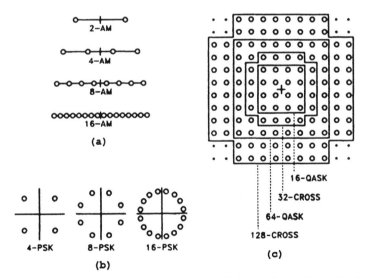

FIG. 55 Signal sets: *a*, one-dimensional amplitude modulation; *b*, two-dimensional phase modulation; *c*, amplitude/phase modulation (PSK = phase-shift keying). (From Ref. 50, reprinted by permission of the Institute of Electrical and Electronics Engineers, © 1987 IEEE.)

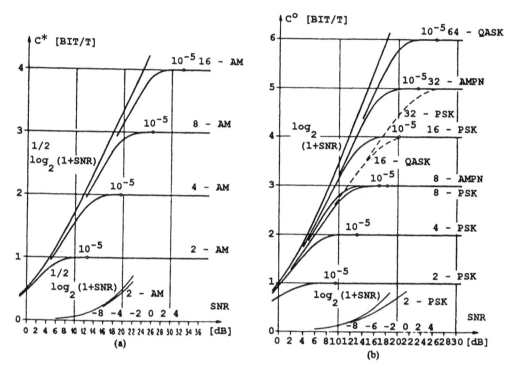

FIG. 56 Channel capacity C^* of bandlimited additive white Gaussian noise channels with discrete-valued input and continuous-valued output: a, one-dimensional modulation; b, two-dimensional modulation. (From Ref. 45, reprinted by permission from the Institute of Electrical and Electronics Engineers, © 1982 IEEE.)

Two-Dimensional Trellis-Coded Modulation

In this section, we illustrate with an example how 2D TCM works. Figure 57 depicts signal sets and trellis diagrams for uncoded 4PSK modulation and coded 8PSK modulation with four trellis states. A trivial, one-state trellis diagram is shown in Fig. 57-b only to illustrate uncoded 4PSK from the viewpoint of TCM. Every connected path through a trellis in Fig. 57 represents an allowed signal sequence. In both the 4PSK and 8PSK systems, starting from any state, four transitions can occur as required to encode two information bits per modulation interval.

The four "parallel" transitions in the one-state trellis diagram of Fig. 57-b for uncoded 4PSK do not restrict the sequences of 4PSK signals that can be transmitted; that is, there is no sequence coding. Hence, the optimum decoder can make independent nearest-signal decisions for each noisy 4PSK signal is $\sqrt{2}$, denoted as d_0. We call it the *free distance* of uncoded 4PSK modulation, to use common terminology of sequence-coded systems. Each 4PSK signal has two nearest-neighbor signals at this distance.

In the four-state trellis of Fig. 57-d for the coded 8PSK scheme, the transitions occur in pairs of two parallel transitions. Figure 57-d shows the numbering

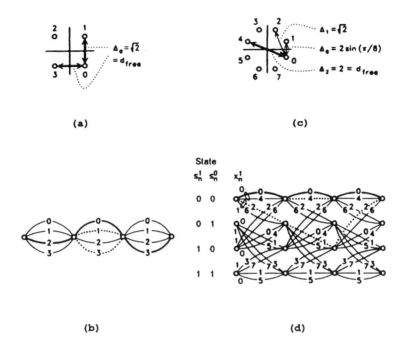

FIG. 57   Example of two-dimensional trellis-coded modulation: *a*, uncoded four-phase modulation for 4 phase-shift keying; *b*, uncoded four-phase modulation for one-state trellis; *c*, four-state trellis-coded eight-phase modulation for redundant 8PSK signal set; *d*, four-state trellis-coded eight-phase modulation (8PSK) four-state trellis. (From Ref. 50, reprinted by permission from the Institute of Electrical and Electronics Engineers, © 1987 IEEE.)

of the 8PSK signals and relevant distances between these signals: $\Delta_0 = 2 \sin (\pi/8)$, $\Delta_1 = \sqrt{2}$, and $\Delta_2 = 2$. The 8PSK signals are assigned to the transitions in the four-state trellis in accordance with the following rules:

1. Parallel transitions are associated with signals with maximum distance between them Δ_2 (8PSK) $= 2$, and the signals in the subsets (0,4), (1,5), (2,6), or (3,7).
2. Four transitions originating from or merging in one state are labeled with signals with a distance of at least Δ_1 (8PSK) $= \sqrt{2}$ between them, that is, the signals in the subsets are (0,4,2,6) or (1,5,3,7).
3. All 8PSK signals are used in the trellis diagram with equal frequency.

Any two signal paths in the trellis of Fig. 57-*d* that diverge in one state and remerge in another after more than one transition have at least the squared distance $\Delta_1^2 + \Delta_0^2 + \Delta_1^2 = \Delta_2^2 + \Delta_0^2$ between them. For example, the paths with signals 0, 0, 0 and 2, 1, 2 have this distance. The distance between such paths is greater than the distance between the signals assigned to parallel transitions, Δ_2 (8PSK) $= 2$, which thus is found as the free distance in the four-state 8PSK code $d_{\text{free}} = 2$. Expressed in decibels, this amounts to an improvement of 3 dB over the minimum distance $\sqrt{2}$ between the signals of uncoded 4PSK modulation. Figure 58 illustrates one possible realization of an encoder modulator for the four-state coded 8PSK scheme.

Soft-decision decoding is accomplished in two steps. In the first step, called *subset decoding*, within each subset of signals assigned to parallel transitions the signal closest to the received channel output is determined. These signals are stored together with their squared distances from the channel output. In the second step, the Viterbi algorithm is used to find the signal path through the code trellis that has the minimum sum of squared distances from the sequence of noisy channel outputs (46). Only the signals already chosen by subset decoding are considered.

The essential points of the Viterbi algorithm are summarized next. Assume that the optimum signal paths from the infinite past to all trellis states at time n are known; the algorithm extends these paths iteratively from the states at time n to the states at time $n + 1$ by choosing one best path to each new state as the "survivor" and "forgetting" all other paths that cannot be extended as the best paths to the new states. Looking backward in time, the surviving paths tend to merge into the same "history path" at some time $n - d$. With a sufficient decoding delay D so that the randomly changing value of d is highly likely to be smaller than D, the information associated with a transition on the common history path at time $n - D$ can be selected for output.

Let the received signals be disturbed by uncorrelated Gaussian noise samples with variance σ^2 in each signal dimension. The probability that at any given time the decoder makes a wrong decision among the signals associated with parallel transitions, or starts to make a sequence of wrong decisions along some path diverging for more than one transition from the correct path, is called the *error-event probability*. At high signal-to-noise ratios, this probability is generally well approximated by (45)

$$Pr(e) \simeq N_{\text{free}} \cdot Q[d_{\text{free}}/(2\sigma)] \tag{71}$$

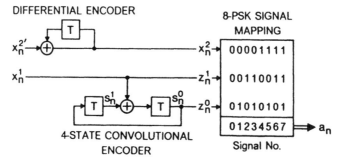

FIG. 58 An encoder for the four-state 8 phase-shift keying code. (From Ref. 50, reprinted by permission from the Institute of Electrical and Electronics Engineers, © 1987 IEEE.)

where $Q(x)$ represents the Gaussian error integral, and N_{free} denotes the (average) number of nearest-neighbor signal sequences with distance d_{free} that diverge at any state from a transmitted signal sequence, and remerge with it after one or more transitions. The approximate formula above expresses the fact that at high signal-to-noise ratios the probability of error events associated with a distance larger than d_{free} becomes negligible.

For uncoded 4PSK, we have $d_{free} = \sqrt{2}$ and $N_{free} = 2$, and for four-state coded 8PSK we find $d_{free} = 2$ and $N_{free} = 1$. Since free distance is found in both systems between parallel transitions, single signal-decision errors are the dominating error events. In the special case of these simple systems, the numbers of nearest neighbors do not depend on which particular signal sequence is transmitted.

Figure 59 shows the error-event probability of the two systems as a function of signal-to-noise ratio. For uncoded 4PSK, the error-event probability is extremely well approximated by Eq. (71). For four-state coded 8PSK, Eq. (71)

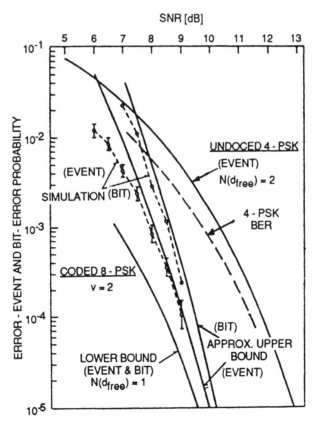

FIG. 59 Error-event and bit-error performance of coded 8 phase-shift keying ($\nu = 2$, minimal encoder) and uncoded 4PSK, 2 bit/T. (From Ref. 45, reprinted by permission of the Institute of Electrical and Electronics Engineers, © 1982 IEEE.)

provides a lower bound that is asymptotically achieved at high signal-to-noise ratios. Simulation results are included in Fig. 59 for the coded 8PSK system to illustrate the effect of error events with distance larger than free distance, of which the probability of occurrence is not negligible at low signal-to-noise ratios.

We have illustrated the principle of 2D TCM by a simple 4state 8PSK scheme. Many other 2D TCM schemes can be found in Refs. 49 and 50. Let us now consider the application of 2D TCM to higher spectral-efficient QAM. Suppose it is required to transmit 7 bits/T (i.e., 128QAM should be used). Using the 2D approach, 256 states in the constellation are needed. Since the constellation size of 128QAM is already very large, it is desirable to reduce it. This leads to the development of multidimensional TCM schemes.

Multidimensional Trellis-Coded Modulation

To send Q information bits per signaling interval T using a rate $m/m + 1$ trellis code with a 2N-dimensional constellation partitioned into 2^{m+1} subsets, m of the NQ information bits arriving in each block of N signaling intervals enter the trellis encoder, and the resulting $m + 1$ coded bits specify which 2N-dimensional subset is to be used. The remaining information bits specify which point from the selected 2N-dimensional subset is to be transmitted. We illustrate this with a simple 16-state code with four-dimensional (4D) rectangular constellation.

A rate 2/3, 16-state code with a 4D rectangular constellation of 2^{15} points is shown in Fig. 60. In Fig. 61, the 192 states (with minimum distance d_0) are partitioned into 4 subsets A, B, C, and D such that with each subset, the minimum distance between states becomes $2d_0$. A 4D state or "type" can be visualized as a pair of 2D states in two consecutive signaling intervals. The 4D constellation is constructed from the 192-point 2D constellation of Fig. 61 and is partitioned into 8 subsets as in Table 10. The three output bits $Y0_n$, $I1_n$, and $I2_n'$ of the trellis encoder are associated with the 4D subsets in accordance with Table 10.

If we denote the current and next states of the trellis encoder as $W1_p W2_p$ $W3_p W4_p$, $p = n$ and $n + 2$, the trellis diagram is as shown in Fig. 62. The association of 4D subsets with the state transitions of Fig. 62 satisfies three requirements: (1) the 4D subsets associated with the transitions leading from a state are different from each other and belong to the same 4D family $U_{i=0}^{3}i$ or $U_{i=4}^{7}i$, and likewise for the 4D subsets associated with the transitions leading to a state; (2) the minimum squared Euclidean distance (MSED) between two allowed sequences of 4D subsets corresponding to two distinct trellis paths is larger than $4d_0^2$, which is the MSED of each 4D subset; and (3) a one-to-one function F that maps each state of the trellis encoder into another state may be defined so that the following statement is valid. Denote X as the 4D subset associated with the transition from a current state i to a next state j, and Y as the 4D subset obtained when X is rotated 90° clockwise. Then Y is associated with the transition from the current state $F(i)$ to the next state $F(j)$. The function F for this code is

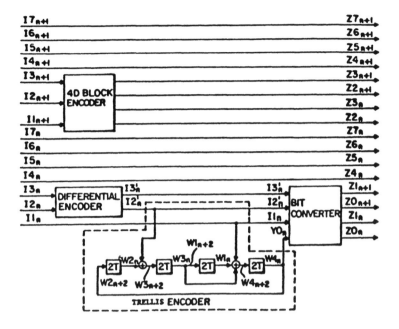

⊕ Exclusive OR

T Signaling Interval

2T Delay Element

FIG. 60 Block diagram of 16-state code with four-dimensional rectangular constellation. (From Ref. 51, reprinted by permission of the Institute of Electrical and Electronics Engineers, © 1987 IEEE.)

$$F: W1_p\,W2_p\,W3_p\,W4_p \to \overline{W1_p}\,W2_p\,\overline{W3_p}\,W4_p \tag{72}$$

where an overbar denotes inversion.

The first requirement guarantees that the MSED between any two allowed sequences of 4D points is $4d_0^2$. Therefore, the asymptotic coding gain of the code over the uncoded 128QAM (see the inner cross-constellation in Fig. 61) is

$$10\,\log_{10}\left(\frac{4d_0^2}{28.0625d_0^2}\,\Big/\,\frac{d_0^2}{20.5d_0^2}\right) = 4.66\,\text{dB}, \tag{73}$$

where $28.0625d_0^2$ is the average power of the 4D constellation, and $20.5d_0^2$ is the average power of 128QAM. This is the largest possible coding gain that can be achieved with the partitioning of the 4D rectangular constellation of Table 10. This coding gain may be viewed as the combination of a gain of 6.02 dB from the trellis code if the 4D constellation were not expanded from 2^{14} to 2^{15} points, and a loss of 1.36 dB due to that expansion. The expansion loss is less than the

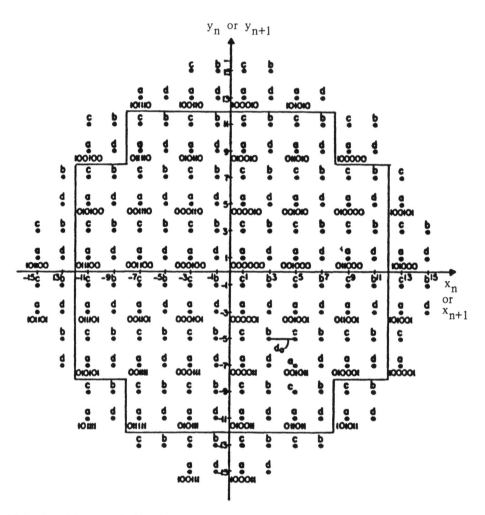

FIG. 61 Diagram of 192-point two-dimensional constellation partitioned into four subsets (number beneath each point equals $Z2_{n+i}\ Z3_{n+i}\ Z4_{n+i}\ Z5_{n+i}\ Z6_{n+i}\ Z7_{n+i}$; $i = 0$ or 1). (From Ref. 51, reprinted by permission of the Institute of Electrical and Electronics Engineers, © 1987 IEEE.)

3-dB loss of a 2D rate $m/m+1$ trellis code as promised by the use of a 4D constellation.

It should be remembered that the asymptotic coding gain can be approached only at high SNR. At low SNR (say, a BER in the range of 10^{-2} to 10^{-3}), the real coding gain could be much less than the asymptotic coding gain. This is true for all TCM schemes because the Viterbi decoder makes longer error events and errors occur in bursts at low SNR.

The second requirement eliminates MSED error events that differ in more than one 4D point from a given sequence of 4D points. The error coefficient of the code thus is minimized to 24 per 4D point (equivalent to 12 per 2D point), which is the number of nearest neighbors to any point in the same 4D subset.

TABLE 10 Eight-Sublattice Partitioning of Four-Dimensional Rectangular Lattice

4D Sublattice (Subset)	$Y0_n$	$I1_n$	$I2'_n$	$I3'_n$	4D Types	$Z0_n$	$Z1_n$	$Z0_{n+1}$	$Z1_{n+1}$
0	0	0	0	0	(A, A)	0	0	0	0
	0	0	0	1	(B, B)	0	1	0	1
1	0	0	1	0	(C, C)	1	0	1	0
	0	0	1	1	(D, D)	1	1	1	1
2	0	1	0	0	(A, B)	0	0	0	1
	0	1	0	1	(B, A)	0	1	0	0
3	0	1	1	0	(C, D)	1	0	1	1
	0	1	1	1	(D, C)	1	1	1	0
4	1	0	0	0	(A, C)	0	0	1	0
	1	0	0	1	(B, D)	0	1	1	1
5	1	0	1	0	(C, B)	1	0	0	1
	1	0	1	1	(D, A)	1	1	0	0
6	1	1	0	0	(A, D)	0	0	1	1
	1	1	0	1	(B, C)	0	1	1	0
7	1	1	1	0	(C, A)	1	0	0	0
	1	1	1	1	(D, B)	1	1	0	1

Source: Ref. 51, reprinted by permission of the Institute of Electrical and Electronics Engineers, © 1987 IEEE.

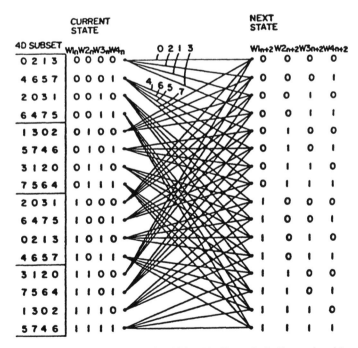

FIG. 62 Trellis diagram of 16-state code of Fig. 60. (From Ref. 51, reprinted by permission of the Institute of Electrical and Electronics Engineers, © 1987 IEEE.)

TABLE 11 Four-Dimensional Block Encoder

$I1_{n+1}$	$I2_{n+1}$	$I3_{n+1}$	$Z2_n$	$Z3_n$	$Z2_{n+1}$	$Z3_{n+1}$
0	0	0	0	0	0	0
0	0	1	0	0	0	1
0	1	0	0	0	1	0
0	1	1	0	1	1	0
1	0	0	1	0	0	0
1	0	1	1	0	0	1
1	1	0	0	1	0	0
1	1	1	0	1	0	1

Source: Ref. 51, reprinted by permission of the Institute of Electrical and Electronics Engineers, © 1987 IEEE.

Taking into account the boundary effect of the finite constellation would reduce this value.

The third requirement guarantees that the code can be made transparent to all phase ambiguities (90°, 180°, and 270°) of the constellation. Since the same 4D subset is obtained when a 4D subset is rotated by 180°, the construction of the trellis code needs to take into account only 90° rotation. The 270° rotation then is taken care of automatically. Detailed bit mapping can be found in Ref. 51.

To summarize, the bit converter and the 4D block encoder in Fig. 60 and Table 11 take the 3 trellis-encoded bits and the 12 remaining uncoded information bits and produce 2 groups of 8 selection bits each, $Z2_p Z3_p Z4_p Z5_p Z6_p Z7_p$ $Z0_p Z1_p$, $p = n$ and $n + 1$. The first group then is used to address a 2D mapping table to obtain the first 2D point. The table may be constructed from Fig. 61 and Table 12. The second group addresses the same 2D mapping table to obtain the second 2D point. The 4D point corresponding to the pair of 2D points is the one selected for transmission.

That the scheme is transparent to all the phase ambiguities of the constellation may be seen as follows. If we translate a sequence of the bit pairs $I3'_n I2'_n$

TABLE 12 Correspondence between $Z0_p Z1_p$ and Four Two-Dimensional Subsets

2D Subset	$Z0_p Z1_p$
A	00
B	01
C	10
D	11

Source: Ref. 51, reprinted by permission of the Institute of Electrical and Electronics Engineers, © 1987 IEEE.

appearing at the inputs of the trellis encoder and the bit converter by the same number of positions (one, two, or three) in a circular sequence 00, 01, 10, 11, then the sequence of 2D points produced by the 4D constellation-mapping procedure will be rotated by 90°, 180°, and 270° clockwise, respectively. Therefore, a differential encoder of the form

$$I3'_n I2'_n = (I3'_{n-2} I2'_{n-2} + I3_n I2_n) \bmod 100_{\text{base 2}} \tag{74}$$

in Fig. 60, and a corresponding differential decoder of the form

$$I3_n I2_n = (I3'_n I2'_n - I3'_{n-2} I2'_{n-2}) \bmod 100_{\text{base 2}} \tag{75}$$

at the output of the trellis decoder will remove all the phase ambiguities of the constellation.

There are two general principles in constructing a trellis code with a multidimensional constellation. The first principle says that the intersubset MSED of the multidimensional subsets associated with transitions originating from each state of the trellis encoder should be kept as large as possible, and likewise for the multidimensional subsets associated with transitions leading to each state.

The second principle says that for each of those phase ambiguities of the constellation such that the multidimensional subsets are not invariant under the corresponding rotations, it should be possible to define a one-to-one function F that maps each state of the trellis encoder into another state so that the following statement is valid. Let X be the multidimensional subset associated with the transition from a current state i to a next state j. Let Y be the multidimensional subset obtained when X is rotated by a number of degrees corresponding to that phase ambiguity. Then Y is the multidimensional subset associated with the transition from the current state $F(i)$ to the next state $F(j)$. The second principle also is used in Ref. 52 for constructing a rotationally invariant 2D trellis code.

Applications and Other Recent Work

Though TCM originally was invented for high-speed voiceband telephone channel modem applications, it has found wide applications in such other areas as satellite communications. For example, an experimental trellis-coded 8PSK (C8PSK) modem was developed successfully to comply with the International Telecommunications Satellite Organization (INTELSAT) specifications for single channel per carrier (SCPC) modems operating at 64 Kbps. Measured results show a 5-dB improvement of the C8PSK modem over the currently used 4PSK modem (48). Generally speaking, for applications with a bit rate in the range of Kbps, an all-digital implementation approach would be an easy task through the use of any of the commercially available digital signal processing (DSP) chips.

For the development of modems operating in the Mbps range, the task would be more difficult but not impossible. As an example, Comsat Laboratories (Clarksburg, MD) has developed a 140-Mbps transmission technology (the standard digital telephone multiplex rate) over INTELSAT satellites' 80-

MHz transponders (53). The technique employs the principles of TCM. The 140-Mbps datastream is converted first to a 180-Mbps stream by a 7/9 convolutional encoder. The 180-Mbps stream is mapped into a trellis-coded 8PSK symbol stream, yielding a symbol rate of 60 MHz. This is to reduce the risk of high-speed implementation. Direct use of trellis-coded 8PSK in this application would require a baud rate of 70 MHz, which at the time of development was thought to be difficult or high risk. The resulting modem achieves a BER as low as those found in fiber-optic circuits, thus allowing satellites to carry the same digital services as the TAT-8 cable.

Trellis codes also have been designed for 1D and 2D signal sets with non-equally spaced ("asymmetric") signals (54). Some modest coding gains compared to schemes with equally spaced signals are achieved when the codes have few states and small signal sets. These gains disappear for larger signal sets and higher code complexity.

Conventional TCM schemes are designed for the Gaussian channel. The design of TCM schemes for Rayleigh-fading channels is the subject of Ref. 55. Noncoherent detection of TCM schemes is another exciting research area (56,57).

Advances in Differentially Coherent Modems

Carrier recovery is required for coherent detection. In a fading channel, coherent detection exhibits high error floors (58). Hence, for burst-mode applications and/or for channels experiencing fading, such as mobile radio, mobile satellite communications, or time division multiple access (TDMA) systems, such incoherent schemes as differentially coherent detection and discriminator detection are preferred because carrier recovery can be avoided. However, the major disadvantage of incoherent detection is an SNR penalty with respect to coherent detection in a Gaussian channel. In this section, we introduce the advances in differentially coherent detection as applied to QPSK and GMSK.

Novel Receiver Structures for Systems Using Differential Detection

The performance of differential detection in an AWGN environment is inferior to coherent detection, stemming from using a noisier reference signal for demodulation. However, some inherent properties of differential detection permits the improvement of BER performance by introducing additional processing at the receiver. For example, majority logic detection (59) and nonredundant error correction (NEC) (60) were proposed. The BER performance can be improved further by introducing new receiver structures (61). We present a detailed description of the new receivers for differential detection of QPSK or differential QPSK (DQPSK).

The block diagram of a DQPSK demodulator is shown in Fig. 63. The signal at the output of the predetection IF filter can be represented as

$$y(t) = \sqrt{2} \cos{(\omega_c t + \theta(t))} + n'(t),$$

$$\left(k - \frac{1}{2}\right)T_s \leq t \leq \left(k + \frac{1}{2}\right)T_s \qquad (76)$$

where T_s is the symbol duration (twice the bit duration), and $\theta(t)$ is the phase for $kT_s - T_s/2 \leq kT_s + T_s/2$.

At the transmitter, the input data to quadrature channels $\{a_k\}$ and $\{b_k\}$ are differentially encoded such that

$$\theta(kT_s) - \theta(kT_s - T_s) = m(k)\frac{\pi}{2} \qquad (77)$$

where

$$m(k) = 0 \quad \text{for } a_k = 1, \quad b_k = 1$$
$$m(k) = 1 \quad \text{for } a_k = 1, \quad b_k = -1$$
$$m(k) = 2 \quad \text{for } a_k = -1, \quad b_k = -1$$
$$m(k) = 3 \quad \text{for } a_k = -1, \quad b_k = 1. \qquad (78)$$

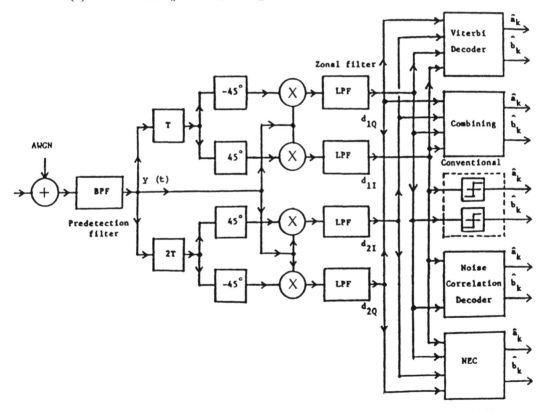

FIG. 63 Block diagram of differential detectors for quaternary phase-shift keying (BPF = band-pass filter; LPF = low-pass filter; NEC = nonredundant error correction). (From Ref. 61, reprinted by permission of the Institute of Electrical and Electronics Engineers, © 1987 IEEE.)

Note that the relationships in Eq. (77) are exact if the combination of the transmit and predetection filters satisfies the Nyquist criterion. For other filter combinations, the phase difference is distorted (i.e., not an integer multiple of $\pi/2$). In a conventional DQPSK detector, the multiplication of the received signal by a T_s second delayed and 45° shifted version of itself produces the decoded in-phase channel I output, that is,

$$d_{1I}(k) = y(k)y(k-1)_{+45°} \tag{79}$$

Ignoring the second harmonic terms,

$$d_{1I}(k) = \cos\left(\theta(k) - \theta(k-1) - \frac{\pi}{4}\right) + n_{1I}(k) \tag{80}$$

where $n_{1I}(k)$ lumps all signal-times-noise and noise-times-noise terms.

Similarly, the 1-bit detector output for the quadrature channel Q is given by

$$d_{1Q}(k) = y(k)y(k-1)_{-45°},$$
$$= \cos\left(\theta(k) - \theta(k-1) + \frac{\pi}{4}\right) + n_{1Q}(k). \tag{81}$$

Note that, in the absence of noise, the output of $d_{1I}(k) = a_k$ and $d_{1Q}(k) = b_k$.

The outputs of the 2-bit detectors, which are used to improve the BER performance, are given by

$$d_{2I}(k) = \cos\left(\theta(k) - \theta(k-2) - \frac{\pi}{4}\right) + n_{2I}(k) \tag{82}$$

and

$$d_{2Q}(k) = \cos\left(\theta(k) - \theta(k-2) + \frac{\pi}{4}\right) + n_{2Q}(k). \tag{83}$$

We use Eqs. (80) to (83) to describe the new receiver structures in the next section.

Combining with Feedback (CWF) for Differential
Quaternary Phase-Shift Keying

To use the 1- and 2-bit detector outputs in estimating the data symbols a_k and b_k, or $m(k)$ equivalently, the maximum-likelihood ratio law is applied:

$$\frac{f(d_1(k), d_2(k)|a_k = 1, \hat{a}_{k-1})}{f(d_1(k), d_2(k)|a_k = -1, \hat{a}_{k-1})} \overset{1}{\underset{-1}{\lessgtr}} 1. \tag{84}$$

This provides us the following relationships:

(1) if $\hat{m}(k-1) = 0$, then

$$\hat{a}_k = \text{sgn}\,(d_{1I}(k) + d_{2I}(k))$$
$$\hat{b}_k = \text{sgn}\,(d_{1Q}(k) + d_2Q(k)).$$

(2) if $\hat{m}(k-1) = 1$, then

$$\hat{a}_k = \text{sgn}\,(d_{1I}(k) - d_{2Q}(k))$$
$$\hat{b}_k = \text{sgn}\,(d_{1Q}(k) + d_{2I}(k)).$$

(3) if $\hat{m}(k-1) = 2$, then

$$\hat{a}_k = \text{sgn}\,(d_{1I}(k) - d_{2I}(k))$$
$$\hat{b}_k = \text{sgn}\,(d_{1Q}(k) - d_{2Q}(k)).$$

(4) if $\hat{m}(k-1) = 3$, then

$$\hat{a}_k = \text{sgn}\,(d_{1I}(k) + d_{2Q}(k))$$
$$\hat{b}_k = \text{sgn}\,(d_{1Q}(k) - d_{2I}(k)). \quad (85)$$

Viterbi Decoder for Differential Quaternary Phase-Shift Keying

For DQPSK, the known quantities are $d_{1I}(k)$, $d_{1Q}(k)$, $d_{2I}(k)$, and $d_{2Q}(k)$ and the quantity to be estimated is $m(k)$ (i.e., the phase difference between two successive symbols). Since the signal has four states, a four-state Viterbi decoder is needed. The maximization problem can be stated as

$$R(k) = \max\,[u_k(0),\, u_k(1),\, u_k(2),\, u_k(3)] \quad (86)$$

where $u_k(j)$ are the metrics for the surviving paths ending in state j; $u_k(j)$ can be expressed as

$$u_k(j) = \max_{q=0,1,2,3}\,(L_k(q,j) + u_k(q)) + 2d_{1I}(k)\cos\left(j\,\frac{\pi}{2} - \frac{\pi}{4}\right)$$
$$- 2d_{1Q}(k)\sin\left(j\,\frac{\pi}{2} - \frac{\pi}{4}\right) \quad (87)$$

where

$$L_k(q,j) = 2d_{2I}(k)\cos\left((j+q)\,\frac{\pi}{2} - \frac{\pi}{4}\right) - 2d_{2Q}k\sin\left((j+q)\,\frac{\pi}{2} - \frac{\pi}{4}\right)$$
$$- \cos\left(q\,\frac{\pi}{2}\right)(d_{1I}(k)d_{2I}(k) + d_{1Q}(k)d_{2Q}(k))$$
$$+ \sin\left(q\,\frac{\pi}{2}\right)(d_{1I}(k)d_{2Q}(k) - d_{1Q}(k)d_{2I}(k)). \quad (88)$$

It can be observed that, for all practical purposes, the last two terms in Eq. (88) can be ignored, resulting in substantial simplifications in the calculation of $L_k(q,j)$. The steps of the algorithm are as follows:

Step 1. Determine the surviving paths at instant k, that is, compute $u_k(0)$, $u_k(1)$, $u_k(2)$, and $u_k(3)$.

Step 2. Select the maximum among $u_k(0)$, $u_k(1)$, $u_k(2)$, and $u_k(3)$.

Step 3. Trace the selected path L steps back and decode $m(k - L)$.

Using the Noise Correlation

In differential detection, the noise samples at successive sampling instants are correlated. In DQPSK, we can take advantage of the noise correlation (NC) to improve the BER performance. The problem can be stated as estimating the most probable transmitted phase difference $m(k)$ when $d_{1I}(k)$, $d_{1Q}(k)$, $d_{1I}(k - 1)$, $d_{1Q}(k - 1)$, and $m(k - 1)$ are known.

Assuming that the random variables $d_{1I}(k)$, $d_{1Q}(k)$, $d_{1I}(k - 1)$, and $d_{1Q}(k - 1)$ are jointly normal, and the noise samplers $n_{1I}(k)$, $n_{1Q}(k)$, $n_{1I}(k - 1)$, and $n_{1Q}(k - 1)$ have zero mean and $2\sigma^2$ variance, then they satisfy the following identities:

$$\overline{n_{1I}(k)n_{1Q}(k)} = 0$$

$$\overline{n_{1I}(k)n_{1I}(k - 1)} = -\overline{n_{1Q}(k)n_{1Q}(k - 1)} = \sigma^2 \sin\left(m(k)\frac{\pi}{2} + m(k - 1)\frac{\pi}{2}\right)$$

$$\overline{n_{1I}(k)n_{1Q}(k - 1)} = \overline{n_{1Q}(k)n_{1I}(k - 1)} = \sigma^2 \cos\left(m(k)\frac{\pi}{2} + m(k - 1)\frac{\pi}{2}\right). \qquad (89)$$

Applying the maximum-likelihood ratio test (MLRT), we find that the in-phase and quadrature signals can be decoded by using the following expressions:

if $\hat{m}(k - 1) = 0$, then $\hat{a}_k = \text{sgn}\,(d_{1I}(k) + A)$, $\hat{b}_k = \text{sgn}\,(d_{1Q}(k) + B)$

if $\hat{m}(k - 1) = 1$, then $\hat{a}_k = \text{sgn}\,(d_{1I}(k) - B)$, $\hat{b}_k = \text{sgn}\,(d_{1Q}(k) + A)$

if $\hat{m}(k - 1) = 2$, then $\hat{a}_k = \text{sgn}\,(d_{1I}(k) - A)$, $\hat{b}_k = \text{sgn}\,(d_{1Q}(k) - B)$

if $\hat{m}(k - 1) = 3$, then $\hat{a}_k = \text{sgn}\,(d_{1I}(k) + B)$, $\hat{b}_k = \text{sgn}\,(d_{1Q}(k) - A)$ (90)

where

$$A = \frac{1}{4}\,(d_{1I}(k)d_{1I}(k - 1) + d_{1I}(k)d_{1Q}(k - 1)$$

$$- d_{1Q}(k)d_{1Q}(k - 1) + d_{1Q}(k)d_{1I}(k - 1))$$

$$B = \frac{1}{4}\,(-d_{1I}(k)d_{1I}(k - 1) + d_{1I}(k)d_{1Q}(k - 1)$$

$$+ d_{1Q}(k)d_{1Q}(k - 1) + d_{1Q}(k)d_{1I}(k - 1)). \qquad (91)$$

The NC technique requires more baseband processing of the received signal than the NEC or CWF technique. However, in this case the number of detectors is reduced from four to two.

The BER performances of DQPSK for five different receiver structures were evaluated by computer simulations (61). The system configuration is shown in Fig. 63. Overall Nyquist characteristics are split equally between the transmit and the predetection IF filter, and the postdetection filter is only for harmonic removal. In a linear AWGN channel, this configuration corresponds to the theoretical performance of conventional DQPSK. The simulation results are shown in Fig. 64. A Viterbi decoder, even with a decoding depth of 1, outperforms the other receivers and provides 1.2-dB improvement over conventional DQPSK at BER = 10^{-4}. The NEC, CWF, and NC all perform equally well and they offer about 0.8-dB advantage over conventional differential detection.

The same principles except for NC also can be applied to differential detection of MSK. Similar improvements in differential MSK (DMSK) can be found in Ref. 61.

Differential Detection of Gaussian Minimum-Shift Keying Using Decision Feedback

GMSK, with its constant envelope and compact spectrum, is an attractive modulation technique for satellite communications and mobile radio applications (11). For the detection of GMSK, a coherent detector (11), a differential detector (62,63), or a limiter discriminator detector can be employed (64).

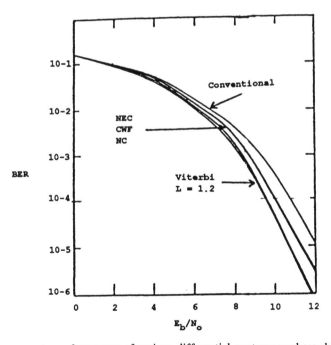

FIG. 64 Bit error rate performances of various differential quaternary phase-shift keying receivers (NEC = nonredundant error correction; CWF = combining with feedback; NC = noise correlation). (From Ref. 61, reprinted by permission of the Institute of Electrical and Electronics Engineers, © 1987 IEEE.)

In this section, we concentrate on the differential detection of GMSK. The application of the conventional 1-bit differential detection to GMSK results in poor performance; thus, the earlier literature focuses on 2-bit differential detection (62,63). The conventional 2-bit differential detection is about 7 dB inferior to the coherent detection of GMSK (62). Hence, there is ample room for improvement by introducing additional processing at the receiver.

The block diagram of a GMSK transmitter is shown in Fig. 65. The transmitter consists of a differential encoder, a Gaussian low-pass filter (GLPF), and an FM modulator. The differential encoder is required for 2-bit and 3-bit differential detectors. The antipodal input symbols to the differential encoder are denoted as a_k and the output symbols as b_k. For the 1-bit differential detection, a differential encoder is not needed (i.e., $a_k = b_k$). For the 2-bit differential detection, the differential encoding rule is given by

$$b_k = -a_k b_{k-1}. \tag{92}$$

Similarly, for the 3-bit differential detection, the differential encoding rule is

$$b_k = -a_k b_{k-1} b_{k-2}. \tag{93}$$

The input to the GLPF is an NRZ sequence. The output of the GLPF can be expressed as

$$s(t) = \sum_{i=-\infty}^{\infty} b_i p(t - iT) \tag{94}$$

where $p(t)$ is the response of the GLPF to a unit amplitude rectangular pulse of duration T, given by (11)

$$p(t) = \frac{1}{2T} \left[Q\left(k_1 B_t T \left(-\frac{t}{T} \right) \right) - Q\left(k_1 B_t T \left(1 - \frac{t}{T} \right) \right) \right]. \tag{95}$$

In Eq. (95), B_t is the 3-dB bandwidth of the GLPF, T is the bit duration and, thus, $B_t T$ is the bandwidth–time product of the transmit GLPF, $k_1 = 7.546$, and $Q(x)$ is the Gaussian integral function. The output of the FM modulator can be expressed as

$$x(t) = A_0 \cos(\omega_c t + \phi(t) + \gamma). \tag{96}$$

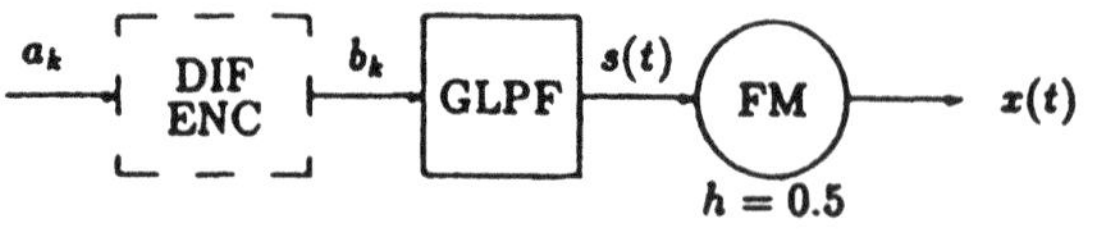

FIG. 65 Block diagram of a Gaussian minimum-shift keying transmitter. The differential encoder is not needed for the 1-bit differential detection (GLPF = Gaussian low-pass filter; FM = frequency modulation). (Courtesy Ref. 65, © 1987.)

In Eq. (96), A_0 is the constant envelope of the signal, ω_c is the carrier frequency, γ is the initial phase (which can be assumed to be zero), and $\phi(t)$ is the excess phase defined by

$$\phi(t) = k_m \int_{-\infty}^{t} s(\tau)d\tau = k_m \sum_{j=-\infty}^{\infty} b_j \int_{-\infty}^{t} p(\tau - jT)d\tau. \qquad (97)$$

The phase change over one symbol interval is

$$\Delta\phi_i = \phi(iT) - \phi(iT - T) = k_m \sum_{j=-\infty}^{\infty} b_j \int_{iT-T}^{iT} p(\tau - jT)d\tau. \qquad (98)$$

The maximum value that the summation term in Eq. (98) can assume is the area under the curve $p(t)$, which is equal to T. For a modulation index $h = 0.5$ and binary input data, $(\Delta\phi_i)_{\max} = \pi/2$; thus, $k_m = \pi/2T$.

Conventional Receivers

The block diagram of the conventional 2-bit differential detector is illustrated in Fig. 66. The output of the 2-bit detector $d_2(t)$ is obtained by multiplying $y(t)$ with a $2T$ delayed version of itself and then low-pass filtering the product, that is,

$$d_2(t) = r(t)r(t - 2T) \cdot \cos\left(k_m \sum_{j=-\infty}^{\infty} b_j \int_{t-2T}^{t} p(\tau - jT)\, d\tau\right) + n_2(t) \qquad (99)$$

where $n_2(t)$ represents all the noise terms.
   At the time instant kT, $d_2(t)$ is given by

$$d_2(kT) = r(kT)r(kT - 2T) \cos\left(\sum_{j=-\infty}^{\infty} b_j V_{k-j}\right) + n_2(kT) \qquad (100)$$

where

$$V_{k-j} = k_m \int_{kT-2T}^{kT} p(\tau - jT)\, d\tau. \qquad (101)$$

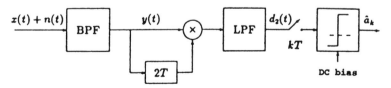

FIG. 66   Block diagram of a conventional 2-bit differential detector (BPF = band-pass filter; LPF = low-pass filter). (Courtesy Ref. 65, © 1987.)

The values of V_i for different B_tT have been tabulated in Table 13, where V_0 and V_1 represent the signal and V_{-3}, V_{-2}, V_{-1}, V_2, V_3, and V_4 are the ISI terms. For any row in Table 13, $\Sigma\ V_i = 180°$. Observe that V_i is almost 0 for $i \geq 4$ and $i \leq -3$. Therefore, we can rewrite Eq. (100) as

$$d_2(kT) = r(kT)r(kT - 2T) \cos (\Delta V_k) + n_2(kT) \qquad (102)$$

where

$$\Delta V_k = b_{k+2}V_{-2} + b_{k+1}V_{-1} + b_kV_0 + b_{k-1}V_1 + b_{k-2}V_2 + b_{k-3}V_3. \qquad (103)$$

For $B_tT = 0.25$, the differential phase angles ΔV_k corresponding to all possible input data combinations have been tabulated in Table 14. Using Table 14, the phase states at the decision instants for the 2-bit differential detector are shown in Fig. 67.

To determine the polarity of the output bit, as a first approximation let us assume that the decision threshold is the y axis. When the phase difference ΔV_k is to the right of the y axis, b_kb_{k-1} is -1; otherwise, b_kb_{k-1} is $+1$. Suppose, for a moment, that the differential encoder in Fig. 65 was omitted (i.e., $a_k = b_k$). Then, the $\hat{b}_k$ can be determined by using the knowledge of the already decoded $\hat{b}_{k-1}$. However, with this approach an error in $\hat{b}_{k-1}$ will affect the subsequent decisions. The differential encoder defined by Ref. 11 circumvents this problem because

$$a_k = -b_kb_{k-1}. \qquad (104)$$

Hence,

$$\hat{a}_k = \text{sgn}\ [d_2(kT)]. \qquad (105)$$

Observe from Fig. 67 that the phase states at the output of the 2-bit differential detector are not symmetrical with respect to the y axis. As a result of this,

TABLE 13   Phase Shifts for V_i of Gaussian Transmit Filter B_tT (in Degrees)

B_tT	V_{-3}	V_{-2}	V_{-1}	V_0	V_1	V_2	V_3	V_4	$\Delta V_{\min}$	$\Delta D_{\min}^{DF}$
0.15	0.3	4.85	26.4	58.45	58.45	26.4	4.85	0.3	–	52.4
0.18	–	2.7	23.9	63.4	63.4	23.9	2.7	–	20.4	73.6
0.2	–	1.7	22.3	66.0	66.0	22.3	1.7	–	36.0	84.0
0.25	–	0.6	18.8	70.6	70.6	18.8	0.6	–	63.6	102.4
0.3	–	0.2	16.2	73.6	73.6	16.2	0.2	–	81.6	114.4
0.4	–	–	12.5	77.5	77.5	12.5	–	–	105.0	130.0
0.5	–	–	10.3	79.7	79.7	10.3	–	–	118.2	138.8
1.0	–	–	5.9	84.1	84.1	5.9	–	–	144.6	156.4
∞	–	–	–	90.0	90.0	–	–	–	180.0	180.0

Source: Courtesy Ref. 65, © 1987.
Note: Phase shifts (in degrees) correspond to signal terms V_0 and V_1 and ISI terms as a function of transmit Gaussian filter B_tT for the 2-bit differential detector. $\Delta V_{\min}$ and $\Delta V_{\min}^{DF}$ are the minimum differential phase angles before and after applying decision feedback, respectively.

TABLE 14 Differential Phase Angles ΔV_k of Two-Bit Detector for Various Input Data Combinations

Bit Combinations					ΔV_k
b_{k-2}	b_{k-1}	b_k	b_{k+1}	State	(in Degrees)
1	1	−1	1	7	37.6
1	−1	1	1	7	37.6
1	1	−1	−1	8	0.0
1	−1	1	−1	8	0.0
−1	1	−1	1	8	0.0
−1	−1	1	1	8	0.0
−1	1	−1	−1	9	−37.6
−1	−1	1	−1	9	−37.6
1	−1	−1	1	10	−103.6
1	−1	−1	−1	11	−141.2
−1	−1	−1	1	11	−141.2
−1	−1	−1	−1	12	−178.8
1	1	1	1	12	178.8
1	1	1	−1	13	141.2
−1	1	1	1	13	141.2
−1	1	1	−1	14	103.6

Source: Courtesy Ref. 65, © 1987.
Note: $B_t T = 0.25$. The contributions of b_{k+2} and b_{k-3} are ignored.

the corresponding eye diagram shown in Fig. 68 is also asymmetrical. To improve the BER performance in this case, it has been proposed to insert a bandpass limiter after the BPF and to apply a DC bias to the threshold comparator (62). This is equivalent to shifting the decision region to the zz' line in Fig. 67.

For the 2-bit detector, the minimum differential phase angle is defined as

$$\Delta V_{\min} = (V_0 + V_1) - 2 \sum_{i \neq 0,1} V_i. \tag{106}$$

The values of $\Delta V_{\min}$ for various $B_t T$s are tabulated in Table 13. From Table 13, we can conclude that the conventional 2-bit differential detection is applicable to systems with $B_t T \geq 0.18$.

Decision Feedback in a Two-Bit Differential Detector

The phase-state diagram (Fig. 67) indicates that a large amount of ISI is inherent in the received signal. In this section, we demonstrate how the effect of ISI can be reduced with the help of "decision feedback." Let us first assume that the effect of b_{k-3} and b_{k+2} is negligible. Observe from Fig. 67 that Phase States 10

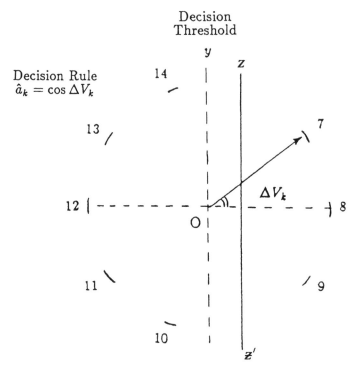

FIG. 67   Phase-state diagram of conventional 2-bit differential detection ($B_tT = 0.25$). (Courtesy Ref. 65, © 1987.)

and 14 are subject to the maximum amount of ISI. If States 10 and 14 are ignored, then the y axis represents the optimal decision threshold. Hence, our objective is to move State 10 clockwise and State 14 counterclockwise. While doing this, we also want to keep the other states as far away from the decision threshold (i.e., y axis) as possible.

To meet these objectives, the phase of the $2T$ delayed signal is shifted by λ degrees. The phase-shifting rule can be formulated as follows:

$$\lambda = \begin{cases} 2\hat{b}_{k-2}V_2 & \text{if } \hat{b}_{k-1} \neq \hat{b}_{k-2} \\ 0 & \text{if } \hat{b}_{k-1} = \hat{b}_{k-2}. \end{cases} \tag{107}$$

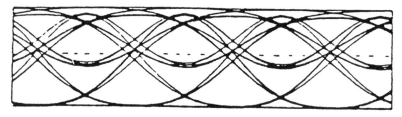

FIG. 68   Eye diagram of a conventional 2-bit differential detector ($B_tT = 0.25$). (Courtesy Ref. 65, © 1987.)

When the effect of b_{k-3} is not negligible,, the required phase shifts are slightly different. Then, we have

$$
\lambda = \begin{cases}
2\hat{b}_{k-2}V_2 + 2\hat{b}_{k-3}V_3 & \text{if } \hat{b}_{k-1} \neq \hat{b}_{k-3} \text{ and } \hat{b}_{k-1} \neq \hat{b}_{k-2} \\
2\hat{b}_{k-2}V_2 & \text{if } \hat{b}_{k-1} = \hat{b}_{k-3} \text{ and } \hat{b}_{k-1} \neq \hat{b}_{k-2} \\
2\hat{b}_{k-3}V_3 & \text{if } \hat{b}_{k-1} \neq \hat{b}_{k-3} \text{ and } \hat{b}_{k-1} = \hat{b}_{k-2} \\
0 & \text{if } \hat{b}_{k-1} = \hat{b}_{k-3} \text{ and } \hat{b}_{k-1} = \hat{b}_{k-2}. \quad (108)
\end{cases}
$$

The phase states for $B_t T = 0.25$ after applying these phase shifts are shown in Fig. 69. Note that the resulting phase states and the corresponding eye diagram shown in Fig. 70 are symmetrical. Comparing Figs. 68 and 70, we observe that after applying decision feedback the eye opening is increased significantly. The minimum differential phase angle after applying decision feedback becomes

$$
\Delta V_{\text{min}}^{\text{DF}} = V_0 + V_1 - 2 \sum_{i < -1} V_i = \Delta V_{\text{min}} + 2(V_2 + V_3 + V_4). \quad (109)
$$

The values of $\Delta V_{\text{min}}^{\text{DF}}$ for different $B_t T$ are given in the rightmost column of Table 13.

The principle outlined above for the 2-bit detector can be extended to 1-bit and 3-bit detectors to improve the performance significantly (65). More important, it is possible to get further improvement by the joint use of more detectors. The structure of the envisioned receiver can be established by using the MLRT (61). The MLRT indicates that the optimal utilization of detectors requires a

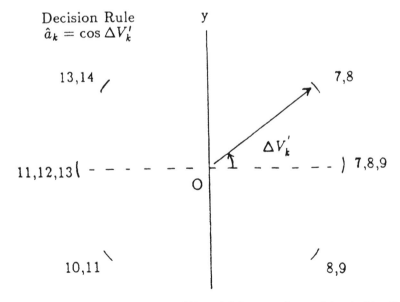

FIG. 69 Phase-state diagram of a 2-bit differential detector after applying decision feedback. (Courtesy Ref. 65, © 1987.)

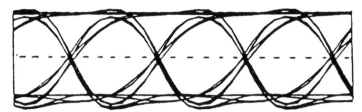

FIG. 70   Eye diagram of a 2-bit differential detector after applying decision feedback. (Courtesy Ref. 65, © 1987.)

nonlinear receiver. In order to reduce the complexity of such an optimal receiver, suboptimum linear receivers can be used (65). Fig. 71 shows an example of combining the outputs of 2-bit and 3-bit detectors with decision feedback. This receiver is denoted as 2+3DF. The decision law of the 2+3DF receiver is given by

$$\hat{a}_k = \text{sgn} [d_2'(kT) - c_2 \hat{a}_{k-1} d_3'(kT)] \tag{110}$$

where $d_2'(kT)$, $d_3'(kT)$ are the sampled outputs of the 2-bit and 3-bit detectors after decision feedback, respectively, and c_2 is the combining coefficient ($c_2 < 1$). Note that the transmitter corresponding to the 2+3DF receiver requires a differential encoder.

The BER performance of various differential GMSK (DGMSK) receivers ($B_tT = 0.25$) obtained by computer simulation is summarized in Fig. 72 (65). We observe that 2+3DF performs the best of the differential detectors. For example, at a BER of 10^{-4} it offers a 1-dB and a 4-dB E_b/N_0 advantage over 1+2DF and 2CBW receivers, respectively. In Fig. 72, we also have plotted the performance of coherent GMSK in a AWGN channel (11). It can be observed that the 2+3DF receiver is only 3 dB away from the coherent receiver, while carrier recovery is avoided.

Related Work

In this section, we select some important research work related to digital modulation.

Effect of High-Power Amplifier (HPA) Nonlinearities on Crosstalk and Performance of Digital Radio Systems

Multilevel QAM systems are very sensitive to nonlinearities of such HPAs as TWT and GaAs FET. It is desirable that TWT and FET be operated as closely as possible to the saturation level to get a maximum output power. However,

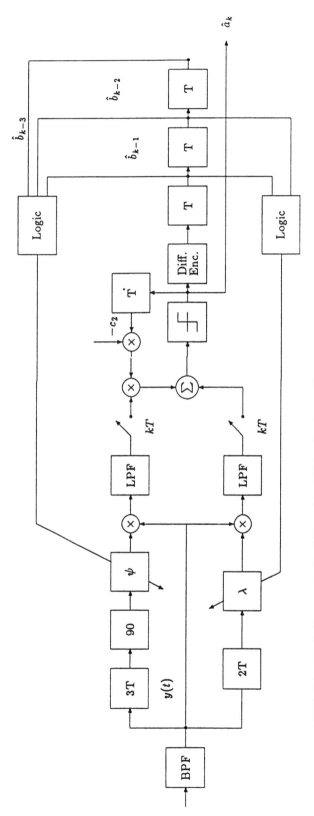

FIG. 71 Block diagram of a 2 + 3DF receiver (BPF = band-pass filter; LPF = low-pass filter). (Courtesy Ref. 65, © 1987.)

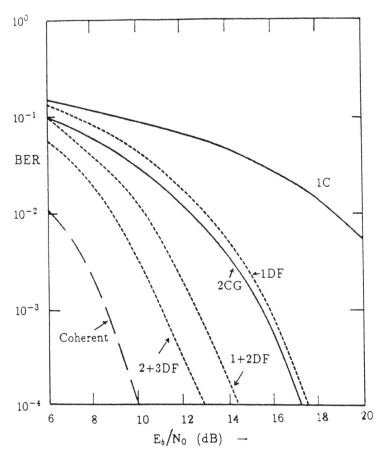

FIG. 72   Bit error rate performance of various differential Gaussian minimum-shift keying receivers (B_tT = 0.25; 1C = conventional one-bit differential detection; 2CG = conventional two-bit differential detection using a Gaussian predetection filter; 2CBW = conventional two-bit differential detection using a Butterworth predetection filter; 1+2DF = combined 1-bit and 2-bit differential detection with decision feedback; 2+3DF = combined 2-bit and 3-bit differential detection with decision feedback). (Courtesy Ref. 65, © 1987.)

the amplitude and phase distortions of HPAs give rise to intermodulation (IM) components and phase distortions in transmitting signals. To limit signal distortions, it is necessary to operate the power amplifiers at average power levels that are significantly below the saturation power or to predistort the signal before amplification.

Quadrature crosstalk is a major problem caused by these distortions (66–70). In this section, the impact of HPA nonlinearities on quadrature crosstalk for spectrally efficient QAM systems is investigated. A computer-simulation model is shown in Fig. 73. To simulate the different extent of nonlinearities for illustrative applications, the measured characteristics of the Hughes 261-H TWT and Fujitsu linearized GaAs FET are considered in the analysis. The measured

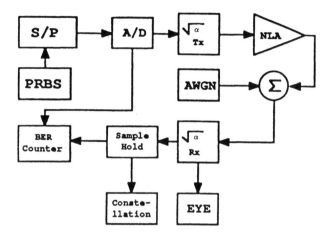

FIG. 73 Block diagram of computer-simulation model; equivalent baseband model is shown (PRBS = pseudorandom binary sequence; BER = bit error rate; AWGN = additive white Gaussian noise; NLA = nonlinear amplifier).

amplitude modulation/amplitude modulation (AM/AM) and amplitude modulation/pulse modulation (AM/PM) characteristics are shown in Fig. 74. From Fig. 74, we note that the AM/AM compression and the AM/PM conversion for TWT are more prominent. There are different representations for expressing the nonlinearity of HPAs (68,69). We use the quadrature nonlinear model as shown in Fig. 75. The functions $P(r)$ and $Q(r)$ are the mathematical expressions of the HPA nonlinearities. They are plotted in Fig. 76 for the HPAs above. Let the input signal of HPAs be

$$u_i(t) = r(t) \cos [\omega_c t + \theta(t)]. \tag{111}$$

The corresponding output signal is written as

$$u_0(t) = r(t)g[r(t)] \cos \{\omega_c t + \theta(t) + \phi[r(t)]\}$$
$$= A[r(t)] \cos \{\omega_c t + \theta(t) + \phi[r(t)]\} \tag{112}$$

where $r(t) = \sqrt{a^2(t) + b^2(t)}$; $\theta(t) = \tan^{-1} [b(t)/a(t)]$; ω_c is the carrier frequency; $a(t)$ and $b(t)$ are the filtered signals after the premodulation LPF of the I and Q channels, respectively; $g[r(t)]$ represents the AM/AM compression of the HPA; and $\phi[r(t)]$ is the AM/PM conversion.

$P(r)$ and $Q(r)$ are related to $A[r(t)]$ and $\phi[r(t)]$:

$$P(r) = A[r(t)] \cos (\phi [r(t)])$$
$$Q(r) = A[r(t)] \sin (\phi [r(t)]) \tag{113}$$

The I and Q channel signals, after demodulation and filtering by a LPF having an impulse response $Hr(t)$, are expressed as

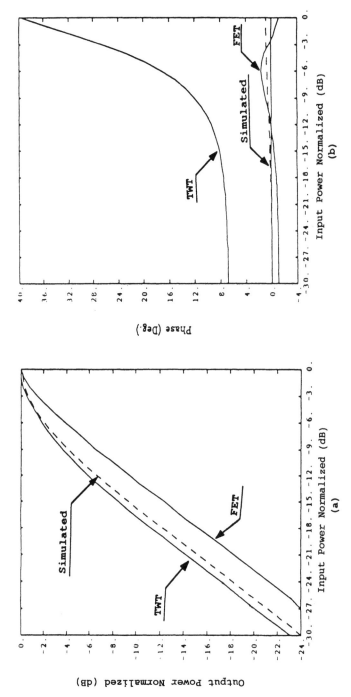

FIG. 74 Measured amplifier characteristics: *a*, AM/AM characteristics; *b*, AM/PM characteristics. (From Ref. 66, reprinted by permission of the Institute of Electrical and Electronics Engineers, © 1988 IEEE.)

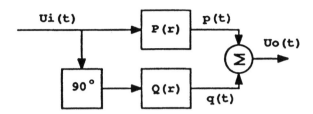

FIG. 75   Quadrature nonlinear model of nonlinear amplifier (NLA).

I channel: $x(t) = r(t) g[r(t)] \cos \{[\theta(t) + \phi[r(t)]\} * Hr(t)$

$= g[r(t)] \{a(t) \cos (\phi [r(t)]) - b(t) \sin (\phi [r(t)])\} * Hr(t)$ (114-a)

Q channel: $y(t) = r(t) g[r(t)] \sin \{\theta(t) + \phi[r(t)]\} * Hr(t)$

$= g[r(t)] \{b(t) (\cos (\phi [r(t)]) + a(t) \sin (\phi [r(t)])\} * Hr(t)$ (114-b)

where $- b(t) \sin (\phi [r(t)])$ and $a(t) \sin (\phi [r(t)])$ represent the crosstalk components introduced by the AM/PM conversion, and * denotes convolution operation.

From Eq. (114), we note that $x(t) \neq 0$ for the I channel when $a(t) = 0$, and $y(t) \neq 0$ for the Q channel when $b(t) = 0$. If we let

$$CI = b(t) \sin (\phi [r(t)])$$

$$CQ = a(t) \sin (\phi [r(t)]) \qquad (115)$$

then CI and CQ are the quadrature crosstalk components of the I and Q channels, respectively. They are all proportional to $\sin (\phi [r(t)])$. The amplitude of crosstalk is caused mainly by AM/PM conversion. In addition, it also depends on the amplitude of the input signals $a(t)$ and $b(t)$. The larger the value of the ϕ $[r(t)]$ function, the more severe the quadrature crosstalk. This, of course, will lead to performance degradation.

Computer-simulated results of CI and CQ versus output backoff (OBO) for 64QAM and 256QAM are plotted in Fig. 77, in which the independent variable OBO is represented in dB. The results show a general tendency that the crosstalk is reduced when OBO increases. Crosstalk characteristics for different HPAs are different. The smaller the variable extent of AM/PM conversion of HPA, the smaller the crosstalk. There are minima for some HPAs when they have a zero phase, such as for the linearized GaAs FET. Actually, it means that the function $Q(r)$ as shown in Fig. 76-b has peaks and the AM/PM conversion approaches zero at certain ranges of OBO.

The dashed lines shown in Fig. 74 represent the simulated characteristics of HPAs. The AM/PM characteristics of the simulated results are chosen to be approximately the same as the measured TWT. However, the AM/PM characteristic has been chosen to be practically ideal, that is, a horizontal line. Crosstalk versus OBO for 64QAM is shown by the upper lines in Fig. 77-a. From these curves, it can be observed that the crosstalk generated by FET is consider-

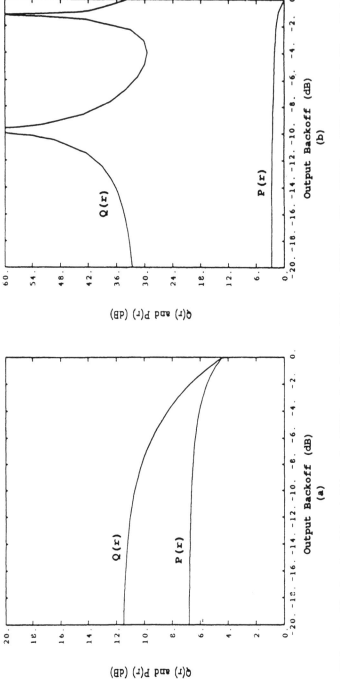

FIG. 76   The function $P(r)$ and $Q(r)$: a, for TWTA; b, for FETA (TWTA = TWT amplifier; FETA = FET amplifier). (From Ref. 66, reprinted by permission of the Institute of Electrical and Electronics Engineers, © 1988 IEEE.)

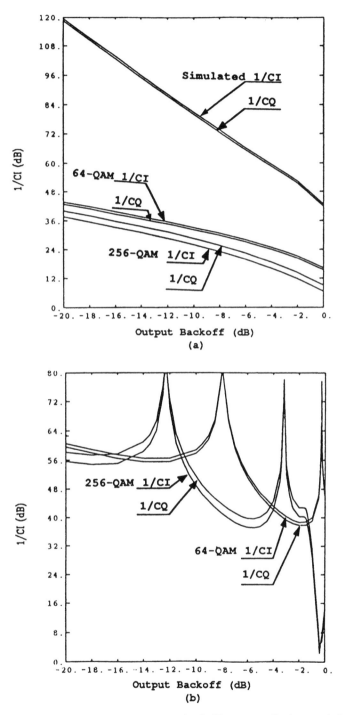

FIG. 77 The quadrature crosstalk versus output backoff: *a*, crosstalk characteristics of TWTA; *b*, crosstalk characteristics of FETA (TWTA = TWT amplifier; FETA = FET amplifier). (From Ref. 66, reprinted by permission of the Institute of Electrical and Electronics Engineers, © 1988.)

ably less than that generated by the TWT. That is to say, the quadrature cross-talk is caused mainly by AM/PM conversion.

The measured power spectrum density of a 64QAM system with an $\alpha = 0.2$ filter in the linear channel is shown in Fig. 78-*a*. In Fig. 78-*b*, the spectrum amplified nonlinearly (having an OBO of 6 dB) is illustrated. In Fig. 78-*b*, the spectral restoration is visible.

The simulated P_e performance for the different HPAs and the various OBOs are shown in Fig. 79. From Fig. 79-*a*, it can be seen that P_e performances for the different HPAs are almost the same when they all operate at OBO = 20

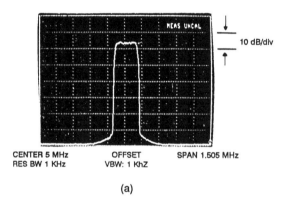

(a)

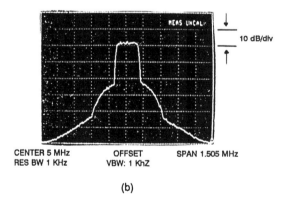

(b)

FIG. 78 Measured power spectrum density of a 64 quadrature amplitude modulation (QAM) system having an $\alpha = 0.2$ filter: *a*, linearly amplified spectrum; *b*, nonlinearly amplified spectrum with spectral restoration visible and having an output backoff of 6 dB. In this experiment, a normalized (low bit rate) system has been used. The transmitted symbol rate is $f_s = 250$ kilobaud (Kbaud). (Source: Ref. 66.)

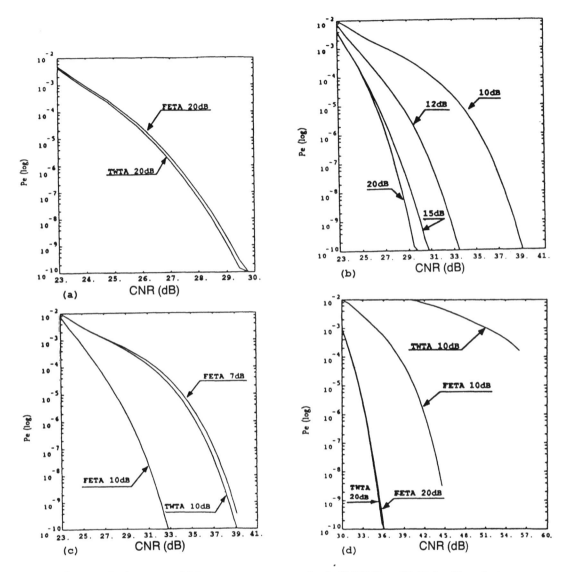

FIG. 79 Performance of *M*-ary systems: *a*, output backoff (OBO) = 20 dB for 64 quadrature amplitude modulation (QAM), α = 0.4; *b*, OBO = 10 dB and 7 dB for 64QAM, α = 0.4; *c*, different OBO of TWTA for 64QAM, α = 0.4; *d*, OBO = 10 dB and 20 dB for 256QAM, α = 0.4 (TWTA = TWT amplifier). (From Ref. 66, reprinted by permission of the Institute of Electrical and Electronics Engineers, © 1988 IEEE.)

dB, namely, a nearly linear situation. A CNR of 27.2 dB for 64QAM at $P_e = 10^{-6}$ is required, for either the TWT or the linearized FET.

On the other hand, the required CNR for various HPAs are different (see Fig. 79-*b*) when two HPAs operate at OBO = 10 dB, α = 0.4. A CNR = 29.2 dB is required for FET at $P_e = 10^{-6}$, but 35.2 dB is required for TWT. This means that the corresponding CNR at $P_e = 10^{-6}$ for TWT degrades 6 dB compared to FET. From the same figure, we also conclude that the P_e performance for TWT operating at OBO = 10 dB and for FET operating at OBO = 7 dB are almost identical. In other words, the effective power of the linearized device is increased about 3 dB.

Figure 79-*c* shows that the more the OBO, the less the required CNR at an equal P_e for the TWT. Figure 79-*d* shows P_e performance of 256QAM for TWTA and FETA, with OBO = 10 dB or 20 dB, α = 0.4. The constellation diagrams of a 256QAM system have been simulated. A linearly amplified constellation is shown in Fig. 80-*a* and a nonlinearly amplified constellation in Fig. 80-*b*.

It is mentioned in the section, "Fundamentals of Pulse Shaping," that filtering is necessary in band-limited systems. Although the Nyquist criterion assures that there is no ISI at the optimum sampling instants, the filtering process causes overshoot in the signal. The overshoot can be as high as 3 to 6 dB, depending on the roll-off factor and can cause significant degradation after nonlinear amplifiers. Some filtering strategies to reduce overshoot are very desirable and are the subject of Ref. 71. In Ref. 71, a reduced overshoot power (ROP) filter instead of the conventional raised-cosine filter is used. For moderate α > 0.4, the ROP filter yields less overshoot, which in turn increases system gain because it allows less backoff in the power amplifier.

Performance of 16-State Superposed Quadrature Amplitude Modulation in a Nonlinearly Amplified Multichannel Interference Environment

For the operation of *M*-ary QAM at the highest power efficiency, parallel-type modulation techniques have been proposed (72,73). To achieve a high spectral efficiency in a multichannel system, it is desirable to transmit signals having a compact power spectrum. A new technique called 16SQAM (superposed QAM) was introduced in Ref. 74 for operation through fully saturated HPAs. In this section, the performance of 16SQAM in a nonlinear multichannel interference environment is analyzed (22) and compared to that of the multiamplitude minimum-shift keying (MAMSK) system (73).

A block diagram of a 16SQAM transmitter is shown in Fig. 81. The hard limited then nonlinearly amplified 16SQAM signal at the transmitter output is represented as

$$p(t) = 2/\sqrt{a^2(t) + b^2(t)}\ [a(t) \cos \omega_c t + b(t) \sin \omega_c t]$$
$$+ 1/\sqrt{c^2(t) + d^2(t)}\ [c(t) \cos \omega_c t + d(t) \sin \omega_c t] \quad (116)$$

where

$$a(t) = \sum_k a_k s(t - kT_s), \quad b(t) = \sum_k b_k s(t - kT_s - T_s/2),$$

$$c(t) = \sum_k c_k s(t - kT_s), \quad d(t) = \sum_k d_k s(t - kT_s - T_s/2) \quad (117)$$

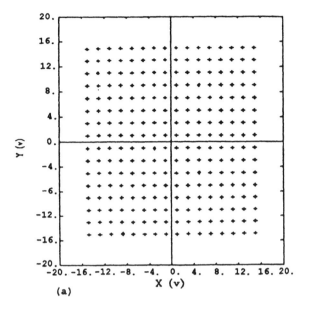

(a)

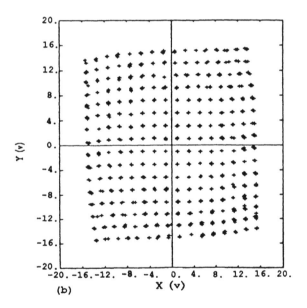

(b)

FIG. 80 Received constellation diagram for TWTA: *a*, output backoff (OBO) = 20 dB, 256 quadrature amplitude modulation (QAM); *b*, OBO = 10 dB, 256QAM (TWTA = TWT amplifier). (From Ref. 66, reprinted by permission of the Institute of Electrical and Electronics Engineers, © 1988 IEEE.)

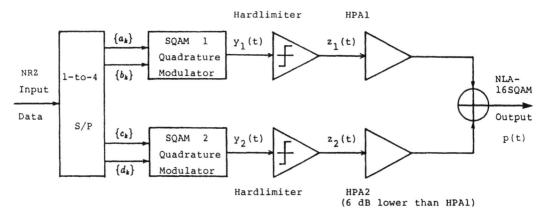

FIG. 81   Block diagram of NLA 16SQAM (superposed quadrature amplitude modulation) transmitter where $p(t) = 2z_1(t) + z_2(t)$ (HPA1 and HPA2 are operated in saturation mode). (Courtesy Ref. 22, © 1988.)

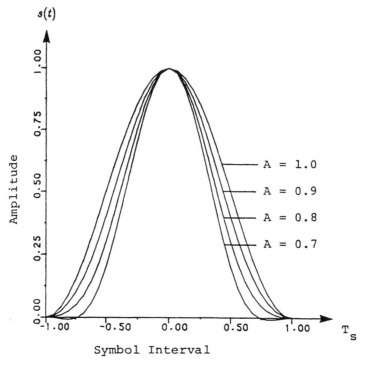

FIG. 82   Impulse response $s(t)$ of 16SQAM (superposed quadrature amplitude modulation) baseband signal processor where $s(t) = (1 + \cos \pi t/T_s)/2 - (1 - A)(1 - \cos 2\pi t/T_s)/2$ (A = amplitude parameter of the 16SQAM signal). (Courtesy Ref. 22, © 1988.)

and $\{a_k\}$, $\{b_k\}$, $\{c_k\}$, and $\{d_k\}$ are 1-to-4, serial-to-parallel, converted input data having a normalized amplitude of $\{\pm 1\}$. T_s is the symbol duration, and ω_c is the carrier angular frequency. The impulse response $s(t)$ of 16SQAM baseband signal processor is defined as

$$s(t) = (1 + \cos \pi t/T_s)/2 - (1 - A)(1 - \cos 2\pi t/T_s)/2 \qquad (118)$$

where A is an amplitude parameter that controls the power spectrum and the envelope fluctuation of the 16SQAM signal (74,75). Examples of the impulse response $s(t)$ are shown in Fig. 82. Figure 83 shows the out-of-band to total power ratios of 16SQAM, MAMSK, and 16QAM signals in a saturated (hardlimited) channel. Note that 16SQAM has the lowest out-of-band power.

A quasianalytical method of computer simulation is used for the evaluation of the system performance. Assumptions used in the simulations are as follows (75):

1. An illustrative data bit rate $f_b = 200$ Mbps (i.e., a symbol rate of $f_s = 50$ Mbaud) is used.
2. Interfering channel signals have the same modulation format as the desired main channel signal.
3. Carrier phases and symbol timings of the interfering channel signals are randomized over $[0,2\pi)$ and $[-T_s/2, T_s/2)$, respectively, to avoid coherence between the main and interfering channel signals.
4. The main and interfering channel transmitters are operated in a fully saturated mode of HPAs. An ideal hardlimiter is used as a first-order approximation of the saturated HPA.

The simulation model of a multichannel 16SQAM (or MAMSK) system is represented in Fig. 84. Figure 85 shows the corresponding frequency allocation in which two interfering adjacent channels are assumed to be spaced equally. The main channel has a carrier frequency f_c, and the adjacent channels have carrier frequencies $f_{\pm 1} = f_c \pm \Delta F$, where ΔF represents a channel spacing.

The performance in an adjacent channel interference (ACI) environment is governed mostly by the channel spacings and the out-of-band powers of the adjacent channel signals. In 16SQAM, the signal wave shape could be optimized for the best P_e performance by varying the amplitude parameter A. This serves to trade off the envelope fluctuations (or ISI) and the spectral spreading due to nonlinear amplification. It is shown in Ref. 22 that 16SQAM with $A = 0.8$ performs best, whereas in the AWGN single channel $A = 0.7$ performed best (74). Thus, in the following evaluations of system performance, a 16SQAM signal with $A = 0.8$ is assumed.

Due to thermal noise and/or ACI effects, a receive filter bandwidth cannot be too wide, nor can it be too narrow due to the ISI effect. In Ref. 22, it has been shown that an optimum filter bandwidth exists near $f_{3dB} = 26$ MHz (or $BT_s = 0.25$).

For this analysis, ACI signals are assumed to have the same power as the main channel signal, and the receive filter bandwidth is optimized in any channel

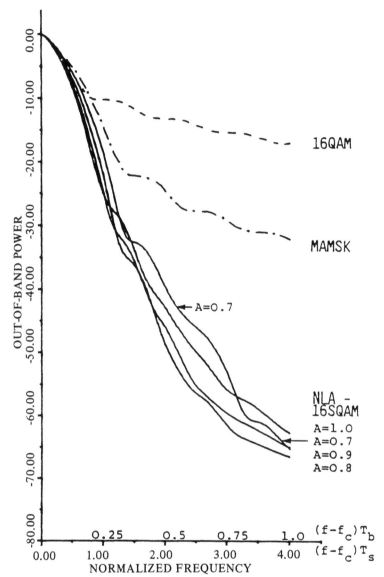

FIG. 83   Out-of-band to total power ratios of NLA 16SQAM (superposed quadrature amplitude modulation), MAMSK (multiamplitude minimum-shift keying) and 16QAM (quadrature amplitude modulation) signals in a hard-limited channel. (Courtesy Ref. 22, © 1988.)

spacing. The P_e performance of 16SQAM in the ACI environment is shown in Fig. 86-*a* for different values of channel spacing. For comparison, P_e performance of MAMSK is shown in Fig. 86-*b*. Note that 16SQAM, owing to its compact power spectrum, outperforms MAMSK. Figure 87 compares the eye diagrams of the demodulated and filtered 16SQAM ($A = 0.8$) and MAMSK signals in the presence of ACI when the illustrative channel spacing is $1.9\,f_s$.

In terrestrial microwave radio links, the signal power often is suppressed

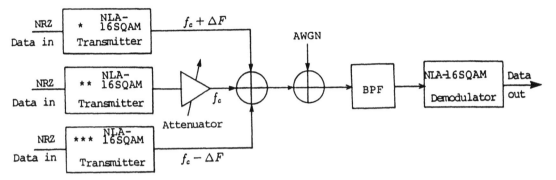

FIG. 84 Simulation model of NLA 16SQAM (superposed quadrature amplitude modulation) multichannel system in the presence of two adjacent interfering channels (* = upper adjacent channel; ** = desired main channel; *** = lower adjacent channel; NRZ = nonreturn to zero; AWGN = additive white Gaussian noise; BPF = band-pass filter). (Courtesy Ref. 22, © 1988.)

due to a fade in the transmission channel. The worst channel condition is when only the desired main channel suffers a flat fade. It is shown in Ref. 22 that a 16SQAM system is much more tolerable to flat fading than the MAMSK system in an ACI environment.

Concluding Remarks

Modern digital modems employ various modulation techniques, from simple BPSK to sophisticated 256QAM. It is expected that more spectral-efficient mo-

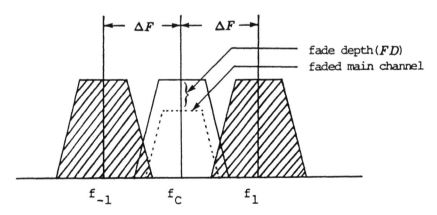

FIG. 85 Frequency allocation of NLA 16SQAM (superposed quadrature amplitude modulation) multichannel system (f_c = main channel carrier frequency; $f_{\pm 1}$ = adjacent channel carrier frequency; $f_{\pm 1} = f_c \pm \Delta F$; ΔF = channel spacing). (Courtesy Ref. 22, © 1988.)

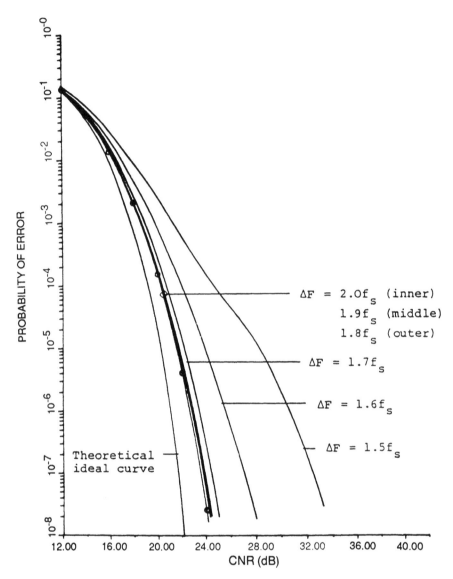

FIG. 86-*a*   Probability of error P_e performance of NLA 16SQAM (superposed quadrature amplitude modulation) ($A = 0.8$) in the adjacent channel interference environment for different values of the channel spacing ΔF. *Note:* Data bit rate is $f_b = 200$ Mbps ($f_s = 50$ Mbaud); equal power channels ($FD = 0$ dB); receive filter; fifth-order Butterworth low-pass filter. (Courtesy Ref. 22, © 1988.)

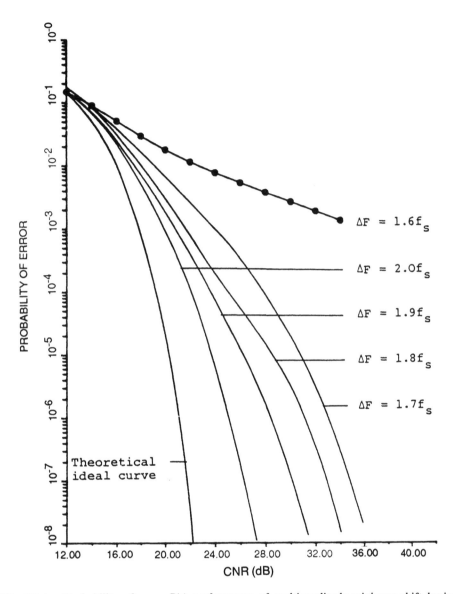

FIG. 86-b Probability of error $P(e)$ performance of multiamplitude minimum-shift keying (MAMSK) in the adjacent channel interference environment for different values of the channel spacing (ΔF). *Note:* Data bit rate is $f_b = 200$ Mbps ($f_s = 50$ Mbaud); equal power channels ($FD = 0$ dB); receive filter; fifth-order Butterworth low-pass filter. (Courtesy Ref. 22, © 1988.)

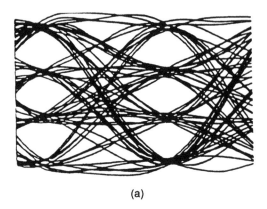

(a)

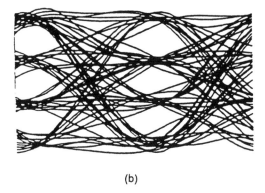

(b)

FIG. 87   Demodulated and low-pass filtered (fifth-order Butterworth) eye diagrams in the non-linearly amplified adjacent channel interference environment (ΔF = channel spacing = $1.9 f_s$; low-pass filter = f_{3dB} = 26 MHz): a, NLA 16SQAM (superposed quadrature amplitude modulation); b, MAMSK (multiamplitude minimum-shift keying). (Courtesy Ref. 22, © 1988.)

dems will be demanded in various applications. A successful operation of modems relies on the marriage of many modem-related techniques. These techniques include:

1. Synchronization techniques (i.e., carrier and symbol timing recovery and frame synchronization)
2. Adaptive equalization, especially necessary for multilevel QAM systems
3. Forward error correction to improve the system BER. There are choices for using the conventional block and convolutional codes, the trellis codes, or a combination of different codes. There are no rules; the choice depends on the different applications.

Each of these areas deserves further study to become a real modem expert. References 4, 5, 26, 43, and 45 are recommended as a good starting point.

Bibliography

Bell Telephone Laboratories Technical Staff, *Transmission Systems for Communications*, Bell Telephone Laboratories, Technical Publications, Winston-Salem, NC, 1971.

Borgne, M., Comparison of High-Level Modulation Schemes for High-Capacity Digital Radio Systems, *IEEE Trans. Commun.*, COM-33:442–449 (1985).

Calderbank, A. R., and Sloane, N. J. A., Four-Dimensional Modulation with an Eight-State Trellis Code, *AT&T Tech. J.*, 64:1005–1018 (May–June 1985).

Cariolaro, G. L., and Tronca, G., Spectra of Block Coded Digital Signals, *IEEE Trans. Commun.*, COM-22:1555–1563 (October 1974).

Deshpande, G. S., and Wittke, P. H., Correlative Encoded Digital FM, *IEEE Trans. Commun.*, COM-29(2):156–162 (February 1981).

Fang, R., and Lee, W., Four-Dimensionally Coded PSK Systems for Combatting Effects of Severe ISI and CCI, *Globcom '83*, 30.4.1–30.4.7.

Forschini, G. J., Contrasting Performance of Faster Binary Signaling with QAM, *Bell Sys. Tech. J.*, 63:1419–1445 (October 1984).

Gersho, A., and Lawrence, V. B., Multidimensional Signal Constellation for Voiceband Data Transmission, *IEEE J. Sel. Areas Commun.*, SAC-2:687–702 (September 1984).

Lender, A., Rogers, R., and Olszanski, H., 4-Bits/Hz Correlative Single-Sideband Digital Radio at 2 GHz, *Proc. ICC*, 5.2.1-5.2.5 (1979).

McNicol, J. D., Barber, S. G., and Rivest, F., Design and Application of the RD-4A and RD-6A 64-QAM Digital Radio Systems, *Proc. ICC*, 646–652 (1984).

Mathiopoulos, P., and Feher, K., Performance Evaluation of a 512-QAM System in Distorted Channels, *IEE Proc.*, 133(2), Part.F, 199–204 (April 1986).

Mazo, J. E., Faster than Nyquist Signaling, *Bell Sys. Tech. J.*, 54:1451–1462 (October 1976).

Rummler, W. D., Digital Radio Outage Due to Selective Fading – Observation vs. Prediction from Laboratory Simulation, *Bell Sys. Tech. J.*, 58:1073–1110 (May–June 1979).

Saito, Y., Komaki, S., and Murotani, M., Feasibility Considerations of High-Level Multi-Carrier System, *Proc. ICC*, 665–671 (1984).

Schonhoff, T. A., Symbol Error Probabilities for M-ary CPFSK: Coherent and Noncoherent Detection, *IEEE Trans. Commun.*, COM-24(6):644–652 (June 1976).

Seo, J.-S., and Feher, K., Tight-Bound Symbol Error Probability of High-Level QAM in the Presence of Residual Amplitude Fluctuation, *Proc. ICC*, 15.4.1–15.4.5 (1988).

Smith, J. G., Odd-Bit Quadrature Amplitude-Shift Keying, *IEEE Trans. Commun.*, COM-23:385–389 (1975).

Tanaka, Y., et al., A 19.2 Kbps High-Speed Voice Data Modem Based on Eight-Dimensional Trellis Coding, *Globcom*, 36.1.1–36.1.5.

Yongacoglu, A., Makrakis, D., and Feher, K., Differential Detection of GMSK Using Decision Feedback, *IEEE Trans. Commun.*, 36(6):641–649 (June 1988).

References

1. Feher, K., *Advanced Digital Communications: Systems and Signal Processing Techniques*, Prentice-Hall, Englewood Cliffs, NJ, 1987.
2. Pahalavan, K., and Holsinger, J. L., Voice-Band Data Communication Modems – A Historical Review, *IEEE Commun.*, 26(1):16–27 (1988).

3. Feher, K., 1024-QAM and 256-QAM Coded Modems for Microwave and Cable System Applications, *IEEE J. Sel. Areas Commun.*, SAC-5:357–368 (April 1987).
4. Lucky, R., Salz, J., and Weldon, E., *Principles of Data Communication*, McGraw-Hill, New York, 1968.
5. Feher, K., ed., *Digital Communications: Satellite/Earth Station Engineering*, Prentice-Hall, Englewood Cliffs, NJ, 1983.
6. Morais, D. H., Sewerison, A., and Feher, K., The Effects of the Amplitude and Delay Slope Components of Frequency Selective Fading on QPSK, Offset QPSK and 8 PSK Systems, *IEEE Trans. Commun.*, Com-27:1849–1853 (1979).
7. de Buda, R., Coherent Demodulation of Frequency-Shift Keying with Low Deviation Ratio, *IEEE Trans. Commun.*, COM-20(3):429–436 (June 1972).
8. Sundberg, C.-E., Continuous Phase Modulation, *IEEE Commun.*, 24(4):25–38 (1986).
9. Osborne, W. P., and Luntz, M. B., Coherent and Noncoherent Detection of CPFSK, *IEEE Trans. Commun.*, COM-22(8):1023–1036 (August 1974).
10. Chung, K. S., General Tamed Frequency Modulation and Its Application for Mobile Radio Communication, *IEEE J. Sel. Areas Commun.*, SAG-2(4):487–497 (July 1984).
11. Murota, K., and Hirade, K., GMSK Modulation for Digital Mobile Telephony, *IEEE Trans. Commun.*, COM-29(7):1044–1050 (July 1981).
12. Feher, K., *Digital Communications: Microwave Applications*, Prentice-Hall, Englewood Cliffs, NJ, 1981.
13. Lender, A., Correlative Techniques and Applications to Digital Radio Systems. In: *Digital Communications: Satellite/Earth Station Engineering* (K. Feher, ed.), Prentice-Hall, Englewood Cliffs, NJ, 1983, pp. 144–182.
14. Kabal, P., and Pasupathy, S., Partial-Response Signalling, *IEEE Trans. Commun.*, COM-23:921–934 (1975).
15. Proakis, J. G., *Digital Communications*, McGraw-Hill, New York, 1983.
16. Forney, G. D., Jr., Maximum-Likelihood Sequence Estimation of Digital Sequences in the Presence of Intersymbol Interference, *IEEE Trans. Inform. Theory*, IT-18:363–378 (May 1972).
17. Eyuboglu, M. V., and Qureshi, S. U. H., Reduced-State Sequence Estimation with Set Partitioning and Decision Feedback, *IEEE Trans. Commun.*, 36(1):13–20 (January 1988).
18. Kretzmer, E. R., Generalization of a Technique for Binary Data Communication, *IEEE Trans. Commun.*, COM-14:67–68 (1966).
19. Godier, I., DRS-8 Digital Radio for Long Haul Transmission, *Proc. ICC*, 5.4-102–5.4-105 (1977).
20. Feher, K., et al., On 6.67 bps/Hz 256-QAM and 225-QPSR Modems for T1/SG Data-in-Voice (DIV) Application, *Proc. ICC*, 15.4.1–15.4.7 (1985).
21. Wu, K. T., and Feher, K., 256-QAM Modem Performance in Distorted Channels, *IEEE Trans. Commun.*, COM-33:487–491 (1985).
22. Seo, J.-S., Power Efficient Multi-ary QAM Systems, Ph.D. thesis, Dept. of Electrical Engineering, University of Ottawa, Ottawa, Canada, 1988.
23. Wu, K. T., and Feher, K., Multi-Level PRS/QPRS above the Nyquist Rate, *IEEE Trans. Commun.*, COM-33:735–739 (July 1985).
24. Wu, K. T., Sasase, I., and Feher, K., Class-IV PRS above the Nyquist Rate, *IEE Proc.*, 135(2), Part F, 183–191 (April 1988).
25. Bennett, W. R., and Davey, J. R., *Data Transmission*, McGraw-Hill, New York, 1965, pp. 121–128.
26. Forney, G. D., Jr., et al., Efficient Modulation for Band-Limited Channels, *IEEE J. Sel. Areas Commun.*, SAC-2:632–647 (September 1984).
27. Brownlie, J. D., and Cusack, E. J., Duplex Transmission at 4800 and 9600 Bit/s

on the General Switched Telephone Network and Their Use of Channel Coding with a Partitioned Signal Constellation (in English), *Proc. Zurich Int. Sem. Digital Commun.*, 113–120 (March 1984).

28. Tahara, M., et al., 6GHz 140 Mbps Digital Radio System with 256-SSQAM Modulation, *Globcom '87*, 38.3.1–38.3.6 (1987).
29. Daido, Y., et al., Multilevel QAM Modulation Techniques for Digital Microwave Radios, *IEEE J. Sel. Areas Commun.*, SAC-5(3):336–341 (April 1987).
30. Davarian, F., Mobile Digital Communications via Tone Calibration, *IEEE Trans. Veh. Technol.*, VT-36:55–61 (May 1987).
31. Simon, M. K., Dual-Pilot Tone Calibration Technique, *IEEE Trans. Veh. Technol.*, VT-35:63–70 (May 1986).
32. Gorog, E., Redundant Alphabets with Desirable Frequency Spectrum Properties, *IBM J. Res. Devel.*, 12:234–240 (May 1968).
33. Kobayashi, H., A Survey of Coding Schemes for Transmission or Recording of Digital Data, *IEEE Trans. Commun.*, COM-19:1087–1100 (December 1971).
34. Aaron, M., PCM Transmission in the Exchange Plant, *Bell Sys. Tech. J.*, 41:99–141 (January 1962).
35. Kaneko, H., and Sawai, A., Feedback Balanced Code for Multilevel PCM Transmission, *IEEE Trans. Commun.*, COM-17:554–563 (October 1969).
36. Nagagome, Y., Amano, K., and Ota, C., A Multi-Level Code Having Limited Digital Sum (English translation), *Electron. Commun. Japan*, 51-A:31–37 (January 1968).37. Hagiwara, S., Sata, N., and Takimasa, A., 1.544 Mbps PCM-FDM
37. Hagiwara, S., Sata, N., and Takimasa, A., 1.544 Mbps PCM-FDM Converters Over Coaxial and Microwave Systems, *Fujitsu Scientific and Technical J.*, 12(3): 1–26 (1976).
38. Kim, D. Y., and Feher, K., New Carrier and Symbol Timing Technique for Digital Mobile Systems, *Proc. of IEEE Vehicular Tech. Conference*, VTC:371–376 (1988).
39. Pierobon, G. L., Codes for Zero Spectral Density at Zero Frequency, *IEEE Trans. Inform. Theory*, IT-30(2):435–439 (March 1984).
40. Kim, D. Y., and Kim, J.-K., A Condition for Stable Minimum-Bandwidth Line Codes, *IEEE Trans. Commun.*, COM-33(2):152–157 (February 1985).
41. Orton, J., and Feher, K., An Improved Channel Coding Algorithm for Low Frequency Spectral Suppression, paper read at 21st Annual Asilomar Conference on Signals, Systems, and Computers, Pacific Grove, CA, November 2–4, 1987.
42. Leung, P.S.-K., and Feher, K., Block-Inversion-Coded QAM Systems, *IEEE Trans. Commun.*, 36(7):797–805 (July 1988).
43. Gallager, R. G., *Information Theory and Reliable Communication*, Wiley, New York, 1968.
44. Ungerboeck, G., and Csajka, I., On Improving Data-Link Performance by Increasing the Channel Alphabet and Introducing Sequence Coding, paper presented at 1976 International Symposium on Information Theory, Ronneby, Sweden, June 1976.
45. Ungerboeck, G., Channel Coding with Multilevel/Phase Signals, *IEEE Trans. Inform. Theory,* IT-28:55–67 (January 1982).
46. Forney, G. D., Jr., The Viterbi Algorithm, *Proc. IEEE*, 61:268–278 (March 1973).
47. Falconer, D. D., Jointly Adaptive Equalization and Carrier Recovery in Two-Dimensional Digital Communication Systems, *Bell Sys. Tech. J.*, 55:317–334 (March 1976).
48. Ungerboeck, G., et al., Coded 8-PSK Experimental Modem for the INTELSAT SCPC System, *Proc. 7th Int. Conf. on Digital Satellite Communications*, ICDS-7: 299–304 (1986).
49. Ungerboeck, G., Trellis-Coded Modulation with Redundant Signal Sets—Part II, *IEEE Commun.*, 25(2):12–21 (February 1987).

50. Ungerboeck, G., Trellis-Coded Modulation with Redundant Signal Sets – Part I: Introduction, *IEEE Commun.*, 25(2):5–11 (February 1987).
51. Wei, L. F., Trellis-Coded Modulation with Multidimensional Constellations, *IEEE Trans. Inform. Theory*, IT-33(4):483–501 (July 1987).
52. Wei, L. F., Rotationally Invariant Convolutional Channel Coding with Expanded Signal Space – Part 1: 180 Degrees, Part II: Nonlinear Codes, *IEEE J. Sel. Areas Commun.*, SAC-2:659–686 (September 1984).
53. Fang, R. J., A Coded 8-PSK System for 140-Mbps Information Rate Transmission over 80-MHz Nonlinear Transponders, *Proc. 7th Int. Conf. on Digital Satellite Communications*, ICDS-7:410–419 (1986).
54. Simon, M. K., and Divsalar, D., *Combined Trellis Coding with Asymmetric MPSK Modulation,* JPL publication 85-24, Jet Propulsion Laboratory, California Institute of Technology, Pasadena, CA, May 1, 1985.
55. Divsalar, D., and Simon, M. K., Multiple Trellis Coded Modulation (MTCM), *IEEE Trans. Commun.*, 36(4):410–419 (April 1988).
56. Makrakis, D., Yongacoglu, A., and Feher, K., A Sequential Decoder for Differential Detection of Trellis Coded PSK Signals, *Proc. ICC*, 1433–1438 (1988).
57. Makrakis, D., Yongacoglu, A., and Feher, K., Interleaving for Differential Detection Using Multiple Detectors, *Proc. of IEEE Vehicular Tech. Conf.*, VTC:18–21.
58. Hirade, K., et al., Error-Rate Performance of Digital FM with Differential Detection in Land Mobile Radio Channels, *IEEE Trans. Veh. Technol.*, VT-28:204–212 (August 1979).
59. Kato, M., and Inose, H., Majority Logic Detection Scheme of Differentially Phase Modulated Waves, *Electron. Commun. Japan*, 55-A(1) (1972).
60. Samejima, S., et al., Differential PSK System with Nonredundant Error Correction, *IEEE J. Sel. Areas Commun.*, SAC-1 (January 1983).
61. Makrakis, D., Yongacoglu, A., and Feher, K., Novel Receiver Structures for Systems Using Differential Detection, *IEEE Trans. Veh. Technol.*, VT-36(2):71–77 (May 1987).
62. Simon, M. K., and Wang, C. C., Differential Detection of Gaussian MSK in a Mobile Radio Environment, *IEEE Trans. Veh. Technol.*, 307–320 (November 1984).
63. Ogose, S., and Murota, K., Differentially Encoded GMSK with 2-Bit Differential Detection, *Trans. IEEE Japan*, 164-B:248–254 (April 1981).
64. Hirono, M., Miki, T., and Murota, K., Multilevel Decision Method for Bandlimited Digital FM with Limiter Discriminator Detection, *IEEE Trans. Veh. Technol.*, 114–122 (August 1984).
65. Yongacoglu, A., On Differential Detection of Digitally Modulated Signals, Ph.D. thesis, Dept. of Electrical Engineering, University of Ottawa, Ottawa, Canada, 1987.
66. Chen, Z., Wang, J., and Feher, K., Effect of Non-Linearities on Crosstalk and Performance of Digital Radio Systems, *IEEE Trans. Broadcasting*, 34(3):336–343 (September 1988).
67. Cahana, D., et al., Linearizer Transponder Technology for Satellite Communication, Part 1 and Part 2, *COMSAT Tech. Rev.*, 15:277–341 (Fall 1985).
68. Bakken, P. M., A New Method for Computing the Properties of the Output Signal of Nonlinear Amplifiers, *IEEE Trans. Commun.*, 67–71 (January 1986).
69. Salen, A. A. M., Frequency-Independent and Dependent Nonlinear Models of TWT Amplifiers, *IEEE Trans. Commun.*, 1715–1720 (November 1981).
70. Kaverhad, M., Multiple FM/FDM Carrier through Nonlinear Amplifiers, *IEEE Trans. Commun.*, 751–757 (May 1981).
71. Wang, J., and Feher, K., Spectral Efficiency Improvement Techniques for Nonlin-

ear Simplified Mobile Radio Systems, *Proc. of IEEE Vehicular Tech. Conf.*, VTC: 629–635.

72. Morais, D. H., and Feher, K., NLA-QAM: A New Method for Generating High Power QAM Signals through Nonlinear Amplification, *IEEE Trans. Commun.*, COM-30:517–522 (March 1982).

73. Weber, W. J., et al., A Bandwidth Compressive Modulation System Using Multi-amplitude Minimum Shift Keying (MAMSK), *IEEE Trans. Commun.*, COM-26 (May 1978).

74. Seo, J.-S., and Feher, K., Bandwidth Compressive 16-State SQAM Modems through Saturated Amplifiers, *IEEE Trans. Commun.*, COM-35:339–345 (March 1987).

75. Seo, J.-S., and Feher, K., Performance of 16-State SQAM in a Nonlinearly Amplified Multichannel Environment, *IEEE Trans. Commun.*, 36(11):1263–1267 (November 1988).

76. Gronemeyer, S. A., and McBride, A. L., MSK and Offset QPSK Modulation, *IEEE Trans. Commun.*, COM-24:809–819 (August 1976).

77. Morais, D. H., Digital Modulation Techniques for Terrestrial Point-to-Point Microwave Systems, Ph.D. thesis, Dept. of Electrical Engineering, University of Ottawa, Ottawa, Canada, 1981.

KUANG-TSAN WU
KAMILO FEHER

Digital Radio Systems

Features of Digital Radio Communication

Radio systems have played an important part in telecommunications for many years because of their excellent performance, low cost, and high reliability. Nearly half of all long-distance telephone calls and almost all commercial television program services now are transmitted through radio-relay networks. More than 3000 repeater stations have been constructed in the United States. In Japan, there are now more than 800 stations, with total trunk transmission routes in excess of 60,000 kilometers. Recently, these radio networks have been undergoing major changes to cope with the development of new telecommunications information sources. With the ongoing shift to digital transmission, radio-relay networks now are making a fresh start and again are playing an important role in the construction of Integrated Services Digital Networks (ISDNs).

The present popularity of digital radio can be attributed primarily to three factors. First, due to their regenerative repeating function, digital radio systems can secure satisfactory transmission quality regardless of transmission distance since they are free of the noises that generally accumulate during repetition. Second, when the transmission capacity is constant, the required transmission power is smaller than that of frequency modulation (FM) systems, thus allowing for easy introduction of solid-state electronics. The smaller power levels minimize interference with other systems and, as a result, the frequency may be shared between satellite and ground communications facilities. Third, through the use of modulation systems in which the carrier-to-noise (C/N) power ratio required for the bit error rate (BER) is rather small, it is possible to realize a system only slightly affected by interference. This facilitates construction of a large number of routes between two places to increase the utility factor of the frequency employed dramatically.

On the other hand, digital radio-relay systems have three technical problems. First, when the system is applied to telephone transmission, the frequency spectrum spread is broader than that of a comparable analog system. Second, when the fluctuation of the propagation property exceeds the allowable value, quality deterioration increases. Third, the systems easily become subject to the effects of nonlinear distortion due to saturation characteristics, which pose no problem for FM systems. These problems have been overcome effectively by recent advances in technology, including multilevel quadrature amplitude modulation (QAM), the development of a number of countermeasure techniques, and the realization of solid-state circuits. Furthermore, since digital radio-relay systems can make use of equipment employed in such existing analog transmission lines as stations, towers, and the like, radio networks can be digitalized promptly and economically.

Digital Microwave Radio Systems

Digitalization of Existing Frequency Modulation Networks

The microwave band is one of the most excellent of all bands and has been utilized extensively in terrestrial radio transmission during the last 40 years. The band has four advantageous features. First, it can transmit very wide band signals because of its extremely high frequency. Second, its propagation characteristic is very good since the microwave frequency is not affected by rain or snow. Very reliable and stable communications networks for line-of-sight transmission can be offered as a result. Third, microwave radio reception rarely is affected by external noise since there are few noise sources in the high-frequency bands. Fourth, since wavelength is very short, high gain can be obtained easily in small-sized antennas, and sufficiently high signal-to-noise (S/N) ratio can be obtained in low-power transmission. With these excellent features, the microwave band has been developed widely and extensively, and microwave radio networks cover the country (1,2).

In existing radio networks, the FM modulation scheme mainly has been used, and great efforts were needed to increase transmission capacity. The most advanced FM systems now in use have a transmission capacity of 3600 telephone channels at a frequency band of 80 megahertz (MHz). In the early 1970s, a great deal of information about digital signals became available, and a demand arose for the establishment of ISDNs through the digitalization of the current analog networks. Since that time, microwave radio has been gaining steadily in importance as a transmission medium for digital communications. The first digital microwave systems utilized quaternary phase-shift keying (QPSK) modulation and achieved a spectral efficiency of less than 2 bits per second per hertz (bps/Hz). However, major technical advances have been made during the last 15 years, and today spectral efficiency has reached a level of 8 bps/Hz with super multilevel modulation (3,4). The most recently developed of the world's digital microwave radio systems are introduced in Table 1.

The outstanding feature of digital transmission is that it makes possible the transmission of such multimedia signals as telephone, facsimile, data, and video. However, three different kinds of hierarchies have been used in the world, and it became difficult to connect networks based on different hierarchies. In 1989, CCITT (International Telegraph and Telephone Consultative Committee) standardized a new synchronous interface, the synchronous digital hierarchy (SDH), as the network node interface on which future network configurations would be based. Figure 1 shows a comparison of existing digital hierarchies and the new SDH standard hierarchy. Currently, radio networks are being reconstructed to match the SDH (5).

Effects of Multipath Fading on Digital Radio

Propagation Characteristics under Deep Fading

Although the microwave band is an excellent band, it has one big drawback — multipath fading (6,7). Multipath fading is caused by the destructive interference among rays arriving at one receiving antenna through different paths.

TABLE 1 Typical Microwave Digital Radio-Relay Systems

System	North America		Japan		CEPT	
	DR6-30-135	TD-90	4/5/6 G 300M	11/15 G 150M	4/6/8/11 G 150M	4/6/8 150M
Frequency bands	6 GHz	4 GHz	4, 5, 6 GHz	11, 15 GHz	4, 6U, 8, 11 GHz	4, 6L, 8 G
Transmission capacity	144.195 Mbps	90 MHz	334.66 Mbps	167.33 Mbps	146.6 Mbps	149.7 Mbps
Modulation	64QAM	64QAM	256QAM	8PSK	16QAM	64QAM
Symbol rate (Mbaud)	24	16.16	13.944	55.777	136.649	24.955
Roll-off factor	0.41	0.31	0.42	0.40	0.35	0.3
Transmitter power	35.7 dBm	18 dBm	27 dBm/32 dBm	33 dBm/30 dBm	30 dBm	32 dBm
Receiver noise	4.0 dB	4.5 dB	4.5 dB	4 dB/5 dB	3 dB/4, 6, 8 GHz 4 dB/11 GHz	3 dB
Typical system gain (BER = 10^{-6})	103 dB	102 dB	94 dB/99 dB	108 dB/103 dB	103 dB	106 dB
Error correction	Optional	Equipped*	BCH Double**	BCH Double	—	Lee Double***
Power control	—	Equipped*	—	—	—	—

*Convolutionary self-orthogonal code.
**BCH Code double forward error correction.
***Lee code double forward error correction.

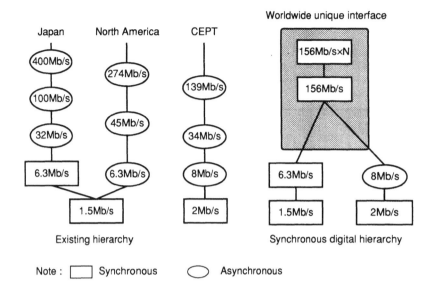

FIG. 1   Comparison of existing digital hierarchies and synchronous digital hierarchy (SDH).

A simple, easy-to-understand, three-ray model can account for most of the distortion effects. A three-ray model is shown in Fig. 2. Transmission characteristics of a three-ray fading model are expressed by the following formula (8–11).

$$H(\omega) = a[1 + \rho e^{-j(\omega\tau - \Theta)}],$$

where a is a coefficient, ρ is the amplitude ratio of the interference ray and the direct ray, and τ and Θ are the delay difference and the phase difference, respectively, between the interference ray and the direct ray.

There are two types of interfering rays. Refractive rays, in which the delay

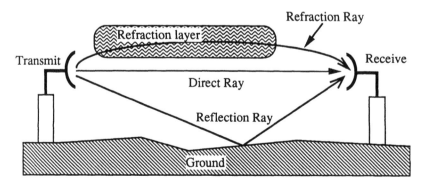

FIG. 2   Three-ray fading model.

time difference between the direct and interfering rays is small, are derived from variations in the refractive index of the atmosphere. Reflective rays, in which the delay time difference between the direct and interfering rays is large, are the result of reflections from water or ground surfaces. If direct and interference rays are received simultaneously by one antenna, the amplitudes of the rays are added at some frequencies and subtracted at others. This results in frequency-selective fading. The spacing between notch frequencies is in inverse proportion to the delay time difference τ, and the notch depth is determined by the amplitude ratio ρ of the interference ray to the direct ray. Frequency response is classified depending on the value of ρ, that is, when ρ is smaller than 1, it is called *minimum phase-shift fading*, and when ρ is larger than 1, it is called *nonminimum phase-shift fading*. The amplitude characteristics of three-ray fading are illustrated in Fig. 3. As can be seen from the figure, the group delay is reversed at the $\rho = 1$ boundary, even though the amplitude characteristics are virtually the same. Two kinds of group-delay characteristics are shown in Fig. 3.

Generation Mechanism of Waveform Distortion Due to Fading

Let us examine the time region with respect to a single pulse to see how the fading causes waveform distortion. Multipath fading occurs when waveforms propagated through different propagation paths interfere with the main wave.

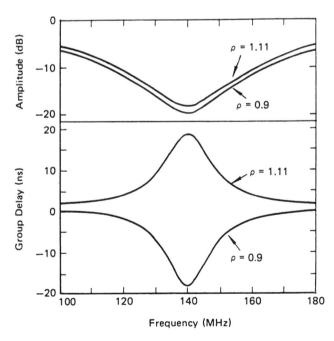

FIG. 3 Two kinds of group-delay characteristics caused by minimum phase-shift fading and nonminimum phase-shift fading.

Figure 4 shows several aspects of the waveform at the code determination timing at which a wave with an amplitude ratio of ρ and a delay time difference of τ overlaps in the reverse phase with the main wave. Even when transmitting or receiving filters are provided, the signal level at the threshold point drops to $1 - \rho$. In actual relay systems, strict band limitation is applied so that the pulses before and after the center point are expanded, making the signal level at the threshold point drop even lower. Under a deep fading, a waveform is severely distorted and finally the polarity of the signal level is reversed, thus causing bit errors.

Fading Countermeasures

The main degradation factors of digital radio communication and their counter-measures are shown in Fig. 5. These factors can be classified as such time-variant factors as multipath fading and rainfall attenuation, and such time-invariant factors as the inherent imperfectness of repeaters. These factors cause waveform distortion, thermal noise, and interference.

Space diversity controlled by a minimum in-band dispersion combiner and adaptive equalization are effective methods of combating waveform distortion (12,13). Against interference noise, interference cancellation techniques, space diversity controlled by an in-phase combiner, and antennas that have a sharp beam or a null for an interference ray direction are effective. Against thermal noise, the techniques of high-power amplification and linearization to increase transmission power, and techniques of diversity reception are useful.

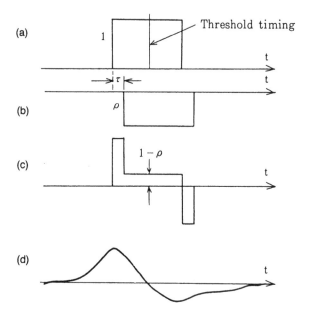

FIG. 4   Generation mechanism of waveform distortion due to fading: a, main wave; b, interference wave; c, composite wave; d, composite wave after passing the filter.

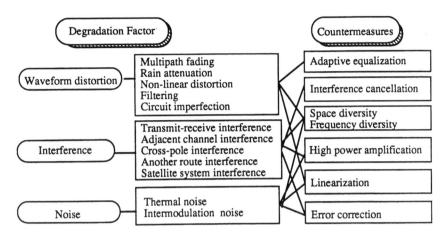

FIG. 5 Digital radio communication degradation factors and countermeasures.

For bands in which the frequency is less than 10 gigahertz (GHz), the fading phenomena is a particularly important problem, and the key to realizing digital microwave radio systems lies in conquering fading degradation. The principal countermeasures employed against fadings in digital microwave radio systems are discussed in the next sections.

Equalizers

Variable-Resonance Adaptive Equalizer. The variable-resonance adaptive equalizer, with its variable-resonance circuit, flattens the frequency characteristics degraded by two-ray fading because it can produce a frequency response opposite to that produced by any two-ray fading. A block diagram of the variable-resonance circuit is shown in Fig. 6. The equalizer circuit is composed of a resonance circuit with a distributed constant line and a varactor diode and, by connecting a positive-intrinsic-negative (PIN) diode, is connected in series to the circuit. The equalizer circuit has two characteristics: (1) the resonance frequency (f_0) can be changed by altering the capacity of the varactor diode, and (2), the sharpness Q can be changed by altering the PIN diode resistance value. This equalizer can compensate for amplitude distortion, but it cannot always compensate for group-delay distortion (12).

The adaptive equalizer performance is measured using a two-ray fading simulator. Both amplitude and group-delay distortions must be compensated for in this system. If amplitude ratio ρ is less than 1, both the amplitude and group-delay distortion can be equalized. If ρ becomes greater than 1, the amplitude distortion is equalized but the group-delay distortion is doubled.

Transversal Equalizer. Deterioration of frequency characteristics due to frequency-selective fading deteriorates the demodulated waveform in the baseband, causing intersymbol interference. The transversal equalizer, which is de-

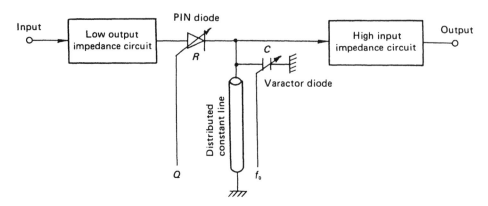

FIG. 6   Variable-resonance equalizer circuit.

signed to remove this intersymbol interference, theoretically can equalize all types of distortion (14).

A three-tap, two-dimensional transversal equalizer is shown in Fig. 7. When a transmitted pulse goes through a fading propagation path, its waveform is distorted and intersymbol interference is added to it at the receiver side. The distorted signals are guided to a tapped delay line, and forward and delayed signals are taken out through the forward tap and the backward tap, respectively. These extracted signals are weighted individually and combined with center tap signals, where weighting is adjusted automatically by using correlation techniques to cancel out the intersymbol interference caused by adjacent pulses.

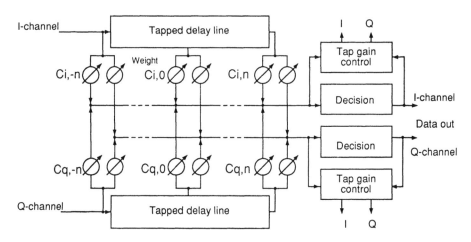

FIG. 7   Configuration of two-dimensional transversal equalizer ($C_{i,j}$ = weight coefficient of the I channel jth tap).

There are two kinds of transversal equalizers. Figure 8-*a* shows a linear transversal equalizer. Its performance is indicated by a signature shown in Fig. 9. A signature is one of the most important indexes used to evaluate equalizer characteristics. When the three-ray model mentioned in the section "Propagation Characteristics under Deep Fading" above, is used, the fading notch frequency is changed by shifting the phase difference Θ, and the limit value ρ capable of obtaining the required BER *Pe* is decided depending on fading notch frequency. The locus given by ρ and τ is called a *signature*, and it determines the resistance of a system to fading. The smaller the area surrounded by the locus, the stronger the system will be against fading.

A linear transversal equalizer can compensate independently for minimum or nonminimum fading, but it needs to have a lot of taps to compensate for steep frequency-selective fading. Figure 8-*b* shows a nonlinear transversal equalizer; its performance is indicated by a signature in Fig. 9 (15). In this case, one side of the signature is completely removed; this means that the nonlinear equalizer works perfectly against minimum phase-shift fading (16,17). The reason is that intersymbol interference is caused by delayed pulses during minimum phase shift fading and they can be completely cancelled out without squeeze out by regenerated pulses that are not corrupted by thermal noise or interference. During nonminimum phase-shift fading, intersymbol interference is caused mainly by forward pulses and equalizing is performed by the forward linear circuit section. Consequently, performance becomes similar to that of a linear transversal equalizer.

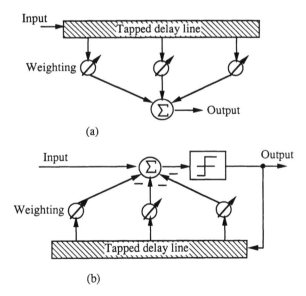

(a)

(b)

FIG. 8 Configuration of transversal equalizers: *a*, linear transversal equalizer; *b*, nonlinear transversal equalizer.

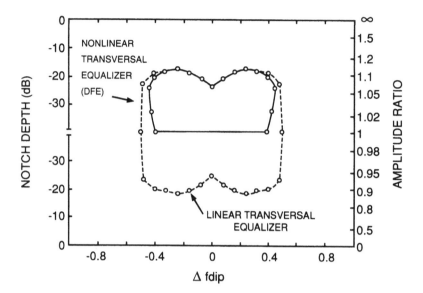

FIG. 9   Equalizer performance for three-path fading model in a 16QAM (quadrature amplitude modulation) system (3 taps) (DFE = decision feedback equalizer).

Diversity Techniques

Diversity techniques are used widely to reduce degradation caused by multipath fading or rain attenuation (18,19). Different antennas, frequencies, polarizations, ray angles, roots, areas, or times are used to obtain different diversity signals. These are referred to as space diversity, frequency diversity, polarization diversity, angle diversity, root diversity, sight diversity, and time diversity, respectively. Frequency utilization efficiency is very important in digital microwave radio; consequently, frequency diversity and polarization diversity rarely are used, but space diversity is used widely.

Space Diversity.   Space diversity has been used for terrestrial radio links for a long time. In FM systems, attenuation is the dominant degradation factor and frequency selectivity of attenuation is not such an important matter. Therefore, an in-phase combiner has been adopted where both received signals are combined so that the combined signal level becomes the maximum. In the in-phase combiner, the phases of the two received signals are made the same so that the level after combining will be maximized, but the improvement in the in-band amplitude dispersion is incomplete. Recently, a new combiner algorithm has been developed for digital microwave radio. It is called a *minimum in-band dispersion combiner* since it can minimize in-band amplitude dispersion (20). It does so by combining interfering rays in antiphase if the two received signals are equal to each other in terms of interfering-ray amplitude. As a result, only the direct-ray component remains after combining and therefore the transmission characteristic's deterioration due to fading can be improved.

Angle Diversity. The existing model of vertical space diversity, based on analog radio experience, indicates that the effect of space diversity increases with the square of the vertical antenna separation. This means that a large vertical antenna separation is required to achieve a large space diversity effect. The analog-radio-based model implies no performance improvement when the vertical separations of the two antennas have different gains and radiation patterns. However, recent studies have clarified that if two antennas have different characteristics, namely, different side lobe patterns or different beam directions, a big improvement can be attained even if the antennas are set on the same platform. This kind of space diversity includes angle diversity, beam-tilt diversity, and antenna pattern diversity (21–24). These diversities are illustrated in Fig. 10.

These diversities are very effective against dispersion-caused outages but are less effective against thermal-noise-caused outages. They are not expected to be very effective against total power fades because their effectiveness is predicated on their ability to prevent simultaneous "deep notches" on the two antennas. In other words, the total power-fade depth within a radio channel is not as sensitive as the deep notch (at a single frequency) to slight variations in the antenna pattern or the antenna pointing angle. Therefore, if the thermal-noise-caused

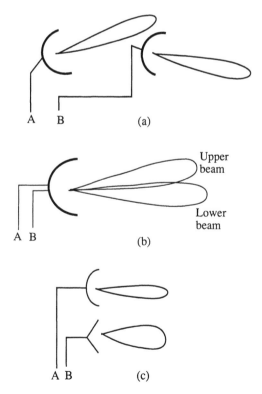

FIG. 10   Diversity techniques depending on antenna characteristics: *a*, angle diversity; *b*, beam-tilt diversity; *c*, antenna pattern diversity.

outages are not negligibly small, then vertical space diversity may provide improvement greater than that provided by antenna pattern diversity or antenna angle diversity.

Frequency Diversity.   Frequency diversity is one of the most vigorous of the countermeasures taken against frequency-selective fading. In frequency diversity, a redundant channel is prepared in a different channel. If a channel signal is faded or distorted by fading, it is switched immediately to a protection channel and can avoid the effect of fading. The conditions under which frequency diversity works well are illustrated in Fig. 11. Digital radio systems are affected strongly by frequency-selective fading. However, diversity very effectively improves frequency-selective fading since frequency response is very different between different channels during fading. If fading is frequency independent, the signal is faded over a wide band and the protection channel signal is faded simultaneously. On the other hand, if fading is frequency dependent, only a few channels are degraded, and the degradation can be avoided by switching to a protection channel (25–27).

The diversity improvement factor depends on the ratio of working channels to protection channels: the smaller the ratio, the larger the gain. Degradation time caused by fading can be reduced by one order for ordinary microwave digital radio systems.

Sophisticated Digital Radio Systems in Microwave Bands

System Design

When designing digital microwave systems, modulation and filtering, channel frequency arrangement, and link design should be considered carefully.

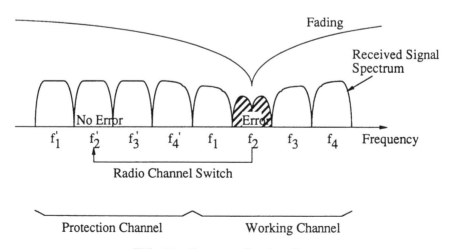

FIG. 11   Frequency diversity effect.

Modulation and Filtering. Transmission capacity should be compatible with existing analog FM systems. To decide the capacity, selection of a modulation scheme and specter shaping becomes very important. The major techniques employed to obtain higher spectrum efficiency are multi-level modulation and narrowband spectrum-shaping techniques. Spectrum efficiencies for various modulation methods are shown in Fig. 12. From the illustration, it is clear that QAM requires less S/N than the phase-shift keying (PSK) modulation scheme. Its signal constellation is very close to the best performance achievable by optimum multisignal point packing in two dimensions (28). QAM has another merit in that high-speed Modem circuitry implementation is more feasible than multi-array PSK modulation. These factors make the use of QAM favorable.

Nyquist cosine roll-off spectrum shaping is one of the superior filtering methods since it can minimize intersymbol interference. It is natural that as the roll-off factor becomes smaller, the spectrum efficiency increases. However, considering the waveform distortion increase due to multipath fading and the difficulty in realizing an accurate roll-off filter, a roll-off factor of 0.5–0.7 is appropriate.

Channel Frequency Arrangement. For long trunk transmission lines, the same frequency used in existing FM radio systems will be used for new digital radio systems. It is very important to coordinate both systems to exist without any mutual interference problems. Figure 13 shows an example of channel frequency arrangement in a digital radio system in the 4-GHz band. In this arrangement, the FM carrier spectrum, which accounts for most of the power of FM modulated signals, is located between digital radio channels; this makes it easy to

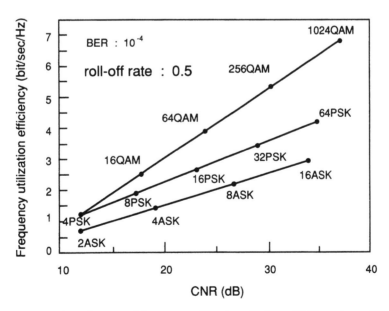

FIG. 12 Modulation schemes and frequency utilization efficiency (PSK = phase-shift keying).

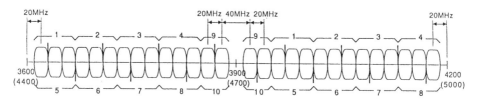

FIG. 13 Example of a channel frequency arrangement.

remove interfering FM carriers by means of filters provided in the digital radio system. A SDH-based system needs three primary careers for each radio channel, therefore, there are eight working channels in 4-GHz band. The remaining four primary careers are used for a protection channel.

Link Design. Public radio trunk transmission lines should follow the 64-kilobits-per-second (kbps), 2500-kilometer (km) hypothetical reference digital path (HRDP) recommended by the International Radio Consultative Committee (CCIR). Recommendation 634-1 decides the performance objective and Recommendation 695 decides the reliability performance for real digital radio-relay links (see Table 2). In designing radio links, the method used to estimate outage probability is very important. Outages caused by fading strongly depend on path characteristics and system parameters (29–31). Various estimation methods have been proposed, and they are quite different from the one used for FM system link design since waveform distortion became the main outage factor in digital radio (32–36). Three principal methods are used: the composite fade margin (CFM) method, the linear amplitude dispersion (LAD) method, and the model parameter simulation method.

256QAM High-Capacity Digital Radio System

In this section, the design of a digital microwave radio system is introduced using the example of a 256QAM system, one of the most advanced current

TABLE 2 Performance Objectives for Real Digital Radio Links

Error performance objectives	BER $> 1 \times 10^{-3}$ for no more than $(L/2500) \times 0.054\%$ of any month; integration time 1 s.
	BER $> 1 \times 10^{-6}$ for no more than $(L/2500) \times 0.4\%$ of any month; integration time 1 min.
	Errored second for no more than $(L/2500) \times 0.32\%$ of any month.
	Residual bit error ratio (RBER) $\leq L \times 5 \times 10^{-9}/2500$
Availability objectives	$A = 100 - (0.3 \times L/2500)\%$

A = availability
L = link length

TABLE 3 Main Parameters of a 300-Mbps 256QAM System

Parameters	Specifications
Frequency bands	4, 5, 6 GHz
Performance objectives	<0.001%/2500 km
Modulation	256QAM
Transmission capacity	312 Mbps
Number of carriers	3 carriers/system
Number of systems (working + protection)	25 + 1
Route capacity	8 Gbps
Frequency arrangement	Co-channel with 20-MHz separation
Clock frequency	13.944 MHz
Roll-off factor	0.42
Transmission output power	27 dBm
Interface	52 Mbps and 156 Mbps
Error correction	BCH double error correction

systems. This system is designed to have a high transmission capacity of 300 megabits per second (Mbps) using a 256QAM modulation scheme and both vertical and horizontal polarization. The main system parameters are given in Table 3. Figure 14 shows a diagram of the system configuration. This system employs three carrier-transmission methods and is referred to as *multicarrier transmission*; it is one of the most effective countermeasures employed to overcome in-band dispersion caused by multipath fading. In a system using three multicarrier transmission methods, for example, permissible amplitude dispersion becomes triple that of a single carrier's transmission capacity.

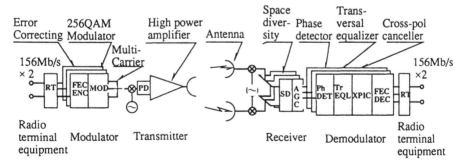

FIG. 14 Configuration of 256QAM digital microwave radio system (RT = radio terminal equipment; FED ENC = forward error correction encoding; MOD = modulator; PD = predistorter; SD = space diversity; AGC = automatic gain control; Ph DET = phase detector; Tr EQL = transversal equalizer; XPIC = cross-pole interference canceler; FEC DEC = forward error correction decoding).

Modulator. The configuration of 256QAM modulation is given in Fig. 15. Eight series of 12.5-Mbps, unipolar, two-level signals are encoded by double forward error correction (FEC) code to reduce the required C/N ratio and to achieve long-term error ratio performance. After four streams are multiplied by the constant coefficients of 1, 1/2, 1/4, or 1/8, respectively, the four manipulated streams are summed. Thus, eight series of 12.5-Mbps signals are converted to two series of 16-level signals. Their bandwidths are limited by the root roll-off filter to shape their spectrum to 42% roll-off. Then they are fed to a quadrature amplitude modulator that consists of orthogonal careers and two amplitude modulators. Eight series of 12.5-Mbps signals then are changed to a 100-Mbps 256QAM modulated signal. In multicarrier systems, three kinds of modulated signals, with frequencies of 110, 130, or 150 MHz, are used and these signals are multiplexed and sent to the transmitter.

Demodulator. The demodulator consists of a coherent phase detector, a transversal equalizer (Tr EQL), a cross-pole interference canceler (XPIC), and an FEC decoder. At the phase detector, an IF (intermediate frequency) modulated signal of 110, 130, or 150 MHz is converted to two 16-level baseband signal streams, and an analog-to-digital (A/D) converter detects eight binary signal pulse streams from a 16-level baseband signal stream. These signal pulse streams are processed to compensate for distortion and cross-polarization interference by a Tr EQL and an XPIC, respectively. The XPIC developed for the 256QAM system is able to achieve an input improvement factor of desired versus undesired (D/U) signal power ratio of up to 16 dB and is able to cancel the interference up to 42 dB (37).

Transmission characteristics can be evaluated by received pulse waveforms. If transmitted pulses are influenced by fading or imperfection of equipment, waveforms are distorted and a system becomes weak against thermal noise or

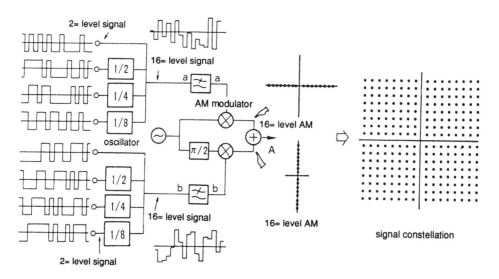

FIG. 15 Principle of 256QAM modulation.

interference noise. The degree of this influence can be seen directly from the eye pattern. In this eye pattern, the pulse waveforms just before the discriminator are drafted on the oscilloscope synchronously with the pulse repetitive frequency overlapping all of the probable waveforms. The difference in voltage between the upper and lower parts is called the *eye aperture*. The larger the eye aperture is, the larger is the allowable noise.

Figure 16 shows the performance of a transversal equalizer. Eyes stained and closed by the fading are equalized, and clear eyes are obtained through the use of the transversal equalizer. This equalizer consists of four LSIs (large-scale integration) and adopts digital signal processing technology.

Transmitting Equipment. Three 256QAM signals are amplified together by a gallium arsenide (GaAs) field-effect transistor (FET) amplifier with a predistortion linearizer after being converted from an IF band to a radio-frequency band. The predistortion linearizer consists of power dividers, circulators, and diode distortion generators, and has very good ability to reduce third-order intermodulation (IM3) noise. Its configuration is shown in Fig. 17 (38).

Receiving Equipment. The receiving signals are amplified by the low noise GaAs FET amplifier (commonly used in each polarization) and then converted to the IF band by a common local oscillator. Space diversity is controlled by a product-phase detection method in each carrier to achieve higher response speed and better effectiveness against fading.

Circuit Switching Equipment. The circuit switching equipment faculties are circuit switching, scrambling, circuit quality supervision, and insertion of pulses used for frame synchronizing, parity checking, supervision, and the control link. The circuit switching is accomplished by individual carrier units (100-Mbps stage) in order to improve the frequency diversity effect. The circuit-switching function can switch the traffic load on a failed channel to a protection channel

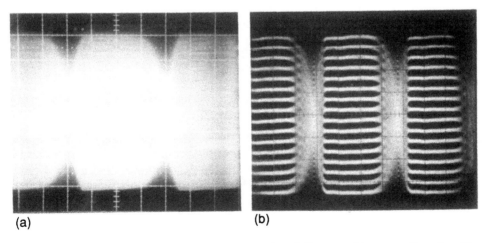

(a)   (b)

FIG. 16 Performance of transversal equalizer against fading: *a*, without equalizer; *b*, with equalizer.

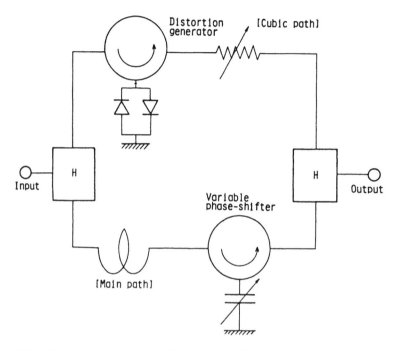

FIG. 17   Configuration of predistortion-type linearizer at microwave band.

without frame-synchronizing loss. Individual carriers for all working channels can be switched freely to an unoccupied carrier in a protection channel.

Figure 18 shows the equipment of the SDH-based 256QAM system. This system was put into nationwide use in Japan in 1989.

Higher Frequency Band Digital Radio Systems

Development of Higher Frequency Band

Frequencies under 10 GHz have been used widely in the progress of radio telecommunications, and as a result it has become difficult to assign bands of sufficient frequency to various radio services. Radio waves are finite and are common assets for use by all humans, and much effort has been expended in attempts to utilize them effectively, such as the development of highly efficient modulation, utilization of both vertical and horizontal polarizations, and improvements in two-dimensional (plane) spectrum utilization efficiency through the use of sharp-beam antennas. However, the demand for frequencies is a significant problem, and solving it will require the utilization of frequency bands above 10 GHz. In order to do so, four issues need to be addressed.

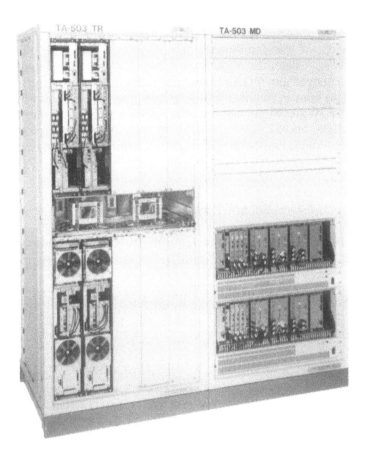

FIG. 18   Equipment of a 256QAM microwave digital radio system.

1. The problem of severe attenuation caused by water vapor and oxygen
2. Restrictions on repeater spacing and lack of economy
3. Accumulation of thermal noise and distortion of repeaters
4. The problem of interference caused by an excessive number of repeater stations

It is hoped that these difficulties can be overcome and that these substantially broadband transmission media can be utilized in large-capacity radio-relay systems. Digital modulation can avoid an accumulation of interference and distortion, high gain is obtained easily for higher frequency band antennas, and the latest remarkable development of solid-state circuits can make the repeaters economical. The early 1960s saw the development of a new quasi-millimeter band and the resultant development of large-capacity digital radio systems used for commercial services in several countries.

Propagation Characteristics above 10 Gigahertz

Radio waves are absorbed or scattered by precipitation like rain or snow. In frequency bands of quasi-millimeter or millimeter wavelength, attenuation is caused mainly by oxygen or water vapor. Figure 19 shows attenuation due to gaseous constituents and precipitation that impede transmission through the atmosphere. As seen in the illustration, oxygen has an isolated absorption line at 118.74 GHz. Water vapor has three absorption lines at frequencies of 22.2 GHz, 183.3 GHz, and 325 GHz. Outside of these absorption frequencies, rain becomes a dominant attenuation factor and determines the repeater spacing for radio links.

Rainfall Attenuation Probability Distribution Estimation

The relationship between attenuation γ_R in decibels per kilometer (dB/km) and rain rate R in millimeters per hour (mm/h) can be approximated by the power law (39)

$$\gamma_R = kR^\alpha \tag{1}$$

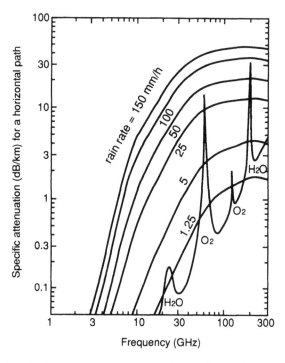

FIG. 19   Attenuation due to gaseous constituents and precipitation (temperature = 20°C; pressure = sea level = 1 atmosphere; water vapor = 7.5 grams/square meter).

Based on the assumption of spherical drops, values of k and α have been calculated at a number of frequencies between 1 and 1000 GHz for several drop temperatures and drop size distributions.

Recent analyses suggest that the rain-rate distribution is better described by a model that approximates a log-normal distribution at low rates, and a gamma distribution at high rates. A practical model that estimates the cumulative distribution of rain rates may be expressed by the equation (40)

$$P(R \geq r) = \frac{\alpha e^{-ur}}{r^{b}} \tag{2}$$

Where α and b are derived from the rainfall rate $R_{0.01}$ exceeded 0.01% of the time and u is a parameter that depends on climate and geographical features. Long-term rain-attenuation statistics can be estimated with Eqs. (1) and (2), taking into account the fact that effective path lengths are affected significantly by heavy rainfall.

Cross-Polarization Discrimination Characteristics during Rainfall

In addition to rainfall attenuation, propagation characteristics during rainfall include cross-polarization discrimination (XPD) deterioration. The main cause of the deterioration is the slanting of raindrops, which is caused by wind. Therefore, the deterioration value in the XPD is considered related to such factors as the rainfall precipitation, raindrop particle distribution, raindrop inclination angle distribution, span length, propagation path inclination, and frequency. A rough estimate of the distribution of XPD can be obtained from a cumulative distribution of the copolarized rain attenuation γ using the equi-probability relation (39)

$$\text{XPD} = U - V(f) \log(\gamma) \, dB \tag{3}$$

where the coefficients U and $V(f)$ are dependent in general on a number of variables and empirical parameters, including frequency f. For line-of-sight paths with small elevation angles and horizontal or vertical polarization, these coefficients may be approximated by

$$U = U_0 + 30 \log f \tag{4}$$

$$V(f) = 20 \tag{5}$$

Sophisticated Digital Radio Systems in Quasi-Millimeter Wave Bands

System Design

In the frequency range above 10 GHz, attenuation due to rain becomes predominant. As a result, there are severe restrictions on repeater spacing when designing systems using this frequency range. Figure 20 shows permissible repeater-spacing frequency dependency, assuming that the circuit outage time due to rain attenuation is 0.1% during an average year. The rapidly decreasing rate becomes rather indefinite above 20 GHz. Permissible repeater spacing decreases from the 50 km required in existing microwave routes to 3 km at 20 GHz. This means that repeater stations can be simplified dramatically. A hypothetical reference circuit is shown in Fig. 21. The standard distance between switching stations is 100 km, with a standard span length of 3 km.

In order to select a modulation scheme, three items must be considered: (1) efficient frequency band utilization, (2) efficient transmitting power utilization, and (3) anti-interference and distortion characteristics. A four-phase PSK modulation and synchronous detection scheme is the most suitable for meeting these requirements.

Repeater equipment should be reliable and economical. These requirements can be fulfilled by employing microwave integrated circuits or monolithic integrated circuits (MICs). These circuits have three advantages: (1) their dimensions are between 1/30 and 1/50 of waveguide circuits; (2) they are very reliable and highly reproducible since they are fabricated by ordinary photolithographic techniques; (3) their production cost is comparatively low since many MICs with the same electrical characteristics can be manufactured simultaneously from one substrate without adjustment.

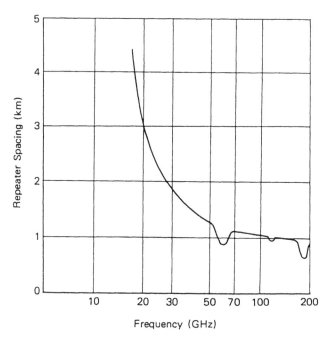

FIG. 20   Permissible repeater spacing in the frequency band above quasimillimeter wave region (circuit outage time due to rain = 0.1% per 2500 km; antenna = 1 mϕ; transmitting power = 200 mW at 200 GHz [−6 dB/octave]).

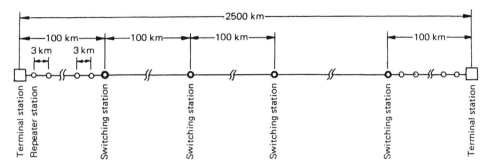

FIG. 21 Hypothetical reference circuit for higher frequency band digital radio systems (link length = 2500 km; switching station spacing = 100 km; standard repeater station spacing = 3 km).

20-Gigahertz Band High-Capacity Digital Radio System

Figure 22 shows an example of a quasi-millimeter system configuration. System parameters are given in Table 4. This system is designed to be capable of carrying 400-Mbps digital signals by 4-phase PSK-coherent detection over a distance of 2500 km using the band of 17.7 ~ 21.2 GHz to meet traffic demands of telephone, data, and video transmissions (40,41). Figure 23 shows a compact repeater station of the type used in this system. The design of repeater stations should take such factors as harmony with the environment, economy, and ease of construction into account because the stations must be erected at about every 3 km along the transmission path. A single-steel-pole structure with a repeater housing on top was chosen for Fig. 23 from among various candidates. The repeaters are encased within a housing, and such maintenance work as replacement of repeater panels can be done inside.

Many MICs have been adopted to increase the compactness of repeater equipment (42,43). FETs were used in this system as high-power and low-noise amplifiers because of their superior performance, simplicity of circuit design, and high efficiency. The contrast between an existing waveguide-type transmitter and the newly developed MIC-type transmitter is shown in Fig. 24.

Future Digital Radio Systems

Combination of Modulation and Coding Techniques

Modulation techniques and coding techniques have been developed individually. In digital radio systems, greater spectrum-efficient modulation is required to increase the transmission capacity of the limited-frequency bands and toward this purpose, highly efficient modulation schemes capable of transmitting 5 bits per hertz have been developed and utilized. However, as the number of modulation levels becomes larger, received signals become susceptible to ther-

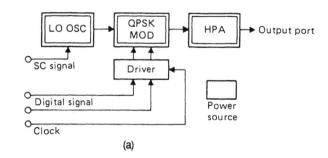

(a)

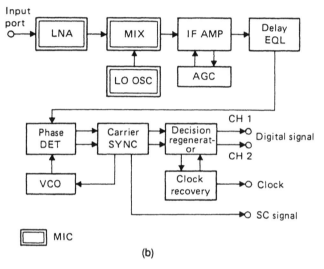

(b)

FIG. 22 Example of quasi-millimeter wave-system configuration: *a*, transmitter; *b*, receiver (HPA = high-power amplifier; VCO = voltage control oscillator).

TABLE 4   Main Parameters of 20-GHz-Band 400-Mbps Digital Radio System

Parameters	20G-D2
Number of systems (working + protection)	8 + 1
Transmission Capacity	400 Mbps/RF channel
	3.2 Gbps/route
Modulation	QPSK
Clock frequency	198.9894 MHz
Unavailability	0.3%/2500 km
Repeater station spacing	6 km (nominal)
Bit error rate of system switching (required C/N)	10^{-4} (11.8 dB)
Equivalent C/N degradation	2 dB
Transmitting power	400 mW (26 dBm)
Noise figure	5.5 dB
Standard antenna	2.4 m (diameter)

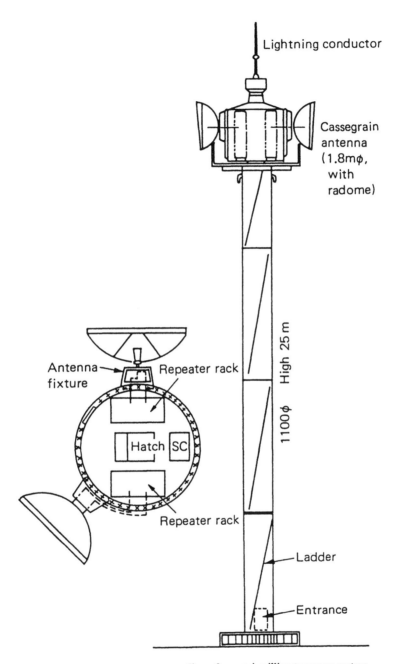

FIG. 23   Repeater tower outline of a quasi-millimeter wave system.

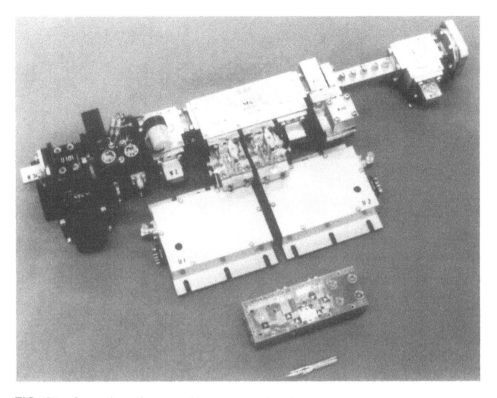

FIG. 24 Comparison of a waveguide-type transmitter (upper) and a transmitter of the monolithic integrated circuit (MIC) type (lower).

mal noise or distortion, and therefore it becomes necessary either to increase
the transmitting power or to equip the system with a transversal equalizer.

In attempting to obtain better steady-state performance or to increase the
margin against thermal noise, coding techniques have been developed. Initially,
these techniques were used in satellite communications, and recently they have
become an indispensable technology in terrestrial radio communications.

Coding techniques have the inherent problem in that redundant bits rapidly
increase with coding gain, and frequency utilization efficiency is decreased
greatly as a result. To attain high coding gain without decreasing frequency
utilization efficiency, detecting and correcting must be done over a great number
of pulses, which results in a marked increase in circuit size and signal throughput
delay.

In multilevel modulation, it is most important to maximize the distance
between adjacent signal points on signal space diagrams, that is, the Euclidean
distance. In contrast, maximizing the Humming distance is most important in
coding technology, and manipulation is carried out in single-dimension space.
Ungerboeck combined both techniques and invented a coded modulation technique, *trellis-coded modulation* (TCM) (44,45). If comparison between a four-
state trellis-coded 8PSK and a conventional 4PSK is made, the former needs 3

dB more C/N than the latter, but the TCM produces 6 dB of gain, so the net result is that TCM is superior to existing techniques by 3 dB (45). TCM was invented only recently and now is being applied only in small-capacity cable transmission. However, it is expected to be used widely in digital radio systems in the near future.

Improvement of Frequency Utilization Efficiency

Cross-polarization and one-frequency repeating are two measures that are expected to bring about improved frequency utilization efficiency in the near future.

Cross-Polarization

Frequency can be utilized in duplicate by utilizing cross-polarized waves. An ordinary antenna has an XPD ratio of 30 to 40 dB, and interference between both polarized waves is of no concern during steady-state propagation. However, XPD is degraded severely during heavy rainfall or during multipath fading and interference is increased (46). Both polarized signals are received simultaneously at a station and a replica of the interference signal can be produced by an orthogonal polarized signal in the transmission of dual polarized waves. In this way, interference can be canceled by adding the replica to the signal concerned (37,47,48).

Degradation caused by rainfall is not difficult to deal with since it does not vary with frequency in the transmission bandwidth, but degradation caused by multipath fading is a rather troublesome problem because it does vary. Therefore, even if a reference signal can be used, it is insufficient to cancel cross-polarized interference by adjusting only the amplitude and phase; they must be adjusted over a wide frequency range. Figure 25 shows one of the most advanced cross-pole-interference cancelers, in which a transversal filter is used to obtain the necessary canceling spectrum.

One-Frequency Repeating

In existing radio-relay systems, different frequencies are used for transmitting and receiving in order to prevent transmitters and receivers at the same station from interfering with each other. If the same frequencies could be used for both transmitting and receiving at the same station, frequency utilization efficiency would be doubled. Figure 26-*a* shows the existing frequency allotment to repeater stations; in this case, the transmitting and receiving frequencies are converted reciprocally. Figure 26-*b* shows the frequency allotment of one-frequency repeating in which the combination of interference routes increases drastically and the use of an interference canceler becomes indispensable. The canceler is required to be able to (1) cancel long-delayed transmitting signals; (2) cancel

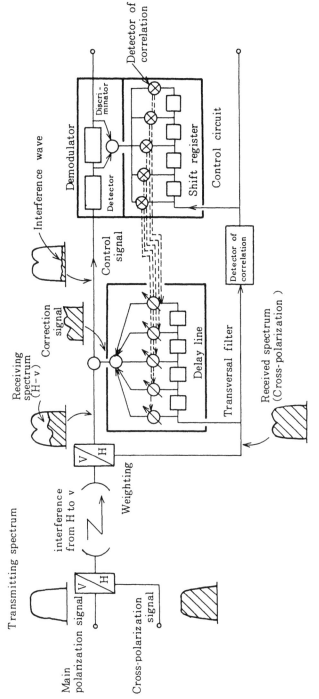

FIG. 25 Cross-pole-interference canceler using transversal filter.

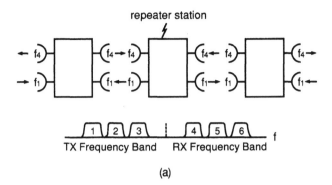

(a)

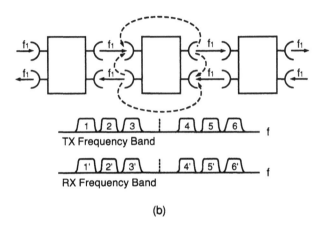

(b)

FIG. 26 Frequency arrangements for digital radio systems: *a*, existing frequency arrangement for dual-frequency repeating; *b*, frequency arrangements for one-frequency repeating (TX = transmitting; RX = receiving).

multiple interference signals; and (3) cancel interference with a dynamic range of 40 dB or more that varies rapidly.

Recently, much progress has been made in the development of a number of such excellent cross-pole-interference cancelers as the transversal-type canceler. These new developments promise to clear the way to full-scale utilization of one-frequency repetition systems.

Flexible Networks

The development of digital radio-relay systems has been pursued in earnest because of the inherently superior feature they possess: thermal noise or interference is not accumulated through the repeater stations of the systems. In digital systems, however, repeater stations become very complicated. In the newest microwave digital radio-relay systems, they also have become very ex-

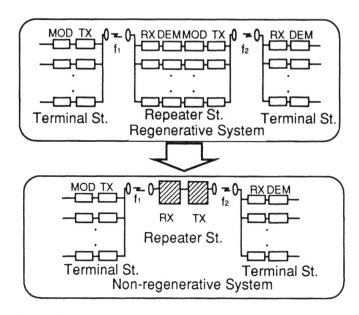

FIG. 27   Comparison of regenerative and nonregenerative repeating system configurations.

pensive with the added features of multilevel modulation, automatic equalization, and FEC. In most cases involving such systems, rainfall heavy enough to cause significant fading will occur only within one repeater span and circuit performance quality of other spans will be affected very little. Therefore, it becomes rather redundant to regenerate and reshape signals at every station. Recently, a hybrid repeating system in which nonregenerative repeating and generative repeating are combined has been proposed and examined (49). Figure 27 shows the configuration of a nonregenerative transmission system in comparison with that of a regenerative system.

Bibliography

Feher, K. *Digital Communications*, Prentice-Hall, Englewood Cliffs, NJ, 1981.
Longton, A. C., DR 18-A High Speed QPSK System at 18 GHz, *ICC Conf. Record*, 18.14–18.17 (June 1976).
Lucky, R. W., Salz, J., and Weldon, E. J. *Principles of Data Communication*, McGraw-Hill, New York, 1967.
Stein, S., and Jones, J. J. *Modern Communication Principles with Application to Digital Signaling*, McGraw-Hill, New York, 1967.

References

1.   Yamamoto, H., et al., Future Trends in Microwave Digital Radio, *IEEE Commun.*, 40–52 (February 1987).

2. Taylor, D. P., and Hartmann, P. R., Telecommunications by Microwave Digital Radio, *IEEE Commun.* 24(8):11–16 (August 1986).

3. Saito, Y. et al., Feasibility Considerations of High-Level QAM Multi-Carrier System, *ICC Conf. Record,* 665–671 (May 1984).

4. Daido, Y. et al., 256 QAM Modem for High Capacity Digital Radio Systems, *GLOBCOM '84,* 547–551 (November 1984).

5. Ichikawa, H., Sango, J., Tanaka, K., and Shindo, Y., Digital Radio System Design with the Network Node Interface, *ICC '89,* 42.4 (June 1989).

6. Giger, A. J., and Barnett, W. T., Effects of Multipath Propagation on Digital Radio, *IEEE Trans. Commun.,* COM-29(9):1345–1351 (September 1981).

7. Komaki, S., Horikawa, I., Morita, K., and Okamoto, Y., Characteristics of a High Capacity 16 QAM Digital Radio System in Multipath Fading, *IEEE Trans. Commun.,* COM-27(12):1854–1861 (December 1979).

8. Rummler, W. D., A New Selective Fading Model: Application to Propagation Data, *Bell Sys. Tech. J.,* 58(5):1037–1071 (May–June 1979).

9. Rummler, W. D., A Statistical Model of Multipath Fading on a Space Diversity Radio Channel, *Bell Sys. Tech. J.,* 61(9):2185–2219 (November 1982).

10. Rummler, W. D., More on the Multipath Fading Channel, *IEEE Trans. Commun.,* COM-29(3):346–352 (March 1981).

11. Rummler, W. D., Coutts, R. P., and Liniger, M., Multipath Fading Channel Models for Microwave Digital Radio, *IEEE Commun.,* 24 (11):30–42 (November 1986).

12. Murase, T., et al., 200Mb/s 16QAM Digital Radio System with New Countermeasure Techniques for Multipath Fading, *ICC Conf. Record,* 46.1.1–5.1 (June 1981).

13. Sari, H., et al., Receiver Techniques for Microwave Digital Radio, *IEEE Commun.,* 24(11):43–54 (November 1986).

14. Lucky, R. W., Automatic Equalization for Digital Communications, *Bell Sys. Tech. J.,* 44:547–588 (April 1965).

15. Araki, M., Time Domain Equalizer Performance in the Non-Minimum Phase Shift Fading Channel, *Trans. IECE Japan,* E66(11):671–677 (November 1983).

16. Taylor, D. P., and Shafi, M., Decision Feedback Equalization for Multipath Induced Interference in Digital Microwave LOS Link, *IEEE Trans. Commun.,* COM-32:267–279 (March 1984).

17. Leclert, A., and Vandamme, P., Non-Minimum Phase Fading Effects on Equalization Techniques in Digital Radio Systems, *GLOBCOM '83* 1(1.2):8–12.

18. Barnett, W. T., Microwave Line-of-Sight Propagation With and Without Frequency Diversity, *Bell Sys. Tech. J.,* 49(8):1827–1871 (October 1970).

19. Vigants, A., Space-Diversity Engineering, *Bell Sys. Tech. J.,* 54(1):103–142 (January 1975).

20. Komaki, S., Tajima, K., and Okamoto, Y., A Minimum Dispersion Combiner for High Capacity Digital Microwave Radio, *IEEE Trans. Commun.,* COM-32(4): 419–428 (April 1984).

21. Lin, E. H., Giger, A. J., and Alley, G. D., Angle Diversity on Line of Sight Microwave Paths Using Dual Beam Dish Antennas, *ICC Conf. Record,* 23.5: 831–834 (1987).

22. Decan, P. M., Berg, J. H., and Evans, M., Aperture Diversity Using Similar Antennas, *ICC Conf. Record,* 23.6.1–23.6.4 (1987).

23. Balaban, P., Sweedyk, E., and Axeling, G., Generalized Angle Diversity: Model and Experimental Results, *ICC Conf. Record,* paper 23.7 (1987).

24. Lin, S. H., Lee, T. C., and Gardina, M. F., Diversity Protections for Digital Radio — Summary of Ten-Year Experiments and Studies, *IEEE Commun.,* 26(2): 51–64 (February 1988).

25. Cellerino, G., Davino, P., and Moreno, L., Frequency Diversity Protection in Digital Radio with Hitless Switching, *ICC Conf. Record*, 1513–1517 (1985).

26. Lee, T. C., and Lin, S. H., More on Frequency Diversity for Digital Radio, *GLOBCOM '85*, 3:1108–1112 (December 1985).

27. Dirner, P. L., and Lin, S. H., Measured Frequency Diversity Improvement for Digital Radio, *IEEE Trans. Commun.*, COM-33(1):106–109 (January 1985).

28. Noguchi, T., Daido, Y., and Nossek, J. A., Modulation Techniques for Microwave Digital Radio, *IEEE Commun.*, 24(10):21–30 (October 1986).

29. Emshwiller, M., Characterization of the Performance of PSK Digital Radio Transmission in the Presence of Multipath Fading, *ICC Conf. Record*, 3:47.3.1–47.3.6 (1978).

30. Anderson, C., Barber, J., and Patel, R., The Effect of Selective Fading on Digital Radio, *IEEE Trans. Commun.*, COM-27:1870–1876 (December 1979).

31. Sakagami, S., and Hosoya, Y., Some Experimental Results on In-Band Amplitude Dispersion and a Method for Estimating In-Band Linear Amplitude Dispersion, *IEEE Trans. on Commun.*, COM-30(8):1875–1888 (August 1982).

32. Greenstein, L. J., and Shafi, M., Outage Calculation Methods for Microwave Digital Radio, *IEEE Commun.*, 25(2):30–39 (February 1987).

33. Jakes, W. C., Jr., An Approximate Method to Estimate an Upper Bound on the Effect of Multipath Delay Distortion on Digital Transmission, *ICC Conf. Record*, 3:47.1.1–47.1.5 (1978).

34. Greenstein, L. J., and Czekaj, B. A., A Polynomial Model for Multipath Fading Channel Responses, *Bell Sys. Tech. J.*, 59(7):1197–1226 (September 1980).

35. Cambell, J. C., and Coutts, R. P., Outage Prediction of Digital Radio Systems, *Electronics Letters*, 18(25/26):1071–1072 (December 9, 1982).

36. Murase, T., and Morita, K., Analysis of Multipath Outage in the Presence of Thermal Noise and Interference on Multi-Level Modulation Systems, *Trans. IECE Japan*, E68(12):844–851 (December 1985).

37. Matsue, H., Ohtsuka, H., and Murase, T., Digitalized Cross-Polarization Interference Canceller for Multilevel Digital Radio, *IEEE Sel. Areas Commun.*, (3):493–501 (April 1987).

38. Imai, N., Nojima, T., and Murase, T., Novel Linearizer Using Balanced Circulators and Its Application to Multilevel Digital Radio System, *IEEE Trans. MTT*, 37(8):1237–1243 (1989).

39. Olsen, R. L., Rogers, D. V., and Hoge, D. B., The aR[b] Relation in the Calculation of Rain Attenuation, *IEEE Trans. Ant. Prop.*, AP-26(2):318–329 (1978).

40. Nakamura, Y., and Yosikawa, T., 20 GHz Digital Radio-Relay System, *ICC Conf. Record*, 88–92 (1977).

41. Yamamoto, H., Kohiyama, K., and Morita, K., 400-Mb/s QPSK Repeater for 20-GHz Digital Radio-Relay System, *IEEE Trans. MTT*, MTT-23(4):334–341 (April 1975).

42. Ogawa, H., Yamamoto, K., and Imai, N., A 26-GHz High-Performance MIC Transmitter/Receiver for Digital Radio Subscriber Systems, *IEEE Trans. Microwave Theory Tech.*, MTT-32:1551–1556 (1984).

43. Yamamoto, K., Ogawa, H., and Imai, N., Design and Performance of Transmitter/Receiver for 20G-D2 System, *Review of ECL*, 32:1083–1091 (1984).

44. Ungerboeck, G., Channel Coding with Multilevel/Phase Signals, *IEEE Trans. Inform. Theory*, IT-28:55–67 (January 1982).

45. Ungerboeck, G., Trellis-Coded Modulation with Redundant Signal Sets, Part I, Part II, *IEEE Commun.*, 25(2):5–21 (February 1987).

46. Cronin, P. H., Dual Polarized Digital Radio Operation in a Fading Environment, *ICC Conf. Record*, 52.5.1–52.5.6 (1980).

47. Namiki, J., and Takahara, S., Adaptive Receiver for Cross-Polarized Digital Transmission, *ICC Conf. Record*, paper 39.6 (1981).
48. Kavehrad, M., Adaptive Cross-Polarization Interference Cancellation for Dual-Polarized M-QAM Signal, *ICC Conf. Record*, paper 29.7 (1983).
49. Watanabe, K., Ohtsuka, H., and Kagami, O., A Non-Regenerative Repeating Digital Microwave Radio System, *GLOBCOM '91*, 1:1812–1816 (December 1991).

TAKEHIRO MURASE

Digital Signal Processing

What Is Digital Signal Processing?

Digital signal processing (DSP) is the mathematical manipulation of a numerical sequence that represents a digital signal. This definition is extremely broad and so are the potential areas for the application of DSP technology. As hardware implementations continue to increase in performance and decrease in price we will see wider use of DSP in such mainstream applications as speech synthesis, speech compression, and image compression and manipulation. Already DSP technology is being used in many communications devices, consumer stereo products, automobiles, and even toys.

In the past, all signal processing was done with analog circuitry. Electrical signals were added, subtracted, mixed, or filtered using entirely analog systems. But with the advent of computers and special-purpose digital technologies, signal processing has entered and flourished in the digital domain. There are many advantages to processing signals with digital rather than analog systems. These advantages include improved quality, reproducibility, predictability, and stability.

Signals processed with digital techniques can retain superior signal quality. Digital filters can be implemented that have very precise transition regions and sharp cutoff responses in the stop bands. In addition, digital filters may have linear phase, unlike analog filters that introduce nonlinear distortions. The linear-phase response is particularly attractive in communications applications. Finally, digital implementation of such complex operations as equalization of telecommunications signals is straightforward.

Digital processing is completely reproducible due to its precise numerical implementation. The processing can be performed many times on the same input set and always will yield the same result: during both development and realization of digital algorithms, system variability can be removed, allowing the engineer to focus on the problem at hand.

DSP systems are entirely predictable. Given that the system is a numerical implementation, the system's behavior can be described mathematically and predicted before the system is realized. Prototype systems can be developed on general-purpose computers and once the design is proven, then the algorithms can be implemented on higher speed special-purpose systems.

Digital processing systems are not vulnerable to the effects of temperature or noise. Analog components typically experience drift over time and temperature. Analog systems also may be susceptible to noise from the power system and other sources. Digital systems are immune to such instabilities.

This article addresses the following aspects of DSP:

- Signal and system fundamentals
- Digitization of signals
- Digital transforms

- Digital filters
- DSP applications
- DSP implementations

A Review of Signals and Linear Systems

Linear system theory provides the mathematical foundation for both analog and digital signal processing. This section provides a brief review of terminology and notations that are utilized in subsequent sections. This section begins with a basic introduction to signals and defines two signals fundamental to DSP theory: the unit step function and the unit impulse (or delta) function. Systems concepts then are introduced, focusing on a specific category of systems, linear time-invariant (LTI) systems. LTI systems possess properties that are fundamental to the operation of many DSP algorithms. Then, the convolution operation is defined and finally the Fourier series and transforms are introduced.

Signals

Signals are used to represent a wide variety of physical phenomena mathematically—ranging from the way the voltage varies across an electrical terminal to describing the acoustic pressure variation caused by speech. Signals can be divided into two distinct types: continuous and discrete. A *continuous-time signal* (Fig. 1) is one that can be represented as a function of one or more independent continuous variables. For example, a musical instrument creates acoustic vibrations that vary continuously with time. A *discrete-time signal* (Fig. 2) can be represented as a function of one or more independent discrete-time variables. Examples of discrete signals might include the daily noontime temperature taken at the local airport or your gross annual income. In the first case, the discrete-time increment is one day; in the second, it is one year.

The unit impulse function is fundamental in linear system analysis and DSP. The unit impulse is derived from the unit step function $u(t)$:

$$u(t) = \begin{array}{ll} 0, & t < 0 \\ 1, & t > 0 \end{array}$$

The unit impulse function is written as $\delta(t)$ and obtained by taking the "derivative" of the unit step function.

$$\delta(t) = \frac{dU(t)}{dt} \qquad (1)$$

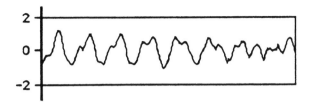

FIG. 1 Continuous-time signal.

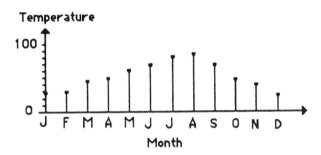

FIG. 2 Discrete-time signal.

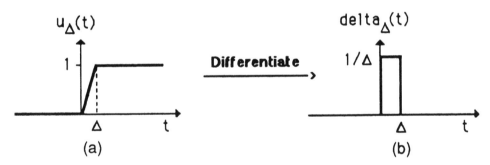

FIG. 3 Approximation of the unit step function and its derivative.

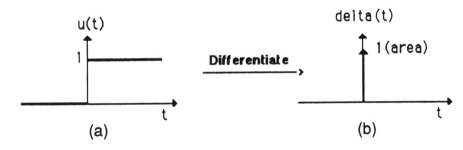

FIG. 4 Unit step function and its derivative.

The unit step function is discontinuous and therefore not differentiable. The unit impulse, however, can be understood in the limiting sense. The approximation of the unit step signal is a continuous form and the resulting derivative is shown in Fig. 3.

As the slope in the region of the approximated discontinuity increases, the width (or duration) of the impulse decreases. As the slope approaches infinity, the pulse duration approaches zero. The area under the pulse always is equal to one (hence the term *unit impulse*). As the width approaches zero, the height approaches infinity (see Fig. 4).

A continuous signal can be expressed as the summation of a series of shifted, weighted impulses spaced infinitely close in time:

$$x(t) = \int_{-\infty}^{\infty} x(\tau) \, \delta(t - \tau) \, d\tau \qquad (2)$$

Systems

A *system* is any process that takes a signal as an input and outputs a signal that relates to that input. A kazoo is an example of a system (Fig. 5). If you blow into the kazoo, the input is the acoustic signal that you create by blowing into the mouthpiece. The output is the resulting music (or noise). The relationship between the input and the output is defined as the *system transformation*.

The continuous input signal to a system is expressed as $x(t)$, indicating that the input is a continuous function of time. The resulting system output $y(t)$ is also a continuous function of time. The system transform often is referred to as $H(t)$.

Linear Time-Invariant Systems

An important class of systems are linear time-invariant (LTI) systems. LTI systems have the properties of linearity and time invariance.

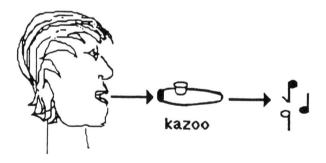

Input x(t) -------> System H(t) ------> output y(t)

FIG. 5 Example of a system.

In regard to linearity, if $x_1(t) \rightarrow y_1(t)$ and $x_2(t) \rightarrow y_2(t)$, then, for a linear system, the inputs may be multiplied by arbitrary constants a and b to produce $ax_1(t) + bx_2(t) \rightarrow ay_1(t) + by_2(t)$.

For time invariance, if $x(t) \rightarrow y(t)$, then $x(t + t_0) \rightarrow y(t + t_0)$, where t_0 is an arbitrary time value. (In a discrete-time system, the notation is changed. The time t is replaced by n, and t_0 becomes n_0 to reinforce the restriction that n and n_0 must be integers in a digital system.)

Since LTI systems are linear, a complex system may be described as the sum of simpler systems. In addition, since the system does not function differently in relation to the time of the arrival of input, system analysis is straightforward. One especially useful property of LTI systems is that their system response to such a complex exponential as $e^{j(\omega + \sigma)t}$ results in an output consisting of the same complex exponential multiplied by an amplitude factor (1, p. 167):

$$e^{j(\omega + \sigma)t} \rightarrow H(j(\omega + \sigma)) * e^{j(\omega + \sigma)t}$$

Hence, if a signal $x(t)$ can be represented as a linear combination of complex exponentials, then the response of an LTI system $y(t)$ can be predicted easily using the principle of superposition. If $x(t) = a_1x_1(t) + a_2x_2(t) + a_3x_3(t)$, then $y(t) = a_1y_1(t) + a_2y_2(t) + a_3y_3(t)$.

Convolution

The response $y(t)$ of an LTI system to a signal $x(t)$ can be characterized by its response to the unit impulse signal. This response is the unit impulse response $h(t)$.

$$y(t) = \int_{-\infty}^{\infty} x(\tau)h(t - \tau)d\tau \tag{3}$$

This equation is the weighted sum of shifted impulse responses to the system. In this case, the weighting factor is the input $x(t)$. Graphically, this equation is represented by reversing the sequence $h(t)$ in time and sliding it to the right, taking infinitesimal time increments. For each time step, the output y is x multiplied by h. The response $y(t)$ is made up of the sum of these products. This process, called *convolution*, often is used to compute the output of a system with known impulse response $h(t)$ to the known input signal $x(t)$.

This integral that defines the system response is referred to as the *convolution integral*. The convolution operation uses the following notation:

$$y(t) = x(t) * h(t) \tag{4}$$

The Fourier Series

The Fourier series is based on the principles of linear system theory. In the early 19th century, Joseph Fourier built upon the work of Euler and Lagrange to conclude that any periodic signal $x(t)$ can be represented as a linear combination of sine and cosine terms:

$$x(t) = \sum_{k=-\infty}^{\infty} a_k \cos(kw_0t) + b_k \sin(kw_0t) \tag{5}$$

Even functions are symmetric about the time-zero axis and can be represented using only the cosine terms. In general, the b_k terms are considered to be zero and the periodic function represented as even.

The terms produced for $k = 1$ and $k = -1$ are considered collectively the fundamental frequency (also the first harmonic) of the signal. The terms produced for $k = \pm 2$, ± 3, and ± 4 are called the second, third, and fourth harmonics, respectively. Each harmonic is an integer multiple of the fundamental frequency. As the sum tends to infinity, the error of the Fourier series approximation will approach zero for finite energy signals. For many signals, the Fourier series representation will be finite.

Intuitively, the ability to represent a periodic signal as a sum of exponentials (sines and cosines) can be understood by considering the principle of superposition. Signal C in Fig. 6 is produced by the linear combination of Signals A and B.

Signals A and B both are sinusoids. When added, they form Signal C. Signal B is the fundamental, or lowest, frequency of C. A is the third harmonic since its frequency is three times the fundamental frequency. Harmonics are always integer multiples of the fundamental frequency. The Fourier series representation of Signal C would be finite because only the first and third harmonics have nonzero amplitudes.

Fourier's introduction of the series expressed the representation as an infinite sum of sines and cosines. It commonly is accepted as more compact notation to use Euler's relations

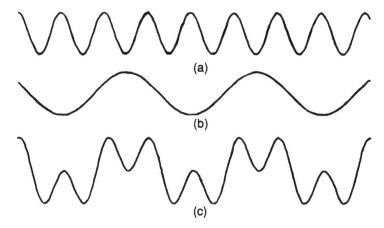

FIG. 6   Addition of two sinusoidal signals (*A* and *B*). Signal *C* is the result.

$$e^{jw_0t} = \cos(w_0t) + j\sin(w_0t) \tag{6}$$

$$\cos(w_0t) = \frac{e^{jw_0t} + e^{-jw_0t}}{2} \tag{7}$$

$$\sin(w_0t) = \frac{e^{jw_0t} + e^{-jw_0t}}{2j} \tag{8}$$

to express the series as an infinite sum of complex exponentials

$$x(t) = \sum_{k=-\infty}^{\infty} a_k e^{jkw_0t} \tag{9}$$

where

$$a_k = \frac{1}{T_0} \int_{T_0} e^{-jkw_0t} \, dt \tag{10}$$

The set of the coefficients a_k comprise the spectrum of the periodic signal.

This representation of a periodic signal as a linear combination of exponentials is especially thought provoking since the response of an LTI system to such a signal may be predicted.

The Fourier Transform

While the Fourier series introduced in the preceding section is a discrete frequency representation of a periodic signal, the Fourier transform is a continuous frequency representation of an aperiodic signal. The Fourier transform is a key ingredient of DSP theory.

The Fourier transform may be understood by considering the Fourier series of a periodic signal as the period of that signal approaches infinity. Recall that the Fourier series representation consists of terms that are harmonics of the fundamental frequency. The lower the fundamental frequency is, the closer together the harmonics will be spaced. As the period of the periodic signal approaches infinity, the spacing of the harmonics approaches zero. An aperiodic signal can be thought of as a periodic signal with an infinite period.

The Fourier transform is expressed as

$$X(w) = \int_{-\infty}^{\infty} x(t)e^{-jwt} \, dt \tag{11}$$

while the inverse Fourier transform is

$$x(t) = \frac{1}{2\pi} \int_{-\infty}^{\infty} X(w)e^{jwt} \, dw \tag{12}$$

The Fourier transform of a rectangular pulse (or "box car") (Fig. 7) is the sinc function (Fig. 8):

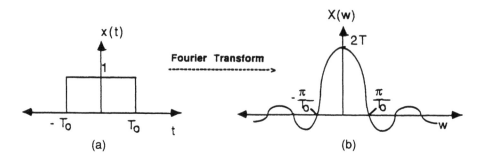

FIG. 7 Rectangular pulse function and its Fourier transform.

$$x(t) = \begin{array}{ll} 1, & |lt| \le T_o \\ 0, & |lt| > T_o \end{array} \quad \xrightarrow[\text{Fourier Transform}]{} \quad X(w) = \frac{\sin{(\pi w)}}{\pi w}$$

Similarly,

$$x(t) = \frac{\sin{(\pi t)}}{\pi t} = \text{sinc } (t) \quad \xrightarrow[\text{Fourier Transform}]{} \quad X(w) = \begin{array}{ll} 1, & |lw| < W \\ 0, & |lw| > W \end{array}$$

Both the box car and the sinc function are aperiodic and hence, as predicted, their Fourier transforms are continuous in the frequency domain.

The Fourier transform of a periodic signal will be a set of discrete spectral components (see Figs. 9–11). The frequency spacing between the lines is determined by the period of the signal being represented. The periodic impulse train is defined as

$$p(t) = \sum_{k=-\infty}^{\infty} \delta (t - kT) \tag{13}$$

The impulse train plays an important role in the discussion of sampling in the next section.

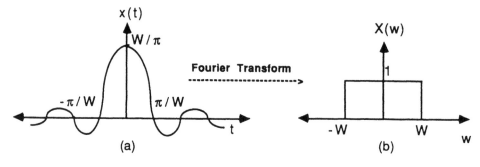

FIG. 8 Sinc function and its Fourier transform.

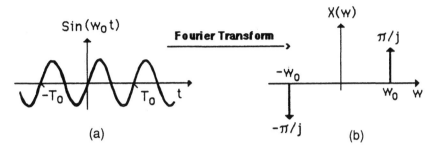

FIG. 9   Sine function and its Fourier transform.

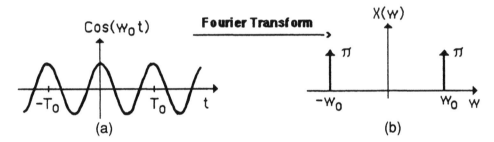

FIG. 10   Cosine function and its Fourier transform.

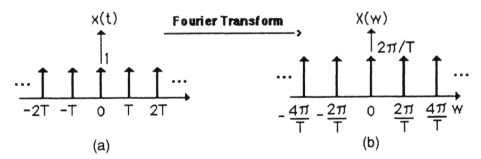

FIG. 11   Pulse train function and its Fourier transform.

Digitization of Signals

DSP builds upon the foundation of linear systems and signals. Applying linear system theory to digital signals creates a set of extremely powerful DSP tools. This section introduces the concept of sampling and explains its mathematical interpretation. The sampling theorem is introduced, and aliasing, which results from violating the sampling theorem, is discussed.

Sampling (Analog-to-Digital and Digital-to-Analog Converters)

The interface between the analog and the digital domain is analog-to-digital (A/D) and digital-to-analog (D/A) circuitry. This circuitry either converts the analog signal to a digital representation or transforms the digital representation to an analog signal. An analog signal is represented as a continuous function over time. A digital representation of an analog signal is a discrete function. It is a quantized representation of the function at evenly spaced time intervals. The process of assigning a discrete quantized value to an analog waveform at a specific point in time is called *sampling* (see Fig. 12) and the collection of data points is referred to as *data samples*. The sampled wave in Fig. 12-*c* is a sequence representing the amplitude of the analog signal at evenly spaced time intervals.

Mathematical Representation of Sampling

Sampling is multiplying the time signal by a series of evenly spaced impulses called an *impulse train*. Each impulse function results in a sample. A sampled

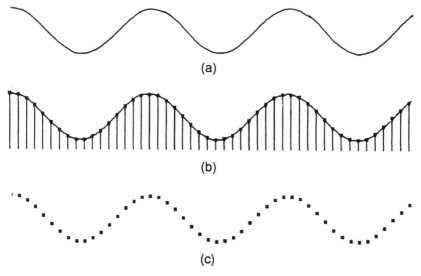

(a)

(b)

(c)

FIG. 12   Sampling of an analog signal: *a*, continuous wave; *b*, sampling the wave; *c*, sampled wave.

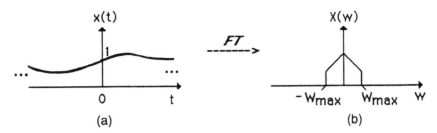

FIG. 13 Band-limited signal in the time and frequency domains.

waveform can be represented as a signal $x(t)$ multiplied by the impulse train $p(t)$

$$X_{\text{sampled}}(t) = x(t)\, p(t)$$

$$= x(t) \sum_{k=-\infty}^{\infty} \delta\,(t - kT) \tag{14}$$

Sampling is equivalent to multiplying the time signal by an impulse train in which the spacing of the impulses corresponds to the sampling period. Multiplication in the time domain corresponds to convolution in the frequency domain:

$$y(t) = p(t)\, x(t) \rightarrow Y(w) = 1/2\pi P(w) * X(w)$$

The signal $x(t)$ is band limited since its highest frequency component is finite. That highest frequency is labeled W_{max} and is shown on the frequency-domain representation of $X(w)$ (Fig. 13). Note that real signals have both positive and "negative" frequency components.

The time impulse train transforms into a frequency impulse train (Fig. 14). The frequency spacing of the frequency-domain impulse train is inversely proportional to the spacing of the impulse train in the time domain. Therefore, the lower the sampling period in the time domain (and the higher the sample rate), the farther apart the impulses will be spaced in the frequency domain. The

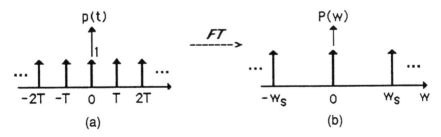

FIG. 14 Impulse train and its Fourier transform.

sampling interval of the pulse train has a period that is labeled T; this period corresponds to a frequency and sampling rate of w_s in the frequency domain.

The sampled time signal is shown in Fig. 15. The frequency-domain representation is derived by convolving the frequency pulse train with the Fourier transform of the original signal. The result is a periodic representation in which the Fourier transform is replicated once for each pulse in the pulse train. The frequency spacing (along the frequency or w axis) is determined by the sampling frequency w_s. Note again that the sampling frequency w_s is the frequency spacing of the impulse train in the frequency domain.

Sampling Theorem

The choice of sampling frequency of an analog signal is dependent on the characteristics of the signal being sampled or digitized. If the signal is band-limited, then the signal can be digitized and then perfectly reconstructed. A band-limited signal, by definition, occupies only a portion or band of the frequency spectrum. The sampling theorem states that the band-limited signal can be reconstructed perfectly if it is sampled at a rate of a least twice the signal's highest frequency. More simply, there must be at least two samples per cycle to reconstruct a periodic signal.

As stated by Oppenheim, Willsky, and Young:

Sampling Theorem:

Let $x(t)$ be a bandlimited signal with $X(w) = 0$ for $|w| > w_m$. Then $x(t)$ is uniquely determined by its samples $x(nT)$, $n = 0, \pm 1, \pm 2 \ldots$ if

$$w_s > 2w_m$$

where

$$w_s = (2\pi)/T.$$

Given these samples, we can reconstruct $x(t)$ by generating a periodic impulse train in which successive impulses have amplitudes that are successive sample values. The impulse train is then processed through an ideal low-pass filter with gain T and cutoff frequency greater than w_m and less than $(w_s - w_m)$. The resulting output signal will exactly equal $x(t)$. (1, p. 519)

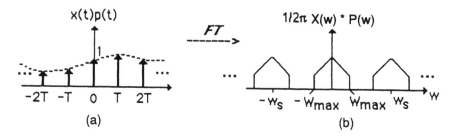

FIG. 15   Sampled time signal and its Fourier transform.

The minimum required sampling frequency also is referred to as the *Nyquist frequency*.

Most real-world signals are band limited naturally. For example, the frequency components of audible music fall below 22.05 kilohertz (kHz). For this reason, compact disks have standardized at a sampling rate of 44.1 kHz. However, a person's voice can be audible and recognizable if sampled at rates significantly less than compact disk rates. The telecommunications industry has standardized at 8-kHz sample rates, providing 4-kHz bandwidth for common voice and modem telephone traffic. In telecommunications applications, low sample rates are desired because the lower the sample rate is, the less data produced and subsequently transmitted.

Aliasing

If the sampling theorem is violated, then the signal's frequencies that are higher than the selected sample rate will not be represented correctly. Instead, they will appear erroneously as lower frequency components. This effect is referred to as *aliasing*.

Aliasing is represented in the frequency domain in Fig. 16. The time impulse train was spaced too widely due to a sampling frequency that was too low. In the frequency domain, the frequency pulse train then is spaced too closely and causes the replicated transforms of the signal to overlap.

To prevent aliasing affects, the analog signal may be filtered before digitization to ensure that the signal is band limited. The filter, often called an anti-alias filter, is a low-pass filter with a cutoff slightly below half the sampling frequency. The analog input signal is processed by the antialias filter first. Then the filtered output may be sampled to convert the signal to its digital representation.

Quantization

An important aspect of any A/D conversion system is the number of bits that are used to represent the voltage level of the sampled waveform. The number

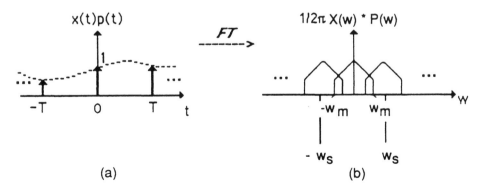

FIG. 16   Effects of aliasing.

of bits refers to the number of binary digits. If a system is an N-bit system, then the number of discrete levels it can represent is $2N$. There is an inherent quantization uncertainty of $\pm \frac{1}{2}$ least significant bit (LSB). This $\pm \frac{1}{2}$ LSB is known as the quantization error and will appear as noise called *quantization noise*.

Digital Transforms

The discussion of digital transforms here makes use of signal and linear system concepts and digitization of signals fundamentals previously introduced. The discrete Fourier transform (DFT), as well as its highly efficient implementation as the fast Fourier transform (FFT) are discussed. Finally, such additional transforms as the Chirp z-transform (CZT) are discussed briefly.

Discrete Fourier Transform

The DFT is analogous to the continuous-time Fourier series. While the Fourier series is the discrete frequency representation of a continuous-time signal, the DFT is the discrete frequency representation of a discrete-time signal.
    Mathematically, the DFT representation is

$$X(k) = \sum_{n=0}^{N-1} x(n)e^{-j(2\pi/N)kn} \tag{15}$$

and the inverse DFT function is (2, p. 14)

$$x(n) = \frac{1}{N} \sum_{k=0}^{N-1} X(k)e^{j(2\pi/N)kn} \tag{16}$$

    The collection of frequency coefficients $X(k)$ is the spectrum of the discrete signal $x(n)$. The coefficients represent the amount of energy that the signal contains at given frequencies.
    The DFT provides a unique, and reversible, representation of periodic- and finite-sequence digital signals. Although the DFT provides an important frequency representation of the signal, the DFT requires so many computations that it is impractical for many applications. A direct implementation of the algorithm requires N^2 multiplications.

Fast Fourier Transform

The FFT is a highly efficient DFT algorithm, especially for sequences with lengths that are a power of 2 (i.e., length equal to 2, 4, 8, 16 . . .). The FFT breaks down the DFT computation into a series of radix-2 DFTs to reduce the number of computations required.

Computing an N-*Point Discrete Fourier Transform*
by Taking Two N/2 *Point Discrete Fourier Transforms*

The N-point DFT algorithm is based on computing an N-point DFT from two $N/2$ point DFTs. The N-point DFT of the finite sequence $x(n)$ defined in Eq. (15) may be written more compactly as

$$X(k) = \sum_{n=0}^{N-1} x(n) \ W^{nk} \tag{17}$$

where $W = e^{-j(2\pi/N)}$ (3, p. 357).

Since W^{nk} is periodic with period N, W often is written as W_N^{nk}. The N-point DFT may be expressed in terms of even and odd n as (3, p. 358)

$$X(k) = \sum_{\substack{n=0 \\ n \text{ even}}}^{N-1} x(n) \ W_N^{nk} + \sum_{\substack{n=0 \\ n \text{ odd}}}^{N-1} x(n) \ W_N^{nk}$$

$$= \sum_{n=0}^{(N/2-1)} x(2n) \ W_N^{2nk} + \sum_{n=0}^{N/2-1} x(2n+1) \ W_N^{(2n+1)k} \tag{18}$$

Alternately, W^2 can be expressed as (3, p. 358)

$$W_N^2 = [e^{j(2\pi/N)}]^2 = e^{j[2\pi/(N/2)]}$$

$$= W_{N/2} \tag{19}$$

If we define $x_1(n)$ and $x_2(n)$ to be the sequences of even and odd terms, respectively, of the input sequence $x(n)$, then the DFT can be written as (3, p. 359)

$$X(k) = \sum_{n=0}^{N/2-1} x_1(n) \ W_{N/2}^{nk} + W_N^k \sum_{n=0}^{N/2-1} x_2(n) \ W_{N/2}^{nk} \tag{20}$$

This N-point DFT can be written as the sum of two $N/2$-point DFTs (3, p. 359):

$$X(k) = X_1(k) + W_N^k X_X(k) \tag{21}$$

Changing the N-point DFT to two $N/2$-point DFTs reduces the number of complex multiplications by a factor of two. The direct DFT calculation of the point sequence would require N^2 complex multiplications, while the calculation of two $N/2$-point DFTs requires 2 times $(N/2)^2$ complex multiplications.

$$2 \, (N/2)^2 = 2 \, (N^2/4) = N^2/2 \qquad (22)$$

Because an N-point DFT can be computed as the sum of two $N/2$-point DFTs, any DFT with a length that is a power of two can be reduced iteratively to a series of radix-2 DFTs. For example, a length 8 DFT may be computed by computing two length 4 DFTs. Each of the length 4 DFTs may be computed by computing two length 2 (radix-2) DFTs.

The Fast Fourier Transform Butterfly Notation

FFT algorithms often are described using a signal flow graph representation as in the diagram below (3, p. 360).

The open circle in the center of the diagram indicates that both addition and subtraction occur. By convention, an arrow accompanied by a multiplier factor denotes the multiplier weight, as seen in the next diagram (3, p. 360).

$$a \xrightarrow{\quad W^k \quad} a \, W^k$$

The multiplier W^k has periodicity of N. A common operation used in the DFT computation is the FFT "butterfly," as shown in Fig. 17 (3, p. 362). N is used to denote the periodicity of W where N equals the length of the original DFT sequence.

Computing Larger Fast Fourier Transforms

A length 8 DFT can be computed by computing 4 radix-2 FFTs. The results of the 4 radix-2 FFTs then are combined using multiple factors called "twiddle

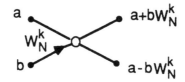

FIG. 17 Fast Fourier transform butterfly notation.

factors" to create the two length 4 DFTs that are combined again using the multiplier factors to generate the length 8 DFT. This process is the FFT. Graphically, this decomposition is illustrated using the butterfly notation shown in Fig. 18 (3, p. 362).

Note that the input order of the time sequence $x(n)$ has been arranged so that the resulting frequency sequence $X(f)$ is in the proper order. This type of prearrangement of the time samples is called a *decimation-in-time FFT*. In essence, the FFT algorithm consists of breaking down the N-point DFT into a series of radix-2 butterflies, the results of which are combined using the twiddle factors. The FFT also may be implemented without prearranging the input sequence. In this case, the resulting output frequency sequence is scrambled. This decimation-in-frequency algorithm is represented in Fig. 19.

Additional Transforms

The FFT is an efficient way to compute the DFT. It is not the only way—in fact, other methods can be more efficient, especially in certain circumstances. Other methods of computing the DFT include the prime factor algorithm (PFA), the doubly odd DFT (odd-time, odd-frequency DFT or O^2DFT), and the odd DFT. While the FFT works best on highly composite numbers (i.e., 16 or 64), the PFA is efficient on a prime number of transformed output points. Or, if the data is real, then the odd DFT may be used. The O^2DFT may be used in the case of symmetrical real data. The odd DFT and the O^2DFT, like the FFT, are computationally simple. Other advanced implementations of the DFT (such as Goertzel's algorithm and the Winograd algorithm) are more complex, but the complexity may be worthwhile for some applications. Still other algorithms, called *partial transforms*, take advantage of the case in which only a subset of the points of the transformed sequence must be computed. Evaluation

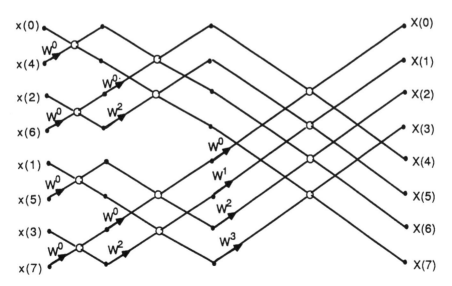

FIG. 18   Decimation-in-time fast Fourier transform.

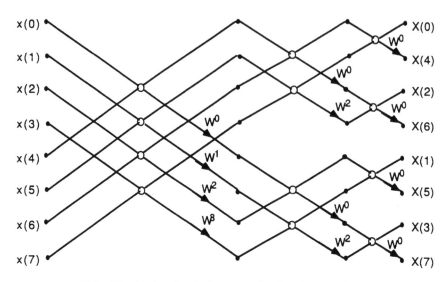

FIG. 19   Decimation-in-frequency fast Fourier transform.

of different algorithms should include assessment of memory required and complexity of operations, not only assessment of the number of required computations.

One additional transform, the Chirp z-transform (CZT), has gained popularity for evaluating the transform of signals with significant sinusoidal components, even though it is not quite as efficient as the FFT. This algorithm provides additional enhancement for spectral lines and may be used to estimate the frequency and amplitude of these small bandwidth signal components better. The CZT may be computed for only the part of the spectrum that is desired, and the number of transform points need not have any relation to the number of time-domain points (unlike the FFT, in which they are required to be the same).

Unlike the transforms previously mentioned that are different implementations of the DFT, the Hilbert transform provides a relationship between the imaginary and real parts of the Fourier transform or the magnitude and phase of the Fourier transform. The Hilbert transform often is used simply to find the real (or imaginary) part of the transform once the imaginary (or real) part is computed.

Digital Filters

Previous sections have discussed the signal and linear system theory upon which DSP techniques are based. Digitization of signals and digital transforms also are introduced in earlier sections. Digital filtering is one of the most powerful DSP functions. This section provides a brief overview of digital filters and then discusses the two chief categories of digital filters: finite impulse response (FIR) and infinite impulse response (IIR). Characteristics and design procedures are summarized. Adaptive filters also are mentioned because of their relevance to DSP. The next section is a discussion of finite-length effects, of interest to anyone using digital filters.

Digital filters change digital signals, for example, by smoothing, predicting, or eliminating unwanted signal components. Digital filters, like digital signals, have many advantages over their analog counterparts. Digital filters do not drift over time or with temperature as analog filters are prone to do. Digital filters can be made to approximate the ideal filter as closely as desired. The digital filter designer can control the accuracy of the filter precisely. Linear-phase characteristics are possible. Usually the difficulty or expense of translating the designed filter into reality is quite reasonable. Digital filters also are better than analog filters in that such complex operations as adaptive filtering are relatively straightforward with a digital implementation.

A digital filter, like an analog filter, typically has a pass band, stop band, and transition band. The pass band refers to the part of the spectrum in which the frequencies are more or less "passed through" the system. The stop-band frequencies are blocked to some extent from appearing in the output signal. The transition region connects the pass-band and the stop-band areas. Filters may have multiple pass bands and stop bands (and therefore transition regions). Note that, unlike an analog filter, a digital filter is specified fully by describing the behavior between zero frequencies and half the sampling frequency w_s. Since $w_s = (2\pi)/T$, then the highest frequency to be described is π/T (see Fig. 20).

Figure 20 illustrates four common filter types: low pass, high pass, band pass, and band stop. The filters shown are "ideal" since (1) the gain of the pass-band regions is exactly one, (2) the transition region's gain is zero, and (3) the stop-band region's gain is exactly zero. These conditions will not be exactly met in a finite-length filter, but a filter (of arbitrary length) may be made that is arbitrarily close to these ideals. In each case, the frequency regions that are passed through the ideal filter undistorted are shaded in Fig. 20. The other regions are blocked by the filter.

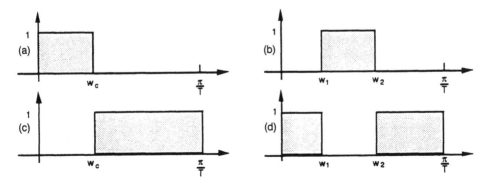

FIG. 20   Frequency responses of ideal filters: *a*, low pass; *b*, band pass; *c*, high pass; *d*, band stop.

A linear digital filter may be defined as a linear combination of past outputs and input signal values:

$$y(n) = \Sigma \, a_k * y(n - 1) + \Sigma \, b_k * x(n) \qquad (23)$$

For this article, we assume that a_k and b_k do not vary with time (that is, they are time-invariant). Time-variant filters are used less frequently. Eq. (23) describes the time behavior of an output signal. We also can examine the pole-zero response form $H(z)$ where $H(z) = y(n)/x(n)$. $H(z)$ often is analyzed to gain information about the stability and steady-state operation of the filter.

The number of coefficients (counting the number of a and b coefficients) gives the number of taps the filter has. This is also known as the *filter order*. This number is used to compare different size filters.

A linear digital filter may be classified as an FIR or IIR filter depending on whether a linear combination of past outputs are used as well as past inputs. FIR filters depend only on past inputs (thus the coefficients a_k are all zero), while IIR filters depend on both past inputs and outputs. Figure 21 illustrates a 5-tap FIR filter, while Fig. 22 shows an IIR filter.

Both FIR and IIR filters have distinct advantages and disadvantages. FIR filters are notable because they may be designed to have a linear-phase characteristic (that is, they introduce no phase distortion in output) and are stable (since no feedback is involved). However, FIR filters generally require more computations, perhaps even a magnitude more computations, than the same performance IIR filter. Also, if the design is transformed, they do not usually keep their linear characteristic. (The exception is the low-to-high-pass transform.) IIR filters are more efficient, but their design is generally more complex and they may have such undesirable characteristics as instability or nonlinear phase.

If computation time is clearly not an issue, then the ease of design of the FIR filters should make them an obvious choice. IIR filters excel particularly when clean, sharp transition regions are needed. Because FIR filters are so much easier to design (and do not have unknown side effects like instabilities or phase distortion), they often are selected. IIR filters may have less delay and therefore can be used in more real-time applications.

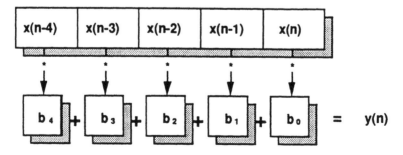

FIG. 21   Five-tap finite impulse-response filter.

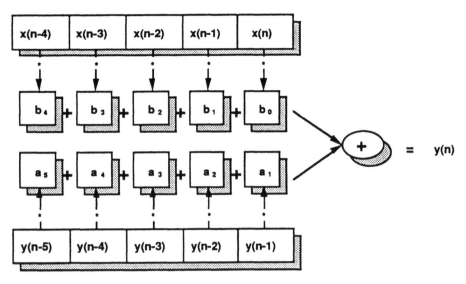

FIG. 22   Infinite impulse-response filter.

Finite Impulse-Response Filters

FIR filters are nonrecursive; that is, only present and past inputs are used. No past outputs are used and there is no feedback as with IIR filters. FIR filters have finite memory since their output depends on only a limited number of past inputs.

There are many different ways to design an FIR filter. One method is the frequency-domain approach. In this method, the desired frequency response is known and plotted in the frequency domain with the desired cutoff and transition areas. The time-domain coefficients corresponding to this filter then are computed. The designer may truncate the coefficients to the number desired. Finally, the coefficients are multiplied by a "window function" to improve the performance of the filter. Window functions that commonly are used include the rectangular, Hamming (and related Hanning), and Kaiser windows. References 3 and 4 provide additional details on these window types.

The final multiplication by the windowing function is important since it smooths out the ripples in the frequency response that were created when the coefficients of the filter with ideal cutoff characters were truncated. (*Gibbs phenomenon* is the term describing the ripples introduced when the coefficients are truncated.) Ripples are not desirable since the rippling is amplified and the signal may be distorted unsatisfactorily if the signal is passed through many such filters. Monotonic filters have the characteristic that they change in only one direction (increasing or decreasing) throughout the entire frequency range. Thus, they have the desirable property of not having ripples.

Another method of optimal FIR linear filter design is the Remez exchange algorithm. The design method utilizes the theory that optimal linear-phase filter function may be written as a Chebyshev approximation $P(e^{jw})$. $P(e^{jw})$ is the sum of r independent, weighted cosine functions with initially unknown weights of alpha:

$$P(e^{jw}) = \sum_{n=0}^{r-1} \text{alpha}(n) \cos(wn) \qquad (24)$$

The filter designer must specify the desired frequency response $D(e^{jw})$, the number of bands, and the filter type. The design relation is formed as

$$W(e^{jw_k}) D(e^{jw_k}) = P(e^{jw_k}) = (-1)^k \partial \qquad \text{for } k = 0 \text{ to } r \qquad (25)$$

where ∂ is the magnitude of the error function. The Remez exchange algorithm solves this set of equations such that the best approximation is found. The DFT then is used to find the weighting values, alpha, for the approximation, from which the filter's impulse response is determined. An excellent discussion of this algorithm, as well as linear programming and other techniques for optimal FIR filter design may be found in Ref. 3. This reference also includes in its appendix Fortran language code to perform optimal filter design and the Remez exchange algorithm in subroutine form.

The designer needs to consider the tradeoff between performance and efficiency. Fewer coefficients, while requiring fewer computations, generally result in a less steep transition region and more ripples. There is a simple relationship between number of coefficients N and performance:

$$N = (\tfrac{2}{3}) * \log\left(\frac{1}{(10 * \partial_p * \partial_s) * (f_s/\Delta f)}\right) \qquad (26)$$

where
    ∂_p = pass-band ripple
    ∂_s = stop-band ripple
    f_s = sampling frequency
    Δf = transition band

Note that this relationship is independent of pass-band width (4, p. 129). This approximation can provide a designer with a rough idea of what to expect for a required filter order.

Infinite Impulse-Response Filters

IIR filters are recursive; that is, their output is computed using both inputs and past output values. One way to design an IIR filter is to use one of the excellent filter design software packages available. The user specifies the filter type and performance desired. The package may use linear programming techniques to arrive at an optimum filter design. One particularly good reference book, *Programs for Digital Signal Processing*, published by the Institute of Electrical and

Electronics Engineers (IEEE) Press, describes and provides the software listing for many DSP functions, including six IIR filter design routines (5). Other programs are available as well. This is quite a bit easier than the traditional IIR design method described next.

A traditional method of IIR filter design is patterned after analog filter design methods. Butterworth, Chebyshev, and elliptic are three widely used types of IIR filters. These filters have long been a mainstay for analog filter design. Since filter designers may be familiar with them, they may choose to create digital IIR filters by constructing the analog filter form, and then transforming it into the digital domain.

Butterworth filters are monotonic filters; they either increase or decrease monotonically throughout the frequency range. A Butterworth filter has the form

$$|H(jw)|^2 = \frac{1}{1 + (w/w_c)^{2N}} \qquad (27)$$

where w_c is the cutoff frequency and N is the order of filter. Figure 23 illustrates the frequency response for Butterworth filters of different order.*

A 10th-order low-pass Butterworth filter appears in Fig. 24. These plots simultaneously show the frequency response of the filter and the effect of filtering on a sinusoidal signal with background noise. With a cutoff frequency of 4 kHz, this 10th-order Butterworth filter effectively filters all input components above 5 kHz in frequency.

Butterworth filters may exceed performance specifications at the lower and upper frequencies. A Chebyshev filter may be appropriate. An IIR filter of the

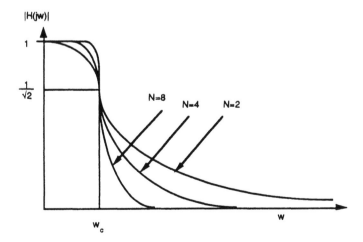

FIG. 23   Butterworth filters.

*The spectral displays of Figs. 23–26 were produced by MacDSP Signal Analysis Software from Spectral Innovations, Inc. (Santa Clara, CA).

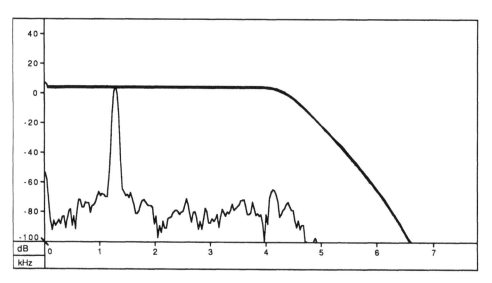

FIG. 24   Tenth-order low-pass Butterworth filter.

Chebyshev form distributes error across a larger range than the Butterworth filter, which generally leads to lower order filter. An approximation having equiripple behavior may be used rather than monotonic design. Chebyshev filters are classified as either Type I, which have equiripple behavior in the pass-band and are monotonic in the stop band, or Type II, which have equiripple behavior in the stop-band and are monotonic in the pass-band.

Chebyshev filters of each type appear in Fig. 25. Again, these figures illustrate the frequency response of the filter overlaid with a filtered sinusoidal signal. Figure 25-*a* has an expanded *y* axis so that the ripples in the pass band may be seen clearly; the monotonic stop-band region is off the scale. Figure 25-*b* depicts a Chebyshev II type filter, as evident by the stop-band ripples. Note that there are half the number of negative spikes in the response as in the filter order.

Elliptic filters are generally even more efficient than Chebyshev filters since elliptic filters are equiripple in both pass band and stop band. Figure 26 shows a 10th-order elliptic filter. The stop-band ripples can be seen clearly. The pass-band ripples, although hidden due to the logarithmic scale, in fact are present.

Although design of the Chebyshev and elliptic filters is more complicated, these figures illustrate the benefits that these filters provide. In particular, the transition regions in Figs. 24–26 may be compared directly. Each of these filter types is the same order (10). However, the elliptic filter shows the steepest transition region, with the Chebyshev II next, followed by the gradual descent of the Butterworth frequency response. Butterworth, Chebyshev, and elliptic filters may be low pass, high pass, band pass, or band stop. Additional information on these analog filters may be obtained from other articles in this encyclopedia.

As mentioned in the beginning of this section, digital filters may be designed from analog filters. However, a causal, stable analog filter may not necessarily

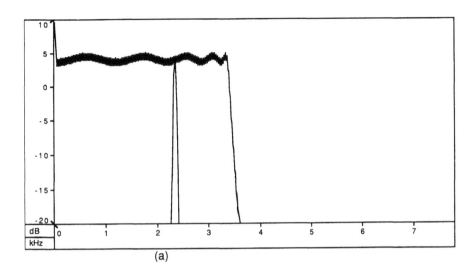

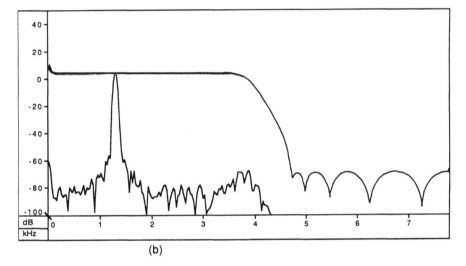

FIG. 25 Low-pass Chebyshev filters: *a*, Type I, 10th order; *b*, Type II, 10th order.

provide a causal, stable digital filter, depending on the design method. The bilinear *z*-transform method provides digital filter stability from a stable analog filter design. The bilinear *z*-transform may be applied only to particular classes of filters, but the common low-pass, high-pass, band-pass, and band-stop filters are included.

The bilinear *z*-transform design methodology is as follows. First the *s*-transform for the comparative normalized low-pass function is determined. There are conversions that allow easy transformations of this normalized low-pass function into a low-pass, high-pass, band-stop, or band-pass filter as desired. Then the *s*-transform function is converted to the *z* domain. The *z*-domain function is inverse *z* transformed to yield the filter coefficients.

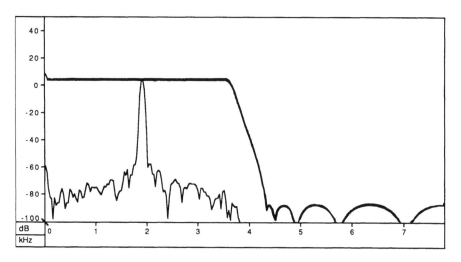

FIG. 26   Tenth-order low-pass elliptic filter.

Adaptive Filters

Besides the categories of IIR and FIR filters, a third type of digital filter warrants discussion: the adaptive filter. Adaptive filters provide many of the functions that nonadaptive filters do, such as prediction, smoothing, and eliminating unwanted signal components. An adaptive filter has a structure that allows the coefficients to change with time, unlike the fixed-coefficient filters described to this point. Like fixed filters, adaptive filters may be either FIR or IIR.

Adaptive filters can be used in real-time, time-varying applications in which the environment is changing slowly. The filter coefficients are updated so that variations in the signal may be tracked. In telecommunications, this type of adaptive filter is used for equalization and noise reduction.

Alternatively, the adaptive filter may self-design a complex structure in a stationary signal environment that would be time consuming for a person to design manually. Once the adaptive filter coefficients converge, these values may be implemented in a fixed filter, if so desired.

Adaptive filters are much easier to implement digitally than by analog means as their structure can be quite complex. A nonrecursive adaptive filter's coefficients are updated using a weighting of the error signal produced from the difference of the adaptive filter's output and a reference signal. Adaptive filter coefficients usually are represented with arrows drawn through them to alert the reader that the coefficients may change (see Fig. 27). Unlike a recursive adaptive filter, a nonrecursive filter has no dependence on past values of the filtered output $y(n)$.

The sampled input signal $x(n)$ is multiplied by the adaptive filter coefficients w_k (often called *weights*) and summed to produce the output $y(n)$ of the filter. (Note that the adaptive filter coefficients w_k have no relation to the FFT multiplier factor W_N^{nk} discussed above.) The adaptive filter's output is compared against a reference signal $r(n)$ to form an error signal $e(n)$. This error is used to

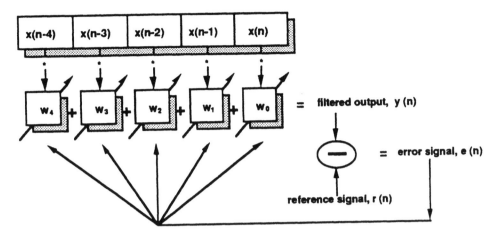

FIG. 27   Nonrecursive adaptive filter.

update the weights. There are many different methods by which the weights may be updated; all aim to minimize the error. Gradient estimation methods or the least mean square (LMS) algorithm put forth by Widrow may be used. The popular LMS algorithm provides a relatively simple, yet effective, technique by which the coefficients may be updated.

Finite-Length Effects in Digital Filters

Our discussion of digital filters (FIR, IIR, and adaptive) would not be complete without presenting some of the concerns facing the digital filter user. Different types of errors may arise during the implementation of digital filters. In high-performance applications in which there is a steep transition region or little ripple in the bands, these effects may become important. The errors may be due to (1) input signal quantization, (2) filter coefficient precision, and (3) round-off or truncation errors. Filter coefficients that may be represented in a computer's memory have limited precision. The coefficients may have either floating or fixed precision. Typically, 8-, 12-, 16-, or 32-bit quantization is used. Input signal quantization was discussed in previous sections.

The rounding or truncation of an IIR filter coefficient may result in poles moving outside the unit circle. This leads to an unstable filter. For this reason, second-order cascaded sections often are used since the stability of each section may be checked more easily than for higher-order sections. If the poles of frequency response in z-plane lie close to the unit circle, then the designer does not want denominator coefficients too close lest these errors push the poles over the unit circle. Quantization effects are dependent on the particular structure in which the IIR filter is realized.

In fixed-point arithmetic, the decimal point stays fixed for all arithmetic operations. There are three common implementations for fixed-point arithmetic: sign and magnitude (used for positive numbers), one's complement, and two's complement (the last two implementations are used for negative numbers). Rounding provides less absolute error than truncation, and generally is preferred in digital filter implementations. An overflow error occurs when the dynamic range is exceeded. Addition and subtraction are straightforward with fixed-point arithmetic; however, multiplication and division are more complex. Although no truncation or rounding errors occur with fixed-point addition, these errors may occur with multiplication.

Floating-point arithmetic, another method of representing numbers, allows greater dynamic range with the same word length as fixed point. Multiplication and division are easy to accomplish. In addition, overflow errors are much less likely to occur. Rounding and truncation errors are present in both floating-point multiplication and addition.

The deadband effect occurs when no input (or constant input) is rounded off and an incorrect steady-state value is given. The addition of a dither signal allows the system to approach the true steady-state value. These small oscillations in the input signal overcome the quantization threshold.

Digital Signal Processing Applications

We now turn our attention to the applications of the DSP tools and techniques presented in the previous sections. The applications of DSP are becoming more widespread and diverse. This article first discusses the breadth of these applications and then focuses on telecommunications applications.

The Breadth of Digital Signal Processing Applications

DSP techniques are being used increasingly in such mainstream applications besides telecommunications as automotives, toys, consumer stereo products, the military, and medicine.

Modern automobiles may use DSP for active suspension. Speech compression and speech synthesis utilize DSP fundamentals in the creation of talking toys. Audio signals produced by modern stereo components often are massaged by digital techniques to produce a more pleasing sound. Medicine also is taking advantage of recent advances in digital image-processing technology, for instance, for more accurate diagnosis of a patient's ailment. Adaptive filters and image processing are discussed briefly below.

Adaptive filters are useful in the same breadth of applications as regular or nonadaptive filters. Adaptive filters commonly are used to eliminate interfering signals. For example, adaptive filters have been used to measure a human fetus' heartbeat more accurately. The heartbeat of a fetus is often hard to hear due to the mother's heartbeat, which is often 2–10 times louder. In an example from Ref. 6, the heartbeat sound from the mother's abdomen provided the input signal for the adaptive filter. The reference signal was the mother's heartbeat measured close to her heart. The fetus's heartbeat was clearly evident in the error signal. Another principal application of adaptive filters is equalization, discussed below in relation to DSP telecommunications signals.

The principles of DSP of one-dimensional signals addressed in this article may be extended to two dimensions. Two-dimensional DSP techniques are used on images in as diverse applications as animation and medical imaging. The motion picture industry is using digital image processing on computerized images to create unusual objects, futuristic environments, backgrounds, and even moving creatures. Image-processing techniques are aiding the medical field greatly. Internal images of the human body may be enhanced as never before to provide evidence of slight fractures, irregular tissue, or unusual features. Although image-processing techniques are often quite computationally intensive, with the advent of more powerful DSP implementations their use will become more commonplace.

Digital Signal Processing in Telecommunications

DSP techniques are used extensively in modern telecommunications networks, which require transmission and switching of signals. Any signal transmitted over a long distance must use repeaters. Digital signals may be transmitted more accurately than analog signals. In analog transmission, an amplifier boosts the signal power and, unfortunately, the noise power. The received signal may have substantial added noise. However, in digital transmission a regenerator receives the signal and retransmits it with minimal introduced noise, thus allowing for much more accurate reception of the original signal at the destination.

DSP often is used in telecommunications for data compression. DSP coding techniques may increase greatly the amount of information that can be sent over a channel. Both speech and data are compressed in our telephone network. One method of compression is to use a nonuniform quantizer when sampling the analog information. This function then is reversed at the receiving end of the transmission. North America and Japan use what is called μ-law data compression, while Europe uses A-law compression.

Data may be compressed through coding such as error-detection codes or error-correction codes (ECCs). Error-detection coding is straightforward. Such additional information as a checksum or cyclic redundancy check (CRC) is sent so that the receiver may distinguish whether errors have occurred in the transmission. Error-detection techniques do not indicate where the errors are — just that they have occurred.

Unlike error-detection methods, ECCs provide information about what piece of data is erroneous and how it should be corrected. Typically, ECC methods are more complex than error-detection methods. Block and convolution codes are the two main types of ECC in use today.

Digital signals also lend themselves to less expensive and potentially more complex multiplexing methods. Analog signals commonly use frequency division multiplexing (FDM). An FDM signal has several signals that occupy distinct frequency bands or regions. Digital signals may be multiplexed in an entirely different manner, time division multiplexing (or TDM). A TDM signal is a digital datastream in which different signals are sent at a higher rate. The digital stream packs several signals at different times within the same stream. Typically, the TDM signal also contains information about how the signals are multiplexed. Current telephone networks utilize both FDM and TDM data. Transmultiplexers are used to convert between the two.

Telephone voice information is band limited to a 4-kHz, or voice-grade, channel. Although the human ear can hear frequencies to 20–25 kHz, a voice-grade channel provides sufficient quality for conversation and efficient transmission. A voice-grade channel is sampled at 8000 samples per second. Each sample is typically 8 bits, so the overall data rate for a voice-grade channel is 64 kilobits per second (Kbps). These channels may transmit modem data as well as voice. Reference 7 contains an excellent discussion of digital transmission.

Modems use DSP technology to transmit compressed data over the narrow 4-kHz telephone bandwidth. Modems have built-in equalizers to receive transmitted information accurately. Equalization is one of the most important implementations of digital adaptive filtering today. Equalizers operate in two phases. In the first (or training) phase, known information is sent to the adaptive filter (or equalizer) so that the adaptive filter coefficients may change to their appropriate values. During the second phase, the equalizer processes the unknown signal (typically, the received signal of interest). This equalizer may compensate for channel distortions that would make the transmission sent unintelligible.

The telephone network is being revamped due to advances in DSP techniques and implementations, as well as the changing nature of consumers' needs. The growth in modem and facsimile traffic mean that more data are being transmitted now than ever before, and the transmission percentage of data to voice has increased substantially over recent years.

The Integrated Services Digital Network (ISDN) currently is being standardized worldwide. The ISDN system will provide one connection in the end-user's home for diverse signals. Since they have a common interface, digital data signals, voice signals, even video signals may be received on the same port. Improved quality, efficiency, and convenience are expected for the end user. While ISDN calls for integration of the delivery of these different services, it does not specify that these signals must be integrated (that is, transmitted over the same path) within the telecommunications network. Regardless, many companies are moving toward integrated transmission.

One certainty remains: DSP will be applied even more widely as our telecommunications systems become all-digital.

Digital Signal Processing Implementations

DSP advances have kept pace with the rapid development of computer technology. As digital hardware implementations continue to become faster, smaller, and more cost effective, the applications of DSP will grow as well. Digital technology that typically is used to implement signal processing algorithms includes:

- general-purpose DSP processors
- single-purpose DSP chips
- DSP coprocessor boards
- general-purpose computers

Both general-purpose and single-purpose DSP processors were developed first in the late 1970s. Since then, microlithography advances have allowed a much larger chip area and a smaller feature size. This combination has boosted the power radically, while lowering the cost of modern DSP chips. In addition, the advent of high-level languages has enabled the chips to be programmed by a wider audience.

General-purpose DSP chips are a rapidly growing technology. DSP chips are microprocessors that have been designed to optimize the implementation of DSP algorithms. They perform the fundamental DSP operation of the multiply-accumulate very quickly. They are generally not as fast as special-purpose hardware implementations, but their programmability makes them very powerful. DSP chips typically are designed to interface easily with analog input and output circuitry. The built-in controls of DSP chips have performed iterative operations efficiently. DSP chips are not wed to a particular operating system like the Reduced Instruction Set Chip (RISC) general-purpose processing chip, which is dependent on the C language and the UNIX™ operating system. Most DSP chips in use today use the Harvard architecture, in which the data and program information are input separately. They are well suited for many real-time signal processing applications. The DSP chip can be programmed in a simple manner, despite a usually complex architecture.

DSP chips may utilize either floating-point or integer (sometimes called fixed-point) arithmetic. Floating-point processors have gained popularity since code may be transported more readily to a floating-point processor from general-purpose computers. Code may be developed on another workstation and migrated to the floating-point processor with minimal difficulty. With a fixed-point implementation, a programmer may need to worry about scaling intermediate results. Successful floating-point processors include Texas Instruments' TMS320C30 and TMS320C40 processors and AT&T's DSP32C family. Most floating-point processors use 32 bits with a 24-bit mantissa with an 8-bit exponent. Popular fixed-point processors include AT&T's DSP16 and Motorola's DSP56000.

DSP operations are both input/output (I/O) and memory intensive. Therefore, both static random-access memory (SRAM) and dynamic random-access memory (DRAM) are used. SRAM is used for the most time-critical operations, while the DRAM, which is cheaper and denser, is utilized in the remaining functions.

The first N$_e$xt computer was introduced with a built-in signal processing chip to offload DSP operations from the main processor.

Using DSP chips together is becoming increasingly popular in high-performance applications. Arrays of DSP chips may be used as accelerators, especially for graphics or image processing. Processing may be divided into portions and distributed to multiple chips. In addition, chips may be used in sequence in a pipelined architecture. This serial structure may be necessary to keep up with the incoming data in a real-time architecture. Modern chips are being designed to work more naturally in the multiprocessor environment. For example, Texas Instruments' TMS320C40 has six built-in communication ports to interface easily with up to six other C40 chips.

A second category of DSP implementations are single-purpose (or special-purpose) DSP chips. Many chip manufacturers now produce sophisticated chips or chip sets that implement such DSP algorithms as an FFT or a correlator. Other chips are used for modems, filters, image compression, speech synthesis, encryption, facsimile transmission, and audio processing. Image-compression chips are an important example of special-purpose DSP chips. Recently image-compression chips have become widely used due to expanding video and graphics capabilities. These chips may use one of two different compression standards that are vying for attention. The Joint Photographers Experts Group (JPEG) standard is used for still images, while the Motion Picture Experts Group (MPEG) standard is used for motion pictures. Image-compression chips will become even more popular as video applications increase.

A DSP chip may form the core engine of the third category of DSP implementations: a coprocessing board. This board often contains the A/D and D/A circuitry needed to process signals in the real world. This coprocessor may fit in one or more slots onto the computer and execute functions while the main processor goes on with its particular tasks. These boards make use of such standard interfaces as the ISAbus for the IBM PC/AT, VME bus, the Nubus (for the Macintosh®), and the Sbus for the Sun Microsystems SPARCstation computer. These coprocessor boards may have arrays of DSP chips in either a parallel or pipelined architecture. The diversity of coprocessor boards available commercially often make it unattractive to develop a custom board.

A key component of the DSP board is not the hardware itself, but the software. The most useful boards have extensive software to make development easy. Fortunately, DSP floating-point compilers have advanced significantly. Improvements in compilers make it less likely that key functions need to be programmed in assembly language. Coprocessor boards often have libraries of commonly used utilities. These subroutines are optimized to perform key functions quite quickly. However, most software developed for one coprocessor board is not readily transferable to another manufacturer's board. Even more advanced software further removes the person from the DSP implementation details. There are graphical interfaces that allow the user to edit, manipulate, and process waveforms in sophisticated ways by just clicking a mouse.

Any general-purpose computer can be programmed to implement DSP algorithms. Usually these machines are used for algorithm development, simulation, or non-real-time signal analysis. Typically, a general-purpose computer is not optimized to execute math functions quickly and has no method of digitizing a live data signal. However, its ease of programmability makes a general-purpose computer an excellent choice for development and simulation. Excellent compilers make it easy to port this code to faster platforms like the general and special-purpose DSP chips and coprocessor boards that we discussed here.

DSP implementation is a rapidly changing field. Once, implementing DSP algorithms would be attempted only by the most serious programmer. Now, such companies as N$_e$xt are building DSP chips right into the main computer board. Advances in chip and electronic technology have fueled the fire of demand for DSP techniques. With cheap, efficient implementations, the applications of DSP are endless.

Acknowledgments: UNIX is a registered trademark of AT&T Bell Laboratories. Macintosh is a registered trademark of Apple Computers, Inc.

Bibliography

Dudgeon, D. E., and Mersereau, R. M., *Multi-dimensional Digital Signal Processing*, Prentice-Hall, Englewood Cliffs, NJ, 1984.

Gabel, A. Robert, & Robert, Richard, *Signals and Linear Systems*, John Wiley and Sons, New York, 1973.

Gold, Bernard, and Rader, Charles M., *Digital Signal Processing*, McGraw-Hill, San Francisco, 1969.

Hamming, R. W., *Digital Filters*, Prentice-Hall, Englewood Cliffs, NJ, 1983.

Oppenheim, A. V., and Schafer, R. W., *Digital Signal Processing*, Prentice-Hall, Englewood Cliffs, NJ, 1975.

Rabiner, Lawrence R., and Rader, Charles M. (eds.), *Digital Signal Processing*, Institute of Electrical and Electronics Engineers Press Selected Reprint Series, IEEE, New York, 1972.

Terrell, T. J., *Introduction to Digital Filters*, Macmillan, New York, 1980.

References

1. Oppenheim, Alan V., Willsky, Alan S., and Young, Ian T., *Signals and Systems*, Prentice-Hall, Englewood Cliffs, NJ, 1983.

2. Burrus, C. S., and Parks, T. W., *DFT/FFT and Convolution Algorithms*, John Wiley and Sons, New York, 1985.

3. Rabiner, Lawrence R., and Gold, Bernard, *Theory and Application of Digital Signal Processing*, Prentice-Hall, Englewood Cliffs, NJ, 1975.

4. Bellanger, M., *Digital Processing of Signals*, John Wiley and Sons, New York, 1987.

5. Institute of Electrical and Electronics Engineers Acoustics, Speech and Signal Processing Society, *Programs for Digital Signal Processing*, IEEE Press, New York, 1979.

6. Widrow, Bernard, and Stearns, Samuel D., *Adaptive Signal Processing*, Prentice-Hall, Englewood Cliffs, NJ, 1985.
7. Saltzberg, B., Basics of Digital Communications. In: *The Froehlich/Kent Encyclopedia of Telecommunications*, Vol. 1 (F. E. Froehlich and A. Kent, eds.), Marcel Dekker, New York, 1990, pp. 487–538.

LISA D. NOBLE
MICHELE D. MATHYS

Digital Speech Processing

What Is Digital Speech Processing?

The technology of digital speech processing encompasses three broad areas of interest: coding, synthesis, and recognition of speech. *Coding* is the process of capturing the speech of one person and processing it for efficient transmission to another person over some type of transmission system (i.e., over a communications channel). The goals of coding include conservation of bandwidth (or bit rate), reduced cost (complexity) of terminal equipment, voice privacy (secure communications) to prevent casual intrusion of a transmission, and, for military applications, encryption so that even well-designed attempts at decoding by anyone but the intended receiver will fail.

Synthesis is the process of converting an ASCII (American Standard Code for Information Interchange) text message to speech to provide a machine with the capability of talking (speaking information) with a person. The goals of synthesis are to provide a broad range of vocabularies, speaking voices, and languages, with high word intelligibility and good quality at low cost in support of a broad range of computer services. Synthesis systems range in complexity from simple computer voice response and announcement systems (which often constitute little more than digital tape recorders) to full text-to-speech systems capable of fluently speaking unrestricted text.

Recognition is the process of extracting meaning from a spoken input so that a person can talk (request information, service, etc.) to a machine. Recognition includes both message decoding and talker recognition for voice access to information. The goals of recognition are to provide a broad, flexible user interface to communications systems for a viable alternative to standard keyboard entry for access to and control of a machine. Within the area of recognition are conventional "command and control systems" in which a small vocabulary of "command" words is all that is required for effective voice access to "control" the machine, conversational systems capable of processing a full speech dialogue to determine and act on the relevant message, and talker recognition systems capable of determining whether a claimed identity is that of the talker who is trying to gain voice access to a machine.

Applications of Speech Coding

The original motivation for research into the coding of speech was the predicted need to conserve bandwidth in the transmission of telephone signals. With the advent of microwave and light-wave transmission, with seemingly unlimited bandwidth, the direct need for coding for long-range transmission has all but disappeared. However, with the advent of developing network services, including mobile telephony, Personal Communication Networks (PCNs), and digital (packetized) transmission, the need for efficient coding of speech within the telephone network is greater than ever. Thus, integrated digital transmission of voice, data, video, facsimile (FAX), and so on, over digital networks such as ISDN (Integrated Services Digital Network) requires coding of speech to optimize use of available facilities.

Another key application of speech coding is the broad area of voice mail and voice messaging in which speech is transmitted through the network and then stored at the receiving end (either in a voice mailbox or in a dedicated local store) for later access and retrieval. Although storage is relatively inexpensive, the lower the bit rate at which speech can be stored, the less expensive will be the resulting service.

Other application areas of speech coding include recorded announcement storage, voice privacy, and message encryption.

Basic Principles of Speech Coding

Figure 1 shows typical examples of speech waveforms. Figure 1-*a* shows a voiced speech section (e.g., a vowel sound) waveform produced by modulating puffs of air (created by the vibrating vocal cords) by the vocal tract shape corresponding to the sounds being spoken. For such voiced waveforms, we see a quasi-periodicity of the signal (over periods of 10s of milliseconds [ms]) as well as a slowly changing waveform character. Figure 1-*b* is the waveform of an unvoiced speech section (e.g., a sibilant sound like "s" or "sh") that has a noiselike character with no periodicity (the vocal cords are not vibrating) and no slowly changing temporal characteristics. Finally, Fig. 1-*c* shows a section of a speech utterance waveform that consists of both voiced and unvoiced sounds.

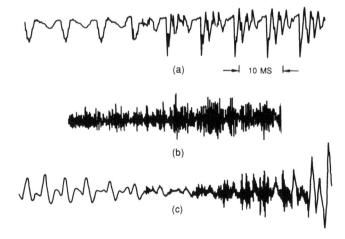

FIG. 1   Typical speech waveforms: *a*, voiced speech; *b*, unvoiced speech; *c*, section of speech with both voiced and unvoiced sounds.

For processing speech signals, the first step is to convert the signal to a digital representation. Digital representations have several advantages, including ease of manipulation (further digital processing), ease of integration with other digital data, relative insensitivity to errors in transmission, and ease of overlaying various levels of security.

The simplest way of converting a speech signal of the type shown in Fig. 1 to a digital representation is to apply the sampling theorem directly. That means we must sample the speech waveform at a rate of twice the highest frequency present in the signal and then digitize the resulting samples to some desired degree of accuracy. For telephone bandwidth speech signals (4-kilohertz [kHz] bandwidth), we need a sampling rate of at least 8000 samples per second and an encoding rate of 16 bits per sample to maintain very high signal-to-noise ratio. Hence, a total of 128,000 bits per second is adequate for a high-quality digital representation of telephone bandwidth speech.

Since one of the goals of speech coding is to provide high-quality speech at low bit rate, to achieve this goal we must exploit one or more of the special properties of speech signals to reduce the bit rate. The properties that can be used include the use of adaptive quantizers (because of the dynamic characteristics of the speech), time-varying filters to exploit both the short-time (within a single period) and long-time (across multiple periods) correlations of the signal, and perceptually based quantization-noise masking to make the "coding noise" inaudible. Figure 2 shows a hierarchy of speech synthesis models that, to varying degrees, attempt to exploit these speech properties to reduce the necessary bit rate for coding of speech.

The *vocoder* (voice coder) model (Fig. 2-*a*) essentially uses a dichotomous excitation model consisting of periodic pulses for voiced speech and random noise for unvoiced speech. A time-varying switch selects between these two

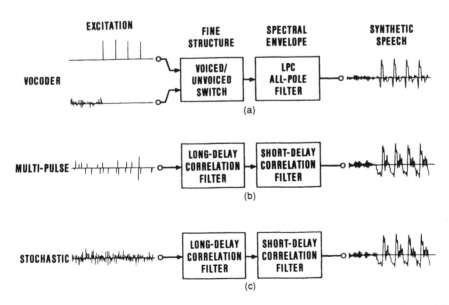

FIG. 2   Hierarchy of speech synthesis models: *a*, the vocoder model; *b*, the multipulse model; *c*, the stochastic model.

sources and sends the resulting signal to a time-varying spectral envelope filter called a linear predictive coding (LPC) all-pole filter that models the short-time correlations in the speech signal.

The multipulse synthesis model (Fig. 2-*b*) integrates the two excitation sources of the vocoder model into a single source that looks noiselike during unvoiced sections and pulselike during voiced sections. The long-delay correlation filter models the periodicity during voiced sections, while the short-delay correlation filter models the vocal tract characteristics of the speech.

The stochastic synthesis model (Fig. 2-*c*) replaces the pulses of the multipulse model with a signal optimally chosen from a stochastically derived codebook of excitations. The long-delay and short-delay correlation filters are the same as for the multipulse model.

To illustrate typical waveforms from the stochastic model, Fig. 3 shows the excitation signal (Fig. 3-*a*), the output of the long-term correlation filter (Fig. 3-*b*) in which the periodicity in voiced sections can be seen, the output of the short-delay correlation filter (the decoded speech) (Fig. 3-*c*), and the original speech signal (Fig. 3-*d*). The high-quality reproduction capability of the stochastic model is seen clearly in this figure.

Attributes of Speech Coders

All (digital) speech coders can be characterized in terms of four attributes: bit rate, quality, delay, and complexity. *Bit rate* is a measure of how many of the

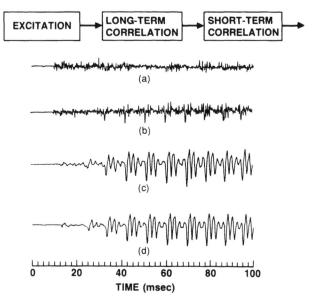

FIG. 3 Typical waveforms from the stochastic model: *a*, excitation waveform; *b*, output of long-term correlation filter; *c*, the decoded speech; *d*, the original speech signal.

special properties of speech signals have been exploited in the coder; the lower the bit rate is, the more reliance on the speech production model is necessary to be accurate in all situations. *Quality* is a measure of degradation of the coded speech signal and can be measured in terms of speech intelligibility and perceived speech naturalness. *Delay* is a measure of the amount of speech signal required to estimate coder parameters reliably for both the encoder and the decoder (overall coder delay is the sum of the encoder delay plus decoder delay). The longer the allowed delay in the coder is, the better the coder can estimate parameters; however, long delays (on the order of 100 ms) are perceived as quality impairments and sometimes even as echo. Finally, complexity is a measure of computation required to implement the coder in digital signal processing (DSP) hardware.

The "ideal" coder has a low bit rate, high quality, low delay, and low complexity. Actual coders make tradeoffs among these four attributes. Rather than discuss this issue further, Figs. 4 and 5 show plots of speech intelligibility as measured in terms of diagnostic rhyme test (DRT) scores, and speech quality as measured in terms of mean opinion scores (MOSs) for a range of speech coders spanning bit rates from 64 kilobits per second (Kbps) to 2.4 Kbps. (Also included in the plots are scores for telephone bandwidth natural speech.) The coders used in these tests included:

μ-law pulse code modulation (PCM) at 64 Kbps
Adaptive differential pulse code modulation (ADPCM) at 32 Kbps
Low-delay code-excited linear prediction (LD-CELP) at 16 Kbps
Code-excited linear prediction (CELP) at 8 Kbps and 4.8 Kbps
Linear predictive coding (LPC10E) at 2.4 Kbps

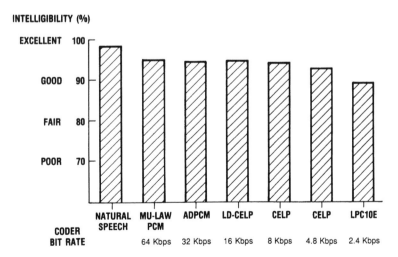

FIG. 4 Speech intelligibility scores in terms of diagnostic rhyme test (DRT) of several coders as a function of bit rate (PCM = pulse code modulation; ADPCM = adaptive differential PCM; CELP = code-excited linear prediction; LD-CELP = low-delay CELP; LPC10E = linear predictive coding).

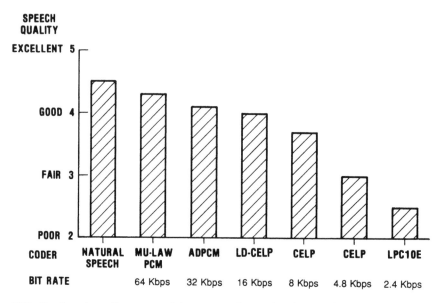

FIG. 5   Speech quality mean opinion scores of several coders as a function of bit rate.

All forms of the CELP coders are stochastic coders of the type discussed in connection with Fig. 2. The LPC10E coder is a U.S. government standard version of a vocoder synthesis model.

It can be seen from Fig. 4 that the intelligibility scores for coders with bit rates of 64 Kbps to 4.8 Kbps are essentially identical, and only slightly lower than that of natural speech. At 2.4 Kbps a further slight degradation in intelligibility also is observed. However, for the most part, all the coders maintain high intelligibility.

In terms of speech quality as measured by listeners who rated the speech on a 5-point scale (excellent = 5 to unacceptable = 1), there are very significant differences among coders. It can be seen in Fig. 5 that the score for natural speech was 4.5 and that all coders with bit rates from 16 Kbps to 64 Kbps achieved scores of 4.0 (good) or higher. Such high MOS scores are considered both necessary and sufficient for network applications of coders in which very high quality is required. At 8 Kbps, the MOS score falls to 3.8, slightly below network quality but quite useful for cellular applications. At 4.8 Kbps and 2.4 Kbps, the MOS scores fall in the range of 2.0–3.0; such coders are acceptable primarily for military applications in which low bit rate is essential for secure communications.

To see the progress made in improving coded speech quality over the past decade (as well as predicted quality improvements over the near term), Fig. 6 shows curves of speech quality versus bit rate as measured in 1980 and 1988 and as predicted for 1992 and 1995. It can be seen that in 1980 the quality began to decline at rates below 32 Kbps; in 1988, the decline began at rates below 16 Kbps; in 1992, it is predicted that the quality decline will begin below 8 Kbps; and in 1995 it is predicted that the quality decline will begin below 4 Kbps.

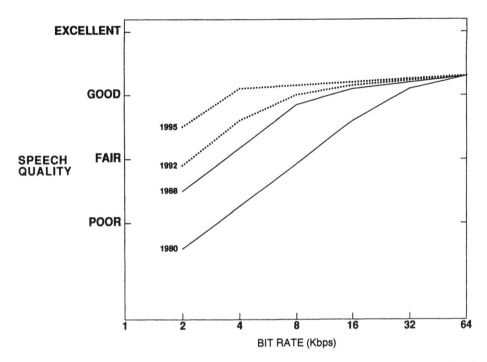

FIG. 6   Curves of speech quality versus bit rate as measured in 1980 and 1988 and as predicted (in 1991) for 1992 and 1995.

Coding Standards

Since the major applications of speech coding are within the telephone network as well as for use in mobile radio and secure voice, it follows that a number of standards have been created around specific coding algorithms. Figure 7 shows a plot of the 3 arenas in which standards have been created: network, mobile radio, and secure voice. The International Telephone and Telegraph Consultative Committee (CCITT) created network standards at 64 Kbps in 1972, at 32 Kbps in 1984, and at 16 Kbps in 1992. The Cellular Telecommunications Industry Association (CTIA) created an 8-Kbps standard for mobile radio in 1990. Finally, U.S. Government National Security Agency (NSA) standards were set for 4.8 Kbps in 1989 and for 2.4 Kbps in 1975 for low-bit-rate secure voice.

Speech Synthesis

The basic goal of research in speech synthesis is to provide a broad range of capability for a machine to speak information (respond) to a user. A simple communications model of a system for voice access to information (a database)

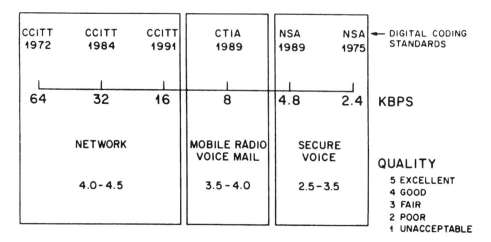

| CCITT 1972 | CCITT 1984 | CCITT 1991 | CTIA 1989 | NSA 1989 | NSA 1975 | ← DIGITAL CODING STANDARDS |

FIG. 7   Three key application areas in which speech-coding standards have been applied.

is shown in Fig. 8. It is assumed that the user has access to a touchtone receiver (TTR) and can enter requests (data) via the TTR keyboard (or via voice input) to a communications interface that transmits the request to a database manager. The requested information is sent back to the user (again through the communications interface) in the form of voice output as this is the only output modality on a standard TTR. Thus, the key issue is how to convert the text equivalent of the database information efficiently to speech for different applications.

There are three factors affecting the way in which a synthesis system is implemented for different tasks: the required quality of the synthetic speech, the range of speaking vocabulary, and the cost (complexity) of the synthesis hardware. Included with the cost factor is the storage costs for words, phrases, and rules, as well as the cost of the speech-generation hardware.

Synthesis Applications

There are two broad classes of synthesis applications: those that require little or no user interaction and those that are highly user interactive. In the first class are a broad range of telecommunications applications such as the Automatic

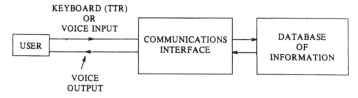

FIG. 8   Model of communications system using speech synthesis to provide spoken messages to the user (TTR = touchtone receiver).

Intercept System (AIS), in which a telephone call dialed to an inactive (or out-of-service) number is intercepted and an appropriate message with details about the problem is spoken back to the user; talking announcements such as time or weather; and entertainment and communications services such as dial-a-joke, dial-a-prayer, and so on. In the second class are the more standard database-access services such as voice banking, stock price quotations, sports scores, flight information, and so on, and services that require the ability to speak such unlimited, unconstrained text as that involved with access to medical or legal encyclopedias, voice FAX, and voice electronic mail.

The applications requiring little or no user interaction generally involve either a small vocabulary of specialized words and phrases or a fixed text that can be spoken and saved in digitally coded form on some digital storage medium. Hence, these applications are often little more than digital tape recordings and require relatively simple hardware for the "synthesis."

The applications requiring a lot of user interaction to determine which information in the database needs to be provided to the user in the form of a spoken response generally require a full text-to-speech (TTS) synthesis system.

Factors Influencing Synthesis Systems

The three factors influencing the use of synthesis systems are the quality of the synthesized speech, the fluency of the spoken output (i.e., the ability to create messages with different vocabularies, emphasis, intonation, speed, etc.), and the complexity of the hardware implementation. Figure 9 shows the status of a range of synthesis systems as a function of these three factors. The "desired performance" of the "ideal" synthesis system is shown in this figure as a point with high quality (the resulting speech sounds quite natural), high fluency (virtually any printed message can be produced with the desired speaking rate and emphasis), and low cost (so that it is cheap enough to integrate into any desired application). Unfortunately, in the real world, there is no practical system with a performance that even comes close to the ideal. Shown in Fig. 9 are three actual classes of systems: announcement machines (such as those used in the AIS application or for voice storage services [VSS]), parametric systems (as exemplified by the Speak and Spell toy introduced by Texas Instruments), and full TTS systems (such as those by Prose, DEC, and Infovox). The announcement machines provide high quality since they essentially are prerecorded and use digitized speech messages, and have low fluency because they only can speak the prerecorded messages or trivial combinations of the words, and they also provide low complexity. The parametric systems provide low-quality speech (highly synthetic sound) with medium fluency and low-to-medium complexity. Full text synthesis systems again provide low quality, but give high fluency (they can say anything in any desired manner) at high complexity.

Technology of Speech Synthesis

A block diagram of a simple voice response system consisting of prerecorded and digitally coded words and phrases is shown in Fig. 10. Based on user

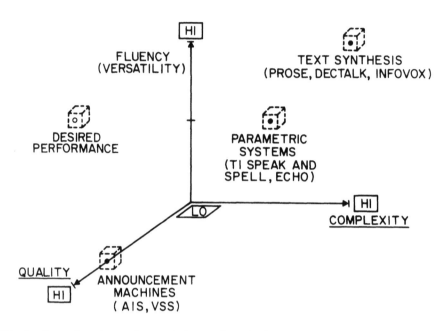

FIG. 9 Examples of several general classes of synthesis systems in regard to quality, fluency, and complexity (AIS = Automatic Intercept System; VSS = voice storage services).

actions (e.g., dialing a disconnected telephone number), a request for a specified sequence of words and phrases is generated and sent to a concatenation device that retrieves from a digital store the coded versions of each of the required vocabulary items, concatenates the vocabulary items for the message, and sends the final result to a decoder that produces the analog speech heard by the user. Thus, for the intercept message, "The number you have dialed, 555–1234, has

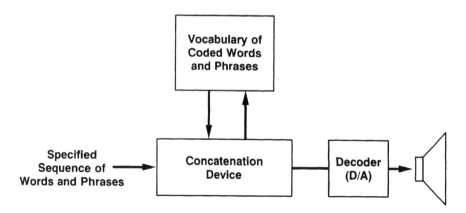

FIG. 10 Block diagram of simple voice response system (D/A = digital to analog).

been disconnected," the voice response system retrieves, in sequence, the phrase "The number you have dialed," followed by each of the digits of the telephone number, followed by the phrase "has been disconnected." (For naturalness, usually several versions of a digit are stored so that a digit at the beginning of a string has a different emphasis than the same digit in the middle or at the end of the string.) Any of the speech-coding techniques discussed here can be used in the voice response system of Fig. 10.

A block diagram of a full TTS system is shown in Fig. 11. The desired message text is an arbitrary ASCII string (usually with appropriate punctuation) so the first task of the system is to convert the text string to a sequence of phonetic symbols indicative of the sounds to be spoken, along with a set of prosody markers indicating the speed of speech, the intonation, and the emphasis on certain words. This "text-to-sound/prosody" conversion involves a combination of dictionary lookup of word pronunciation and rules for exceptions, unusual cases, and for generating durations and emphasis of sounds. To illustrate how difficult this text analysis problem can be, consider the sentence:

Dr. Wojtech from Wynmore Dr. in St. Louis took a job with DEC at the Hawthorne St. laboratory in Gloucester, MA.

The ASCII text "Dr." has to be pronounced "doctor" at its first occurrence and "drive" at its second occurrence. Similarly, the text "St." is pronounced "saint" in the city name but is "street" in the address. The acronym "DEC" should be pronounced as the word "deck" and not as the individual letters. The abbreviation "MA" should not be pronounced as the word "ma" but instead as the state "Massachusetts." Finally, the names "Wojtech," "Wynmore," "Hawthorne," and "Gloucester" all must be pronounced properly, even though it is unlikely any of them will be in a pronouncing or an exceptions dictionary. Other key problems in text analysis include proper determination of part of speech, determination of locations of embedded clauses, dangling prepositions, and so on.

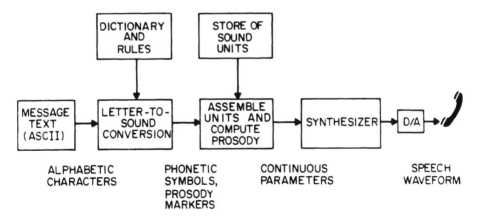

FIG. 11 Block diagram of full text-to-speech synthesis system.

Once the appropriate phonetic symbols and prosody markers have been obtained, the next step in the TTS process is to assemble the appropriate speech units and compute the required prosody (pitch/duration) contour for the message. A store of sound units is required. Creation of an appropriate set of synthesis units is both time consuming and difficult as these units must be robust to different phonetic environments, yet must be rich enough to disambiguate sound combinations that are different in minimal ways. Experience shows that inventories of from 2000 to 4000 dyad/polyad units (dyads are spectral representations of time slices from 2-phone sequences, polyads are spectral representations of time slices from sequences of 3 phones or more) are required for good-quality synthesis.

The final steps in the TTS process are synthesis from spectral parameters appropriate to the sequence of synthesis units and digital-to-analog (D/A) conversion of the resulting speech to render it useful for transmission back to the user.

Text-to-Speech System Performance

The only proper way to judge TTS performance is to listen to one or more paragraphs of speech produced by the system. Without this capability, the next best way of describing TTS performance is in terms of intelligibility scores and quality MOSs. The best TTS systems achieve word intelligibility scores of close to 97% (natural speech achieves 99% scores); hence, the intelligibility of TTS is almost that of natural speech. MOS scores for the best TTS systems are in the 3.0–3.5 range, indicating that the quality is in the fair-to-good range. Finally, the computation necessary to support full TTS systems is about 2 million instructions per second (MIPS) with about 6 megabytes (MB) of memory required for units, rules, and program code.

Speech Recognition

The goal of speech recognition is to provide enhanced access to machines via voice commands. The idea of "enhanced" access is a most important one as, for most applications, there are viable alternatives to voice control, including keyboards, touch panels, mice, and so on. Thus, for voice technology to be of value means that the voice interface to the machine is a natural one in which voice input is a reasonable way of requesting information, and the interface performs reliably (with high accuracy) and robustly for all users and in all environments. Figure 12 shows a block diagram of a model of voice access to and control of a machine. In order to access a database of information, the user is assumed to speak commands in the form of either an isolated word sequence or as a sequence of words drawn from a small vocabulary (e.g., digits). We refer to this second form of spoken input as connected word sequences. Recognition of the spoken input is based on whole-word patterns; hence, the

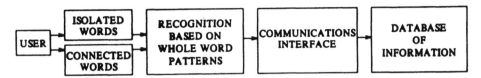

FIG. 12 Communications model of database retrieval system using speech recognition to transmit database requests for information.

output of the recognizer is either the appropriate command word or a recognized sequence of words. A communications interface then is used to access a database for the appropriate information, which is transmitted back to the user using TTS messages.

There are a wide range of factors that influence performance of speech-recognition systems, including the following:

1. *Speaking format:* Whether the system looks for isolated word and/or phrase commands or can recognize fluent speech in the form of connected word sequences or, eventually, unconstrained speech

2. *Degree of speaker dependence:* Whether the system must be trained to the speech of individual users (so-called speaker-dependent [SD] systems) or can work properly with users who never (or seldom) have seen or used the system (so-called speaker-independent [SI] systems)

3. *Vocabulary size and complexity:* The range of vocabulary words and phrases to which the system responds. There are many interesting and useful tasks requiring small- to moderate-size vocabularies (e.g., digit strings, simple commands from menus); however, ultimately we will need to be able to recognize upward of 50,000 words for such tasks as voice dictation, natural language access, control of databases, and so on.

4. *Task constraints:* As the vocabulary size for recognition grows, the number of possible combinations of words to be recognized can grow exponentially. Hence, some form of task constraint in the form of formal syntax (which words can follow other words) and formal semantics (which sentences make sense for the current status of the task transaction) is required to make the recognition task more manageable and to provide higher recognition task accuracy.

5. *Cost, method of implementation:* It will be seen that speech recognition by machine often is computationally quite expensive (upward of 1 gigaflop/second is required for real-time operation for some problems). Hence, a limiting factor is often what can be done with reasonable, but limited, computational resources.

Applications of Speech Recognition

Applications of speech-recognition technology fall into two broad areas: telecommunications and business applications. In the telecommunications area, some representative applications include

1. *Expanded use of rotary phone for menu-based services:* Such services currently are unavailable without a touchtone phone. In addition, even for users with access to touchtone phones, the voice recognition interface can be more attractive than the standard button-pushing alternative because the service names are spoken rather than having to push buttons associated with the service. Thus, for access to different parts of a department store, it would be more natural to speak the words "hardware" or "furniture" rather than to remember to push Button 3 for "hardware" or Button 5 for "furniture."

2. *Repertory dialing:* Voice dialing of telephone numbers and names provides the opportunity for hands-free, eyes-free control and use of a telephone. This is especially important for mobile telephony when the eyes and hands usually are tied up with the process of driving and controlling an automobile.

3. *Catalog ordering:* Most catalogs consist of letter and number codes, often with a great deal of redundancy built into the codes. Hence, ordering items from a catalog by voice is a natural way of interacting with the database of items associated with the catalog.

In the business area, some representative applications include

1. *Data entry for filling out forms and the like:* Such applications are highly repetitive and generally are performed by a small staff of people who can afford to train the system to recognize individual word patterns. Typically, vocabularies for this application are small to moderate in size (e.g., from 10 to 200 words).

2. *Keyboard replacement or expansion:* Here the recognition task is to replace sequences of keystrokes with a single voice command (a voice macro) or to replace the keyboard entirely with spoken input.

3. *Database access:* The recognition task is to query the database to determine specific information contained within the database. Hence, an airline's reservation system could be queried to determine available flights between specified cities, flight costs, type of aircraft, and so on.

Recognition Technology

Although a wide variety of approaches to speech recognition has been studied, the most popular (and successful) approach has been one based on standard pattern-recognition technology. This approach is illustrated in Fig. 13. Basically, the system uses a set of word and/or phrase patterns created using a pattern-training program. These patterns can be typical spectral patterns of words, averages of spectral patterns of words across different talkers, or even sophisticated statistical models that include spectral mean and spectral variance statistics derived over the time duration of the word.

The way in which isolated word speech recognition is carried out once the vocabulary patterns have been created and stored is to record the input speech

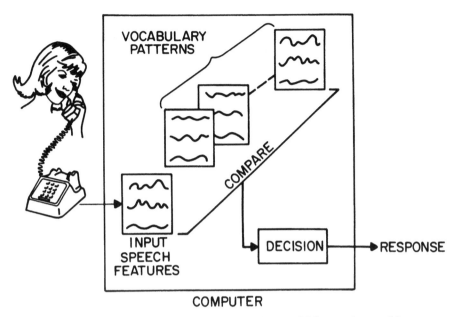

FIG. 13   Block diagram of pattern-recognition model for word recognition.

features (i.e., the unknown spectral pattern) and to compare them against each of the stored vocabulary patterns. The pattern that best matches the input speech features is determined and, if the match is close enough, the decision box provides a response consistent with the recognized word. If the match is not sufficiently close, no decision is made and the user either can repeat the word or choose an alternative way of making the request to the system.

A key problem in comparing speech features from an unknown input against those of stored reference patterns is the problem of time variability, that is, one version of a word can be spoken considerably different from another version of the same word. This means that some form of time alignment and normalization of the two patterns is required to determine the similarity of the spectral features of the two words. Such a procedure is illustrated in Fig. 14. The left of the illustration shows two versions, "Reference" and "Test," of the word "seven." The overall durations are different for the two versions; the Reference version is 30 frames or 450 ms, and the Test version is 35 frames or 525 ms. However, even after normalizing out the overall duration difference, the internal lineups are quite nonlinear, as seen from the lack of lineup of the two sets of vowel peaks.

The solution to the time-alignment problem is to determine a nonuniform alignment path so that the time scales of the Reference and Test patterns are aligned optimally, that is, the highest degree of overall similarity exists for the two word patterns. We show the reference pattern (rotated by 90 degrees) in the middle of Fig. 14, and the test pattern is redrawn at the bottom right. A procedure called dynamic time warping (DTW), a form of dynamic programming optimization, is used to determine the time-alignment path shown at the

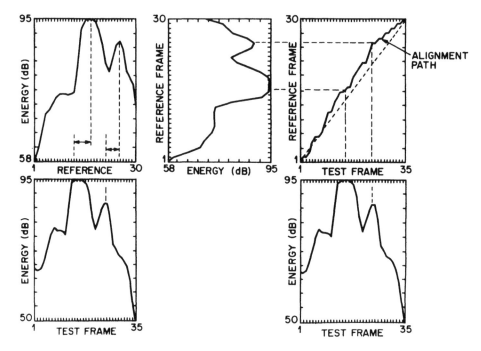

FIG. 14   The problem of time aligning a pair of word utterances. The left side of the illustration shows energy contours of the two words to be time aligned and the right side shows the time-alignment process.

upper right of Fig. 14. In the upper-right-hand frame of Fig. 14 it can be seen that the time-aligned vowel regions (in fact, the whole word) are now in excellent alignment. Figure 15 illustrates this point quite dramatically. Figure 15-*a* shows the linear normalization time alignment of the two contours; Fig. 15-*c* shows the nonlinear time alignment as determined by the DTW match. The degree of improvement in the alignment is indeed dramatic for this example.

The concept of aligning an unknown test pattern against a reference pattern is extended readily from single words or phrases to sequences of words. The way in which this is accomplished, in concept, is by concatenating reference patterns to form longer sequences and matching these longer sequences against the input patterns. In practice, very efficient procedures have been developed to accomplish this task both for connected word and large vocabulary fluent speech recognition.

Performance of Speech Recognizers

Table 1 summarizes, based on laboratory evaluation, the performance of speech recognizers for a wide range of conditions. The technology used is divided into the three commonly defined areas: isolated word, connected word, and fluent speech. The task and the syntax provide constraints that often significantly improve recognizer performance by eliminating from consideration recognition candidates that otherwise would be potential errors. The mode is either SD, in

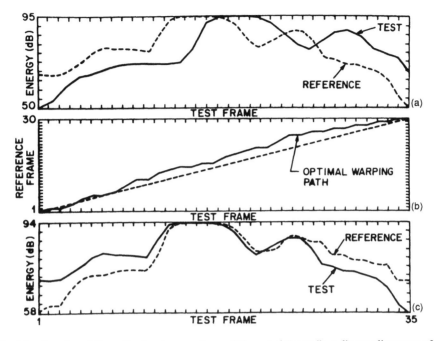

FIG. 15 Results of time aligning two versions of the word "seven": *a*, linear alignment of the two utterances; *b*, optimal time-alignment path; *c*, nonlinearly aligned patterns.

TABLE 1 Performance of Several Speech-Recognition Systems

Technology	Task	Syntax	Mode	Vocabulary	Word Error Rate (%)
Isolated words	None	None	SD	10 Digits	0
				39 Alphadigits	4.5
				1109 Basic English	4.3
			SI	10 Digits	0.1
				39 Alphadigits	7.0
				129 Airline words	2.9
Connected words	Digit strings	Known-length string	SD	10 Digits	0.1
			SI	11 Digits	0.2
	Airline reservations	Finite-state grammar (perplexity = 4)	SD	129 Airline words	0.1
Fluent speech	Naval resource management	Finite-state grammar (perplexity = 60)	SI	991 Words	4.5

which the recognizer is trained separately for each talker, or SI with a single global training set. (For SI evaluations, all talkers used were not part of the training set.) The measure of recognizer performance is the word error rate (in percent) for the given vocabulary, task, and syntax.

As seen in Table 1, for isolated word recognition results are given with no specified task or syntax. This is because very high performance is achieved for a wide range of vocabularies without the added task and syntax constraints. This performance includes a 0.1% word error rate for digits (SI mode), a 2.9% word error rate for 129 airline words, and even a 7.0% word error rate for the spoken alphabet plus digits vocabulary (including the highly confusing e-set consisting of the letters B, C, D, E, G, P, T, V, Z). All error rates are for local telephone recordings.

For connected word recognition, results are presented for a standard vocabulary of digits (10 digits for zero to nine, 11 digits include the pronunciation "oh" in addition to zero), and for a vocabulary of airline words used in an airline's reservation task. It can be seen that word error rates for digit strings are quite low (0.1% for SD mode, 0.2% for SI mode). Similarly, by using a highly constrained syntax for the airline system (perplexity of 4 means that, on average, about 4 words could follow every word in each spoken sentence), the word error rate is very low for this task.

The current state of the art in fluent speech recognition is shown in the bottom line of Table 1. For this task, the naval resource management task, a ship's database can be queried using a 991-word vocabulary and a syntax of perplexity 60 (i.e., a significantly larger syntax than used for the airline's task). The system, when evaluated in an SI mode, has a word error rate of about 4.5%.

Speaker Verification

The basic problem of speaker verification is to decide whether or not an unknown speech sample was spoken by the individual whose identity was claimed. This problem is similar to that of speech recognition in which the problem is to normalize out, in some sense, the individual speaker and extract the message content of the speech. Here, the problem is to normalize out, in some sense, the message content and extract information about the individual speaker. Because of the similarities of these two problems, the processing for speaker verification is similar (with some small differences) to that of speech recognition. The diagram of Fig. 13 applies, except for a speaker-verification system, the set of voice patterns, obtained from a suitable training procedure, for different talkers is stored instead of the voice patterns for different words. The customer, attempting to be verified, provides a voice sample, which is converted to a set of speech features, and a claimed identity. The speech features of the customer are compared, using a time-alignment procedure similar to the one used for speech recognition, to the voice pattern corresponding to the claimed identity and, if a suitable match is obtained, the identity claim is verified.

Speaker-Verification Applications

The major area of application for speaker verification is in access control to information, credit, banking, machines, computer networks, private branch exchanges (PBXs), and even premises. Thus, the concept of a "voice lock" that prevents access until the appropriate speech by the authorized individual(s) is "heard" is made a reality by speaker-verification technology.

Integrated Speaker-Verification System

Figure 16 shows a block diagram of an integrated speaker-verification system in which the customer wishing to be verified provides a claimed identity (to access the appropriate stored voice pattern), the spoken phrase suitable to the verification system, and the transaction requested. A comparison of the spoken phrase with the appropriate stored voice pattern provides a comparison score. Depending on the transaction requested, the decision to accept or reject the identity claim is made and sent back to the customer via a computer speech-answer-back system. Thus, for banking transactions, a much lower degree of match would be required to check an account balance than would be required to withdraw funds.

Performance of Speaker-Verification Systems

Table 2 gives a summary of the performance achieved in laboratory evaluations of a speaker-verification system. The particular system that gave the results shown in Table 2 used digit sequences for both training and testing (in particular, 7-digit test utterances were used). The performance scores shown in the

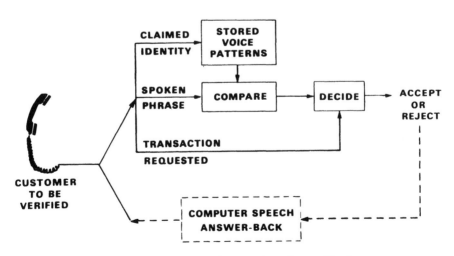

FIG. 16 Block diagram of an integrated speaker-verification system.

TABLE 2 Performance of a Speaker-Verification System Using Connected Digit Sequences as Input Strings

Adaptation	Text	
	Independent	Dependent
Without	3.0%	0.8%
With	2.2%	0.3%

table are equal error rate scores (in percent), which means that these are the scores for a decision threshold set (experimentally, based on measured scores for each talker) so that the true customer is rejected the same percentage of the time an average imposter gets accepted. Shown are results for both *text-independent trials* (those in which the customer can speak any arbitrary 7-digit sequence and the machine does not know the digits), and for *text-dependent trials* in which the machine instructs the customer as to which digit sequence to speak. Clearly, the performance in the text-dependent mode is far superior to that of the text-independent mode because the machine can exploit the known dynamics and spectral content of the speech utterance precisely in making its decision.

Also shown in the table are results without and with adaptation to the changing talker characteristics (individual speaker voice patterns) over time. The best performance of 0.3% equal error rate is achieved with text-dependent mode and with adaptation; the loss in performance without adaptation (0.8% equal error rate) is significantly smaller than the loss in performance in the text-independent mode (2.2% equal error rate). Hence, the extra information obtained from foreknowledge of the spoken string is significantly greater than that obtained from adaptation to the changes in the talker's speech patterns.

Long-Term Challenges

Each of the speech-processing technologies has its own strengths and areas in which more research is required in order to be applicable more universally. In speech coding, the challenge is to achieve high-quality encrypted coding at data speeds in the 2.4–4.8 Kbps range. For speech synthesis, the long-term goal is to improve voice quality and increase flexibility by providing a wide range of voice styles and voice characteristics for a wide range of languages. For speech recognition, the challenge is to improve performance and increase robustness for all recognition modes including isolated word, connected word, and continuous (large vocabulary) speech recognition. Finally, for speaker verification, the goal is to be able to build adaptive talker models based on small amounts of text-independent training, that perform well even for short input strings.

Acknowledgment: The author gratefully acknowledges the invaluable contributions of James Flanagan, Bishnu Atal, Nikil Jayant, Rich Cox, Peter Kroon, Joe Olive, and Aaron Rosenberg in providing ideas, figures, data, and inspiration for the material contained in this article.

Bibliography

Speech Coding

Atal, B. S., Speech Processing Based on Linear Prediction. In: *Encyclopedia of Physical Science and Technology*, Vol. 13, Academic Press, New York, 1987, pp. 219–230.

Chen, J. H., Cox, R. V., Lin, Y. C., Jayant, N. S., and Melchner, M. J., A Low-Delay CELP Coder for the CCITT 16 Kb/s Speech Coding Standard, *IEEE J. Sel. Areas Commun.*, 10(2):830–849 (June 1992).

Jayant, N. S., and Noll, P., *Digital Coding of Waveforms*, Prentice-Hall, Englewood Cliffs, NJ, 1984.

Kroon, P., and Atal, B. S., Strategies for Improving the Performance of CELP Coders at Low Bit Rates, *Proc. IEEE Intl. Conf. Acoustics, Speech, and Signal Proc.*, 151–154 (April 1988).

Schroeder, M. R., Atal, B. S., and Hall, J. L., Optimizing Digital Speech Coders by Exploiting Masking Properties of the Human Ear, *J. Acoust. Soc. Am.*, 66:1647–1652 (1979).

Speech Synthesis

Allen, J., Synthesis of Speech from Unrestricted Text, *Proc. IEEE*, 64:422–433 (1976).

Allen, J., Hunnicutt, S., and Klatt, D. H., *From Text to Speech: The MIT Talk System*, Cambridge University Press, Cambridge, England, 1987.

Coker, C. H., A Model of Articulatory Dynamics and Control, *Proc. IEEE*, 64:452–459 (1976).

Flanagan, J. L., Computers that Talk and Listen: Man–Machine Communication by Voice, *Proc. IEEE*, 64:405–415 (1976).

Flanagan, J. L., and Rabiner, L. R., *Speech Synthesis*, Dowden, Hutchinson and Ross, Stroudsberg, PA, 1973.

Holmes, J. N., Mattingly, I. G., and Shearine, J. N., Speech Synthesis by Rule, *Language and Speech*, 7:127–143 (1964).

Klatt, D. H., Review of Text-to-Speech Conversion for English, *J. Acoust. Soc. Am.*, 82(3):737–793 (September 1987).

Olive, J. P., and Liberman, M. Y., A Set of Concatenative Units for Speech Synthesis. In: *Speech Communication Papers Presented at the 97th Meeting of the Acoustical Society of America* (J. J. Wolf and D. H. Klatt, eds.), American Institute of Physics, New York, 1979, pp. 515–518.

O'Malley, M. H., Larkin, D. K., and Peters, E. W., Beyond the Reading Machine: What the Next Generation of Intelligent Text-to-Speech Systems Should Do for the User, *Proc. Speech Tech. '86*, 216–219 (1986).

Pisoni, D. B., Nusbaum, H. C., and Geene, B. G., Perception of Synthetic Speech Generated by Rule, *Proc. IEEE*, 73:1665–1676 (1985).

Speech Recognition

Bahl, L. R., Jelinek, F., and Mercer, R., A Maximum Likelihood Approach to Continuous Speech Recognition, *IEEE Trans. Pattern Analysis and Machine Intelligence*, PAMI-5(2):179–190 (1983).

Lee, K. F., *Automatic Speech Recognition, the Development of the SPHINX System*, Kluwer Academic Publishers, Boston, 1989.

Ney, H., The Use of a One Stage Dynamic Programming Algorithm for Connected Word Recognition, *IEEE Trans. Acoustics, Speech, and Signal Proc.*, ASSP-32(2): 263–271 (1984).

Rabiner, L. R., A Tutorial on Hidden Markov Models and Its Applications to Speech Recognition, *Proc. IEEE*, 77(2):257–286 (1989).

Rabiner, L.R., and Levinson, S. E., Isolated and Connected Word Recognition – Theory and Selected Applications, *IEEE Trans. Commun.*, COM-29(5):621–659 (1981).

Reddy, D. R., Speech Recognition Machine: A Review, *Proc. IEEE*, 64:502–531 (1976).

Weibel, A., and Lee, K. F. (eds.), *Readings in Speech Recognition*, Morgan Kaufman, San Mateo, CA, 1990.

Zue, V. W., The Use of Speech Knowledge in Automatic Speech Recognition, *Proc. IEEE*, 73(11):1602–1615 (1985).

Speaker Verification

Rosenberg, A. E., Automatic Speaker Verification: A Review, *Proc. IEEE*, 64:475–487 (1976).

Rosenberg, A. E., and Soong, F. K., Evaluation of a Vector Quantization Talker Recognition System in Text Independent and Text Dependent Modes, *Computer, Speech, and Language*, 22:143–157 (1987).

Soong, F. K., Rosenberg, A. E., Juang, B. H., and Rabiner, L. R., A Vector Quantization Approach to Speaker Recognition, *AT&T Tech. J.*, 66:14–26 (1987).

LAWRENCE R. RABINER

Digital Switching Systems

Introduction

Digital time-division switching systems for circuit-switched communications of voice and other voiceband signals are described in this article. These systems are designed to interface with digital time-division transmission systems, and these transmission and switching systems together comprise the two major network elements of circuit-switched telecommunications digital networks. References 1, 2, and 3 comprise a bibliography on digital switching systems; they provide extensive details on the history of digital switching systems. Packet-switched networks and their switching system nodes are covered elsewhere in this encyclopedia, but the convergence of these two techniques in common networks to handle voice and data is discussed below in the recent developments section of this article.

Digital networks take advantage of modern integrated circuit technology to provide the transport and delivery of voice messages that are more resistant to noise, distortion, and variability in loss when compared to analog networks; this is similar to broadcast radio transmission, which has provided superior listening qualities using frequency modulation (FM) instead of amplitude modulation (AM) schemes. Although the analogy is not an exact one, it is no accident that the superior sound qualities of digital networks and FM radio broadcast networks both require more of a constrained resource: the bandwidth available in the transmission media to generate and convey the more robust formats.

Digital time-division transmission systems were introduced commercially in the early 1960s and became practicable from a reliability standpoint with the introduction of such discrete solid-state devices as transistors, diodes, and magnetic cores.

Digital switching had to wait another 15 years for the advances in increased speed and decreased cost, size, and power levels that came with integrated circuits to make these systems commercially practicable. (The adjectives "time-division" and "circuit-switched" usually are dropped for brevity here in describing digital switching and transmission systems in accordance with common usage, but are implied, except as otherwise noted.)

Introducing digital switching nodes in the network eliminated the need for costly signal format transformations required when interconnecting digital transmission and analog switching systems (see Fig. 1).

Figure 1-*a* depicts the telecommunications network as it appeared around 1960. Telephones were connected over metallic loops to analog local switching systems (L_A). All the trunk circuits were either metallic or radio links with repeaters as required. The analog toll switching systems (T_A) completed the picture. The most modern of these systems, in the Bell System's network, were the 4A and 5 crossbar systems for local and toll switching, which used multiple control circuits known as *markers* since they hunted and selected idle network

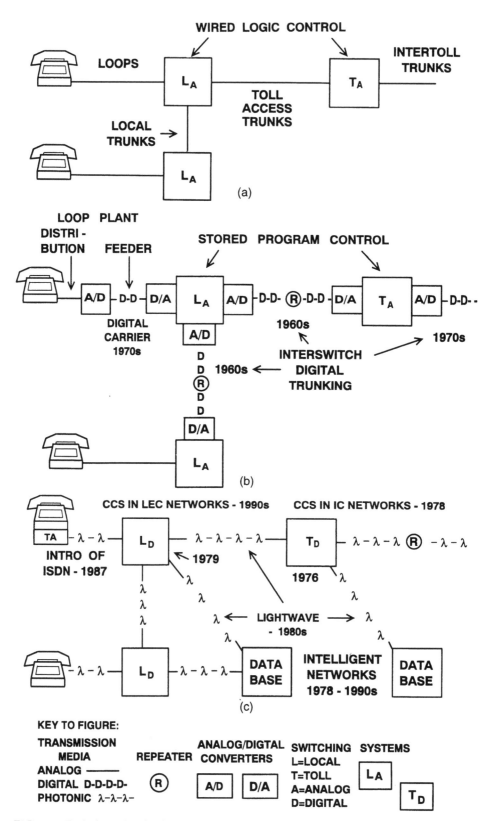

FIG. 1 Evolution of switching networks: *a*, 1960s analog network; *b*, transition to digital transmission (1962–1975); *c*, digital switching, lightwave transmission, and Integrated Services Digital Network (ISDN).

paths by the use of special voltages to "mark" the paths for connection. The markers comprised thousands of complex relay contact logic circuitry and could provide services as sophisticated as Centrex (4).

Figure 1-*b* depicts the first steps toward the Bell System's digital network, beginning with the transmission plant's changeover to digital pulse code modulation (PCM) trunks. This occurred first in short-haul transmission networks in the local networks with a T1 carrier, and then with the use of digital coaxial cable and radio systems for long-haul transmission and with digital loop electronics in the feeder portion of the loop plant between the local switching systems and the final (distribution) portion of the loop to customer telephones (5).

Such stored program control systems as the 1 and 1A Electronic Switching System (1ESS^{TM} and 1AESS^{TM}) switches (with analog space-division network fabrics) in the Bell network began to appear in the local and toll networks, and the 4A crossbar systems had electronic controls added to modernize their control of traffic routing in the long-distance network (6).

The conversion between analog and digital was performed in a remote terminal; this is indicated in Fig. 1-*b* by the boxes in the loop marked A/D (analog to digital) and D/A (digital to analog). Note that to use these digital transmission systems, the signals had to be demultiplexed and converted back to analog before they could be switched by the local and toll switches. Each such transition required this interface. The symbols (D-D) indicate the digital transmission, and occasional circles enclosing an (R) indicate a digital repeater.

Figure 1-*c* represents the last part of the conversion of the North American switched network, beginning with the introduction of toll digital switches (T_D), followed (in approximate order of introduction) by high-speed digital signaling, which has evolved to today's common channel signaling (CCS) (6), local digital switching (L_D), the beginning of centralized databases using the CCS network, and the introduction of Integrated Services Digital Network (ISDN), which has carried digital communications all the way to the user's ISDN telephone, including the personal computer (PC).

The symbols λ-λ-λ in Fig. 1-*c* represent lightwave transmission. Note that the appearance of repeaters has been reduced; this represents the superior loss and distortion characteristics of fiber-optic transmission systems.

Once introduced, the changeover from analog to hybrid to all digital networks has proceeded rapidly; in some cases, some very large national networks have been largely converted to the digital mode in less than 15 years, a rapid switch when one considers the accelerated replacement of large amounts of complex and costly network elements. Several synergistic technological developments have helped stimulate this change and enabled the tens of billions of dollars of investment needed to convert existing networks to the digital mode, a prudent choice for telephone administrations around the world. The technological developments are CCS, which began commercial operation in 1978, ISDN, which began significant deployment in the early 1990s, and fiber-optic transmission, which has grown from very specialized applications in the early 1980s to the media of choice in almost every part of modern networks in the 1990s.

CCS (7) provides a high-speed, fraud-resistant means of signaling among switching nodes in a network. These capabilities alone have stimulated the mod-

ernization of long-distance networks in the United States, which, perhaps seren-
dipitously, accompanied the introduction of digital switching systems to replace
older systems and allow substantial expansion of the network while at the same
time reducing its overall complexity by an order of magnitude and reducing
operating costs substantially. Long-distance calls could be completed with this
signaling network across the United States within 1 to 2 seconds instead of the
10 or more necessary with earlier techniques.

The second force supporting the conversion to integrated digital networks,
the concept of ISDNs, carries the digital signaling techniques directly to the
telephones and other customer premises equipment to provide the new capabilities beginning to be realized in the 1990s. Digital switching systems provide a
natural base for evolution to this technology; analog systems can be made to
provide ISDN functionality by the addition of adjuncts. However, the ineffi-
ciencies of the analog/digital interface persist with the conversion of analog
systems to ISDN. With ISDN, digital networks can accommodate voice, data,
and video and provide additional signaling means directly to users of the new
telephones that provide this kind of service (8).

Even before the introduction of ISDN, network services based on CCS,
known generically as CLASS^{SM} (at one time an acronym for Custom Local-Area
Signaling Service), were tried and now are introduced as forerunners of ISDN.
Out-of-band signaling using CCS allows the introduction of digital control signals that operate over analog or digital links to subscribers (9).

A third element of support for digital networks has been the commercializa-
tion of fiber-optic systems to increase vastly the bandwidth available in terrestrial transmission systems and relieve demands on microwave radio systems—a
fixed resource that otherwise would have been a limiting factor in the development of present-day digital networks. The fiber-optic systems also open the
door for wideband (1.5–2.0 megabits per second [Mbps]) and broadband (150–
600 Mbps) digital networks that are expected to begin appearing in public telecommunications networks in the 1990s.

Historical Background

Digital encoding and transmission of information is truly ancient, going back
to the times of the Grecian Era when bonfires were lit to signal military intelligence over many kilometers at the speed of light. Smoke signals used by the
American Indians also used a robust "on or off" to convey prearranged or
"encoded" information; a more recent singular example is Paul Revere's famous
"one if by land, two if by sea" signal. In the early days of electrical communication, Morse's encoding of signals for telegraphy began commercial application
in the 1860s.

The concept of the digital encoding and decoding of voice signals for use in
multiplex transmission system appeared in a French patent granted to A. H.
Reeves as noted in an article by Delovaine and Reeves (10) of the International
Telephone and Telegraph Company (ITT) in 1938, but was not implemented

commercially until 1962 when PCM transmission systems were introduced in the United States (11).

The development of practical digital transmission systems helped plant the seeds for digital telephone switching systems; perhaps this is most notable in research efforts carried out at AT&T's Bell Laboratories, where the construction and operation of an experimental system known as ESSEX was reported by H. E. Vaughn and others in 1959 (12). Further work on digital telephone switching systems was reported at the International Switching Symposia by the British, French, and Japanese as well as the North Americans in 1960, 1966, 1969, and 1972 (13).

An analog time-division switching system was introduced in the United States in 1963, but the first circuit-switched digital time-division switching system placed into commercial service was France's E-10 system, which provided local service in 1970. (The E-10 system was reintroduced in 1975 to include full stored program control, and is now known as the E-10B.) Six years later, two companies in the United States, AT&T and TRW-Vidar, introduced their digital switching systems to provide long-distance or toll service. The Vidar switch was designed to provide both local and toll service in small installations limited to a few thousand lines or trunks. The AT&T switch was designed to carry the backbone of the long-distance traffic in the United States and was therefore a very large system capable of serving up to 100,000 trunks and completing some 500,000 calls per hour. The AT&T and French systems mentioned here were still in production in 1991 and together serve tens of millions of customer lines and long-distance trunks throughout the world (1,2).

Since the mid-1970s, numerous other manufacturers have designed and introduced digital switching systems into commercial service, largely companies in Canada, France, Germany, Japan, Sweden, and the United States. Most of these systems not only are serving in their country of origin, but are providing service in many countries throughout the world. The development of a local or long-distance digital switching system with a full set of features and supporting operations and maintenance capabilities has a price tag of on the order of $1 billion, and several manufacturers have not survived the competition to participate in worldwide markets. Changes in regulatory policies are sweeping around the globe to eliminate the captive markets and vertical integration within telecommunications administrations, and several major suppliers already have disappeared from the market or have merged with other companies in order to survive.

Fundamental Concepts of Digital Time-Division Switching

Voice communications carries almost all of its intelligence in a relatively narrow band of continuously variable signals ranging from several hundred hertz (Hz) to four kilohertz (kHz); this fact has led to the development of analog transmission and switching systems in the United States and elsewhere to convert this range of audible signals into electrical signals for ubiquitous communication. (It is interesting to note that Alexander Graham Bell was pursuing the quantizing or encoding of analog signals with his work on a "harmonic telegraph," which led to the invention of the analog device, the telephone.)

The concept of reducing analog signals into periodic samples of the voice waveform taken at a rate of 8 kHz and then quantizing and encoding those samples into binary format is described for PCM and digital transmission in many articles and texts (3,5,11). These same references also describe the interleaving or multiplexing of these signals that allows a number of speech signals to be transmitted over a single pair of wires (or a single radio channel) to conserve equipment and reduce capital and operating costs.

The digitally encoded signals that have been subjected to noise and distortion can be reconstructed periodically so that the information can be reconstituted into clean sets of ones and zeros. This periodic refreshment of the signals essentially eliminates the progressive quality degradation of the signals transported in analog networks.

By demultiplexing and decoding the binary signals, a replica of the original voice waveform can be directed to the telephone receiver and the listener's ear. The scheme of multiplexing, transmission, and demultiplexing of digitally encoded signals has been extended to achieve digital switching.

Time-Division Switching Using Digital Multiplexers and Demultiplexers

A time-division switch can be constructed with a digital multiplexer, a digital demultiplexer, and a short bus to interconnect the outputs of the multiplexer to the inputs of the demultiplexer. A control memory is added to rearrange the multiplexing sequence. By altering the contents of the control memory, inputs can be connected in an arbitrary fashion to selected outputs.

As shown in Fig. 2, by simply rearranging the order of multiplexing (alternatively, the order of demultiplexing) the information appearing on the input channels can be selected arbitrarily on the output channels. The rearrangements are accomplished by changing the contents of a control memory that associates the individual input of the multiplexer with a particular time-slot t_i in a repeating frame of n time-slots (t_1, t_2, . . . t_n). (It is assumed that the time-slot relationship for the demultiplexer is fixed in Fig. 2 always to select C, D, . . . , to be associated with the first, second, . . . , time-slots. It also is assumed that each time-slot carries a full 8-bit sample of the encoded signals from such inputs as A and B.)

In the first example (i), the control memory is loaded with the sequence A, B, . . . , and input A is sampled at t_1 and delivered to the demultiplexer that connects the digital stream to output C. During time-slot t_2, input B is multiplexed to the digital stream and the demultiplexer connects the output of the stream to output D. By rearranging the time-slot memory as in (ii), B is connected to the first time-slot and the output is delivered to C, A is delivered to D, and so on. Finally, by repeating A one or more times in the multiplexing control memory as shown in (iii), it will be multiplexed two times and "broadcast" to both C and D.

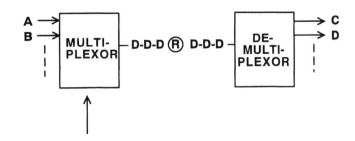

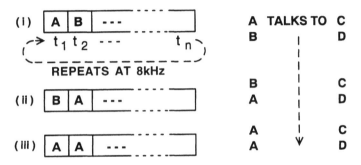

FIG. 2 Time-division bus.

The concept can be extended to as many channels as can be multiplexed practically. The number is limited by the bandwidth capability of the electronics doing the multiplexing and rearranging. Also, one can note that the demultiplexing process can be rearranged so that both *C* and *D* hear the signals from *A* and thus selective broadcasting is possible.

The next section describes the use of a time-division bus in an early time-division switching system.

Time-Division Switching in Analog Format

Observe that rearranging of the encoded information could be done with the analog samples without first resorting to binary encoding and decoding to obtain time-division switching of analog signals. Such techniques, along with sophisticated means to minimize the loss of signal amplitude in switching only brief samples of the analog signals (using a technique invented by G. Svala of L. M. Ericsson), led to the commercial introduction of such a switching system in 1963 in the United States. AT&T's 101ESS™ system was chosen for application in private branch exchanges (PBXs) or private automatic branch exchanges (PABXs), and no attempt was made to gain any synergies with digital time-division transmission systems. It therefore could interconnect more voice channels on a single pair of wires but had to convert to either pure analog or pure digital format in order to be able to interface with commercially available transmission systems. And, importantly, the analog samples were just as subject to noise and distortion as the original analog voice signals. Advances in technology for switching and digital transmission systems foreclosed further development of this approach.

Time-Slot Interchange Switching Techniques

Time-division bus switches are limited in bandwidth and hence the number of channels that can be switched; this applies to digital switches as it does to the analog time-division systems just described. The time-division bus, as shown in Fig. 2 with its multiplexers and demultiplexers, includes the conversion between spatially separated signals and their multiplexed counterpart in the time-division domain. Such an arrangement assumes that separate incoming and outgoing lines or trunks are provided for each voice channel to be switched.

To be able to switch both larger groups of inputs and incoming and outgoing signals in a time-multiplexed format required the invention of several new switching techniques. One of these, the time-slot interchanger (TSI), which is found in almost all of today's digital time-division switching systems, is described next.

Figure 3 depicts a TSI. A repeating stream of signals represented by $A, B, \ldots, N$, is presented to the input. During each of the time-slots t_i the control memory directs the information contained in the incoming stream to a memory location a_j according to its own contents. In the example depicted, during time-slot t_1 the input A is directed to location a_2, and during t_2 the input B to location a_1, and so on until the frame again repeats. The memory read-out is directed

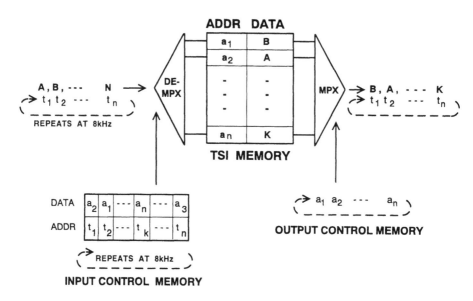

FIG. 3　Time-slot interchanger (TSI).

to the output of the TSI in the ordered sequence t_1, t_2, . . . , and thereby delivers the outputs $B, A, . . . , K$. Just as described above for the time-division bus, the control memory has reordered the sequence of information, in this case in time-division format at the input and output.

The simplistic diagram does not include the possibility of race conditions that could occur in attempting simultaneously to enter and remove information stored in the same location of the time-slot memory; this is a design detail that can be handled in several different ways. For example, the race condition can be avoided by adjusting the relative timing and pulse widths of the t_i driving the input and output control memories.

Note that information entered in a particular memory element may not be read out or removed until almost a full cycle has gone by, typically 125 microseconds (μs) for TSIs operating in synchronism with T1 digital transmission systems. (The time frame usually is determined by the connecting transmission system; in this case, it is assumed to be the 24 channels of the North American Digital Signal Level 1 [DS-1] digital PCM standard.)

As a consequence of the variable delay in switching of the different channels, inputs arriving in one time frame may leave in different time frames. This is not a problem if the input signals (e.g., A and B) are unrelated conversations. However, if there is intent to group A and B into a coherent 128-kilobits per second (Kbps) signal, the actions of the TSI as described here may corrupt the coherence randomly. To maintain such coherence, known as *time-slot integrity*, takes more complex TSI logic and may result in further signal delays.

TSIs can be multiplexed and cascaded to build larger switching fabrics to handle more inputs and extend the effective bandwidth of digital switching systems. However, as more TSIs are cascaded, the cumulative delays are increased and, unless controlled in large networks, create excessive overall delay and echo that will impair speech and data communications to an unacceptable level. Other switching techniques, discussed in the section, "Architectures for Digital Switching Systems" below, are used to achieve overall traffic capacity and delay objectives required for national and global digital telecommunications networks.

Bilateral Networks and Their Control

Also note that the simple example above for the time-division bus and the TSI indicates A and B are talking and C and D are listening. Most practical digital circuits are unilateral, so a corresponding copy of the system must be made to allow C and D to be heard by A and B, and the contents of the control memories must be coordinated correspondingly.

Two techniques to achieve two-way connections are possible. In the first, a second copy of the network fabric is slaved to the control of the first to provide the dual speech paths. In the second, a single network is designed but the control must map the selected path through the space and time stages of the network fabric into a second set of terminals and interstage links to provide the second speech path. Note that, in the second technique, the second connection may not be a simple "mirror" of the first connection in order to avoid interconnection conflicts. In this case, the network interconnection and the control means must be designed to operate a simple translation algorithm to convert the connections of the first path into the correct links of the second path.

Synchronism of Digital Networks

Just as for digital transmission systems, the switching systems of digital networks must be designed to maintain synchronism with the other network elements to minimize the introduction of excessive impairments in the signal being conveyed. Very accurate electronic clocks are embedded in the control circuitry of digital switching systems to ensure that the digital waveforms do not lose valid information or acquire "extra bits" to distort the information being carried. Techniques are employed to allow slower or faster network elements to absorb some of these drifts over time, and the occasional loss of a bit usually results in faint clicks or pops being heard in the reconstructed voice signals. The consequences of lost bits are more serious for the transmission of data, and error-checking techniques must be employed to detect these errors. These problems and techniques for avoiding or remedying them are well understood in the design of digital transmission systems, and the concepts and techniques also have been incorporated in the design of digital switching systems.

Minimization of Delays in Digital Switching Systems on Overall Network Performance

As signals pass through the stages of a digital time-division system's network fabric, they will encounter delays and must be retimed before leaving the switch on outgoing digital transmission systems. This means, for example, that a maximum added delay of 125 μs can be encountered in each TSI stage in a digital switching system. These delays, when accumulated over a number of network elements, create undesirable delays in speech transmission and confusing echoes. Echo cancelers are provided in long-distance toll networks to overcome or minimize these problems. Buffering and retiming may be required within various stages of the switching fabric of a digital switching system, but there are constraints on the total allowable delay incurred in both local and long-distance digital switching systems within a digital network to meet an overall delay requirement.

Interface between Analog Switching Systems and Customer Premises Equipment

The interface between a central-office local switching system and the telephone lines it serves is the *line circuit*. The line circuit conveys power, speech, and signaling information between the telephone and the switching system and also must protect the switching system from the exposure to interfering and hazardous signals in the many kilometers between the telephone and the switch. The design of the line circuit can be crucial to the economic advantage of a switching system. Each component added or saved in the line circuit can be significant in the overall cost of the switching system. If a digital switching system serves analog customer lines, then it has to perform the sampling/integrating, encoding/decoding, and companding of analog-to-digital/digital-to-analog conversion for presentation to or reception from the digital switching system. In the first-generation digital switching systems appearing from the late 1970s to the mid-1980s, the per-line cost of performing these and other interface functions was high, and some of these systems required space-division concentration switching of analog signals in the switching network fabric before the conversion process was performed.

Other conversions and protections were needed for interfacing with the rugged environment required and presented by the traditional pair of telephone wires connecting the customer's telephone to the local switching system. Battery and ringing signals transmitted through an analog switching system could not be handled directly in a digital switching system, and the traditional protection circuitry sufficient for metallic-based switching systems would not be adequate for digital switching systems. Induced power voltages and lightning strikes have tremendous energies that could destroy relatively delicate microcircuitry found in digital switching system network fabrics. Combinations of developing more rugged microcircuits along with more effective voltage and current protection circuitry have been required.

Figure 4 illustrates the critical elements involved in the interface between analog customer lines and a digital switching system that have led to the mnemonic acronym BORSCHT (battery, overvoltage, ringing, signaling, CODEC [coder-decoder], hybrid, test). The acronym serves to point out the critical areas of change in design that tend to penalize digital switching systems in their consideration for replacement of their analog forerunners.

Figure 4-*a* summarizes the issues at hand. The protector circuit (O = overvoltage), the first element of the interface between the telephone loop and the switch, has a stringent job to ensure that lightning, power crosses, and induced voltages from power lines sharing telephone poles or underground ducts will not damage very-large-scale-integrated (VLSI) line and network fabric circuitry.

Second, the high-voltage, high-current signals for office battery (B), 100-volt, 20-Hz ringing (R), dial-pulse line signaling (S), and test (T) access cannot be passed through the switching fabric as in analog switching systems. Special bypass arrangements are needed to provide these functions, and tend to add significantly to the cost and size of the line circuit equipment.

Third, Fig. 4-*a* shows the interface between the 2-wire bilateral analog communication between the telephone and the switch and the 4-wire unilateral signaling presently required in digital switching systems. Hybrid (H) circuits are required to split and merge the 2-wire and the 4-wire modes.

Finally, the coding and decoding of digital bit streams into continuous analog voltages and currents is required. The combined function is known as a CODEC. The design of microcircuits specifically for the line interface applications and the economies of scale in building line interface equipment for tens of millions of telephone lines has helped reduce the relative cost of the BORSCHT function (1–3).

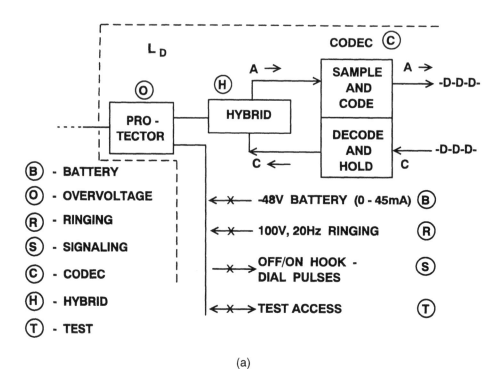

(a)

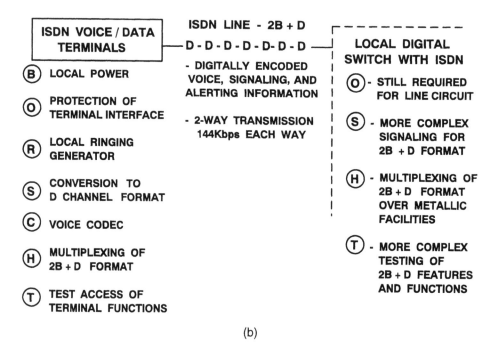

(b)

FIG. 4 Line interfaces in digital switching systems: *a*, BORSCHT interfaces; *b*, extension of BORSCHT functions to ISDN terminals.

With the advent of ISDN, new line interfaces have been developed with more complex control circuitry to handle the two digital voice channels (B channels—each such channel comprising 2-way digital voice or data channels operating at 64 Kbps in each direction) and a new signaling channel (the D channel comprises an out-of-band 2-way digital signaling means operating at 16 Kbps in each direction) to provide the $2B + D$ format. ISDN signaling of two one-way digital bit streams over a pair of wires between the switch and the telephones it serves has been a challenge. Indeed, the standards for this signaling stabilized after some substantial introduction of commercial switching and terminal equipment, which had to be reworked (8).

In addition, some of the control interfaces and other functions in the switching system line circuit represented by BORSCHT have been modified and/or moved to the ISDN telephone set premises. Figure 4-b shows the migration and demonstrates several new issues. First, the office battery may not be sufficient to power the integrated circuit interface at the customer premises. Therefore, the telephone connection is at risk in the face of a power outage. This can be minimized by having the switching system provide battery power and allow minimal services to continue in the face of loss of power at the customer's location.

Second, there are now delicate VLSI circuits at the location of the telephone, and protection now is required at both ends of the pair of wires connecting the telephone to the central office. The extensive utilization of VLSI in these digital and ISDN interfaces now places more logic in individual line interface circuits or a customer's telephone than could be found in entire first-generation electronic telephone switching systems that served tens of thousands of analog telephone sets.

Third, the interface between the ISDN line and the ISDN telephone set can be either "T" or "U". This implies that functionality either can be owned by the customer (U) or the telephone company (T), and some ISDN station sets may require terminal adapters (TAs) to convert from the 2-wire U interface to the 4-wire T interface (8). The different interfaces are due to regulatory differences between the United States and Europe. This creates an additional challenge for the designers of "world-class" digital switching systems to serve this market diversity.

Architectures for Digital Switching Systems

Digital switching systems represent a second generation of electronic telephone switching systems. The first generation is represented by the introduction of a centralized, computerized control of a network fabric for the actual interconnection of customer lines, trunk circuits to other switches, and to service circuits that provide signal sources and detection means to provide dial tones, detect dialing signals from telephones, and provide busy tones, ringing signals, and so on.

The higher cost of providing a number of control equipment units operating at millisecond speeds for each logic decision is outweighed by that of the lesser amount of control equipment operating at submicrosecond speeds. Consequently, the focus was largely on high-speed, highly reliable, special-purpose data processors that featured very large programs and databases that could not have their data mutilated or erased easily by the vagaries of nature or errors in operation, administration, or maintenance of the switching system. An important advantage of the centralized control was that it did not interact with other distributed controls and compete for resources in carrying out its tasks. However, the single high-speed control ultimately would become limited in the total number of tasks required per unit time as the number of lines and trunks served increased and the number of special services it could provide continued to increase.

In central-office environments, the large size of most systems required that the network fabric be metallic (crossbar or reed switches, e.g.) since electronic elements were too expensive and too delicate to carry the required signals. It should be noted that, in this regard, new central-office telephone switching systems must interface with existing network elements for decades and therefore must cope with the requirements imposed by interfaces to older existing systems. The first generation of electronic switching systems introduced in the United States in the 1960s and 1970s had to interwork with equipment that had been placed in service as much as 50 years previously.

A characteristic of space-division circuit-switching systems is that they are designed and engineered for a certain level of traffic. These systems are designed to handle statistical peak loads in which, during most busy hours of communication, not more than 10% to 20% of the telephones would be engaged in calls at any given instant. During only a few days, or during such unusual events as earthquakes or other major catastrophes, would the traffic offered to the switch surge beyond the calculated estimates.

To design space-division systems to handle such rare occurrences of unusually high traffic would be far too costly in terms of the size of the network fabric and the number of switching crosspoints required. For traffic offered to a switching system (or a network of systems) that exceeds the engineered levels, calls trying to enter the system that already is operating at its engineered capacity are turned back and are said to be "blocked." Typical space-division network fabrics are referred to as blocking networks. Only special space-division switching systems, usually very small in total capacity, have been equipped with nonblocking network fabrics. Digital time-division switching systems can be designed much more easily and economically to be nonblocking, or very nearly so, with substantial savings in operation and administration. Some advantages of the nonblocking systems are noted in the next sections.

The introduction of digital switching systems has had much more impact on the network fabric and the line and service circuits interconnected by that fabric, but there has been significant architectural impact on the design of the control as well. First, a look at the fabric and interconnecting circuits is undertaken, followed by a discussion of the trends in the control portion of digital switching systems.

Time-Multiplexed Switch

TSI switching capacity is presently limited to a few hundred interconnections. Since there is a need for tens of thousands of simultaneous interconnections in large digital switching systems, a single TSI is impractical. One can concatenate and interconnect TSIs to build a larger network fabric out of smaller pieces, but the cost of buffering and delays suggests the use of other topologies. The time-multiplexed switch (TMS) (Fig. 5) provides an alternate element that uses to advantage the reduced demand on bandwidth of the familiar space-division matrix switch or crossbar employed in metallic switching fabrics. It brings together a number of multiplexed digital streams and allows signals in the same time-slot to be transferred to different output streams, a spatial switching arrangement much like the traditional space-division matrix or crossbar switch. However, the TMS has an advantage: with each new time-slot the interconnection of inputs to outputs can be rearranged instantaneously — again through the judicious use of control memories, just as previously noted for the TSIs.

Since it does not reorder the time-slots of incoming signals, the retiming delays of a TMS can be minimized. A TMS can be interspersed with TSIs and other TMSs to build larger switching system interconnection capacities. For example, the 4ESS™ switch employs two stages of TSIs with two interim stages of TMSs to attain a nonblocking network of 100,000 terminals, far larger than practicably achievable with space-division switching systems.

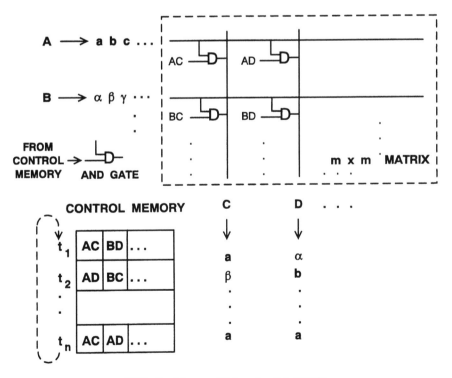

FIG. 5 Time-multiplexed switch (TMS).

It is also worth noting that nonblocking network fabrics reduce the administrative costs of balancing traffic over the network fabric as additional circuits are added to the switch. Nonblocking space-division networks of any useful size in central-office switching applications would be prohibitively expensive, but nonblocking fabrics are quite common in digital switching systems. In blocking networks, the addition of lines and trunks has to be distributed over a number of such switching fabric units as TSIs or TMSs, and existing connections may have to be rearranged to accommodate the additions, but nonblocking networks do not require such rearrangements, greatly reducing administrative expenses in adding circuits over time.

Switching System Networks Using Stages of Time-Slot Interchangers and Time-Multiplexed Switches

As already noted, by concatenating stages of TSIs and TMSs, one can increase the number of time-multiplexed streams by using the time-division shifting of the TSIs and the spatial interconnectivity of the TMSs. The network fabrics of commercial digital switching systems employ various interconnections of these kinds of elements to achieve the fabric design required to provide both small and large systems and, very importantly, the ability to expand or "grow" an installed switching system from small to large as demands increase over the working life of such systems. To ignore or minimize such growth requirements in the design imposes an economic disadvantage on that switching system.

A notational scheme to indicate different concatenations of TSIs and TMSs was developed by A. E. Joel, Jr., and used to describe different topological implementations of the various digital switching systems. The notation denotes a stage of TSIs as T (time) and a stage of TMSs as S (space). Thus, a T-S-T network consists of an array of TSIs connecting to an array of TMSs that in turn connect to an array of TSIs (Fig. 6) (1). It is interesting to note that no one configuration is necessarily the best since many different topologies have been implemented and are enjoying commercial success in the telecommunications marketplace. The overall goal is to achieve a design that is practical for as wide a range of applications as deemed suitable for the market.

Expansion and Modularity of Digital Switching System Network Fabrics

As part of the design consideration for large ranges of digital switching system networks applications, switching networks generally provide modular network elements (usually TSIs and TMSs) that permit a range of 10 to 1 or more in size of the number of lines and/or trunks served. The ability to enlarge a system without having to take it out of service and replace it is a major consideration for telephone administrations.

The lack of the ability to expand an in-service system forces telephone administrations to buy too many smaller systems with more control and less-efficient trunking and networking costs than necessary or to buy switching systems with unused capacity over a longer period of time. Both of these choices

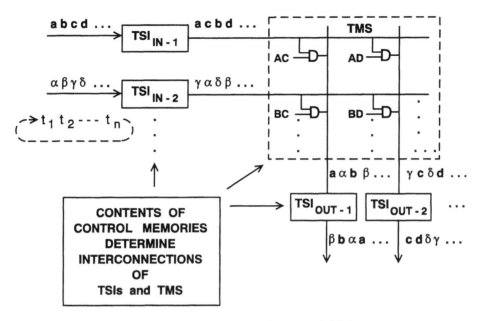

FIG. 6   Time-space-time (TST) network fabric.

will result in more costly networks over the long term, and unnecessarily raise the cost of telecommunications to the end users.

Another factor in the design of a good switching system's network fabric within a digital switching system is the ability to use those modules to distribute the system over a substantial geographic area. Since the links between digital network modules are time-multiplexed signals and digital systems can provide direct digital interfaces to transmission links, such systems allow more economic distribution of modules over a distance of tens of kilometers as compared to those with space-division network fabrics.

Trends toward More Distributed Switching in Local Digital Switching Systems

Again, fiber optics play a role in distributing the functionality of switching systems. The traditional small rural switching systems known as Community Dial Offices (CDOs) are being replaced by remote multiplexing systems and remote modules of the larger centralized switching systems. The previously mentioned direct digital interfaces and the increased reliability and decreased costs of fiber-optic systems have made this alternate means of replacement more attractive. This tends to reduce the overall cost of the switching network and brings the powerful operations, administration, and maintenance advantages of large central switching systems to the remote locations for which the expense of providing full-time craftspersons would be prohibitive.

Distribution of switching modules and their controls in a digital switching network permits a telephone administration to replace CDOs with remote modules of larger switching systems with the advantages previously noted. For example, in the United States, before the advent of digital switching systems there were more than 5000 such systems in the Bell Operating Companies and a similar number in the other operating companies. Figure 7 shows an example of such an arrangement employed in local digital switching systems such as AT&T's 5ESS® switching system. In Fig. 7, the remote modules are arranged to complete calls between its subscribers without necessarily having to extend the connection through the central switch module.

By extending and distributing the centralized control judiciously to these individual modules, more processing capability can be attained. Experience has shown that much of the control can be distributed except for those calls that require special service capability or must make interconnections to lines or trunks in other switch modules. This allows the "stand-alone" approach to work for the remote modules that serve as replacements for CDOs. If the central switch fails or is cut off from the remote module, that module can continue to serve its community of interest, including local police, fire, and medical services, in spite of the loss of central control or connectivity. Figure 8 emphasizes this point. By distributing the control, the amount of control and memory required is well within the reach of microprocessor-based systems, and the switching system cost and performance benefit from the ever-increasing advances in microprocessor and memory chips to be found in the present generation of desktop computers and workstations.

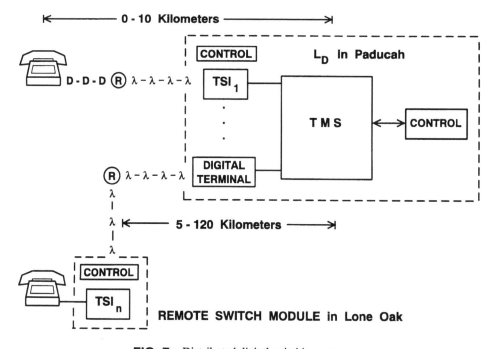

FIG. 7   Distributed digital switching systems.

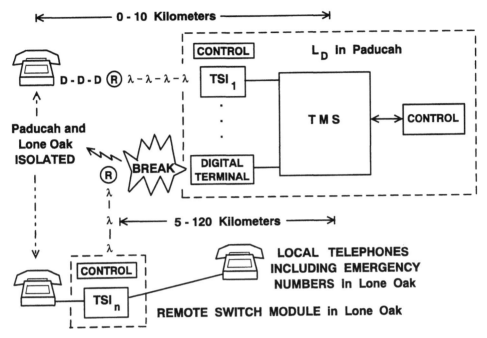

FIG. 8 Local survivability with distributed control.

Recent Developments and Some Future Trends

Beyond 64 Kilobits per Second

With the introduction of ISDN, the ability to switch two or more 64-Kbps channels on a point-to-point basis from end user to end user has been launched in the telephone network. Combining several B channels to obtain a higher bit rate of communication has applicability to high-quality audio, data, and video applications. Larger groupings of such 64-Kbps channels as 384 Kbps (B channels) and 1.5 megabits per second (Mbps) are particular instances that provide capability for compressed video communications. However, almost all present digital switching systems are based on connecting only 64-Kbps channels and do not preserve the format required for multiple channel conveyance of information. For example, the TSIs in digital switching systems placed in service in the 1970s and 1980s do not preserve the relative time order of the grouped channels in the switching process. This destroys the grouping relationship.

Nontrivial redesigns of the digital switching system network fabrics are necessary to overcome this limitation. Maintaining time order of the individual segments when passing through TSIs requires additional memory and increased delay of transit time of digital signals through the switching system. The maintenance of this order is known as *time-slot integrity*. Newer systems will have to either switch more bits per time-slot or add hardware specifically to maintain the time order of time-slots.

Modifications are beginning to appear that augment present-day 64-Kbps switching systems to handle several specific multiples, 128 Kbps on ISDN Basic Rate Interfaces (BRIs) and 384 Kbps over Primary Rate Interfaces (PRIs). More general $N \times 64$-Kbps solutions should be forthcoming.

New Network Elements: Long-Term Connections

Both analog and digital switching systems generally are designed to set up and take down calls on a call-by-call basis. The connections in local exchanges usually are established in a matter of hundreds of milliseconds or less, after the customer has completed dialing the last digit. Once the call has ended, the connection is taken down almost immediately and "forgotten" (except for charging records retained for billing purposes) by the switching systems and the networks in which they are embedded. Tandem and toll switching systems operate the same way. There are longer-term connections that need to last for weeks, months, or years and for which establishment and removal can be made by placing verbal or written orders with telephone system administrators. One application is the provision of special high-quality audio connections for broadcast-quality voice circuits between locations served by switching nodes. Another application is "foreign exchange" lines that permit a customer to reach a business by placing what appears to be a local call, but the call actually completes over a private facility to a distant local exchange actually serving that business. The business pays for the service to extend free calling to its customers over a larger area.

These special services require special processes and equipment to provision, establish, and monitor them. Semipermanent connections made on mainframes or distribution frames have been the means of connecting switching systems, transmission facilities, lines, and trunks. These same frames are used to connect these special services. In this way, switching equipment is bypassed and not tied up for such connections, and there is no risk of losing connections because of switching system outages.

Using a switching system for semipermanent connections was not practical with space-division technology; semipermanent or "nailed-up" connections are becoming practical with digital switching systems and provide an alternate means for providing special services. Time-division switching systems can provide a lower degree of blocking for the same cost compared to space-division systems, and hence provide the difference for nail-up.

Other means of aggregating and rearranging transmission facilities between switching nodes appeared in the early 1980s to reduce the need for manually rearranging connections among digital transmission systems. A special class of digital switching systems known as Digital Cross-Connect Systems (DCSs) were introduced. These new network elements allowed remotely controlled rerouting of individual 64-Kbps channels and rearrangement of groups of channels at ever-increasing aggregate bit rates to meet changing traffic needs. DCSs serve to meet another part of the special service needs and have taken advantage of the relative simplicity of rearranging digital bit streams instead of providing comparable functionality with analog multiplexed transmission systems. To some extent, DCSs complement or compete with the notion of nailed-up connections within digital switching systems, but contribute to an overall more efficient network as evidenced by their rapid introduction into large telecommunications networks.

DCSs allow the remultiplexing of digital channels at 64 Kbps and up within higher bandwidth digital transmission multiplex rates from 1.5 Mbps to 2 gigabits per second (Gbps) and beyond. However, DCSs can maintain, but not create, time-slot integrity.

Another use of multiple-bit-rate facilities, known in the United States as *fractional T1*, allows business customers to lease facilities at multiples of 64 Kbps but below the full 1.5-Mbps rate of a T1 facility. This offers intermediate discounts on multiple 64-Kbps channels. Like DCSs, fractional T1 service only can maintain time-slot integrity.

Wideband Networks: Packet-Switching Systems and Frame Relay Transmission Operating at 1.5–2.0 Megabits per Second

Packet switching has its own approach to handling data traffic which has been different enough in its characteristics to warrant different networks with different switching system designs and capabilities. The possibility of integrating circuit- and packet-switched networks into high-speed packet-switching systems has been proposed and studied for many years. One objective is to gain the efficiencies of one network instead of two. Voice, data, and low-bit-rate video could be fitted into the 1.5–2.0-Mbps Digital Signal Level 1 (DS-1) digital transmission streams of North American and European networks, and some systems have been introduced to implement this mix of traffic. The issue of grouping channels and maintaining time-slot integrity is avoided.

Frame relay, a transmission system designed for wideband networks, is the first digital multiplexing to use a rate of 1.5 Mbps (long-established DS-1 transmission technology) in the United States; it uses X.25 protocols but includes certain changes in addition to the higher speed of operation to provide faster network operation.

First, frame relay does away with error control within the network so that lengthy packets do not have to be stored at each switching node and checked for transmission errors before sent on to the next link. End-user equipment must provide the capability of checking for errors and requesting retransmission. Frame relay relies on the inherently much lower error rate in modern digital networks to allow end-user error checking to be practical.

Second, customers of geographically dispersed data networks must designate their expectations of data traffic intended for different destination nodes in their end-point equipment. The network has some capability of reducing the bit-rate of information transmitted for voice or video traffic to continue to maintain no loss of transmission in the face of short-term increases or bursts of data traffic for which bit error rates must be much lower. But user constraint still is required in order not to overload the network and interfere with the traffic of others. Enforcement of such constraints will confine the impairment of traffic largely to nonabiding subscribers. Third, the switching of variable-length packets has been retained for efficiency in overall network performance.

Frame relay operates with "packet-stuffing" techniques that fill in the gaps of asynchronous data traffic with empty packets introduced to maintain the frame transmission rate of 1.5 or 2.0 Mbps. The empty packets are discarded at the receiving terminals. Private and public networks are being introduced, mostly for high-speed data transmission as opposed to integrated voice–data networks. The speed of transmission in wideband networks is estimated to be approximately 10 times faster than predecessor X.25 networks that operate at tens of kilobits per second.

Switching beyond 1.5–2.0 Megabits per Second: Broadband Integrated Services Digital Network

Several generations of digital transmission systems have increased the capacity from tens of megabits per second to several gigabits per second, largely because of the introduction of fiber-optic systems. This has led to designs and plans for future digital switching systems that can interconnect and switch digital bit streams running at 150 Mbps to more than 600 Mbps. The regime of broadband communications has been defined as that going above the 1.5–2.0 Mbps of wideband digital networks described above. These extremely high rates of network interconnectivity have arisen in planning for Broadband ISDN (BISDN). The vision of BISDN includes much more than higher bit rates; the ability to serve a wide range of bandwidths simultaneously, ranging from voice and slow-speed data to very high speed data and high-resolution, full-motion video (e.g., high-definition television [HDTV]) is a part of the proposed extension of ISDN.

With these bandwidths, the challenge to handle, smoothly and efficiently, two different kinds of traffic in the wideband regime has become a much more severe challenge in the broadband regime. One now has to consider a range from 64 Kbps for voice to 150 Mbps for high-quality video (neither of which can tolerate extensive delays or noticeable variability in those delays) and highly bursty multimegabit-per-second data that can tolerate delay variations but, as noted in the previous section, requires much lower bit error rates.

As in wideband networks, node-by-node error checking is dropped, and packet stuffing is utilized to maintain synchronism in the face of lulls in the characteristic burstiness of data communication. Also, there is a need for customers to limit the burstiness of their traffic and understand that exceeding those limits will cause impairment of communication.

A significant departure in evolving from wideband to broadband is to accept the loss in efficiency by returning to fixed-length packets and thereby returning to a more traditional design of circuit-switched digital time-division fabrics. Broadband switching fabrics use the TMS and TSI techniques to direct dozens of bytes (or octets) as the unit of switched signal instead of the single 8-bit byte signal switched in the present generation of digital switching systems. The fixed-length packet allows a considerable simplicity and speed of operation that more than compensates for the potentially higher efficiency of operation heretofore achieved with variable-length packet-switching fabrics. Broadband switching systems not only operate at much higher speeds but also use a number of switching techniques not found in traditional digital switching systems.

Broadband Circuit-Switching Systems

Handling tens of megahertz of analog signals, small, specialized space-division switching systems and networks serving broadcast television have been operating for 30 years in the United States. Several experimental broadband networks of switching and transmission systems were deployed in the 1980s in Germany (BIGFON), France (Biarritz), and Japan (INS). These systems used both traditional analog and digital switching techniques and pushed the design of VLSI switching chips to achieve the higher speed of operation.

Broadband Systems with Variable Bandwidth Switching Capabilities

"Smart" or self-routing switching network fabrics have been a longtime dream that actually was realized in at least one trial system by AT&T in the early 1960s and revived again in the 1980s and 1990s. This revival depends on the ability to realize economically very complex switching fabric controls that can operate at gigabit-per-second rates on the information to be processed.

Self-routing networks known as *banyan networks*, made of two-by-two crosspoint switches and their interstage connection links, have been advanced and studied for several decades. (They were so named because of their complexly interwoven links that interconnect stages of crosspoint switches, reminiscent of the intertwined roots of the banyan tree.) The simple elements — two-by-two crossbar-type switches — allowed engineers to imagine and construct on paper switching networks that would be hybrid circuit and packet time-division digitally controlled switches. Each time-slot would carry a fixed-length signal of digitally encoded information, but each fixed-length packet included a "header" of bits that steered the signal through the network fabric bit by bit from one stage of the fabric to the next to arrive ultimately at the intended output port of the switch.

Figure 9 shows how the crossbar switch works in the banyan network. Note that it is a time-division switch with control logic associated with each switching element to "decide" whether the signal passes in the cross or bar mode of operation. (The figure shows the ability to process two incoming signals and have them emerge as they arrived or interchange their order in leaving the two-by-two switching element.) If two inputs arrive at the two-by-two switch that require different modes, then one signal will have to be blocked and control logic will have to decide which signal will be passed. (Signals in the header portion of the arriving packets can be used to arbitrate, e.g., such continuous data as voice or video could preempt a competing data packet.) The banyan networks have certain rules for the interconnection of stages so that one could develop arbitrarily large networks in a recursive fashion. Buffering could be supplied at the input of a banyan switch, at the output stages, or distributed within the fabric at each switching element to allow for the random simultane-

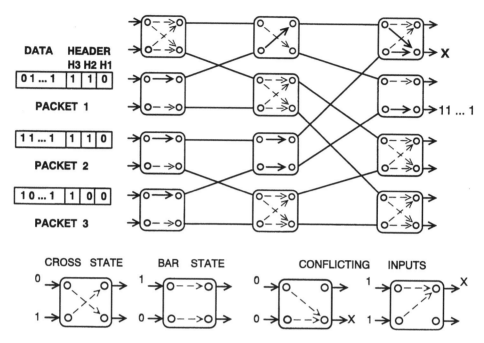

FIG. 9   Self-routing banyan network.

ous arrival of packets from different sources intended for the same destination. The richness of VLSI capabilities allows one to consider these options as reasonable design choices. Just as in the case of space-division networks, one can design network topologies that require more two-by-two switching elements but always will avoid collisions of packets competing for cross and bar connections within the network fabric.

Alternatively, the internal operation of the fabric can be made to operate several times faster than the rate of arriving and departing time-slots also to reduce or eliminate blocking in the network. This technique, known as *fast packet switching*, gains logical simplicity at the expense of the net bandwidth capacity achievable. Again, the ability to provide nonblocking network fabrics using time-division techniques has proved to be less costly than that needed for space-division networks. But, the need for buffering of packets competing for the same destination remains an issue.

Another network topology, named after its proposer, Batcher, and resembling the process of shuffling a deck of playing cards, also provided a way to build networks allowing the ability to replicate incoming packets from stage to stage and therefore to introduce the selective broadcasting of incoming signals to selected output ports. Such an arrangement was described in 1984 by Huang and Knauer using both the Batcher and banyan topologies and came to be known as a member of the class of Batcher–banyan networks (14). This proposal included the idea of allowing different bit-rate information to be passed through the same switch—the early dream of BISDN. Its original intent was for implementation using very high speed photonic circuits, and working models using VLSI chip sets were built and reported.

Other architectures blossomed during this period using a variety of architectures and techniques, including the use of the frequency-hopping techniques employed in radio transmission systems and code-division multiplexing, both forms of spread-spectrum signaling techniques (15). (It is interesting to note that these two techniques have been employed widely in analog military communications systems to provide resistance to enemy detection of intelligence and jamming, a deliberate interference of radio communication instigated by adversaries in military actions.)

For circuit-switched connections, networks such as the Batcher–Banyan are one branch of possible network fabric architecture; memory-based systems are another. (High-speed electronic and photonic techniques continue to be active contenders for implementation in the mid-1990s.)

In any event, the trend in switching for the early 1990s is toward high-speed time-division circuit switching using packet techniques, not only to control the routing of signals, but also to carry other information relating to facility selection and routing treatment.

Asynchronous Transfer Mode: A Resolution for Broadband Integrated Services Digital Network

The concept of the asynchronous transfer mode (ATM) originated in France, and much work has been stimulated in other International Telegraph and Telephone Consultative Committee (CCITT) member countries on ATM network and switching system design to meet the goals of BISDN. In addition to the techniques noted above for wideband and broadband switching systems, ATM added a new element to handle a mixture of smooth and bursty traffic more efficiently. A companion article on ATM by the author of the first published paper on ATM appears in this encyclopedia (16).

The ATM network uses fixed-length packets and synchronous transmission as described above, but ATM does away with the concept of having a fixed time-slot in each periodically repeating frame of a fixed number of time-slots dedicated to a particular communication channel. As more packets are needed, they are added to the stream of traffic; their headers indicate their destination. Such packets occupy the next available time-slot and, as long as they do not crowd out competing packets, the switching systems can sort out the packets for delivery to the next stage or to the intended receiving port. There is no fixed frame of reference of repeating time-slots. It is in that sense that ATM is considered asynchronous.

High- and low-bit-rate traffic can coexist and, as long as the network is engineered properly and the sources constrain the extent of the burstiness of their offered traffic, such systems will provide the means for delivery. Priority indicators in packet headers can indicate relative urgency (for the needs of synchronism) in order of delivery.

As in the case of wideband, compression of video signals is assumed even at the high bit rates in order to accommodate HDTV, and various sophisticated techniques are proposed that allow the occasional shedding of bits in video transmission in order to accommodate bursts in data traffic. Flow control and error control will be kept at the edges of broadband networks. The maintenance of balanced traffic, so that Customer A cannot hog the network to the inadvertent detriment of Customer B's traffic, is an issue.

SONET (Synchronous Optical Network) transmission standards have been proposed by Bell Communications Research (Bellcore) in the United States to interwork with ATM switching systems and represent the North American standard to build a broadband network of switching nodes and interconnecting links for the future (17). This standard has been accepted within CCITT, and both ATM switching systems and SONET transmission systems are in the early stages of commercial introduction in the early and mid-1990s.

Acknowledgments: Thanks are due to Amos E. Joel, Jr., for broadening my background in telecommunications switching in the years that I worked for him and afterward during his "retirement" to nearly 10 years of consultancy with AT&T. Getting the figures done using a personal computer could not have been accomplished in any timely fashion without the expert guidance of Major D. L. Blosser, USMC Ret. It has been a pleasure to work with both Amos and Don on a number of switching projects at AT&T Bell Laboratories. CLASS is a service mark of Bellcore. 1ESS, 1AESS, and 4ESS are trademarks, and 5ESS is a registered trademark, of AT&T.

Glossary

ASYNCHRONOUS TRANSFER MODE (ATM). Format for transmission and switching for BISDN; channels provided as high as 600 Mbps.

BROADBAND. Term applied to networks and switching and transmission elements operating at rates higher than 1.5–2.0 Mbps.

COMMON CHANNEL SIGNALING (CCS). High-speed out-of-band technique first introduced into the AT&T network in 1978 as Common Channel Interoffice Signaling System No. 6 (CCIS6) and evolved through CCITT standards to become a worldwide signaling standard in both local and interexchange networks.

COMMUNITY DIAL OFFICE (CDO). A small local switching system serving a few hundred to a few thousand lines, usually in a rural area and generally operating unattended.

CODER/DECODER (CODEC). Dual-function circuit to encode and decode between digital and analog signals.

CUSTOM LOCAL-AREA SIGNALING SERVICE (CLASS). Interlocal switching features based on switches being served by CCS networks.

DIGITAL CROSS-CONNECT SYSTEM (DCS). Specialized high-speed SPC-controlled digital cross-connects that allow individual channels or groups of channels to be added or dropped within higher-bit-rate digital multiplex transmission rates.

HIGH-DEFINITION TELEVSION (HDTV). Next-generation television under development in Japan, the United States and elsewhere; can be compressed to 150-Mbps signal and carried over BISDN transmission channels and switches.

INTEGRATED SERVICES DIGITAL NETWORK (ISDN).

INTERNATIONAL SWITCHING SYMPOSIUM (ISS). Premier worldwide conference dedicated to switching systems and held approximately every two years since 1960.

LOCAL ACCESS AND TRANSPORT AREA (LATA). The formal name for an area to be served by local exchange companies. Interexchange carriers provide transport between LATAs, and the local exchange carriers such as the Bell Operating Companies and GTE provide local service within each of the more than 500 LATAs in the United States.

LOCAL-EXCHANGE CARRIER (LEC). Local telephone companies that serve traffic within the LATAs.

PRIVATE AUTOMATIC BRANCH EXCHANGE (PABX). In Europe, PBXs in which the word "automatic" is employed to distinguish from manual switchboards or operator positions.

PRIVATE BRANCH EXCHANGE (PBX). In the United States, small to large automatic switching systems that serve the group of telephones at a business location and provide groups of lines or trunks to be extended to local or toll switching systems. Usually includes an attendant position to screen incoming and outgoing calls.

STORED PROGRAM CONTROL (SPC). The method of controlling a telecommunications switching system with one or more data processors; first introduced commercially in 1965 and found in all new systems being placed in operation in the network.

SYNCHRONOUS OPTICAL NETWORK (SONET). Bellcore standard for broadband transmission of digital signals that facilitates the insertion and extraction of various subrate groups within the overall digital bit stream.

TIME MULTIPLEXED SWITCH (TMS). A crossbar type of switching element used in time-division switching networks in which the control of input-to-output connections are changed at microsecond or submicrosecond speeds from one time slot to the next.

TIME SLOT INTERCHANGER (TSI). A switching element used in time-division switching networks in which the sequence of input signals assigned to individual time slots is reordered at the output of the TSI.

Bibliography and References

The first three references serve also as bibliography since they were written to cover the subject of this article in an extensive fashion. Brief descriptions of the contents of these three are included.

1. Joel, A. E., Jr. (ed.), *Electronic Switching: Digital Central Office Systems of the World.* Institute of Electrical and Electronics Engineers (IEEE) Press, New York, 1982.

This text contains selected papers from IEEE and other publications on digital switching systems around the world as of the publication date. Includes several tutorial papers on digital switching by Joel and M. R. Aaron, followed by papers describing virtually all first-generation digital switching systems worldwide. Extensive references and appendices are provided.

2. Chapuis, R. J., and Joel, A. E., Jr., *Electronics, Computers and Telephone Switching,* Elsevier Science Publishing, Amsterdam, 1990.

   This book covers the period from 1960 to 1985 (and actually 4–5 years beyond) of electronic telephone switching systems and includes much material of a historical and philosophical nature as well as providing technical insights into the design of switching systems.

3. McDonald, J. C. (ed.), *Fundamentals of Digital Switching*, 2d ed., Plenum Press, New York, 1990.

   A tutorial text in which the editor and several of the contributors are designers of digital switching systems in commercial service as of 1992. Used as a source text in seminars on this topic.

4. Richter, M. J., Centrex. In: *The Froehlich/Kent Encyclopedia of Telecommunications*, Vol. 2 (F. E. Froehlich and A. Kent, eds.), Marcel Dekker, New York, 1991, pp. 385–389.
5. Rey, R. F. (ed.), *Engineering and Operations in the Bell System*, 2d ed., AT&T Bell Laboratories, Murray Hill, NJ, 1977.
6. Joel, A. E., Jr., *A History of Engineering and Science in the Bell System — Switching Technology (1925–1975)*, Bell Telephone Laboratories, Murray Hill, NJ, 1982.
7. Goldberg, Richard R., Common Channel Signaling. In: *The Froehlich/Kent Encyclopedia of Telecommunications*, Vol. 3 (F. E. Froehlich and A. Kent, eds.), Marcel Dekker, New York, 1992, pp. 163–181.
8. Stallings, William, *ISDN: An Introduction*, Macmillan, New York, 1989.
9. Bell Communications Research, *LATA Switching Systems Generic Requirements*, FR-NWT-000064, Bell Communications Research, Piscataway, NJ, 1992.

   Twelve modules provide technical details of operations and features of CLASS services.

10. Deloraine, E. M., and Reeves, A. H., The 25th Anniversary of Pulse Code Modulation, *IEEE Spectrum*, 2(5):56–63 (May 1965).
11. O'Neill, E. F., *Engineering and Science in the Bell System — Transmission Technology (1925–1975)*, AT&T Bell Laboratories, Murray Hill, NJ, 1985.
12. Vaughan, H. E., Integrated Communication, *Bell Sys. Tech. J.*, 38(4):909–932 (July 1959).
13. Lucas, Pierre, World Developments in Electronic Switching, *Commutation and Transmission*, 2:7–78 (1979).
14. Huang, Alan, and Knauer, Scott, Starlite: A Wideband Digital Switch, *Globecom '84*, 5.3.1–5.3.5 (1984).
15. Dixon, Robert C., *Spread Spectrum Systems*, 2d ed., John Wiley and Sons, New York, 1984.
16. Coudreuse, Jean-Pierre, Communications Using the Asynchronous Transfer

Mode. In: *The Froehlich/Kent Encyclopedia of Telecommunications*, Vol. 4 (F. E. Froehlich and A. Kent, eds.), Marcel Dekker, New York, 1992, pp. 123–164.

17. Bell Communications Research, SONET Transport Criteria, FR-NWT-000919, 1992, *Transport Systems Generic Requirements*, FR-NWT-000440, 1992, Bell Communications Research, 1992.

FRANK F. TAYLOR

Distortion and Noise in
Telephone Systems   (see Analog and Digital Distortion
in Speech and Data Transmission)

Distributed Computing

Introduction

A distributed computer system (DCS) is a collection of computers connected by a communications subnet and logically integrated in varying degrees by a distributed operating system (DOS) and/or distributed database system. Each computer node may be a uniprocessor, or multiprocessor, or multicomputer. The communications subnet may be a widely geographically dispersed collection of communication processors or a local-area network. Typical applications that use distributed computing include E-mail, teleconferencing, electronic funds transfers, multimedia telecommunications, command-and-control systems, and support for general-purpose computing in industrial and academic settings. The widespread use of DCSs is due to the price–performance revolution in microelectronics, the development of cost-effective and efficient communication subnets (due to the merging of data communications and computer communications) (1), the development of resource-sharing software, and the increased user demands for communication, economical sharing of resources, and productivity.

A DCS potentially provides significant advantages, including good performance, good reliability, good resource sharing, and extensibility (2,3). Potential performance enhancement is due to multiple processors and an efficient subnet, as well as avoiding contention and bottlenecks that exist in uniprocessors and multiprocessors. Potential reliability improvements are due to the data and control redundancy possible, the geographical distribution of the system, and the ability for hosts and communication processors to perform mutual inspection. With the proper subnet, DOS (4), and distributed database (5), it is possible to share hardware and software resources in a cost-effective manner, increasing productivity and lowering costs. Possibly the most important potential advantage of a DCS is extensibility. *Extensibility* is the ability to adapt easily to both short- and long-term changes without significant disruption of the system. Short-term changes include varying workloads and host or subnet failures or additions. Long-term changes are associated with major modifications to the requirements or content of the system.

DCS research encompasses many areas, including local- and wide-area networks, DOSs, distributed databases, distributed file servers, concurrent and distributed programming languages, specification languages for concurrent systems, theory of parallel algorithms, theory of distributed computing, parallel architectures and interconnection structures, fault-tolerant and ultrareliable systems, distributed real-time systems, cooperative problem solving techniques of artificial intelligence, distributed debugging, distributed simulation, distributed applications, and a methodology for the design, construction, and maintenance of large, complex distributed systems. Many prototype DCSs have been built at university, industrial, commercial, and government research laboratories, and production systems of all sizes and types have proliferated. It is impossible to survey all distributed computing system research. An extensive survey and bibliography would require hundreds of pages. Instead, this article focuses on two important areas: distributed operating systems and distributed databases.

Distributed Operating Systems

Operating systems for distributed computing systems can be categorized into two broad categories: network operating systems (NOSs) and distributed operating systems (6).

Consider the situation in which each of the hosts of a computer network has a local operating system that is independent of the network. The total of all the operating system software added to each host in order to communicate and share resources is called a *network operating system*. The added software often includes modifications to the local operating system. NOSs are characterized by being built on top of existing operating systems, and they attempt to hide the differences between the underlying systems (Fig. 1).

Consider an integrated computer network in which there is one native operating system for all the distributed hosts. This is called a *distributed operating system* (Fig. 2). Examples of DOSs include the V system (7), Eden (8), Amoeba (9), the Cambridge distributed computing system (10), Medusa (11), Locus (12), and Mach (13). Examples of real-time DOSs include Maintainable Real-Time System (MARS) (14) and Spring (15). A DOS is designed with the network requirements in mind from its inception and it tries to manage the resources of the network in a global fashion. Therefore, retrofitting a DOS to existing operating systems and other software is not a problem. Since DOSs are used to satisfy a wide variety of requirements, their various implementations are quite different. Note that the boundary between NOSs and DOSs is not always clearly

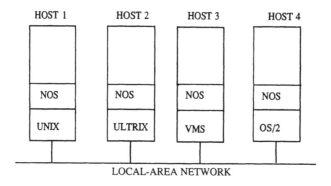

NATIVE OPERATING SYSTEMS - UNIX,ULTRIX,VMS,OS/2

FIG. 1   Network operating systems.

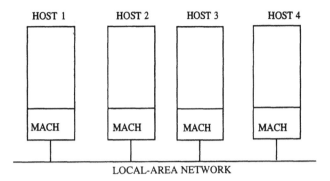

FIG. 2 Distributed operating system for Mach.

distinguishable. In this article, we primarily consider DOS issues for six catego-ries: process structures, access control and communications, reliability, hetero-geneity, efficiency, and real time. A similar breakdown is applied to distributed databases.

Process Structures

The conventional notion of a *process* is an address space with a single execution trace through it. Because of the parallelism inherent in multiprocessing and distributed computing, we have seen that recent operating systems are support-ing the separation of address space (sometimes called a *task*) and execution traces (called *threads* or *lightweight processes*) (13,16). In most systems, the address space and threads are restricted to reside on a single node (a uniproces-sor or multiprocessor). However, such systems as Ivy, the Apollo domain, and Clouds (17) support a distributed address space (sometimes called a *distributed shared memory*; see Ref. 18 for a summary of issues involved with distributed shared memory), and distributed threads executing on that address space. Re-gardless of whether the address space is local or distributed (19), there has been significant work done on the following topics: supporting very large but sparse address spaces, efficiently copying information between address spaces using a technique called *copy on write* in which only the data actually used get copied (20), and supporting efficient file management by mapping files into the address space and then using virtual memory techniques to access the file.

At a higher level, DOSs use tasks and threads to support either a procedure or an object-based paradigm (21). If objects are used, there are two variations: the passive and active object models. Because the object-based paradigm is so important and well suited to distributed computing, we present some basic information about objects and then discuss the active and passive object para-digms.

A *data abstraction* is a collection of information and a set of operations defined on that information. An *object* is an instantiation of a data abstraction. The concept of an object usually is supported by a kernel, which also may define a primitive set of objects. Higher-level objects then are constructed from more primitive objects in some structured fashion. All hardware and software resources of a DCS can be regarded as objects. The concept of an object and its implications form an elegant basis for a DCS (6,11). For example, such distributed systems' functions as allocation of objects to a host, moving objects, remote access to objects, sharing objects across the network, and providing interfaces between disparate objects are all "conceptually" simple because they all are handled by yet other objects. The object concept is powerful and easily can support the popular client-server model of distributed computing.

Objects also serve as the primitive entity supporting more complicated distributed computational structures. One type of distributed computation is a process (thread) that executes as a sequential trace through passive objects, but does so across multiple hosts (Fig. 3). The objects are permanent but the execution properties are supplied by an external process (thread) executing in the address space of the object. Another form of a distributed computation is to have clusters of objects, each with internal, active threads, running in parallel and communicating with each other based on the various types of interprocess communication (IPC) used. This is known as the *active object model* (Fig. 4). The cluster of processes may be co-located or distributed in some fashion. Other notions of what constitutes a distributed program are possible, that is, object invocations can support such additional semantics as those found in database transactions.

The major problem with object-based systems has been poor execution time performance. However, this is not really a problem with the object abstraction itself, but is a problem with inefficient implementation of access to objects. The most common reason given for poor execution time is that current architectures are ill suited for object-based systems. Another problem is choosing the

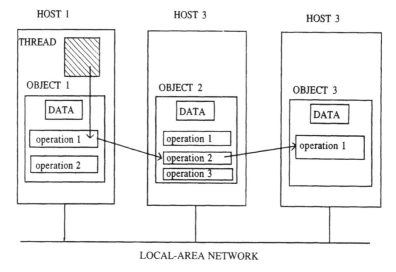

FIG. 3 Distributed computation (thread through passive objects).

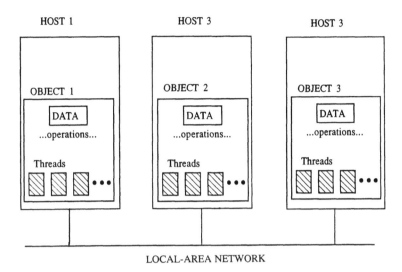

FIG. 4   Active objects.

right granularity for an object. If every integer or character and their associated operations are treated as objects, then the overhead is too high. If the granularity of an object is too large, then the benefits of the object-based system are lost.

Access Control and Communications

A distributed system consists of a collection of resources managed by the DOS. Accessing the resources must be controlled in two ways. First, the manner used to access the resource must be suitable to the resource and requirements under consideration. For example, a printer must be shared serially, local data of an object should not be shareable, and a read-only file can be accessed simultaneously by any number of users. In an object-based system, access can be controlled on an operation-by-operation basis. For example, a given user may be restricted to using only the insert operation on a queue object while another user may be able to access the queue using both insert and remove operations. Second, access to a resource must be restricted to a set of allowable users. In many systems, this is done by an access control list, an access control matrix, or capabilities. The Massachusetts Institute of Technology's (MIT's) Athena project developed an authentication server called Kerberos based on a third-party authentication model that uses private-key encryption (22).

Communication has been the focus of much of the work in DOSs, providing the glue that binds logically and physically separated processes (23,24). Remote procedure calls (RPCs) extend the semantics of a programming language's procedure calls to communication across nodes of a DCS. Lightweight RPCs have been developed to make such calls as efficient as possible (25). Many systems also support general synchronous and asynchronous send-and-receive primitives with semantics that are different and more general than RPC semantics. Broadcasting and multicasting are also common primitives found in DOSs and they provide useful services in achieving consensus and other forms of global coordination. For systems with high reliability requirements, reliable broadcast facilities might be provided (26).

Implementing communications facilities is done either directly in the kernels of the operating systems as in the V system (7) or as user level services as in Mach (13). This is a classical tradeoff between performance and flexibility. Intermediate between these approaches lies the x-kernel (23), in which basic primitives required for all communications primitives are provided at the kernel level and protocol-specific logic is programmable at a higher level.

Reliability

While reliability is a fundamental issue for any system, the redundancy found in DCSs make them particularly well suited for the implementation of reliability techniques. We begin the discussion on reliability with a few definitions.

A *fault* is a mechanical or algorithmic defect that may generate an error. A fault may be permanent, transient, or intermittent. An *error* is an item of information that, when processed by the normal algorithms of the system, will produce a failure. A *failure* is an event at which a system violates its specifications. *Reliability* then can be defined as the degree of tolerance against errors and faults. Increased reliability comes from fault avoidance and fault tolerance. *Fault avoidance* results from such conservative design practices as using high-reliability components and nonambitious design. *Fault tolerance* employs error detection and redundancy to deal with faults and errors. Most of what we discuss in this section relates to the fault tolerance aspect of reliability.

Reliability is a complex, multidimensional activity that simultaneously must address some or all of the following: fault confinement, fault detection, fault masking, retries, fault diagnosis, reconfiguration, recovery, restart, repair, and reintegration. Furthermore, distributed systems require more than reliability, that is, they need to be dependable. Dependability is the trustworthiness of a computer system and it subsumes reliability, availability, safety, and security. Such system architectures as Delta-4 strive for dependability (27). We cannot do justice to all these issues in this short article. Instead, we will discuss several of the more important issues related to reliability in DOSs.

Reliable DOSs should support replicated files, exception handlers, testing procedures executed from remote hosts, and avoid single points of failure by a combination of replication, backup facilities, and distributed control. Distributed control could be used for file servers, name servers, scheduling algorithms, and other executive control functions. Process structure, how environment information is kept, the homogeneity of various hosts, and the scheduling algorithm may allow for relocatability of processes. IPC might be supported as a reliable RPC (28,29) and also provide reliable atomic broadcasts as is done in ISIS (30). Reliable IPC would enforce "at least once" or "exactly once" semantics depending on the type of IPC being invoked, and atomic broadcasts guarantee that either all processes to receive the message indeed will receive it, or none will. Other DOS reliability solutions are required to avoid invoking processes that are not active, to avoid the situation in which a process remains active but is not used, and to avoid attempts to communicate with terminated processes.

ARGUS, a distributed programming language, has incorporated reliability concerns explicitly into the programming language (31). It does this by supporting the idea of an atomic object, transactions, nested actions, reliable RPCs, stable variables, guardians (modules that service node failures and synchronize concurrent access to data), exception handlers, periodic and background testing procedures, and recovery of a committed update if the present update does not complete. A distributed program written in ARGUS potentially may experience deadlock. Currently, deadlocks are broken by timing out and aborting actions.

Distributed databases make use of many reliability features such as stable storage, transactions, nested transactions (32), commit and recovery protocols (33), nonblocking commit protocols (34), termination protocols (35), checkpointing, replication, primary/backups, logs/audit trails, differential files (36), and time-outs to detect failures. Operating system support is required to make these mechanisms more efficient (12,37–39).

One aspect of reliability not stressed enough in DCS research is the need for robust solutions, that is, the solutions must assume an unreliable network explicitly, tolerate host failures, network partitionings, and lost, duplicate, out of order, or noisy data. Robust algorithms sometimes must make decisions after reaching only approximate agreement or by using statistical properties of the system (assumed known or dynamically calculated). A related question is at what level should the robust algorithms, and reliability in general, be supported? Most systems attempt to have the subnet ensure reliable, error-free data transmission between processes. However, according to the end-to-end argument (40), such functions placed at the lower levels of the system often are redundant and unnecessary. The rationale for this argument is that since the application has to take into account errors introduced not only by the subnet, many of the error-detection and recovery functions can be provided correctly and completely only at the application level.

The relationship of reliability to the other issues discussed in this article is very strong. For example, object-based systems confine errors to a large degree, define a consistent system state to support rollback and restart, and limit propagation of rollback activities. However, if objects are supported on a distributed shared memory, special problems arise (41). Since objects can represent unreliable resources (such as processors and disks), and since higher-level objects can be built using lower-level objects, the goal of reliable system design is to create reliable objects out of unreliable objects. For example, a stable storage can be created out of several disk objects and the proper logic. Then, a physical processor, a checkpointing capability, a stable storage, and logic can be used to create a stable processor. One can proceed in this fashion to create a very reliable system. The main drawback is potential loss of execution time efficiency. For many systems, it is just too costly to incorporate an extensive number of reliability mechanisms. Reliability also is enhanced by proper access control and judicial use of distributed control. The major challenge is to integrate solutions to all these issues in a cost-effective manner and produce an extremely reliable system.

Heterogeneity

Incompatibility problems arise in heterogeneous DCSs in a number of ways and at all levels (42). First, incompatibility is due to the different internal formatting schemes that exist in a collection of different communications and host processors. Second, incompatibility also arises from the differences in communications protocols and topology when networks are connected to other networks via gateways. Third, major incompatibilities arise due to different operating systems, file servers, and database systems that might exist on a (set of) network(s).

The easiest solution to the general problem of incompatibility for a single DCS is to avoid the issue by using a homogeneous collection of machines and software. If this is not practical, then some form of translation is necessary. Some earlier systems left this translation to the user. This is no longer acceptable.

Translation done by the DCS system can be done at the receiver host or at the source host. If it is done at the receiver host, then the data traverse the network in their original form. The data usually are supplemented with extra information to guide the translation. The problem with this approach is that at every host there must be a translator to convert each format in the system to the format used on the receiving host. When there exist n different formats, this requires the support of $(n - 1)$ translators at each host. Performing the translation at the source host before transmitting the data is subject to all the same problems.

There are two better solutions, each applicable under different situations: an intermediate translator or an intermediate standard data format. First, an intermediate translator accepts data from the source and produces the acceptable format for the destination. This usually is used when the number of different types of necessary conversions is small. For example, a gateway linking two different networks acts as an intermediate translator. Second, for a given conversion problem, if the number of different types to be dealt with grows large, then a single intermediate translator becomes unmanageable. In this case, an intermediate standard data format (interface) is declared, hosts convert to the standard, data are moved in the format of the standard, and another conversion is performed at the destination. By choosing the standard to be the most common format in the system, the number of conversions can be reduced.

At a high level of abstraction, the heterogeneity problem and the necessary translations are well understood. At the implementation level, a number of complications exist. The issues are precision loss, format incompatibilities (e.g., minus zero value in sign magnitude and one's complement cannot be represented in two's complement), data type incompatibilities (e.g., mapping of an upper/lowercase terminal to an uppercase only terminal is a loss of information), efficiency concerns, the number of locations of the translators, and what constitutes a good intermediate data format for a given incompatibility problem.

As DCSs become more integrated, one can expect that both programs and complicated forms of data might be moved to heterogeneous hosts. How will a program run on this host, given that the host has different word lengths, different machine code, and different operating system primitives? How will database relations stored as part of a CODASYL model database be converted to a relational model and its associated storage scheme? Moving a data structure object requires knowledge about the semantics of the structure (e.g., that some of the fields are pointers and these have to be updated upon a move). How should this information be imparted to the translators, what are the limitations, if any, and what are the benefits and costs of having this kind of flexibility? In general, the problem of providing translation for movement of data and programs among heterogeneous hosts and networks has not been solved. The main problem is ensuring that such programs and data are interpreted correctly at the destination host. In fact, the more difficult problems in this area largely have been ignored.

The Open Systems Foundation (OSF) distributed computing environment (DCE) is attempting to address the problem of programming and managing heterogeneous DCSs by establishing a set of standards for the major components of such systems. This includes standards for RPCs, distributed file servers, and distributed management.

Efficiency

DCSs are meant to be efficient in a multitude of ways. Resources (files, compilers, debuggers, and other software products) developed at one host can be shared by users on other hosts, limiting duplicate efforts. Expensive hardware resources also can be shared, minimizing costs. Such communications facilities as RPC, electronic mail, and file transfer protocols also improve efficiency by enabling better and faster transfer of information. The multiplicity of processing elements also might be exploited to improve response time and throughput of user processes. While efficiency concerns exist at every level in the system, they also must be treated as an integrated system-level issue. For example, a good design, the proper tradeoffs between levels, and the paring down of overambitious features usually improves efficiency. In this section, however, we concentrate on discussing efficiency as it relates to the execution time of processes (threads).

Once the system is operational, improving response time and throughput of user processes (threads) is largely the responsibility of scheduling and resource management algorithms (43–50) and the mechanisms used to move processes and data (7,17,19,51). The scheduling algorithm is related intimately to the resource allocator because a process will not be scheduled for the central processing unit (CPU) if it is waiting for a resource. If a DCS is to exploit the multiplicity of processors and resources in the network, it must contain more than simply n independent schedulers. The local schedulers must interact and cooperate, and the degree to which this occurs can vary widely. We suggest that a good scheduling algorithm for a DCS will be a heuristic that acts like an "expert system." This expert system's task is to utilize the resources of the entire distributed system effectively given a complex and dynamically changing environment. This is illustrated in the discussion that follows. In the remainder of this section, when we refer to the scheduling algorithm, we are referring to that part of the scheduler (possibly an expert system) that is responsible for choosing the host of execution for a process. We assume that there is another part of the scheduler that assigns the local CPU to the highest priority ready process.

We divide the characteristics of a DCS that influence response time and throughput into system characteristics and scheduling algorithm characteristics. *System characteristics* include the number, type, and speed of processors, caches, and memories, the allocation of data and programs, whether data and programs can be moved, the amount and location of replicated data and programs, how data are partitioned, partitioned functionality in the form of dedicated processors, any special-purpose hardware, characteristics of the communications subnet, and such special problems of distribution as no central clock and the inherent delays in the system. A good scheduling algorithm would take the system characteristics into account. *Scheduling algorithm characteristics* include the type and amount of state information used, how and when that information is transmitted, how the information is used (degree and type of cooperation between distributed scheduling entities), when the algorithm is invoked, adaptability of the algorithm, and the stability of the algorithm (52,53).

The type of state information used by scheduling algorithms in making scheduling decisions includes queue lengths, CPU utilization, amount of free memory, estimated average response time, or combinations of various information. The type of information also refers to whether the information is local or networkwide information. For example, a scheduling algorithm on Host 1 could use queue lengths of all the hosts in the network in making its decision. The amount of state information refers to the number of different types of information used by the scheduler.

Information used by a scheduler can be transmitted periodically or asynchronously. If asynchronous, it may be sent only when requested (as in bidding), it may be piggybacked on other messages between hosts, or it may be sent only when conditions change by some amount. The information may be broadcast to all hosts, sent to neighbors only, or to some specific set of hosts.

The information is used to estimate the loads on other hosts of the network in order to make an informed global scheduling decision. However, the data received are out of date and even the ordering of events might not be known (54). It is necessary to manipulate the data in some way to obtain better estimates. Several examples are (1) very old data can be discarded, given that state information is timestamped, a linear estimation of the state extrapolated to the current time might be feasible; (2) conditional probabilities on the accuracy of the state information might be calculated in parallel with the scheduler by some monitor nodes and applied to the received state information; (3) the estimates can be some function of the age of the state information; or (4) some form of (iterative) message interchange might be feasible.

Before a process actually is moved, the cost of moving it must be accounted for in determining the estimated benefit of the move. This cost is different if the process has not yet begun execution instead of already being in progress. In both cases, the resources required also must be considered. If a process is in execution, then environment information (e.g., the process control block) probably should be moved with the process. It is expected that in many cases the decision will be not to move the process.

Schedulers invoked too often will produce excessive overhead. If they are not invoked often enough, they will not be able to react fast enough to changing conditions. There will be undue start-up delay for processes. There must be some ideal invocation schedule that is a function of the load.

In a complicated DCS environment, it can be expected that the scheduler will have to be quite adaptive (53,55). A scheduler might make minor adjustments in weighing the importance of various factors as the network state changes in an attempt to track a slowly changing environment. Major changes in the network state might require major adjustments in the scheduling algorithms. For example, under very light loads there does not seem to be much justification for networkwide scheduling, so the algorithm might be turned off, except for the part that can recognize a change in the load. At moderate loads, the full-blown scheduling algorithm might be employed. This might include individual hosts refusing all requests for information and refusing to accept any process because it is too busy. Under heavy loads on all hosts, it again seems unnecessary to use networkwide scheduling.

A bidding scheme might use both source- and server-directed bidding (49). An overloaded host asks for bids and is the source of work for some other hosts in the network. Similarly, a lightly loaded host may make a reverse bid, that is, ask the rest of the network for some work. The two types of bidding might coexist. Schedulers could be designed in a multilevel fashion with decisions being made at different rates (e.g., local decisions and state information updates occur frequently, but more global exchange of decisions and state information might proceed at a slower rate because of the inherent cost of these global actions).

A classic efficiency question in any system is what should be supported by the kernel, or more generally by the operating system, and what should be left to the user? The trend in DCS is to provide minimal support at the kernel level, for example, support objects, primitive IPC mechanisms, and processes (threads). Then, other operating system functions are supported as higher-level processes. On the other hand, because of efficiency concerns, some researchers advocate putting more in the kernel, including communications protocols, real-time systems' support, or even supporting the concept of a transaction in the kernel. This argument never will be settled conclusively since it is a function of the requirements, type of processes running, and so on.

Of course, many other efficiency questions remain. These include those related to the efficiency of the object model, the end-to-end argument, locking granularity, performance of remote operations, improvements due to distributed control, the cost effectiveness of various reliability mechanisms, efficiently dealing with heterogeneity, hardware support for operating system functions (56), and handling the input/output (I/O) bottleneck via disk arrays of various types (RAID [redundant arrays of inexpensive disks] 1 through RAID 6). Efficiency, therefore, is not a separate issue but must be addressed for each issue in order to result in an efficient, reliable, and extensible DCS. A difficult problem to be addressed is exactly what is acceptable performance given that multiple decisions are being made at all levels and that these decisions are being made in the presence of missing and inaccurate information.

Real-Time Distributed Operating Systems

Such real-time applications as nuclear power plants and process control are inherently distributed and have severe real-time and reliability requirements. These requirements add considerable complication to a DCS. Examples of demanding real-time systems include the Electronic Switching System (ESS) (59), REBUS (60), and Software Implemented Fault Tolerance (SIFT) (61). ESS is a software-controlled electronic switching system developed by the Bell System for placing telephone calls. The system meets severe real-time and reliability requirements. REBUS is a fault-tolerant distributed system for industrial real-time control, and SIFT is a fault-tolerant flight control system. Generally, these systems are built with technology that is tailored to these applications because many of the concepts and ideas used in general-purpose distributed computing are not applicable when deadlines must be guaranteed. For example, RPCs, creating tasks and threads, and requesting operating system services all are done in a manner that ignores deadlines, causes processes to block at any time, and only provides reasonable average case performance. None of these things are reasonable when it is critical that deadlines are met. In fact, many misconceptions exist when dealing with distributed real-time systems (62). However, significant new research efforts now are being conducted to combat these misconceptions and to provide a science of real-time computing in the areas of formal verification, scheduling theory and algorithms (63,64), communications protocols (65,66), and operating systems (15,67,77). The goal of this new research in real-time systems is to develop predictable systems even when the systems are highly complex, distributed, and operate in nondeterministic environments (69).

Other distributed real-time applications such as airline reservation and banking applications have less severe real-time constraints and are easier to build. These systems can utilize most of the general-purpose distributed computing technology described in this article, and generally approach the problem as if real-time computing were equivalent to fast computing, which is false. While this seems to be common practice, some additions are required to deal with real-time constraints, for example, scheduling algorithms may give preference to processes with earlier deadlines, or processes holding a resource may be aborted if a process with a more urgent deadline requires the resource. Results from the more demanding real-time systems also may be applicable to these soft real-time systems.

Distributed Databases

Database systems have existed for many years, providing significant benefits in high availability and reliability, reduced costs, good performance, and ease of sharing information. Most are built using what can be called a database architecture (5,70) (see Fig. 5). The architecture includes a query language for users, a data model that describes the information content of the database as seen by the users, a schema (the definition of the structure, semantics, and constraints on the use of the database), a mapping that describes the physical storage structure used to implement the data model, a description of how the physical data will be accessed, and, of course, the data (database) itself. All of these components then are integrated into a collection of software that handles all accesses to the database. This software is called the *database management system* (DBMS). The DBMS usually supports a transaction model. In this section, we discuss various transaction structures, access and concurrency control, reliability, heterogeneity, efficiency techniques, and real-time distributed databases.

Transaction Structures

A *transaction* is an abstraction that allows programmers to group a sequence of actions on the database into a logical execution unit (71–84). Transactions either commit or abort. If the transaction successfully completes all its work, the transaction commits. A transaction that aborts does so without completing its operations and any effect of executed actions must be undone. Transactions have four properties, known as the ACID properties: atomicity, consistency, isolation, and durability. *Atomicity* means that either the entire transaction completes or it is as if the transaction never executed. *Consistency* means that

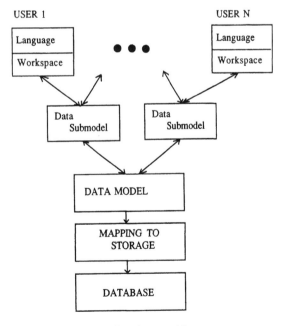

FIG. 5 Database architecture.

the transaction maintains the integrity constraints of the database. *Isolation* means that even if transactions execute concurrently, their results appear as if they were executed in some serial order. *Durability* means that all changes made by a committed transaction are permanent.

A distributed database is a single logical database, but it is distributed physically over many sites (nodes). A distributed database management system (DDBMS) controls access to the distributed data supporting the ACID properties but with the added complication imposed by the physical distribution. For example, a user may issue a transaction that updates various data that, transparent to the user, physically resides on many different nodes (75). The software that supports the ACID properties of transactions then must interact with all these nodes in a manner that is consistent with the ACID properties. This usually requires supporting remote node slave transactions, distributed concurrency control, recovery, and commit protocols, and so on. There have been many papers and books written about these basic aspects of DDBMSs (5,71,72,76,77). Rather than discussing these basic issues, we will discuss some of the extended transaction models that have been developed recently.

The traditional transaction model, while powerful because of its ability to mask the effects of failures and concurrency, has shortcomings when applied to such complex applications as computer-aided design, computer-aided software engineering, DOSs, and multimedia databases. In these applications, there is a need for greater functionality and performance than can be achieved with traditional transactions. For example, two programmers working on a joint programming project may wish for their transactions to be cooperative and to see partial results of the other user, rather than being competitive and isolated, properties exhibited by the traditional transaction model. Also, traditional transactions only exploit such very simple semantics as read-only and write-only semantics of the transactions in order to achieve greater concurrency and performance.

Many extended transaction models have been proposed to support the greater functionality and performance required by complex applications. These include nested transactions (32,78), multilevel transactions (79,80), cooperative transactions (81), compensating transactions (82), recoverable communicating actions (83), split transactions (84), and sagas (85,86). Each of these transaction models has different semantics with respect to visibility, consistency, recovery, and permanence in their attempt to be useful for various complex applications. As an example, a nested transaction is composed of subtransactions that may execute concurrently. Subtransactions are serializable with respect to siblings and other nonrelated transactions. Subtransactions are failure atomic with respect to their parent transaction. A subtransaction can abort without necessarily causing the parent to abort. The other extended models have such features as relaxing serializability or failure atomicity, may have structures other than hierarchical ones, and may exhibit different abort dependencies. The relationship and utility of these models currently are being explored. In this regard, there has been the development of a comprehensive and flexible framework called ACTA (A formal Characterization of TransActions) used to provide a formal description and a reasoning procedure for the properties of all these extended transaction models (87).

There is another dimension along which traditional transactions have changed for complex applications. Initially, traditional transactions were considered as performing simple read or write operations on the database. However, there has been a merger of ideas from object-based programming and database systems, resulting in object-based databases (88–90). Here, transactions and extended transactions perform higher-level operations on objects that reside in the database. Using object-based databases provides more support for complex applications than having to work with simple read and write operations.

Access and Concurrency Control

Most access control in distributed databases is supported by the underlying operating system. In particular, it is the operating system that verifies a database user and controls which user is allowed to read or write various data. Users typically are grouped so that each group has the same rights. Maintaining and updating the various rights assigned to each group is nontrivial, and is exacerbated when heterogeneous databases are considered.

In a database system, multiple transactions will be executing in parallel and may conflict over the use of data. Protocols for resolving data-access conflicts between transactions are called concurrency control protocols (76,91–94). The correctness of a concurrency control protocol usually is based on the concept of serializability (77). *Serializability* means that the effect of a set of executed transactions (permitted to run in parallel) must be the same as some serial execution of that set of transactions. In many cases and at all levels, the strict condition of serializability is not required. Relaxing this requirement usually can result in access techniques that are more efficient in the sense of allowing more parallelism and faster execution times. However, due to space limitations, we do not discuss these extensions here; see Refs. 95 and 96 for pertinent information.

Three major classes of concurrency control protocols are locking, timestamp ordering (97), and validation (also called the optimistic approach [92]). *Locking* is a well-known technique. In the database context, the most common form of locking is called two-phase locking. Reference 77 provides a description of this protocol.

Timestamp ordering is an approach in which all accesses to data are time-stamped and then some common rule is followed by all transactions in such a way as to ensure serializability (94). This technique can be useful at all levels of a system, especially if timestamps already are being generated for such other reasons as to detect lost messages or failed hosts. A variation of the timestamp-ordering approach is the multiversion approach. This approach is interesting because it integrates concurrency control with recovery. In this scheme, each change to the database results in a new version. Concurrency control is enforced on a version basis and old versions serve as checkpoints. Handling multiple versions efficiently is accomplished by differential files (36).

Validation is a technique that permits unrestricted access to data items (resulting in no blocking and hence fast access to data), but then checks for potential conflicts at the commit point. The commit point is the time at which a transaction is sure that it will complete. This approach is useful when few conflicts are expected because access is very fast and, if most transactions are validated (due to the few conflicts), then there is also little overhead due to aborting any nonvalidated transactions. Most validation protocols assume a particular recovery scheme and also can be considered to integrate concurrency control and recovery.

Reliability

Recovery management in distributed databases is a complex process (33,34,72). Recovery is initiated due to such problems as invalid inputs, integrity violations, deadlocks, and node and media failures. All of these faults, except node failures, are addressed by simple transaction rollback. Node failures require much more complicated solutions. All solutions are based on data redundancy and make use of stable storage where past information concerning the database has been saved and can survive failures.

It is the recovery manager component of database systems that is responsible for recovery. Generally, the recovery manager must operate under six scenarios.

1. Under normal operation, the recovery manager logs each transaction and its work, and at transaction commit time checks transaction consistency, records the commit operation on the log, and forces the log to stable storage.
2. If a transaction must be rolled back, the recovery manager performs the rollback to a specific checkpoint or completely aborts the transaction (essentially a rollback to the beginning of the transaction).
3. If any database resource manager crashes, the recovery manager must obtain the proper log records and restore the resource manager to the most recent committed state.
4. After a node crash, the recovery manager must restore the state of all the resource managers at that node and resolve any outstanding distributed transactions that were using that node at the time of the crash and were not able to be resolved due to the crash.
5. The recovery manager is responsible for handling media recovery (such as a disk crash) by using an update log and archive copies or copies from other nodes if replicated data is supported by the system.
6. It is typical that distributed databases have recovery managers at each node that cooperate to handle many of the previous scenarios. These recovery managers may fail. Restart requires reintegrating a recovery manager into the set of active recovery managers in the distributed system.

Actually supporting all of the above scenarios is difficult and requires sophisticated strategies for what information is contained in a log, how and when to write the log, how to maintain the log over a long time period, how to undo transaction operations, how to redo transaction operations, how to utilize the archives, how and when to checkpoint, and how to interact with concurrency control and commit processing.

Many performance tradeoffs arise when implementing recovery management (e.g., how often and how to take checkpoints). If frequent checkpoints are taken, then recovery is faster and less work is lost. However, taking checkpoints too often significantly slows the "normal" operation of the system. Another question arises in how to log the information needed for recovery efficiently. For example, many systems perform a lazy commit in which all the log records are created at commit time but only pushed to disk at a later time. This reduces the guarantees that the system can provide about the updates, but it improves performance.

Heterogeneity

As defined above, a *distributed database* is a single logical database physically residing on multiple nodes. For a distributed database system, there is a single query language, data model, schema, and transaction management strategy. This is in contrast to a *federated database system*, which is a collection of autonomous database systems integrated for cooperation. The autonomous systems may have identical query languages, data models, schemas, and transaction management strategies, or they may be heterogeneous along any or all of these dimensions. The degree of integration varies from a single unified system constructed with new models for query languages, data models, and so on, built on top of the autonomous components, to those systems with minimal interaction and without any unified models. These last-mentioned systems are called *multidatabase systems*. Federated systems arise when databases are developed independently, then need to be highly integrated. Multidatabase systems arise when individual DBMSs wish to retain a great degree of autonomy and only provide minimal interaction.

Heterogeneity in database systems also arises from differences in underlying operating systems or hardware, from differences in DBMSs, and from semantics of the data itself. Several examples of these forms of heterogeneity follow.

Operating systems on different machines may support different file systems, naming conventions, and IPC. Hardware instruction sets, data formats, and processing components also may be different at various nodes. Each of these may give rise to differences that must be resolved for proper integration of the databases.

If DBMSs have different query languages, data models, or transaction strategies, then problems arise. For example, differences in query languages may mean that some requests become illegal when issued on data in the "other" database. Differences in data models arise from many sources, including what structures each supports (e.g., relations versus record types) and how constraints are specified. One transaction strategy might employ two-phase locking with after-image logging, while another uses some form of optimistic concurrency control.

Other problems arise from semantics of the data. Semantic heterogeneity is not well understood and arises from differences in the meaning, definition, or interpretation of the related data in the various databases. For example, course grades may be defined on an alphabetical scale in one database and on a numerical scale in another database. How do we resolve the differences when a query fetches grades from both databases?

Solutions to the heterogeneity problem are usually difficult and costly both in development costs in dollars and in run-time execution costs. Differences in query languages usually are solved by mapping commands in one language to an equivalent set of commands in the other, and vice versa. If more than two query languages are involved, then an intermediate language is defined and mappings occur to and from this intermediate language. Mappings must be defined for the data model and schema integration. For example, an integrated schema would contain the description of the data described by all component schemas and the mappings between them. It also is possible to restrict what data in a given database can be seen by the federation of databases. This is sometimes called an *export schema*.

Solutions also must be developed for query decomposition and optimization, and global transaction management in federated database systems. The complexities found here are beyond this article. Interested readers should see Refs. 98–100.

Efficiency Techniques

Distributed database systems make use of many techniques for improving efficiency including distributed and local query optimization, various forms of buffering, lazy evaluation, disk space allocation tailored to access requirements, the use of mapping caches, parallelism in subtransactions, and nonserializability. Parallelism in subtransactions and nonserializability are mentioned above so these issues are not discussed in this subsection. It often is necessary for operating systems that support databases to be tailored specifically for database functions and, if not, then the support provided by the operating system usually is inefficient. See Ref. 38 for a discussion of this issue.

To obtain good performance in distributed databases, query optimization is critical. Query optimization can be categorized as either heuristic, in which ad hoc rules transform the original query into an equivalent query that can be processed more quickly, or systematic, in which the estimated costs of various alternatives are computed using analytical cost models that reflect the details of the physical distributed database. In relational databases, JOIN, SELECTION, and PROJECTION are the most time consuming and frequent operations in queries and hence most query optimizers deal with these operations. For example, a JOIN is one of the most time consuming relational operators because the resultant size can be equal to the product of the sizes of the original relations. The most common optimization is to perform PROJECTIONS and SELECTIONS first to minimize intermediate relations to be JOINED. In a distributed setting, minimizing the intermediate relations also can have the effect of minimizing the amount of data that needs to be transferred across the network. Specialized query optimization techniques also exist for statistical databases (101) and memory-resident databases (102).

Supporting a database buffer (sometimes called a *database cache*) is one of the most important efficiency techniques required (103,104). The database buffer manager acts to make pages accessible in main memory (reading from the disk) and to coordinate writes to the disk. In doing the reading and writing, it attempts to minimize I/O and to do as much I/O in a lazy fashion as possible. One example of the lazy I/O would be that at transaction commit time only the log records are forced to the disk, the commit completes, and the actual data records are written at a later time when the disk is idle or if forced for such other reasons as a need for more buffer space. Part of the buffer manager's task is to interact with one log manager by writing to the log and to cooperate with the recovery manager. If done poorly, disk I/O's for logging can become a bottleneck.

Disk space allocation is another important consideration in attaining good database performance. In general, the space allocation strategy should be such that fast address translation from logical block numbers to physical disk addresses can occur without any I/O, and the space allocation should be done to support both direct and sequential access.

The full mapping of relations through multiple intermediate levels of abstractions (e.g., from relations to segments, to OS files, to logical disks, to extents, to disk blocks) down to the physical layer must be done efficiently. In fact, one should try to eliminate unnecessary intermediate layers (still retaining data independence), and use various forms of mapping caches to speed up the translations.

Real-Time Distributed Databases

Real-time transaction systems are becoming increasingly important in a wide range of applications. One example of a real-time transaction system is a computer-integrated manufacturing system in which the system keeps track of the state of physical machines, manages various processes in the production line, and collects statistical data from manufacturing operations. Transactions executing on the database may have deadlines in order to reflect, in a timely manner, the state of manufacturing operations or to respond to the control messages from operators. For instance, the information describing the current state of an object may need to be updated before a team of robots can work on the object. The update transaction is considered successful only if the data (the information) is changed consistently (in the view of all the robots) and the update operation is done within the specified time period so that all the robots can begin working with a consistent view of the situation. Other applications of real-time database systems can be found in program trading in the stock market, radar tracking systems, command-and-control systems, and air-traffic-control systems.

Real-time transaction processing is complex because it requires an integrated set of protocols that not only must satisfy database consistency requirements but also operate under timing constraints (105–108). The algorithms and protocols that must be integrated include CPU scheduling, concurrency control, conflict resolution, transaction restart, transaction wakeup, deadlock, buffer management, and disk I/O scheduling (107,109–113). Each of these algorithms or protocols should address the real-time constraints directly.

To date, work on real-time databases has investigated a centralized, secondary storage real-time database (114–116). As is usually required in traditional database systems, work so far has required that all the real-time transaction operations maintain data consistency as defined by serializability. Serializability may be relaxed in some real-time database systems, depending on the application environment and data properties (108,117,118), but little actual work has been done in this area. Serializability is enforced by using a real-time version of either the two-phase locking protocol or optimistic concurrency control. Optimistic concurrency control has been shown to perform better than two-phase locking when integrated with priority-driven CPU scheduling in real-time database systems (119–121).

In addition to timing constraints, in many real-time database applications each transaction imparts a value to the system that is related to its criticalness and to when it completes execution (relative to its deadline). In general, the selection of a value function depends on the application (63). To date, the value of a transaction has been modeled as a function of its criticalness, start time, deadline, and the current system time. Here, criticalness represents the importance of transactions, while deadlines constitute the time constraints of real-time transactions. Criticalness and deadline are two characteristics of real-time transactions and they are not necessarily related. A transaction that has a short deadline does not necessarily have high criticalness. Transactions with the same criticalness may have different deadlines and transactions with the same deadline may have different criticalness values. Basically, the higher the criticalness of a transaction, the larger is its value to the system. It is important to note that the value of a transaction is time variant. A transaction that has missed its deadline will not be as valuable to the system as it would be if it had completed before its deadline. Other important issues and results for real-time distributed databases include the following:

- In a real-time system, I/O scheduling is an important issue with respect to the system performance. In order to minimize transaction loss probability, a good disk scheduling algorithm should take into account not only the time constraint of a transaction, but also the disk service time (111).
- Used for I/O, the earliest deadline discipline ignores the characteristics of disk service time and therefore does not perform well except when the I/O load is low.
- Various conflict-resolution protocols that directly address deadlines and criticalness can have an important impact on performance over protocols that ignore such information.
- How can priority inversion (this refers to the situation in which a high-priority transaction is blocked due to a low-priority transaction holding a lock on a data item) be solved (122,123)?
- How can soft real-time transaction systems be interfaced to hard real-time components?
- How can real-time transactions themselves be guaranteed to meet hard deadlines?

- How will real-time buffering algorithms have an impact on real-time optimistic concurrency control (124)?
- How will semantics-based concurrency control techniques have an impact on real-time performance?
- How will the algorithms and performance results be affected when extended to a distributed real-time system?
- How can correctness criteria other than serializability be exploited in real-time transaction systems?

Summary

DCSs began in the early 1970s with a few experimental systems. Since that time, tremendous progress has been made in many disciplines that support distributed computing. The progress has been so remarkable that DCSs are commonplace and quite large. For example, the Internet has over 500,000 nodes on it. This article has discussed two of the areas that played a major role in this achievement: distributed operating systems and distributed databases. For more information on distributed computing, Refs. 1–6, 62, 72, and 125 are good books and surveys. As mentioned in the introduction, many areas of distributed computing could not be covered in this article. One important area omitted is such distributed file servers as NFS (network file server) and Andrew. For more information on these and other distributed file servers, see the survey article in Ref. 126.

Acknowledgments: I enthusiastically thank Panos Chrysanthis and Krithi Ramamritham for their valuable comments on this work.

Glossary

ACCESS CONTROL LIST. A model of protection in which rights are maintained as a list associated with each object. *See also* Access control matrix; Capability list.

ACCESS CONTROL MATRIX. A model of protection in which rows of the matrix represent domains of execution and columns represent the objects in the system. The entries in the matrix indicate the allowable operations each domain of execution can perform on each object.

ACTIVE OBJECT MODEL. An object that has one or more execution activities (e.g., threads) associated with it at all times. Operation invocations on the object use these resident threads for execution.

ASYNCHRONOUS SEND. An IPC primitive in which the sending process does not wait for a reply before continuing to execute. *See also* Synchronous send.

ATOMIC BROADCAST. A communications primitive that supports the result that either all hosts (or processes) receive the message or none of them see the message. *See also* Broadcasting.

ATOMICITY. A property of a transaction such that either the entire transaction completes or it is as if the transaction never executed. *See also* Transaction.

BIDDING. A distributed scheduling scheme that requests hosts to provide information in the form of a bid as to how well that host can accept new work.

BROADCASTING. Sending a message to all hosts or processes in the system. *See also* Multicasting.

CAPABILITY LIST. A model of protection in which rights are maintained as a list associated with each execution domain. *See also* Access control list; Access control matrix.

CLIENT-SERVER MODEL. A software architecture that includes server processes that provide services and client processes that request services via well-defined interfaces. A particular process can be both a server and a client process.

CONSISTENCY. A property of a transaction that means the transaction maintains the integrity of the database. *See also* Transaction.

COPY-ON-WRITE. Data is delayed from being copied between address spaces until the destination actually requires it, or the sender needs to modify it.

DATA ABSTRACTION. A collection of information (data) and a set of operations on that information.

DIFFERENTIAL FILES. A representation of a collection of data as the difference from some point of reference. Used as a technique for storing large and volatile files.

DISTRIBUTED COMPUTING ENVIRONMENT (DCE). A computing environment that exploits the potential of computer networks without the need to understand the underlying complexity. This environment is to meet the needs of end users, system administrators, and application developers.

DISTRIBUTED OPERATING SYSTEM (DOS). A native operating system that runs on and controls a network of computers.

DISTRIBUTED SHARED MEMORY. The abstraction of shared memory in a physically nonshared distributed system.

DURABILITY. A property of a transaction that means all changes made by a committed transaction are permanent. *See also* Transaction.

FEDERATED DATABASE SYSTEM. A collection of autonomous database systems integrated for purposes of cooperation.

HARD REAL TIME. Tasks have deadlines or other timing constraints and serious consequences could occur if a task misses a deadline. *See also* Soft real time.

ISOLATION. A property of a transaction that means even if transactions execute concurrently, their results appear as if they were executed in some serial order. *See also* Serializability; Transaction.

LAZY EVALUATION. A performance-improvement technique that postpones taking an action or even a part of an action until the results of that action (subaction) actually are required.

LIGHTWEIGHT PROCESS. An efficiency technique that separates the address space and rights from the execution activity. Most useful for parallel programs and multiprocessors.

LIGHT REMOTE PROCEDURE CALL. An efficiency technique to reduce the execution time cost of RPCs when the processes happen to reside on the same host.

MULTICASTING. Sending a message to all members of a defined group. *See also* Broadcasting.

NESTED TRANSACTION. A transaction model that permits a transaction to be composed of subtransactions that can fail without necessarily aborting the parent transaction.

NETWORK OPERATING SYSTEM (NOS). A layer of software added to local operating systems to enable a distributed collection of computers to cooperate.

NETWORK PARTITIONING. A failure situation in which the communications network(s) connecting the hosts of a distributed system have failed in such a manner that two or more independent subnets are executing without being able to communicate with each other.

OBJECT. An instantiation of a data abstraction. *See also* Data abstraction.

PASSIVE OBJECT. An object that has no execution activity assigned to it. The execution activity gets mapped into the object upon invocation of operations of the object.

PROCESS. A program in execution including the address space, the current state of the computation, and various rights to which the program is entitled.

RAID (*redundant arrays of inexpensive disks*). Enhance I/O throughput and fault tolerance.

REAL-TIME APPLICATIONS. Applications in which tasks have specific deadlines or other timing constraints such as periodic requirements.

RECOVERABLE COMMUNICATING ACTIONS. A complex transaction model to support long and cooperative nonhierarchical computations involving communicating processes.

REMOTE PROCEDURE CALL (RPC). A synchronous communication method that provides the same semantics as a procedure call, but it occurs across hosts in a distributed system.

SAGAS. A complex transaction model for long-lived activities consisting of a set of component transactions that can commit as soon as they complete. If the saga aborts, committed components are compensated.

SERIALIZABILITY. A correctness criterion that states that the effect of a set of executed transactions must be the same as some serial execution of that set of transactions.

SOFT REAL TIME. In a soft real-time system, tasks have deadlines or other timing constraints, but no serious complications occur if a deadline is missed. *See also* Hard real time.

SPLIT TRANSACTIONS. A complex transaction model in which the splitting transaction delegates the responsibility for aborting or committing changes made to a subset of objects it has accessed to the split transaction.

SYNCHRONOUS SEND. An IPC primitive in which the sending process waits for a reply before proceeding. *See also* Asynchronous send.

TIMESTAMP ORDERING. A concurrency control technique in which all accesses to data are timestamped and then some common rule is followed to ensure serializability. *See also* serializability.

THREAD. Represents the execution activity of a process. Multiple threads can exist in one process.

TRANSACTION. An abstraction that groups a sequence of actions on a database into a logical execution unit. Traditional transactions have four properties: atomicity, consistency, isolation, and durability. *See also* Atomicity; Consistency; Isolation; Durability.

VALIDATION. A concurrency control technique that permits unrestricted access to data items, but then checks for potential conflicts at the transaction commit point.

Bibliography

Ball, J., Feldman, J., Low, J., Rashid, R., and Rovner, P., RIG, Rochester's Intelligent Gateway: System Overview, *IEEE Trans. Software Eng.*, SE-2(4) (December 1980).

Birrell, A., Levin, R., Needham, R., and Schroeder, M., Grapevine: An Exercise in Distributed Computing, *Commun. ACM*, 25:260–274 (April 1982).

Cheriton, D., Goosen, H., and Boyle, P., Paradigm: A Highly Scalable Shared Memory Multicomputer Architecture, *IEEE Computer* (February 1991).

Dion, J., The Cambridge File Server, *ACM Operating Sys. Rev.* (October 1980).

Kearns, J. P., and DeFazio, S., Diversity in Database Reference Behavior, *Performance Evaluation Rev.*, 17(1) (May 1989).

Lamport, L., Shostak, R., and Pease, M., The Byzantine Generals' Problem, *ACM Trans. Programming Language and Sys.*, 4(3) (July 1982).

Shrivastava, S. K., On the Treatment of Orphans in a Distributed System, *Proc. 3d Symposium Reliability Distributed Sys.* (October 1983).

Sturgess, H., Mitchell, J., and Isreal, J., Issues in the Design and Use of Distributed File System, *ACM Operating Sys. Rev.* (July 1980).

Theimer, M., Lantz, K., and Cheriton, D. R., Preemptable Remote Execution Facility for the V-System, *Proc. 10th Symposium on OS Principles*, 2–12 (December 1985).

References

1. Abeysundara, B., and Kamal, A., High Speed Local Area Networks and Their Performance: A Survey, *ACM Computing Surveys*, 23(2) (June 1991).

2. Davies, D. W., Holler, E., Jensen, E., Kimbleton, S., Lampson, B., LeLann, G., Thurber, K. J., and Watson, R. W., Distributed Systems — Architecture and Implementation. In: *Lecture Notes in Computer Science*, Vol. 105, Springer-Verlag, Berlin, 1981.

3. Enslow, P., What is a Distributed Data Processing System? *IEEE Computer*, 11 (January 1978).

4. Goscinski, A., *Distributed Operating Systems: The Logical Design*, Addison-Wesley, Sydney, Australia, 1991.

5. Ozsu, M., and Valduriez, P., *Principles of Distributed Database Systems*, Prentice-Hall, Englewood Cliffs, NJ, 1991.

6. Stankovic, J. A., A Perspective on Distributed Computer Systems, *Trans. Computers*, C-33(12):1102–1115 (December 1984).

7. Cheriton, D., and Zwaenepoel, W., Distributed Process Groups in the V Kernel, *ACM Trans. Computer Sys.*, 3(2) (May 1985).

8. Lazowska, E., Levy, H., Almes, G., Fischer, M., Fowler, R., and Vestal, S., The Architecture for the Eden System, *Proc. 8th Annual Symposium on Operating System Principles* (December 1981).

9. Mullender, S. J., van Rossum, G., Tanenbaum, A. S., van Renesse, R., and van Staveren, H., Experiences with the Amoeba Distributed Operating System, *Commun. ACM*, 33(12) (December 1990).

10. Needham, R. M., and Herbert, A. J., *The Cambridge Distributed Computing System*, Addison-Wesley, London, 1982.

11. Ousterhout, J., Scelza, D., and Sindhu, P., Medusa: An Experiment in Distributed Operating System Structure, *Commun. ACM*, 23 (February 1980).

12. Popek, G., et al., LOCUS, A Network Transparent, High Reliability Distributed System, *Proceedings 8th Symposium Operating System Principles*, 14–16 (December 1981).

13. Rashid, R., Threads of a New System, *UNIX Review*, 37–49 (August 1986).

14. Kopetz, H., Damm, A., Koza, A., Mulazzani, M., Schwabl, W., Senft, C., and Zainlinger, R., Distributed Fault Tolerant Real-Time Systems: the Mars Approach, *IEEE Micro*, 9(1):25–40 (February 1989).

15. Stankovic, J., and Ramamritham, K., The Spring Kernel: A New Paradigm for Real-Time Systems, *IEEE Software*, 8(3):62–72 (May 1991).

16. Rashid, R., Tevanian, A., Young, M., Golub, D., Baron, R., Black, D., Bolosky, W., and Chew, J., Machine Independent Virtual Memory Management for Paged Uniprocessor and Multiprocessor Architectures, *IEEE Trans. Computers*, 37(8) (August 1988).

17. Dasgupta, P., LeBlanc, R., Ahamad, M., and Ramachandran, U., The Clouds Distributed Operating System, *IEEE Computer*, 24(11):34–44 (November 1991).

18. Nitzberg, B., and Lo, V., Distributed Shared Memory: A Survey of Issues and Algorithms, *IEEE Computer*, 24(8):52–60 (August 1991).

19. Felten, E., and Zahorjan, J., Issues in the Implementation of a Remote Memory Paging System, University of Washington, Seattle, TR 91-03-09, March 1991.

20. Fitzgerald, R., and Rashid, R., Integration of Virtual Memory Management and Interprocess Communication in Accent, *ACM Trans. Computer Sys.*, 4(2) (May 1986).

21. Chin, R., and Chanson, S., Distributed Object Based Programming Systems, *ACM Computing Surveys*, 23(1) (March 1991).

22. Needham, R. M., and Schroeder, M., Using Encryption for Authentication in Large Networks of Computers, *Commun. ACM*, 21(12):993–999 (December 1978).

23. Hutchinson, N., and Peterson, L., The x-Kernel: An Architecture for Implementing Network Protocols, *IEEE Trans. Software Engineering*, 17(1) (January 1991).

24. Rashid, R. F., and Robertson, G. G., Accent: A Communication Oriented Network Operating System Kernel, *Proc. 8th Symposium Operating System Principles* (December 1981).

25. Bershad, B. N., Anderson, T. E., and Lazowska, E. D., Lightweight Remote Procedure Call, *ACM Trans. Computer Sys.*, 8(1):37–55 (February 1990).

26. Joseph, T., and Birman, K., Reliable Broadcast Protocols, TR 88-918, Cornell University, Ithaca, NY, June 1988.

27. Powell, D., et al., The Delta-4 Distributed Fault Tolerant Architecture, Laboratoire d'Automatique et d'Analyse des Systemes, Report No. 91055, February 1991.

28. Nelson, B. J., Remove Procedure Call, Xerox Corporation, Tech. Rep. CSL-81-9, May 1981.

29. Shrivastava, S. K., and Panzieri, F., The Design of a Reliable Remote Procedure Call Mechanism, *Trans. Computers*, C-31 (July 1982).

30. Birman, K., Replication and Fault-Tolerance in the ISIS System, *ACM Symposium on OS Principles*, 19(5) (December 1985).

31. Liskov, B., and Scheifler, R., Guardians and Actions: Linguistic Support for Robust, Distributed Systems, *Proc. 9th Symposium on Principles of Programming Languages*, 7–19 (January 1982).

32. Moss, J.E.B., Nested Transactions: An Approach to Reliable Distributed Computing, Ph.D. thesis, Massachusetts Institute of Technology, Cambridge, MA, April 1981.

33. Skeen, D., and Stonebraker, M., A Formal Model of Crash Recovery in a Distributed System, *IEEE Trans. Software Eng.*, SE-9(3) (May 1983).

34. Skeen, D., Nonblocking Commit Protocols, *Proc. ACM SIGMOD* (1981).

35. Skeen, D., A Decentralized Termination Protocol, *Proc. 1st IEEE Symposium on Reliability in Distributed Software Database Sys.* (1981).

36. Severance, D. G., and Lohman, D. G., Differential Files: Their Application to the Maintenance of Large Databases, *ACM Trans. Database Sys.*, 1(3) (September 1976).

37. Gray, J., Notes on Database Operating Systems. In: *Operating Systems: An Advanced Course*, Springer-Verlag, Berlin, 1979.

38. Stonebraker, M., Operating System Support for Database Management, *Commun. ACM*, 24:412–418 (July 1981).

39. Weinstein, M., Page, T., Livezey, B., and Popek, G., Transactions and Synchronization in a Distributed Operating System, *ACM Symposium on OS Principles*, 19(5) (December 1985).

40. Saltzer, J. H., Reed, D. P. J., and Clark, D. D., End-to-End Arguments in System Design, *Proc. 2d Intl. Conf. Distributed Computing Sys.* (April 1981).

41. Wu, K., and Fuchs, W., Recoverable Distributed Shared Virtual Memory, *IEEE Trans. Computers*, 39(4) (April 1990).

42. Bach, M., Coguen, N., and Kaplan, M., The ADAPT System: A Generalized Approach towards Data Conversion, *Proc. VLDB* (October 1979).

43. Anderson, T., Bershad, B., Lazowska, E., and Levy, H., Scheduler Activations: Effective Kernel Support for the User-Level Management of Parallelism, TR 90-04-02, University of Washington, Seattle, October 1990.

44. Chou, T., and Abraham, J., Load Balancing in Distributed Systems, *IEEE Trans. Software Eng.*, SE-8(4) (July 1982).

45. Chu, W., Holoway, L., Lan, M., and Efe, K., Task Allocation in Distributed Data Processing, *IEEE Computer*, 13:57–69 (November 1980).

46. Efe, K., Heuristic Models of Task Assignment Scheduling in Distributed Systems, *IEEE Computer*, 15 (June 1982).

47. Mirchandaney, R., Towsley, D., and Stankovic, J., Analysis of the Effects of Delays on Load Sharing, *IEEE Trans. Computers*, 38(11):1513–1525 (November 1989).

48. Stankovic, J. A., Bayesian Decision Theory and Its Application to Decentralized Control of Job Scheduling, *IEEE Trans. Computers*, C-34 (January 1985).

49. Stankovic, J. A., and Sidhu, I. S., An Adaptive Bidding Algorithm for Processes, Clusters and Distributed Groups, *Proc. 4th Intl. Conf. Distributed Computing* (May 1984).

50. Towsley, D., Rommel, G., and Stankovic, J., Analysis of Fork-Join Program Response Times on Multiprocessors, *IEEE Trans. Parallel and Distributed Sys.*, 1(3):286–303 (July 1990).

51. Vaswani, R., and Zahorjan, J., Implications of Cache Affinity on Processor Scheduling for Multiprogrammed, Shared Memory Multiprocessors, TR 91-03-03, University of Washington, Seattle, March 1991.

52. Casavant, T., and Kuhl, J., A Taxonomy of Scheduling in General Purpose Distributed Computing Systems, *Trans. Software Eng.*, 14(2): (February 1988).

53. Stankovic, J. A., Simulations of Three Adaptive Decentralized Controlled, Job Scheduling Algorithms, *Computer Networks*, 8(3):199–217 (June 1984).

54. Lamport, L., Time, Clocks, and the Ordering of Events in a Distributed System, *Commun. ACM* (July 1978).

55. Mirchandaney, R., Towsley, D., and Stankovic, J., Adaptive Load Sharing in Heterogeneous Distributed Systems, *J. Parallel and Distributed Computing*, 9: 331–346 (September 1990).

56. Anderson, T., Levy, H., Bershad, B., and Lazowska, E., Interaction of Architecture and OS Design, Department CS, TR, University of Washington, Seattle (August 1990).

57. Katz, R., Gibson, G., and Patterson, D., Disk System Architectures for High Performance Computing, *Proc. IEEE*, 77(12) (December 1989).

58. Patterson, D., Gibson, G., and Katz, R., A Case for Redundant Arrays of Inexpensive Disks (RAID), *Proc. ACM SIGMOD* (July 1988).

59. Barclay, D., Byrne, E., Ng, F., A Real-Time Database Management System for No. 5 ESS, *Bell Sys. Tech. J.*, 61(9) (November 1982).

60. Ayache, J., Courtiat, J., and Diaz, M., REBUS, A Fault Tolerant Distributed System for Industrial Control, *IEEE Trans. Computers*, C-31 (July 1982).

61. Wensley, J., Lamport, L., et al., SIFT: Design and Analysis of a Fault Tolerant Computer for Aircraft Control, *Proc. IEEE*, 1240–1255 (October 1978).

62. Stankovic, J., Misconceptions about Real-Time Computing: A Serious Problem for Next Generation Systems, *IEEE Computer*, 21(10):10–19 (October 1988).

63. Locke, C. D., Best-Effort Decision Making for Real-Time Scheduling, Ph.D. dissertation, Carnegie Mellon University, Pittsburgh, 1986.

64. Ramamritham, K., Stankovic, J., and Shiah, P., Efficient Scheduling Algorithms for Real-Time Multiprocessor Systems, *IEEE Trans. Parallel and Distributed Computing*, 1(2):184–194 (April 1990).

65. Arvind, K., Ramamritham, K., and Stankovic, J., A Local Area Network Architecture for Communication in Distributed Real-Time Systems, *Real-Time Sys. J.*, 3(2):113–147 (May 1991).

66. Zhao, W., Stankovic, J., and Ramamritham, K., A Window Protocol for Transmission of Time Constrained Messages, *IEEE Trans. Computers*, 39(9):1186–1203 (September 1990).

67. Schwan, K., Geith, A., and Zhou, H., From Chaos(Base) to Chaos(Arc): A Family of Real-Time Kernels, *Proc. Real-Time Sys. Symposium*, 82–91 (1990).

68. Tokuda, H., and Mercer, C., Arts: A Distributed Real-Time Kernel, *ACM Operating Sys. Rev.*, 29–53 (July 1989).

69. Stankovic, J., and Ramamritham, K., What is Predictability for Real-Time Systems—An Editorial, *Real-Time Systems J.*, 2:247–254 (December 1990).

70. Date, C. J., *An Introduction to Database Systems*, Addison-Wesley, Reading, MA, 1975.

71. Bernstein, P. A., Hadzilacos, V., and Goodman, N., *Concurrency Control and Recovery in Database Systems*, Addison-Wesley, Reading, MA, 1987.

72. Kohler, W., A Survey of Techniques for Synchronization and Recovery in Decentralized Computer Systems, *ACM Computing Surveys*, 13(2) (June 1981).

73. Lampson, B., Atomic Transactions. In: *Lecture Notes in Computer Science*, Vol. 105, Springer-Verlag, Berlin, 1980, pp. 365–370.

74. Spector, A. Z., and Schwarz, P. M., Transactions: A Construct for Reliable Distributed Computing, *ACM Operating Sys. Rev.*, 17(2) (April 1983).

75. Stonebraker, M., and Neuhold, E., A Distributed Database Version of INGRES, *Proc. Berkeley Workshop on Distributed Data Management and Computer Networks*, 19–36 (1977).

76. Bernstein, P. A., Shipman, D. W., and Rothnie, F. B., Jr., Concurrency Control in a System for Distributed Databases (SDD-1), *ACM Trans. Database Sys.*, 5(1): 18–25 (March 1980).

77. Bernstein, P., and Goodman, N., Concurrency Control in Distributed Database Systems, *ACM Computing Surveys*, 13(2) (June 1981).

78. Pu, C., Replication and Nested Transactions in the Eden Distributed System, Ph.D. thesis, University of Washington, Seattle, 1986.

79. Moss, J.E.B., Griffeth, N., and Graham, M., Abstraction in Recovery Management, *Proc. ACM SIGMOD Intl. Conf. Management of Data*, 72–83 (May 1986).

80. Badrinath, B., and Ramamritham, K., Performance Evaluation of Semantics-Based Multi-Level Concurrency Control Protocols, *Proc. ACM SIGMOD*, 163–172 (May 1990).

81. Korth, H., Kim, W., and Bancilhon, F., On Long-Duration CAD Transactions, *Information Sciences*, 46(1–2):73–107 (October–November 1988).

82. Korth, H., Levy, E., and Silberschatz, A., Compensating Transactions: A New Recovery Paradigm, *Proc. 16th VLDB Conf.*, 95–106 (August 1990).

83. Vinter, S., Ramamritham, K., and Stemple, D., Recoverable Actions in Gutenberg, *Proc. 6th Intl. Conf. Distributed Computing Sys.*, 242–249 (May 1986).

84. Pu, C., Kaiser, G., and Hutchinson, N., Split-Transactions for Open-Ended Activities, *Proc. 14th Intl. Conf. VLDB*, 26–37 (September 1988).

85. Garcia-Molina, H., Gawlick, D., Klein, J., Kleissner, E., and Salem, K., Modeling Long-Running Activities as Nested Sagas, *IEEE Technical Committee on Data Engineering*, 14(1):14–18 (March 1991).

86. Garcia-Molina, H., and Salem, K., SAGAS, *Proc. ACM SIGMOD Intl. Conf. Management of Data*, 249–259 (May 1987).

87. Chrysanthis, P., and Ramamritham, K., ACTA: A Framework for Specifying and Reasoning about Transaction Structure and Behavior, *Proc. ACM SIGMOD Intl. Conf. Management of Data*, 194–203 (May 1990).

88. Batory, D., GENESIS: A Project to Develop an Extensible Database Management System, *Proc. International Workshop on Object-Oriented Database Systems*, 207–208 (September 1986).

89. Carey, M., et al., The Architecture of the EXODUS Extensible DBMS. In: *Readings in Database Systems*, Morgan Kaufmann, 1988, pp. 488–501.

90. Maier, D., Stein, J., Ottis, A., and Purdy, A., Development of an Object-Oriented DBMS, *Proc. Object-Oriented Programming Systems, Languages, and Applications*, 472–482 (October 1986).

91. Agrawal, R., Carey, M. J., and Livny, M., Concurrency Control Performance Modeling: Alternatives and Implications, *ACM Trans. Database Sys.*, 12(4) (December 1987).

92. Kung, H. T., and Robinson, J. T., On Optimistic Methods for Concurrency Control, *ACM Trans. Database Sys.*, 6(2) (June 1981).

93. LeLann, G., Algorithms for Distributed Data-Sharing Systems which Use Tickets, *Proc. 3rd Berkeley Workshop on Distributed Databases and Computer Networks* (1978).

94. Thomas, R. H., A Majority Consensus Approach on Concurrency Control for Multiple Copy Databases, *ACM Trans. Database Sys.,* 4(2):180–209 (June 1979).

95. Weihl, W., Specification and Implementation of Atomic Data Types, Ph.D. thesis, Massachusetts Institute of Technology, Cambridge, MA, March 1984.

96. Weihl, W., Commutativity-Based Concurrency Control for Abstract Data Types, *Trans. Computers*, 37(12):1488–1505 (December 1988).

97. Rosenkrantz, D. J., Stearns, R. E., and Lewis, P. M., System Level Concurrency Control for Distributed Database Systems, *ACM Trans. Database Sys.*, 3(2) (June 1978).

98. Litwin, W., Mark, L., and Roussopoulos, N., Interoperability of Multiple Autonomous Databases, *ACM Computing Surveys*, 22(3) (September 1990).

99. Sheth, A., and Larson, J., Federated Database Systems for Managing Distributed, Heterogeneous, and Autonomous Databases, *ACM Computing Surveys*, 22(3) (September 1990).

100. Thomas, G., et. al., Heterogeneous Distributed Database Systems for Production Use, *ACM Computing Surveys*, 22(3) (September 1990).

101. Ozsoyoglu, G., Matos, V., Meral Ozsoyoglu, V., Query Processing Techniques in the Summary-Table-by-Example Database Query Language, *ACM Trans. Database Sys.*, 14(4):526–573 (1989).

102. Whang, K., and Krishnamurthy, R., Query Optimization in a Memory-Resident Domain Relational Calculus Database System, *ACM Trans. Database Sys.*, 15(1): 67–95 (March 1990).

103. Effelsberg, W., and Haerder, T., Principles of Database Buffer Management, *ACM Trans. Database Sys.*, 9(4) (December 1984).

104. Sacco, G. M., and Schkolnick, M., Buffer Management in Relational Database Systems, *ACM Trans. Database Sys.*, 11(4) (December 1986).

105. Dayal, U., et. al., The HiPAC Project: Combining Active Database and Timing Constraints, *ACM SIGMOD Record* (March 1988).

106. Dayal, U., Active Database Management Systems, *Proc. 3d Intl. Conf. Data and Knowledge Management* (June 1988).

107. Huang, J., Stankovic, J., Towsley, D., and Ramamritham, K., Experimental Evaluation of Real-Time Transaction Processing, *Proceedings Real-Time System Symposium* (December 1989).

108. Lin, K. J., Consistency Issues in Real-Time Database Systems, *Proc. 22d Hawaii International Conf. System Sciences* (January 1989).

109. Buchmann, A., et al., Time-Critical Database Scheduling: A Framework For Integrating Real-Time Scheduling and Concurrency Control, Data Engineering Conf. (February 1989).

110. Carey, M. J., Jauhari, R., and Livny, M., Priority in DBMS Resource Scheduling, *Proc. 15th VLDB Conf.* (1989).

111. Chen, S., Stankovic, J., Kurose, J., and Towsley, D., Performance Evaluation of Two New Disk Scheduling Algorithms for Real-Time Systems, *Real-Time Sys.*, 3(3) (September 1991).

112. Chen, S., and Towsley, D., Performance of a Mirrored Disk in a Real-Time Transaction System, *Proc. 1991 ACM SIGMETRICS* (May 1991).

113. Sha, L., Rajkumar, R., and Lehoczky, J. P., Concurrency Control for Distributed Real-Time Databases, *ACM SIGMOD Record* (March 1988).

114. Abbott, R., and Garcia-Molina, H., Scheduling Real-Time Transactions, *ACM SIGMOD Record* (March 1988).

115. Abbott, R., and Garcia-Molina, H., Scheduling Real-Time Transactions: A Performance Evaluation, *Proc. 14th VLDB Conf.* (1989).

116. Abbott, R., and Garcia-Molina, H., Scheduling Real-Time Transactions with Disk Resident Data, *Proc. 15th VLDB Conf.* (1989).

117. Son, S. H., Using Replication for High Performance Database Support in Dis-

tributed Real-Time Systems, *Proc. 8th Real-Time Systems Symposium* (December 1987).

118. Stankovic, J. A., and Zhao, W., On Real-Time Transactions, *ACM SIGMOD Record* (March 1988).

119. Haritsa, J. R., Carey, M. J., and Livny, M., On Being Optimistic About Real-Time Constraints, *Principles of Distributed Computing* (1990).

120. Haritsa, J. R., Carey, M. J., and Livny, M., Dynamic Real-Time Optimistic Concurrency Control, *Proc. 11th Real-Time Systems Symposium* (December 1990).

121. Huang, J.. Stankovic, J. A., Ramamritham, K., and Towsley, D., Experimental Evaluation of Real-Time Optimistic Concurrency Control Schemes, *Proc. VLDB* (September 1991).

122. Huang, J., Stankovic, J., Towsley, D., and Ramamritham, K., Priority Inheritance Under Two-Phase Locking, *Proc. Real-Time Systems Symposium* (December 1991).

123. Son, S. H., and Chang, C. H., Priority-Based Scheduling in Real-Time Database Systems, *Proc. 15th VLDB Conf.* (1989).

124. Huang, J., and Stankovic, J., Real-Time Buffer Management, COINS TR 90-65, University of Massachusetts, August 1990.

125. Tanenbaum, A., and van Renesse, R., Distributed Operating Systems, *ACM Computing Surveys*, 17(4):419–470 (December 1985).

126. Levy, E., and Silberschatz, E., Distributed File Systems: Concepts and Examples, *ACM Computing Surveys*, 22(4) (December 1990).

JOHN A. STANKOVIC

Divestiture Impact on Local Telephone Rate Policy in the United States: Diffusion of Local Measured Service, 1977–1985

Introduction

Traditionally, political scientists and economists have viewed regulatory policy subsystems as those in which private-interest groups capture their regulators (1–5). Capture tendencies seem most stark in utility regulation, especially in the United States. However, in the past decade these captured policy subsystems* have been opened up to public and outside pressures, especially those emanating from organized public-interest research groups (PIRGs) and increased intervenor activity in regulatory proceedings (6). In the United States, electric utility regulation was the first area to feel these new pressures as the oil embargoes of the 1970s led to rapidly increasing electricity prices. In contrast, the telecommunications regulatory subsystem remained relatively closed until the late 1970s and early 1980s, when two changes occurred.

First, economists and regulators in the late 1970s altered their views about regulation (7). Regulation began to be seen as socially inefficient and costly, in part because the true costs of service were hidden from users, often in the form of cross-subsidies, cost averaging, and flat-rate charges. They suggested that user-sensitive pricing schemes should replace the older rate forms. In the telecommunications area, these often are known as local measured (or metered) service (LMS), in which the timing, duration, and distance of a call are assessed to determine the price to consumers. Such a change in pricing policy disrupted the old alignments of these once-closed policy-making subsystems in the United States.

A second change, the divestiture of AT&T,† reinforced the movement to LMS. Prior to divestiture, a complex system of cross-subsidies existed: from long distance to local, urban to rural, and businesses to residents. With divestiture, the long-distance-to-local cross-subsidy was reduced, with plans for complete elimination. Long distance was arguably the most profitable service that AT&T provided, and demand for it grew at a heady rate. Further, technological innovations reduced the real cost of a long-distance call over the years (8). These savings were passed on to the local residential customer in the form of lower local rates.

Divestiture threatened the long-distance-to-local cross-subsidy. Easier entry by independent telephone companies into the long-distance market, which in-

---

*In the terminology to follow, the phrase *closed policy subsystem* is used to refer either to iron triangles or regulatory capture. In contrast, the term *open policy subsystems* refers to issue networks.

†On August 24, 1982, Judge Harold Greene issued the Modification of Final Judgment to the divestiture agreement between AT&T and the Department of Justice. Divestiture would become fully implemented on January 1, 1984.

creased competition there, placed pressures on regulators to remove the cross-subsidy. However, this revenue loss would affect the local companies greatly, forcing them to seek replacement revenues, perhaps by raising local rates. Some policy advocates suggested that one way to cap local prices would be to offer LMS. Such a user-sensitive pricing scheme would allow users to budget their demand and bring capacity and demand more into balance.

The divestiture of AT&T threatened to force local telephone prices steeply upward. One option to raising rates is to change the pricing structure from a flat-rate system to a user-sensitive pricing scheme (e.g., LMS). Proponents of this policy course argued that LMS would help suppress rates by reducing telephone company operating costs. First, people only would consume the amount of the system that they needed. For instance, it is estimated that the average residential customer uses the telephone only 15 minutes a day (9). Flat-rate systems, which allow unlimited calls, often require the erection and maintenance of excess capacity, thereby increasing the cost of the telephone network and ultimately requiring increased company revenues. Second, LMS would lessen the likelihood of system bypass. If local rates rose and/or long distance was assessed to subsidize local systems, some heavy users might find it more profitable to build their own networks. This would decrease local revenues further as the wealthiest users, mostly big business, bypassed the local network. Consequently, costs to the customer would rise, perhaps forcing even more people out of the system. For all of these reasons, some felt that LMS systems provided one way out of the local company revenue dilemma. Thus, both the rise of new economic theory and the divestiture of AT&T led to the movement to increase LMS offerings in the United States. These twin threats lead to a crisis in the policy subsystem and LMS became one solution.

Focusing on the case of state telephone pricing policy, this article raises the theoretical question, Can closed policy subsystems respond to change effectively? I argue that the theories of capture and iron triangles poorly account for change and adaptation in policy subsystems. Rather, divestiture and new economic theory converted the once-closed policy subsystem into an issue network. Former antagonists forged coalitions and one-time nonparticipants became active agents of change and policy making. As a result, policy was redirected, and as we will see, formerly favored rural areas lost policy advantages to urban centers. Thus, this article focuses on the adoption and diffusion of an innovation in telephone pricing policy.

Change, Capture, and Issue Networks

Interest-group capture theory can be viewed as a subset of iron triangle theory, which holds that a coalition composed of interest groups, bureaus, and legislative committees makes policy within its realm. Further, the coalition, based on the self-interest of each actor in the triangle, is stable and mutually reinforcing. Hence, the triangle tends to resist change and often is powerful enough to withstand attacks from the outside, especially from the executive.

One source of the triangle's political power comes from the fact that it tends to make policy hidden from public view. And since triangle policy focuses on the concerns of a narrow interest, the public rarely cares about any particular triangle's activity.

Interest-group capture theory is built upon the same foundation, except that the politicians allow the interest group to dictate policy to the bureaucrats. In both situations, the policy subsystem is static; policy is made to support the status quo. In capture, in fact, regulations creating barriers to protect the regulated interest group from outside competitors are the major policy goal. If any stimulus for change exists, it comes from the outside, and often is motivated technologically (10).

Changes in politics beginning in the 1960s altered many triangles and captured systems, converting them into issue networks. First, the great increase in the number of interest groups often led to intense conflict among competing interests. Second, the rise of new issues cut across the old boundaries that once separated distinct policy subsystems, leading to intersubsystem conflicts. Third, the rise of public advocacy and PIRGs added pressure to the closed policy subsystems. These public advocates helped transform once-hidden policy-making processes into objects of public scrutiny (11). These factors lead to breakdowns in the formerly stable and static policy subsystems, resulting in more fluid issue networks.

Issue networks, in contrast, have three characteristics (12). First, issue network participation is broader and may include chief executives, PIRGs, public opinion, the media, and perhaps even actors from other subsystems. Second, policy making is fluid and alignments among participants may change as the issue of concern changes and/or evolves. Alignments among participants therefore may be highly unstable and unpredictable. Third, these fluid issue networks may not be change resistant; they actually may promote change. The instability of alignments, the evolving nature of issues, and the changing intersection of issue boundaries all may lead to new alignments. Hence, internal as well as external factors may lead to change.

Support and Opposition to the Adoption of Local Measured Service

In this section, we discuss the alignment of support and opposition to the adoption of LMS. The theory of closed policy subsystems suggests that its members will be hostile to the adoption of LMS as this new policy represents a change in their interrelationships. If we consider regulatory policy to be a highly conflictual and complex policy arena, then the creation of a closed policy subsystem represents a resolution to conflict among these interests.

The basic interests to be balanced here are the telephone service providers and customers. The service providers require enough revenue to operate profitably, and those profit levels must be substantial enough to attract investment capital. Customers, on the other hand, prefer lower prices, reliable service, and easy access. In essence, the job of the state regulators has been to balance "fair rates of return" to the telephone companies with "universal service" to customers, especially residents.

If the old subsystem is still in control of telecommunications policy, we expect to find support for these two hypotheses:

H1: *The members of the closed subsystem will resist the diffusion of LMS.*

H2: *Nonmembers of the closed system will not affect the diffusion of LMS.*

An important implication of H1 is that all subsystem actors will agree in their resistance to LMS. If the subsystem is captured, only the telephone companies will affect policy outcomes significantly.

If, however, telecommunications policy in the United States has become an issue network, we expect to find support for these hypotheses:

H3: *The members of the closed system will conflict in their support/opposition to the adoption of LMS.*

H4: *Nonmembers of the closed subsystem will affect the adoption of LMS significantly.*

The discussion that follows presents the lineup of support and opposition to LMS that follows from the issue network perspective.

Closed Policy Subsystem Actors

Local Operating Companies

The local operating companies see the adoption of LMS as an opportunity to resolve the threat to their financial health that divestiture posed. First, the cross-subsidy from long-distance operations was reduced, with plans for eventual elimination. Second, large users began to bypass or planned to bypass the local system as local rates increased or threatened to increase and the operation of a "private" long-distance network became desirable and feasible economically.

With these potential losses of revenue, the local operating companies either had to increase local rates and/or reduce operating expenses. One study of Washington State estimated that if 125 businesses bypassed the local public network, the telephone system could lose 20% of its revenues and local rates subsequently would have to be raised $3 a month to make up the shortfall (13).

Some argue that user-sensitive pricing schemes may help reduce local company operating expenses. Since most flat-rate systems do not price discriminate between peak and nonpeak hours, peak hours become loaded and nonpeak periods become comparatively underutilized. Yet, the system requires enough capacity to meet the peak hours' demand.

User-sensitive pricing would lead to more cost-conscious behavior among customers, who probably would restrict their use at times and/or shift their use to nonpeak hours, when prices would be lower. Such a shift also would lower the necessary system capacity to handle demand, spreading demand more evenly throughout the day. This kind of change alone would lower operating costs by reducing the size of plant investment required to meet demand.

Some feel that overcapacity inherent in flat-rate schemes has lead many subscribers, whose demand for telephone service is limited, to overpay for service. One study reports that 7% of local telephone subscribers do not use the local loop in any particular month, and that 17% of local users account for over half of the use of the network. But all flat-rate subscribers pay the same. Hence, underusers are subsidizing overusers. The same study contends that 65% of local users are overcharged under flat-rate billing systems because of their restrained calling habits. Presumably, they would benefit by user-sensitive pricing if their calling habits did not change (13).

Thus, the local operating companies argue that user-sensitive pricing, of which LMS is one variety, not only would restrain their operating costs, but would help keep customer prices from rising at accelerated rates. Moreover, bypass would be somewhat countered, providing the local systems with larger revenue bases and, consequently, residents would benefit.

Business Users

For many of the same reasons mentioned above, business users should favor the adoption of LMS pricing schemes. If local rates rose, bypass might become more attractive to heavy users of long-distance services. As stated earlier, such bypass would reduce local system revenues, requiring even greater local price increases. Public outcry may lead to a greater cross-subsidy from business to residents. However, not all businesses are heavy long-distance users nor can all take advantage of bypass economically. They would be subject to redistributive politics without being able to exit. Without the exit–bypass recourse, these anchored businesses would have to voice their needs and try to influence the political system. Hence, to keep off the potential burden of redistribution, these businesses should support LMS.

Regulators

Unlike the local operating companies, state regulators generally have been slow to support LMS. (This does vary by state, as some state Public Utility Commissions [PUCs], like New York, were in the forefront of change, but they seem to be the exception.) We argue here, in fact, that the rise of these new issues in telecommunications uncoupled the PUCs from their "capturing" interests. If capture adequately describes the telecommunications subsystem in the age of divestiture, there would be no reason to refer to any subsystem actor other than the telephone companies. However, conflict between the PUCs and the telephone companies erupted over the issue of LMS in many states.

Yet, while the PUCs are not unaware of or unsympathetic to the concerns of the local operating companies, their environment is more complex. They are concerned particularly with the universal service goal. While the poor may be affected greatly by LMS, the most severe impacts perhaps will be felt among the PUCs' most vocal popular constituency, rural users. Noll, for instance, argues that statewide rate averaging, an integral element of flat-rate pricing, was in effect a rural subsidy (14).* In Noll's view, universal service is a public rationale for subsidy to rural residents.

However, some feel that user-sensitive pricing would price telephone service beyond the means of most nonurban dwellers. While it is not clear what the exact impact would be, one study estimates that rates for rural users in Alaska would rise to $200 a month (15). While this may be an extreme example, it illustrates the rhetoric of antidivestiture, anti-LMS advocates and the attendant fears of regulators, who would take the heat from the public and politicians if rates to rural residents increased so precipitously.

Further, in an area of uncertainty, regulators may be resistant to wide-scale change. Regulators, for instance, are uncertain about the true effects of LMS plans, how much rates would rise, and for whom. Coupling this with the fact that even where residents are offered LMS they tend not to opt for it (Ref. 13 estimates that 90% of consumers use flat-rate pricing plans even when LMS is available), regulators may question the wisdom of the policy. Last, flat-rate pricing is an easy policy to police, especially compared to LMS (13), where complications arise with inspection of meters that measure customer usage. We expect that the PUCs will oppose the adoption of LMS.

Legislative Committees

Legislative committees will feel pressures similar to those of the PUCs, and thus may be relatively resistant to LMS options. In particular, the geographic bias of committee representation, whereby rural areas are more likely to be over-represented on committees and subcommittees with jurisdiction over telecommunications policy, may insulate the committee politicians partially from the local operating companies, especially the Bell Operating Companies, which do not provide much service in rural areas. However, when the interests of the local operating companies and rural users were balanced, as was the case prior to divestiture, the legislative committees did not need to choose sides.

Yet, once conflict in the subsystem erupted, committee preferences shifted toward rural interests and away from telephone company interests. In so shifting, the legislative committees may have lost influence within the legislature because the legislature as a whole was more interested in protecting the larger urban interests. By trying to preserve their individual careers, the reason for the shift, these committee overseers may have had to cede policy-making influence to their chambers.

---

*Why, however, did the rural subsidy not diminish with the reapportionments of the 1960s? Noll suggests that "the answer would seem to be the fortuitous happenstance of explosive growth in telecommunications technology and the demand for services. With all prices falling, even during inflation, and new technologies coming on-line at a rapid rate, this particular rural subsidy (from urban to rural users) was not a salient political issue—at least, not until the mid-1980s" (pp. 181–182).

As long as telecommunications policy was not controversial, the legislative chambers allowed the committee freedom to guide policy. And, if this meant telephone company capture of the subsystem, such an arrangement also was acceptable. As soon as public interest and concern heightened, the chambers of the legislature saw that, without intervention, policy preferences would be implemented. Hence, in the new environment we are not likely to find legislative committees with much independent impact on policy. But, if any impact is detected, it should oppose the adoption of LMS.

The New Issue Network Actors

The theory of issue networks argues that not only should we find conflict among those actors who made up the old closed policy subsystem, but others who were not active and/or influential in the predivestiture subsystem now would become active. Of greatest concern for present purposes are PIRGs, governors, and the legislative chambers.

Public-Interest Groups

A great degree of opposition to LMS comes from PIRGs. PIRG arguments against LMS are legion. Among the most common arguments are the following (13):

- LMS would further confuse consumers already confused by deregulation and divestiture. Confusion about how to estimate the cost of a telephone call under LMS would be especially sharp.
- Charitable organizations, already heavily dependent on telephones to generate contributions, would be hit hard. Increasing their costs of operation would ripple, affecting those who depend on these organizations: the poor, the elderly, and the infirm.
- People already socially isolated, such as the poor and the elderly, who depend heavily on the telephone for social contact, may be isolated increasingly as they curtail their telephone use to save money.
- Small businesses that rely heavily on telephones to generate and conduct business would be hit hard. As rates under LMS would be greatest during peak business hours, small and marginal businesses would be burdened greatly.
- Rates would increase, not decrease, under LMS, as expensive metering devices would have to be installed. This may make the overall operation of an LMS system more expensive than retaining flat rate service. Use of those meters would invade user privacy.

If the important cleavage in the postdivestiture age is between rural and urban interests, why do the PIRGs, with their heavily urban support base, oppose LMS? Part of the answer must be that PIRGs see the issue in big business versus consumer terms, and they viewed the adoption of LMS as a sop to big business. Also, PIRGs may view the interests of the urban poor and rural areas as coinciding.

One possible impact of PIRGs on telecommunications policy may have been to try to convert an urban–rural conflict to one between consumers and big business. If successful, not only would the political alignment in the states look quite different, but incentives for politicians to support LMS would be diminished as well. Therefore, we are likely to find that the stronger the PIRGs, the less likely that LMS will diffuse across the states' rate classes.

Legislatures

As we argue above, the legislative chambers generally should support LMS. This results from their responsiveness to urban interests. Teske found that state legislatures increasingly have become active and important in the adoption of user-sensitive pricing schemes (16). Insofar as the state legislatures are more well apportioned than the PUCs, we should find them, with less of a rural bias than the PUCs, somewhat more supportive of LMS. (The theory, however, does not indicate any party-related association, hence we do not use a party variable in this analysis. The same holds for the governors' parties, discussed next.)

Governors

Governors also may have a keen impact on the adoption of LMS. It is likely that governors will be even more supportive of LMS than the legislature, owing to their even greater independence from rural areas. Further, a governor's position as the leading state politician may provide the state executive with a vantage point from which to counter PIRG arguments against LMS. Gubernatorial political support for LMS should be most evident in the most urbanized states where rural voters play a smaller part in state politics.

Other Factors

Diffuse environmental and economic pressures also may be important. A major concern of state-level telecommunications policy has been the maintenance of universal service. The more that LMS may threaten or enhance that goal, the less likely it is LMS policies will be adopted. If, as we argue, politicians at the state level see LMS as part of a rural–urban conflict and also as a solution to the specter of rising consumer prices, then we are likely to see LMS adopted where threatened populations are large. Two such conditions may be when poverty levels are high and when comparatively few people are telephone subscribers. Hence, we expect that as poverty increases and penetration decreases, adoption of LMS will accelerate. Further, if the urban–rural split is the issue of great significance, then the more urbanized a state is, the more likely that LMS adoption will move apace.

Finally, two conditions of the states may make them more or less hospitable to LMS. We do not expect southern states to adopt this policy as swiftly as nonsouthern states, as much comparative state research on the adoption of innovations has found. Also, a state's level of innovativeness may affect its ability to adopt this innovation. We employ Walker's measure of state innovativeness to measure this capability (17).

Results of the Analysis

Trends

Our interest is whether or not it is a trend to offer LMS as an option in any of four designated rate classes: urban businesses, urban residents, small-town businesses, and small-town residents. Few states require LMS among residents (though businesses in the largest metropolitan areas of the largest states often are required to use LMS and, in a handful of cities, like New York, residents are offered only LMS). (We use small-town instead of rural rates because our focus is on telephones in the territories of the Bell Operating Companies. This helps insure comparability. Bell rarely serves rural areas, but does serve small towns that are adjacent to larger cities.)

Table 1 illustrates the pattern of LMS use for the four rate classes. LMS systems were not foreign to the United States even before AT&T's divestiture. Urban businesses often had the option. In 1977, fully 88% of the United States (excluding Alaska and Hawaii, where Bell companies do not operate) and the District of Columbia allowed businesses the LMS option. This figure stayed quite steady across the years. Thus, state regulators knew something about the LMS option, its level of popularity, and the kind of revenues that it could generate.

However, Table 1 also makes another point: the diffusion of LMS to the

TABLE 1   Percentages of States with Local Measured Service, By Year

	1977	1978	1979	1980	1981	1982	1983	1984	1985
Small-town residents	18	18	22	27	49	52	59	MD	68
Urban residents	37	37	45	51	63	75	78	MD	96
Small-town businesses	27	29	33	33	49	52	63	MD	72
Urban businesses	88	84	86	84	90	96	92	MD	91

Adapted from National Association of Regulatory Utility Commissioners, Bell Operating Companies Exchange Service Telephone Rates, NARUC, Washington, DC, 1977–1985.
The urban business figures fluctuate because of minor differences in the number of states reporting each year.

other three customer classes began before divestiture took effect and even before
the AT&T court case concluded. From 1977 to 1981, LMS options for urban
U.S. residents grew from 37% to 63%. The percentages for small-town residents
and businesses went from 18% to 49% and from 27% to 49%, respectively.
Diffusion continued through 1985, with 96% of the states offering the option
to urban residents, 68% to small-town residents, and 72% to small-town businesses.

We can see the diffusion patterns from another perspective. In 1977, the
average number of LMS options offered to these three customers groups (excluding urban businesses) was 0.82. By 1981, it increased to 1.61, almost double,
and by 1985 it rose to 2.27, again an absolute increase of almost the same
magnitude. LMS seemed to diffuse at about the same rate before divestiture as
after.

We can use regression analysis to inspect the diffusion rate patterns more
systematically. For each of the four LMS policy options, I created a dependent
variable, defined as the percentage of states per year allowing the LMS option.
I then regressed these dependent variables on an equation that includes three
terms: a time counter (1977 = 1, 1978 = 2, . . . , 1985 = 9), a dummy variable for the pre- and post–Modified Final Judgment (MFJ) periods (1977–1981
= 0, 1982–1985 = 1), and a dummy variable for the full implementation of
divestiture (1977–1983 = 0, 1984–1985 = 1). The equation for the urban business LMS does not contain a significant variable. As the trends in Table 1 show,
this LMS option is relatively invariant over time. However, for each of the other
three LMS options, one finds the same results: the only significant variable is
the time counter. Neither the MFJ or the divestiture dummies have any impact
(see Table 2). Hence, we can assume safely that divestiture's events did not
affect the rate of diffusion of the policy—it began diffusing before divestiture,
continuing along the same course at the same rate as divestiture took hold.
These patterns suggest that LMS was used to deal with postdivestiture problems
because it was a policy option already gaining popularity among regulators. If
divestiture had any impact, it was to justify the further spread of LMS.

Yet, in 1985 considerable state variation still exists. While most states offer
the service to urban businesses, of the other 3 customer groups, 3 states do not
offer the service to any, 11 states offer it to 1 group, 5 states offer it to 2
groups, and 30 states offer it to all 3 customer classes. The dependent variable
used in the tables here indicates how many of the three customer groups (urban
residents, small-town residents, and small-town businesses) offered LMS per
state per year from 1977 to 1985. Thus, each state–year case can take on a value
from 0 to 3, representing the number of customer classes offered the LMS
option, the LMS index.

Regression Analysis

Table 3 presents the results of a pooled cross-sectional time series of the closed
subsystem and issue network variables for the adoption of LMS, the *LMS
index*. Definitions of variables used are presented in Table 4.

In tables, *b,* SE, *t,* and Sign. refer to statistics associated with each substan-

TABLE 2   An Aggregate Regression View of the Diffusion of Local Measured
Service, 1977–1985

Variable		Small-Town Residential	Urban Residential	Small-Town Business	Urban Business	LMS Index
Counter	b	8.59	13.81	10.21	23.39	.33
	SE	.83	1.65	1.24	5.02	.003
	t	10.23	8.39	8.23	4.66	9.51
MFJ	b	−.31	−13.27	−8.85	−58.02	−.24
	SE	7.02	13.77	10.38	42.03	.28
	t	−.004	−.96	−.85	−1.38	−.82
Divestiture	b	−8.96	−15.03	−11.02	−61.47	−.42
	SE	7.93	15.57	11.74	47.53	.32
	t	−1.13	−.97	−.94	−1.29	−1.31
R^2		.93	.77	.78	−55.99	.85
Adjusted R^2		.90	.67	.70	−78.79	.79
n		8	8	8	8	8

The dependent variables are the number of states offering LMS service to each customer rate class
each year. The LMS index is the average number of LMS offerings per state per year for the
following three rate classes: small-town residential, urban residential, and small-town business.
Counter is $1977 = 1, \ldots, 1985 = 9$.
MFJ is $1977–1981 = 0, 1982–1985 = 1$.
Divestiture is $1977–1983 = 0, 1984–1985 = 1$.

tive, that is, independent variable. The regression coefficient b tells us how
much of a unit change in the dependent variable, the LMS index, is caused by a
one-unit change in the independent variable. SE is the independent variable's
standard error. Dividing b by its SE produces the t value, a measure of statistical
significance. Rules of thumb usually insist that the t value be greater than 2.00,
though values nearing that range also at times are considered important. Sign.
is the significance level, and informs us about the probability that the t value
was not a chance occurrence. The closer the significance is to .00, the less
likelihood there is that it occurred by chance; usually, one insists that the significance passes or is near the .05 level. As an example, inspect the Bell variable in
Table 3. The left-hand b column tells us that a 1% increase in the size of the
Bell System leads to an increase of .03 units on the LMS index. Further, the SE
is small, producing a t value much larger than the 2.00 marker (it is 6.01). As
the significance level is near .00, we can assume safely that the impact did not
happen by chance.

Results of the pure closed subsystem model are presented in the left-hand
columns of Table 3. Of the substantive independent variables included in the
equation, four clearly reach conventional levels of statistical significance, and
another almost does. Further, R^2, the percentage of the variance explained, is
significant and healthy at .52, indicating an equation that performs well. Upon
closer inspection, however, results appear murkier as many of the significant
variables point in the wrong direction and/or contradict related variables.

In all, support for the closed subsystem theory is mixed. The pointing of

TABLE 3   Regression Results of Impact of Closed Subsystem and Issue Network on the Adoption of Local Measured Service

	Pure Model				Interest Variables Added			
	b	SE	t	Sign.	b	SE	t	Sign.
Closed subsystem								
Constant	−1.49				−5.42			
LRULE	.005	.05	.10	.92	−.05	.06	−.88	.38
CONIN	.33	.14	2.32	.02	.03	.15	.22	.82
TIME	−.25	.14	−1.82	.07	−.36	.15	−2.39	.02
BELL	.03	.004	6.01	.00	.01	.007	1.57	.12
COMPLEX	.03	.05	.60	.55	−.01	.06	−.21	.83
EMPLOY	.0008	.0003	2.11	.03	−.0009	.0006	−1.69	.09
LICEN	−.09	.03	−2.54	.01	−.16	.05	−3.36	.00
BUSIPCT	.06	.03	2.15	.03	.08	.05	1.62	.11
Issue network								
LSAL	—				.25	.09	2.66	.01
LARM	—				.31	.17	1.78	.09
EXARM	—				−.05	.21	−.23	.82
GPOWER	—				.07	.03	2.19	.03
GRASS	—				.13	.15	.89	.38
PROXY	—				−.30	.17	−1.81	.07
CCAUSE	—				−.0005	.02	−.03	.98
POV	—				.14	.04	3.86	.00
PENET	—				−.006	.02	−.30	.76
SOUTH	—				−.50	.24	−2.08	.04
ELECTION	—				−.81	.24	−3.36	.00
INNOV	—				1.34	1.22	1.09	.28
URBAN	—				−.007	.007	−.92	.36
R^2/adjusted R^2	.52/.48				.61/.56			
F/n	12.63/376				11.91/376			

Significant state and year dummies respectively are 1978, 1979, 1980, 1981, 1982, 1983, 1985, and Connecticut, Delaware, Georgia, Iowa, Kentucky, Maine, Minnesota, Montana, Ohio, Rhode Island, South Dakota, Texas, Virginia, West Virginia, Wisconsin.

TABLE 4 Definitions of Variables Used in the Analysis

LRULE	Legislative committee rule review; 0 = no rule review power; 1 = review authority must be given to a committee; 2 = review authority also must be general, that is, not limited to a specific agency or program; 3 = review authority has to include the possibility of authoritative action on agency regulations. Source: Ref. 19.
CONIN	PUC is covered under conflict of interest provisions; 1 = covered; 0 = not covered. Source: Ref. 18.
TIME	Time regulations between leaving the PUC and beginning employment in a regulated firm; 1 = time regulations; 0 = no time regulations. Source: Ref. 18.
BELL	Percentage of telephones in the Bell network. Source: Ref. 18.
COMPLEX	Number of the following that the PUC possesses: (1) a research section or permanent research staff; (2) a research library; (3) a separately staffed public information office; (4) whether or not administrative law judges and hearing examiners are full time employees; (5) whether the staff is covered under civil service or merit protection; (6) whether data processing and computer equipment are used.
EMPLOY	Number of employees in the PUC. Source: Ref. 18.
LICEN	Number of the following licensing powers that the PUC possesses: (1) to initiate telephone service; (2) to lay transmission lines; (3) to lay distribution lines; (4) to operate other types of plants; (5) to allow abandonment of facilities and services; (6) to issue indeterminate permits; (7) to allocate unincorporated territory among the telephone utilities. Source: Ref. 18.
BUSIPCT	Percentage of state's telephones that are for business as opposed to residential. Source: Ref. 21.
LSAL	Legislative salaries, annual, corrected for inflation. Source: Ref. 19.
LARM	PUC is an arm of the legislature; 0 = not an arm; 1 = an arm. Source: Ref. 18.
EXARM	PUC is an arm of the executive; 1 = is an arm; 0 = is not an arm. Source: Ref. 18.
GPOWER	Index of gubernatorial power. Source: Ref. 20.
GRASS	Grassroots advocacy activity level; 1 = High; 0 = Low. Source: Ref. 6, p. 40.
PROXY	Proxy advocacy activity level; 1 = High; 0 = Low. Source: Ref. 6, p. 40.
CCAUSE	Number of members of Common Cause per 10,000. Source: Ref. 22, p. 52.
POV	Percentage of the state's population living below the poverty level. Source: Ref. 23.
PENET	Percentage of households in the state with telephone service. Source: Ref. 23.
SOUTH	South; 1 = Confederate state; 0 = not a Confederate state.
ELECTION	If the PUC commissioners are elected or not; 1 = elected; 0 = not elected. Source: Ref. 18.
INNOV	Walker's index of state innovativeness.
URBAN	Percentage of the state's population living in urban areas. Source: Ref. 23.

coefficients in contradictory directions undermines the closed subsystem model as an explanation of the adoption of LMS. One possible reason for these incorrect signs may be model underspecification. We deal with this possibility by adding the issue network variables into the equation, as seen in the bottom portion of Table 3. If the pattern of the closed subsystem variables remains as is, but some of the issue network variables are found to be significant predictors of LMS adoption, then we also must reject the pure closed subsystem theory for this case.

The right-hand columns of Table 3 add the PIRG, political, and diffuse interests variables that the issue network theory suggests. Initially, three points of interest arise. First, this model significantly improves upon the subsystem model in pure statistical terms. The explained variance increases by 9% to .61. An F test confirms that this increase is statistically significant ($F = 3.90$, $p > .01$). Second, the subsystem variables of the pure equation decline in overall importance. Now, only two are statistically significant: time restrictions on finding employment in a regulated industry after leaving the PUC and licensing power. Both point in the same direction, as the issue network theory suggests — they retard adoption. Third, when we inspect the coefficients for the issue network variables, we find a number reaching statistical significance with the expected signs. These are the PIRG variable, proxy advocacy; if PUC regulators are not elected; legislative and gubernatorial variables; the poverty rate; and if states are not southern.

In all, the results speak more strongly toward accepting the issue network perspective than the more traditional closed policy subsystem one. First, conflict among subsystem participants is noted, especially that between the special interests and the PUCs. While the special interests favored adoption of the user-sensitive pricing scheme, PUCs did not. Second, actors from outside the closed subsystem affected the adoption of LMS. Further, inclusion of issue network variables boosts the explained variance considerably.

Of equal significance, governors and state legislative chambers have an impact. This is somewhat counter to Teske, who only found legislatures important, but did not include gubernatorial variables in his analysis (16). Further, while these peak-level politicians had an impact on adoption behavior, subsystem politicians in the legislative committees did not. This indicates that peak-level politicians may have restricted the influence of subsystem politicians, another sign that an issue network has developed. Lack of committee impact is consistent with a capture perspective, but capture also would hold that peak-level politicians would not have much impact. That they do counters capture theory.

The Rural Bias to Telephone Policy Revisited

An important theme in the discussion above was the rural bias of policies during the predivestiture era. The adoption of LMS shows how much the old, closed policy subsystem has been disrupted. Participants of the old system have begun to disagree over the direction of telephone policy. In particular, the local operating companies and the regulators now disagree over the shape and look of new policy in the era of divestiture. The local operating companies prefer some kind of user-sensitive pricing mechanism, while the bureaucrats, fearing a backlash from rural residents, have resisted too speedy acceptance of such a policy course.

Further, while urban residents may find that divestiture has pushed up their telephone bills, they may find relative benefits from a user-sensitive scheme. Under such a scheme the urban to rural cross-subsidy (in the form of statewide rate averaging) may recede. In effect, while LMS may cause urban rates to rise, they may not rise as fast as if under a flat-rate statewide rate averaging scheme.

However, with the greater electoral importance of urban versus rural voters, we must ask, Has policy generally become less rurally beneficial? Has the urban-to-rural subsidy even for more traditional flat-rate service moderated, as urban residents began to feel the impact of actual or impending rate increases? Previous data has revealed the spread of LMS programs across most of the states to most telephone users. Does this spread mean that policy is taking a decidedly more urban, less rural turn?

We can investigate this question. One measure of relative urban–rural rates might take the following form:

rural flat rates/urban flat rates

We should note that this ratio is not a good representation of the actual cross-subsidy. It does not take into account marginal costs of supplying service to urbanites versus rural residents, nor does it account for the aggregate amount of money transferred from one group to the other. However, on a political level it has appeal because voters are likely to relate their rates to the rates that others pay.

But, although we do not possess good data on rural rates (our data are exclusively Bell rates, and Bell does not service rural areas), we do possess good rate data on small-town rates for Bell service. We can use such data to examine the relative rates of urban to small-town areas. Urban residents not only pay more for their service than rural residents, but they pay more than small-town residents, as well. Thus, we replace the rural flat-rate term in the formula above with the small-town flat rate that corresponds to the rate class for which the LMS data were taken.

Table 5 indicates that the relative small-town to urban flat rate has grown. In 1977, the series witnessed a ratio of .68. In 1985, the ratio had grown to .75, to the disadvantage of small-town residents. In other words, relative to urban residents, small-town residents saw their telephone rates grow 7% faster. More significantly, these changes occurred subsequent to divestiture. From 1977 to 1982, the ratio of small-town to urban rates hardly changed. However, the first calendar year after the issuance of Judge Greene's MFJ, 1983, saw the ratio step up over 2%, to almost .71, and by 1985 it had jumped almost another five points, to slightly over .75. This trend clearly implicates divestiture as a causal agent and superficially underlies the theme of increasingly less rurally biased telephone policy as divestiture matured.

Again, multiple regression analysis can be used to understand better the

TABLE 5 Change in the Small-Town to Urban Flat-Rate Ratio, 1977–1985

Year	Average	Standard Deviation	n
1977	.68	.095	48
1978	.66	.100	48
1979	.67	.107	47
1980	.68	.096	47
1981	.67	.113	47
1982	.68	.119	47
1983	.71	.127	49
1984	.68	.176	4
1985	.75	.136	47
Grand Mean	.69	.095	384

process of change associated with the ratio measure. We use the same variables that were employed in the LMS equation except for the Walker innovation index and the percentage of business telephones, which are not relevant here. This rate policy is not an innovation, but suggests only changes in existing policy. Hence, there is no need for the innovativeness index. Further, while business may have an indirect interest in the residential cross-subsidy, it has no direct interest in how much urban residents pay compared to nonurbanites. Results are presented in Table 6. Variables should be interpreted as for Table 3.

There is strong indication that the antirural factions associated with the adoption of LMS also have had an impact on relative rates. The left-side columns of Table 6 present the results of an equation that only includes the closed subsystem variables, and the left-side columns present the full equation results.

The closed subsystem variables poorly explain the ratio. Only two of the seven variables measuring subsystem participants had an impact. A better statistical fit is found for the issue network theory. First, equation performance improves: the R^2 increases from .49 to .64, a boost of 15%. An F test demonstrates that this is a statistically significant increase ($F = 7.04$, $p > .01$). Further, we find a number of the subsystem nonparticipants with an impact from these relative rate changes: Both governors and legislatures have impact, as does the PIRG variable proxy advocacy, the election of PUC regulators, the size of the Bell System, conflict-of-interest regulations for PUC regulators, and institutional resources of the PUCs.

One finding of interest among the closed subsystem variables in the fuller equation is that they all point in the expected direction and highlight the conflict between the Bell Telephone companies and the PUCs, something unheard of before divestiture. This is another indication of the dissolution of the old closed subsystem, the lack of private-interest-group capture, and the rise of telecommunications issue networks in the United States.

Last, as we also may expect, subsystem factors loomed more importantly in the rate equation than the one for the diffusion of LMS. Subsystems have a

TABLE 6 Determinants of the Small-Town to Urban Flat-Rate Ratio, 1977–1985

	Pure Model				Interest Variables Added			
	b	SE	t	Sign.	b	SE	t	Sign.
Closed subsystem								
Constant	$.07$				$.13$			
LRULE	-2.54^{-3}	8.70^{-4}	-2.83	$.00$	-1.89^{-3}	8.77^{-4}	-2.16	$.03$
CONIN	1.64^{-3}	2.20^{-3}	$.75$	$.46$	-6.11^{-3}	2.38^{-3}	-2.57	$.01$
TIME	1.51^{-3}	2.42^{-3}	$.62$	$.53$	1.31^{-3}	2.20^{-3}	$.60$	$.55$
BELL	-2.60^{-5}	7.40^{-5}	$-.35$	$.73$	2.11^{-4}	$.89^{-4}$	2.36	$.02$
COMPLEX	$.30^{-3}$	$.94^{-3}$	$.32$	$.75$	$.57^{-3}$	$.90^{-3}$	$.58$	$.56$
EMPLOY	$-.31^{-4}$	$.90^{-5}$	-3.48	$.00$	$-.21^{-4}$	$.11^{-4}$	-1.88	$.06$
LICEN	$-.90^{-4}$	$.55^{-3}$	$-.16$	$.87$	$-.15^{-2}$	$.67^{-3}$	-2.21	$.03$
URBAN	-2.34^{-4}	$.81^{-4}$	-2.90	$.00$	$-.46^{-4}$	$.83^{-4}$	$-.55$	$.58$
Issue network								
LSAL	—				$-.34^{-3}$	1.31^{-3}	$-.26$	$.80$
LARM	—				$.01$	$.24^{-2}$	4.85	$.00$
EXARM	—				$.64^{-2}$	$.29^{-2}$	2.18	$.03$
GPOWER	—				$-.59^{-3}$	$.42^{-3}$	-1.41	$.16$
GRASS	—				$.30^{-2}$	$.24^{-2}$	1.29	$.20$
PROXY	—				$.59^{-2}$	$.24^{-2}$	2.50	$.01$
CCAUSE	—				$-.41^{-3}$	$.30^{-3}$	-1.35	$.18$
POV	—				$.70^{-2}$	$.49^{-3}$	1.42	$.16$
PENET	—				$-.89^{-3}$	$.31^{-3}$	-2.84	$.00$
SOUTH	—				$.63^{-3}$	$.35^{-3}$	1.81	$.07$
ELECTION	—				$-.93^{-2}$	$.32^{-2}$	-2.93	$.00$
R^2/adjusted R^2	$.49/.45$				$.64/.59$			
F/n	$11.24/376$				$14.22/376$			

The ratio is defined as small-town residential flat rates/urban residential flat rates. Significant state dummy variables are California, Georgia, Kansas, Kentucky, Maine, Minnesota, Missouri, New Jersey, New York, North Carolina, Ohio, Oregon, Utah, Virginia, Washington, and Wisconsin, and year dummies are from 1977 to 1985. Variable definitions are presented in Table 4.

harder time responding to innovations than making adjustments to already existing policy, hence, the lesser impact of the subsystem on LMS and consequently the greater relative impact of the issue network. However, we also see the infection of divestiture and LMS on rates. More than just incremental adjustments in rate policy were being made. The subsystem was breaking apart there, as well, though perhaps at a slower rate. In other words, there may be a lag effect whereby the effects of divestiture and LMS interceded into rate policy only after those new forces rooted into the policy process and fashioned an issue network.

We can demonstrate further the connection between these two policies. States that are more likely to adopt LMS are also more likely to shift rate-making policies in an urban direction, as correlational evidence attests. The Pearson product moment correlation between adoption of LMS and the rate ratio is .275, which is significant at the .01 level. That the correlation is no larger signifies some differences: the differences between innovative and incremental policies, the greater impact of subsystem actors on rate than LMS decisions, and the possible lag of impact from the LMS arena to the rate arena. Still, the significant correlation indicates a relationship between the two policies.

Conclusion: Capture, Networks, and Changing Telephone Policy in the United States

Two themes dominate this article. One concerns the changing structure of the telephone regulatory subsystem in the United States, the break-up of the closed policy subsystem and its replacement with a larger, more diffuse, and conflictual issue network. The second theme looks at the changing policy direction of state telephone regulation, specifically, the altering balance between urban and nonurban interests. In this conclusion, we tie these two themes together more closely and try to speak to the issue of the future of state telephone regulatory policy.

The predivestiture policy subsystem was not only a classically closed subsystem, but it favored rural interests over urban ones. The urban/nonurban division was not an issue of great concern because telephone rates in urban areas had been falling since the 1940s. The source of the rate declines was the long-distance cross-subsidy of local service. This subsidy could be maintained as long as the cost of long-distance service declined and demand grew, allowing long-distance-generated revenues to be transferred to local users.

Such a system benefited everyone except those who wanted to enter the lucrative long-distance markets that AT&T monopolized. Two joint factors provided incentives for prospective market entrants: technological innovations and profits. Technological innovations, especially microwave communications, allowed the possibility of offering long-distance communications without having to duplicate the expensive wiring system that AT&T had constructed over the years. Also, growing demand and high profits surely attracted entrepreneurs into the market.

However, once divestiture allowed these new entrants into AT&T's former long-distance monopoly, competitive forces among the long-distance providers threatened the long-distance-to-local cross-subsidy. These newer, cheaper technologies would allow heavy users to build their own networks, thereby bypassing the existing telephone network if they found that the common carriers were charging more than these heavy users would be willing to bear. Hence, competitive pressures and bypass threatened the long-distance-to-local cross-subsidy.

With this cross-subsidy removed, the delicate balance among the telephone, business, urban, and nonurban interests at the state level started to unravel. This has led to the break-up of the closed subsystem and its replacement by the issue network described above. The rise of LMS is not just an innovative policy, but also one that signaled the destruction the old urban/nonurban status quo.

What are the implications of these findings for future state telephone regulatory policies? It seems evident that policy may continue down the path described above. User-sensitive schemes may become more popular, as will greater equity in prices between urban and nonurban customers.

However, underlying these themes are venerable issues of social equity that may arise and may retard the push to allow market forces to determine telephone pricing. The old AT&T monopoly was based upon a delicate balance of social interests. One interest was AT&T's. By reducing competition, AT&T could secure its financial future. And while regulation would cap AT&T's profits by disallowing monopolistic predatory pricing, "reasonable" profits would be granted to AT&T. To make such a deal, AT&T had to offer something to the public. What they offered was universal access at a reasonable price. And since the cost of service is in part a function of the cost of plant and equipment plus the use made on that plant, service to nonurban areas would be costly. To connect rural areas to the network required extensive plant and wiring. Further, low population density made it unlikely that these costs could be recovered from rural users. To offer universal access, rural areas would have to be subsidized. Thus, the new age of telephone (de)regulation threatens service to nonurban areas. In all likelihood, while the cross-subsidy to rural interests may decline, it will not cease. Social equity and political pressure from nonurban areas probably will insure that.

Two trends likely will develop, however. First, more creative and innovative pricing solutions will be developed. Lifeline services probably will become one attractive option. Perhaps the more important long-term trend will relate to technological development and the provision of more inexpensive ways to offer communications to nonurban users. While nonurban service probably always will be more expensive than urban service because of low population density, nonurban service still represents a huge market. Just noting that GTE's service to nonurban areas alone makes it among the largest companies in the nation reveals the extent of the nonurban market. Such a market surely will be of interest to someone. In the meantime, telephone policy in the United States will be controversial as the states try to adapt to the realities in the postdivestiture era.

Bibliography

Agnew, C., and Romeo, A., *Telecomm. Policy*, 5:273–288 (1981).

Aufderheide, P., *J. Comm.*, 37:81–96 (1987).

Bailey, E. E., *Ec. J.*, 96:1–17 (1986).

Barnett, W. P., and Carroll, G. R., *Admin. Sci. Q.*, 32:400–421 (1987).

Baughcum, A., and Faulhaber, G., *Telecommunication Access and Public Policy*, Ablex, Norwood, NJ, 1984.

Bickers, K., *The Problem of Governance and Institutional Change in the American Telecommunications Industry, 1876-1984*, American Political Science Association, 1986.

Berg, S., and Pacey, P., *Telecomm. Policy*, 6:308–314 (1982).

Bolter, W., Duvall, J., Kelsey, F., and McConnaughey, J., *Telecommunications Policy for the 1980s: The Transition to Competition*, Prentice-Hall, Englewood Cliffs, NJ, 1984.

Brock, G., *The Telecommunications Industry*, Harvard University Press, Cambridge, MA, 1981.

Brooks, J., *Telephone: The First Hundred Years*, Harper and Row, New York, 1975.

Brown, C. L., *Telecomm. Policy*, 7:91–95 (1983).

Cain, R. M., *Comm/ent: Hasting J, of Comm. and Entertain. Law*, 10:1–32 (1987).

Coll, S., *The Deal of the Century: The Breakup of AT&T*, Atheneum, New York, 1986.

Copeland, B., and Severn, A., *Yale J. Reg.*, 3:53–85 (1985).

Cornell, N. W., Pelcovits, M. D., and Brenner, S. R., *Regulation*, 7:37–42 (1983).

Crandall, R. W., *Amer. Ec. Assoc. Papers and Proc.*, 78:323–327 (1988).

Crandall, R. W., *Regulation*, 12:28–33 (1988).

Danielian, N. R., *AT&T: The Story of Industrial Conquest*, 1939. Reprint, Arno Press, New York, 1974.

de Fontenay, A., Hoffberg, M., Shugard, M., White, R., *Telecomm. Policy*, 11:45–57 (1987).

Denious, R. D., *Telecomm. Policy*, 10:259–267 (1986).

Derthick, M., and Quirk, P., *The Politics of Deregulation*, Brookings, Washington, DC, 1985.

Drucker, P. F., *Public Interest*, 76:3–27 (1984).

Eberle, D. C., and Williamson, L., *Public Utilities Fortnightly*, 122:20–28 (1988).

Egan, B. L., and Weisman, D. L., *Telecomm. Policy*, 10:164–176 (1986).

Evans, D. (ed.), *Breaking Up Bell*, North-Holland, New York, 1983.

Evans, D., and Heckman, J., *Amer. Ec. R.*, 74:620 (1984).

Face, H. K., *Public Utilities Fortnightly,* 121:27–31 (1988).

Faulhaber, G., *Telecommunications in Turmoil: Technology and Public Policy,* Ballinger, Cambridge, MA, 1987.

Fischer, C. S., *J. Social Hist.*, 21:5–26 (1987).

Fuhr, J. P., *J. Policy Analysis and Manag.*, 5:583–590 (1986).

Fuhr, J. P., *Telecomm. Policy*, 10:193–194 (1986).

Garnet, R., *The Telephone Enterprise: The Evolution of the Bell System's Horizontal Structure*, 1876–1909, Johns Hopkins, Baltimore, 1985.

Griffin, J. M., and Mayor, T. H., *J. Law and Ec.*, 30:465–487 (1987).

Hillman, J. J., *Northwestern Law R.*, 79:1183–1234 (1984–1985).

Horwitz, R. B., *Critical Studies in Mass Comm.*, 3:119–154 (1986).

Irwin, M., and Ela, J., *Telecomm. Policy*, 5:24–32 (1981).

Johnson, E., *Telecomm. Policy*, 10:57–67 (1986).

Johnson, L., *Competition and Cross-Subsidization in the Telephone Industry,* Rand Corporation, Santa Monica, CA, 1982.

Jost, B. S., and Tollin, L. A., *Public Utilities Fortnightly,* 121:32–35(1988).

Kahn, A., *Yale J. Reg.,* 1:139–157 (1984).

Kahn, A., and Shew, W., *Yale J. Reg.,* 4:191–257 (1987).

Kaserman, D., and Mayo, J., *J. Policy Analysis and Manag.,* 6:84–92 (1986).

Kaserman, D., and Mayo, J., *Policy St. J.,* 15:395–413 (1987).

Kaserman, D., and Mayo, J., *Public Utilities Fortnightly,* 122:18–27 (1988).

Langdale, J., *Telecomm. Policy,* 6:283–299 (1982).

Lipartito, K., *J. Ec. History,* 48:419–421 (1988).

Lipartito, K., *J. Ec. History,* 49:323–336 (1989).

MacAvoy, P., and Robinson, K., *Yale J. Reg.,* 1:1–42 (1983).

MacAvoy, P., and Robinson, K., *Yale J. Reg.,* 2:225–267 (1985).

Mahan, G., *The Demand for Residential Telephone Service,* Graduate School of Business Administration, Michigan State University, East Lansing, MI, 1979.

Mansell, R., *J. Ec. Issues,* 20:145–164 (1986).

McKenna, R., *Admin. Law R.,* 49:43–59 (1987).

McKenna, R., *Fed. Comm. Law J.,* 37:32–43 (1985).

Meyer, J., et. al., *The Economics of Competition in the Telephone Industry,* Oelgeschlager, Gunn, and Hain, Cambridge, MA, 1980.

Miller, E. S., *Public Utilities Fortnightly,* 118:14–18 (1986).

Miller, J. (ed.), *Telecommunications and Equity: Policy Research Issues,* Elsevier, New York, 1986.

Noam, E. (ed.), *Telecommunications Regulation Today and Tomorrow,* Harcourt Brace Janovich, New York, 1983.

Noam, E., *Vanderbilt Law R.,* 36:949–983 (1983).

Noll, A. M., *J. Comm.,* 37:73–80 (1987).

Noll, M., *Telecomm. Policy,* 11:161–78 (1987).

Noll, R., *Amer. Ec. Assoc. Papers and Proceedings,* 75:52–56 (1985).

Orton, B., *Policy St. J.,* 16:134–145 (1987).

Pacey, P., and Singell, L., *Telecomm. Policy,* 8:249–255 (1984).

Pepper, R., and Brotman, S., *J. Comm.,* 37:64–72 (1987).

Powers, E., *Policy St. J.,* 16:146–159 (1987).

Renshaw, E., *Amer. Ec. R.,* 75:515–518 (1985).

Robinson, K., *J. Policy Analysis and Manag.,* 5:572–583 (1986).

Rohlfs, J. H., *Bell J. Ec. and Manag.,* 5:16–37 (1974).

Rowland, W. D., *J. Comm.,* 32:114–136 (1982).

Ryan, W. J., *Telecomm. Policy,* 5:136–148 (1981).

Schiller, D., *Telematics and Government,* Ablex, Norwood, NJ, 1982.

Schwartz, L. B., *J. Comm.,* 34:180–187 (1984).

Simon, S., *After Divestiture: What the AT&T Settlement Means for Business and Residential Telephone Service,* Knowledge Industry Publications, White Plains, NY, 1985.

Smith, G., *The Anatomy of a Business Strategy: Bell, Western Electric, and the Origins of the American Telephone Industry,* Johns Hopkins, Baltimore, 1985.

Soloman, P. J., *Telecomm. Policy,* 2:146–157 (1978).

Taylor, L., *Telecommunications Demand: A Survey and Critique,* Cambridge University Press, New York, 1980.

Temin, P., with Galambos, L., *The Fall of the Bell System,* Cambridge University Press, New York, 1987.

Temin, P., and Peters, G., *Willamette Law R.,* 21:199–223 (1985).

Temin, P., and Peters, G., *Amer. Ec. R.: Papers and Proceedings,* 75:326 (1985).

Trauth, E., Trauth, D., and Huffman, J., *Telecomm. Policy,* 7:111–120 (1983).

Trebing, H., *J. Ec. Issues,* 20:613–632 (1986).

Tunstall, J., *Telecomm. Policy,* 9:203–214 (1985).

Tunstall, W. B., *Disconnecting Parties: Managing the Bell System Break-Up: An Inside View,* McGraw-Hill, New York, 1985.

Vietor, R., and Davidson, D., *J. Policy Analysis and Manag.,* 5:3–22 (1985).

von Auw, A., *Heritage and Destiny: Reflections on the Bell System in Transition,* Praeger, New York, 1983.

Wanders, J., *The Economics of Telecommunications,* Ballinger, Cambridge, MA, 1987.

Wanders, J., *Telecomm. Policy,* 11:243–246 (1987).

Wanders, J., and Egan, B., *Telecomm. Policy,* 10:33–40 (1986).

Worthy, P. M., *Public Utilities Fortnightly,* 123:9–13 (1989).

References

1. Bernstein, M., *Regulating Business by Independent Commission,* Princeton University Press, Princeton, NJ, 1955.

2. Huntington, S. P., *Yale Law J.,* 61:467–509 (1952).

3. Stigler, G., *Bell J. Ec. and Manag.,* 2:3–21 (1971).

4. Peltzman, S., *J. Law and Ec.,* 19:211–240 (1976).

5. Posner, R., *Bell J. Ec. and Manage. Sci.,* 5:335–358 (1974).

6. Gormley, W., *The Politics of Public Utility Regulation,* University of Pittsburgh Press, Pittsburgh, 1983.

7. Derthick, M., and Quirk, P., *The Politics of Deregulation,* Brookings, Washington, DC, 1985.

8. Crandall, R. W., and Flamm, K., *Changing the Rules: Technological Change, International Competition, and Regulation in Communication,* Brookings, Washington, DC, 1989.

9. Dizard, W. P., *The Coming Information Age: An Overview of Technology, Economics, and Politics,* 3d ed., Longman, New York, 1989, p. 133.

10. Hammond, T., and Knott, J., *J. Politics,* 50:3–30 (1988).

11. Heclo, H., Issue Networks and the Executive Establishment. In: *The New American Political System* (A. King, ed.), American Enterprise Institute, Washington, DC, 1978, pp. 87–115.

12. Berry, J. M., *The Interest Group Society,* 2d ed., Scott Foresman, Glenview, IL, 1989, pp. 164–195.

13. Vietor, R., *Strategic Management in the Regulatory Environment: Cases and Industry Notes,* Prentice-Hall, Englewood Cliffs, NJ, 1989.

14. Noll, R. G., State Regulatory Responses to Competition and Divestiture in the Telecommunications Industry. In: *Antitrust and Regulation* (R. E. Grieson, ed.), Lexington, Lexington, MA, 1986, pp. 181–182.

15. Sweet, D. D., and Hexter, K. W., *Public Utilities and the Poor: Rights and Responsibilities,* Praeger, New York, 1987, p. 94.

16. Teske, P. E., *Implementing Deregulation: The Political Economy of Post-Divestiture State Telecommunications Regulation,* American Political Science Association, 1987.

17. Walker, J. L., Innovation in State Politics. In: *Politics in the American States,* 2d ed. (H. Jacob and K. Vines, eds.), Little Brown, Boston, 1971, pp. 354–387.

18. National Association of Regulatory Utility Commissioners, *Annual Report on Utility and Carrier Regulation*, National Association of Regulatory Utility Commissioners, Washington, DC, 1977–85.

19. Council on State Government, *Book of the States*, Council on State Government, Lexington, KY, 1977–85.

20. Beyle, T., Governors. In: *Politics in the American States*, 4th ed. (V. Gray, H. Jacob, and K. Vines, eds.), Little, Brown, Boston, 1983, pp. 180–221.

21. U.S. Federal Communications Commission, *Statistics on Common Carriers*, U.S. Government Printing Office, Washington, DC, 1977–85.

22. MacFarland, A., *Common Cause*, Chatham House, Chatham, NJ, 1984.

23. U.S. Department of Commerce, *Statistical Abstract of the United States*, U.S. Government Printing Office, Washington, DC, 1977–85.

JEFFREY E. COHEN

Divestiture of AT&T (see The Bell System;
Computer Inquiry I, II, and III—Computers and Communications: Convergence, Conflict, or Policy Chaos?)

Divestiture of Bell Operating Companies: Technical Challenges and Achievements

January 1, 1984, ushered in a new era in telecommunications in the United States. On that date, the AT&T Company, which had provided telephone service to about 80% of telephone subscribers, was divided into eight completely separate and independent telephone companies. Seven of these were restricted to the provision of local service within local access and transport areas (LATAs). The eighth, which retained the name AT&T, was allowed to continue the manufacture of communications equipment and to compete with other long-distance carriers in providing service between LATAs. This new financial and network structure resulted from a negotiated settlement of an antitrust suit against AT&T, which had been moving forward ponderously for nearly a decade since its filing by the U.S. Department of Justice on November 20, 1974 (1).

Planning for the technical implementation of this divestiture agreement began as soon as the broad outlines of the network plan became known in January 1982, seven months before the consent decree was signed on August 24, 1982, and nearly two years in advance of the actual divestiture. As a matter of fact, consideration of the technical issues had to affect the December 16, 1982, Plan of Reorganization (POR), which spelled out the details of how the human, financial, and physical resources of AT&T would be divided into eight parts (2). The underlying concern was continuing a high level of service to telephone customers, as well as maintaining the economic viability of each of the resulting business entities.

At the time that the technical planning began, it was not yet clear how many local service companies there would be or how many LATAs each would serve. However, it was soon established that there would be 7 regional companies serving a total of 161 LATAs, an average of about 23 per company. (This total later was reduced to 157 by a court-ordered consolidation of 8 LATAs into 4.) Under the divestiture agreement, these LATAs were to become subnetworks of the national network, which only could be interconnected by completely separate companies serving as inter-LATA carriers. (These most commonly are referred to now as interexchange carriers.) With a few exceptions at LATA boundaries allowed by the court, the seven regional telephone companies were restricted to handling intraexchange and interexchange customer traffic completely within the LATAs.

The Bell System before divestiture had been made up of 24 local operating telephone companies. Close technical ties to AT&T allowed a high level of integration of the network facilities and a sharing of responsibilities in their operation. Since AT&T had only a minor financial interest in two of the operating companies, Southern New England Telephone and Cincinnati Bell, these were separated from AT&T apart from the divestiture agreement. The terms of divestiture were applicable to the remaining 22, forcing the intimate technical planning and operating arrangements to be replaced by agreements appropriate for separate corporate entities. This meant a system of clearly defined interfaces and independent responsibilities for engineering, operation, administration, and maintenance of individually owned networks.

Among the things that had to be accomplished were the implementation of equal-access arrangements so that all long-distance carriers could compete on an equal basis without significant technical differences or subsidies, the allocation of network facilities to intra-LATA and inter-LATA functions, and the similar separation of the network operations databases and operations support centers and systems. The first of these was the most difficult to accomplish because of the changes that were required in the existing network switching equipment.

Equal-Access Considerations

The divested Bell Operating Companies (BOCs) were committed by the agreement approved by the court to provide exchange access service equal in type and quality to that provided to AT&T to all interexchange carriers (3). Exchange access service meeting this requirement began on September 1, 1984, with the goal of implementing the capability in all 157 LATAs by September 1, 1986. Behind this ambitious implementation program was a major task, namely, the design and introduction of software into the telephone switching systems to provide the new switching and transmission features needed for equal access (4). This major challenge was met in the shortest time intervals ever for such a large-scale software deployment.

Actually, the technical plan for equal access had its roots in the negotiations involving AT&T and the competing interexchange carriers in meetings run by the Federal Communications Commission (FCC). The matter under discussion, the provision of Exchange Network Facilities for Interstate Access (ENFIA), also responded to anticipated legislation, in particular Senate Bill S.898. The divestiture agreement converted a commitment on the part of the Bell System to implement equal-access arrangements from high priority to a definitely committed schedule of deployment. It also brought about more rapid closure on the dialing features to be provided than had been possible in the negotiations among the parties involved to that time.

The technical planning also had to contend with the already existing constraints that resulted from the prior series of computer inquiries conducted by the FCC (5). The conditions the FCC had imposed on AT&T relative to the provision of services that involved the storage and processing of customer information continued to apply to all eight entities resulting from the divestiture agreement.

An important equal-access requirement is that telephone subscribers be able to make a per-call specification of the interexchange carrier to be used. This feature requires modification of switching systems to accept and use a three-digit carrier designation in call disposition. Dialing these extra digits on every inter-LATA call would be a burden for those who choose to use the same carrier for all such calls. Therefore, equal-access switching systems have been modified to implement a presubscribed carrier for each customer. All calls that involve the presubscribed carrier are dialed the same way AT&T calls were dialed before divestiture. This technical feature has resulted in very aggressive and on-going competition to be the carrier of choice for residences and places of business.

The choice of a different carrier still can be made on a per-call basis by dialing "10" and the 3-digit carrier designation. The presubscription approach has turned out to be a very sensible way to minimize the effects of divestiture on the telephone-using public while providing the benefits of competition among all carriers. Figure 1 shows typical dialing patterns required for directly billed and operator-handled calls to the same and different number plan areas (NPAs) within LATAs and to those NPAs where competition among interexchange carriers has been established. Variations in dialing patterns among jurisdictions are the result of different rules of competition that apply.

Prior to divestiture, the local BOC billed for all calls completed on the Bell System network whether those calls were local or long distance. The local company recorded the calling number, the time of origination and the time of termination at the originating end of the call. This information was processed together with applicable tariffs at automatic message accounting centers. Revenues then were divided depending on what resources were used in completing the calls.

Under the terms of divestiture, the BOCs were required to offer billing service like that provided to AT&T to all competing interexchange carriers. However, they also were required to forward the calling number on each call so that the competing carriers would be able to provide their own billing service. This feature made it possible to discontinue the manual entering of caller identification numbers on calls carried by AT&T competitors that billed their own calls.

The challenge was to add this calling number information to the string of

	SAME NPA	DIFFERENT NPA
PRESUBSCRIBED IXC OR INTRA-LATA CARRIER	NXX-XXXX	1-NPA-NXX-XXXX 0-NPA-NXX-XXXX
PER-CALL CARRIER DESIGNATION	10XXX-NXX-XXXX	10XXX-1-NPA-NXX-XXXX 10XXX-0-NPA-NXX-XXXX

FIG. 1 Dialing patterns for access to carriers providing competing services (LATA = local access and transport area; XXX in 10XXX = interexchange carrier designation; NXX = local central-office code; NPA = numbering plan area code; 1-NPA = direct-dialed long-distance call; 0-NPA = operator-handled long-distance calls; XXXX = last 4 digits of the called number).

called number digits without introducing perceptible delay in the set up of the call. The ingenious solution is to begin the transmission of the calling number as soon as the chosen carrier and destination office code are known. This improvement in automatic number identification (ANI) begins high-speed transmission of the calling station number before the last four digits are entered manually by the customer.

The network plan for the LATAs is based on a two-level hierarchy, consisting of end offices and access tandems, that provides exchange access quality for all interexchange carriers equal to that provided to AT&T (6). As shown in Fig. 2, connections to the point of termination (POT) for a specific interexchange carrier may be made directly from end offices as well as from access tandems if there is enough traffic to justify a dedicated trunk group. Overflow from such direct trunk groups goes to the access tandem over trunk groups shared among all interexchange carriers. All of the calls collected by the access tandem are sorted and routed to the appropriate carrier for completion.

Calls via the access tandems include an extra link. There was initially a concern that this might favor the dominant carrier, AT&T, since more of its traffic was routed directly from end offices. To make the transmission performance for these two categories of connections equal, the tandem inter-LATA connecting trunks were assigned zero transmission loss.

Based on computer modeling of the performance of telephone networks (7), it was possible to determine whether there were likely to be any perceptible

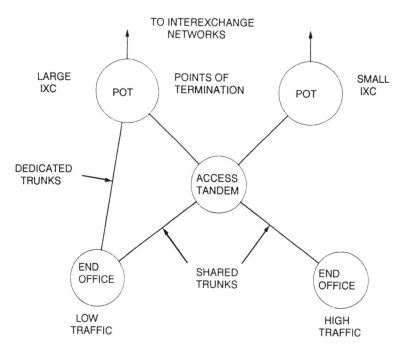

FIG. 2 Equal-access network design (POT = point of termination).

differences when all of the significant impairments on local and long-distance trunks were taken into account. The customer opinion model used for this purpose was the Long Toll Model (8), based on the opinions of over 10,000 Bell System customers who had just completed calls over long toll connections. The model predicted that the performance would meet the underlying objective of equal quality from the customers' point of view. This has been borne out in actual experience.

Partitioning of the Network

The divestiture agreement called for dividing the then-existing Bell System network into 157 LATA networks and one inter-LATA network. This required the resolution of some difficult questions of ownership of those network facilities used to support both local and long-distance traffic. Since joint ownership of a switching system or a transmission facility was not allowed, ownership went in the direction of the dominant use. As a result of this rule, 80% of the large 4-ESS™ toll switching systems, all of the traffic service position system (TSPS) operator service position systems, and the extensive common channel signaling (CCS) interconnecting the higher levels of the switching hierarchy went to AT&T (9).

Each BOC had to develop a new network plan for carrying the intra-LATA traffic between exchanges that had been carried by those 4-ESS switches now solely owned by AT&T. This was in addition to implementing inter-LATA access as shown in Fig. 2. In many cases, these functions have been combined in existing or new tandem switching systems. The necessary rearrangements and additions could not be accomplished overnight, and the divestiture agreement allowed for a transitional period.

In the case of operator services, the BOCs lost all of the technical facilities and the operators required to handle operator-assisted traffic within the LATAs. Initially, all "dial-0" calls went to AT&T operators, who provided this function under contract. Over a period of about five years, the BOCs added modern operator service positions to their digital tandem switching systems and took back the operator-handled traffic and the operator work force as well. Now customers dial "00" to reach interexchange operators to handle credit and dialing problems. For operator assistance on collect, person-to-person, and credit card calls a single "0" in place of the "1" before the NPA code in the dialing codes shown in Fig. 1 suffices.

At the time of divestiture, connection of end offices to AT&T's network of CCS had just begun. The signaling arrangements and protocols did not provide for interconnection to multiple interexchange carriers, so the program of deployment of CCS had to be abandoned for that reason alone. Other issues emerged that further delayed the availability of CCS. Among them was the court ruling that every LATA must have its own signal transfer point (STP) to avoid allowing the BOCs to transport signaling messages across LATA boundaries. Now, all BOCs have begun to deploy CCS within their LATAs to support Intelligent Network and Integrated Services Digital Network (ISDN) services.

Nationwide interconnection with interexchange carriers will happen soon in support of ISDN services.

Operations Support Systems

Major AT&T operations support systems affected by divestiture were Plug-in Inventory Control System with Detailed Continuing Property Records (PICS/DCPR) and Trunk Integrated Record Keeping System (TIRKS™). PICS/DCPR was well established at the time of divestiture as the means for tracking plug-in equipment and for managing materials. Part of the PICS system is a detailed investment database supporting the accounting records for all central-office equipment (not just plug-in equipment). These records, compiled in 19 separate installations and containing a total central-office investment base of $53 billion, had to be sorted into intra- and inter-LATA segments by the effective date of divestiture, January 1, 1984. Each new company then could begin its postdivestiture life with a proper set of investment records for regulatory and financial reporting purposes.

Major new software features were required in PICS/DCPR to bring about this separation. Both the development work and the population of the new databases were accomplished on time and with minimal disruption of regular BOC functions. Furthermore, a completely functioning PICS system was replicated for AT&T to manage the equipment added to its inventory by the intra-/inter-LATA split.

The changes in TIRKS necessary to support divestiture were more extensive and continued for two years past the divestiture date. The BOCs had used 21 separate TIRKSs to support their circuit-provisioning activities. The databases of each system contained the detailed records of about one million circuits. These data had to be combined with data resident in other automated systems to build a complete record of the circuits in inventory. After this was done, a determination was made as to which new operating entity would have ownership and control, and points of interface for each circuit were established. Finally, the data were partitioned either to be retained by the BOC or handed off to the new TIRKS deployed by AT&T.

Once the ownership and control of the network was split in this way, the network planning and service-provisioning processes also had to be modified. Prior to divestiture, the design of a switch-to-switch trunk had been the responsibility of a single company, regardless of the equipment ownership. This was one of the advantages of corporate integration of the local and long-distance portions of the Bell System. In the postdivestiture environment, each entity must own some part of an interconnecting trunk and has the responsibility for engineering that portion. Obviously, these trunk segments must fit together and meet the demanding requirements of the equal-access transmission plan. This created the need for new design methods (10). Characteristics that had been implicit in old designs had to be made explicit requirements in new designs. New codes of trunks now have been sufficiently well specified so that two companies can design their segments separately and have the completed trunk meet the overall performance objectives. Software modifications have been made to TIRKS to support the new trunk design process.

Delayed Network Progress

As mentioned earlier, divestiture interrupted the progress being made toward a complete nationwide stored program controlled network linked by a CCS network. The driving force for deployment, 800 service, was assigned to AT&T by the divestiture agreement along with the database and the CCS network required for support. By that decision, 800 service became an interexchange service, with blocks of 800 numbers assigned to the competing interexchange carriers as they deployed their own databases. The BOC LATA networks simply routed 800 calls to the appropriate interexchange carrier based on the 800–NXX portion of the number. Not only did this relegate the BOCs simply to providing an access service with limited revenue potential, but it made it impossible for customers to transfer their prestige numbers from one competing carrier to another. For these reasons, the BOCs started from scratch to implement the advanced network features required for 800 and other Intelligent Network services on an intra-LATA basis. A simplified view of a network capable of providing Intelligent Network services on both an intra-LATA and interexchange access basis is shown in Fig. 3.

The BOC plan requires the local operating company databases to have infor-

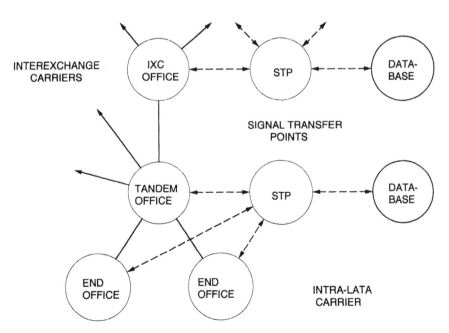

FIG. 3   A network architecture for 800 and other Intelligent Network services (STP = signal transfer point).

mation on every 800 customer, including knowledge of the interexchange carrier providing service for that specific 10-digit number. This allows customers very broad flexibility in deciding which carriers will carry their traffic from specific regions of the country. The mechanized facilities and management procedures for collecting, storing, protecting, and distributing this sensitive customer information have been put into place. Of course, this same network capability also allows the BOCs to compete for the portion of the 800 traffic that originates and terminates within a LATA. It took more than five years from the time of divestiture for this to happen, even on a limited basis. While the technical issues were significant, the political issues were even tougher. The leading interexchange carriers, already well established in providing 800 service, had little incentive to cooperate in the deployment of BOC database systems that might lead to an erosion of their business. In the end, the divestiture court ordered that deployment was in the best interest of the customers. However, this was with the requirement that CCS be in place to offset the time delay inherent in querying the BOC databases on a 10-digit basis.

There were some technical benefits of the deferred deployment. In particular, it was possible to provide Intelligent Network software capabilities in the new BOC database systems beyond those required for 800 service alone. Also, the CCS protocol used is a newer version, International Telegraph and Telephone Consultative Committee (CCITT) System 7, which is better suited to the long-term evolution of networks than the older System 6 originally used by AT&T.

Despite the wide use of System 7 within local and Interexchange networks, it has not become the means of signaling across LATA boundaries yet. The deployment of a nationwide and ubiquitous ISDN requires that signaling protocol and political problems finally be solved and that the interconnection of CCS finally be accomplished. Hopefully, the ISDN imperative will usher in the era of full interconnection of all switching offices with high-speed CCS.

ISDN has been an internationally supported goal for nearly two decades. ISDN is not a service, but the capability of providing public switched connections that are digital from end to end and that can support a wide spectrum of services. The deployment of both digital switching systems and digital transmission systems make this capability possible, but agreements at every network interface are necessary to make ISDN realizable. Of course, real and useful services together with appropriate customer premises equipment also must exist to make deployment economically attractive and likely to happen on a widescale basis.

The impact of divestiture on ISDN is debatable. On one hand, the segmentation of the network created more interfaces and made agreement on those interfaces very difficult to achieve. On the other hand, the rest of the world has not made much more progress in the deployment of ISDN than has the United States. All that has been accomplished in the United States to date is the deployment of proprietary islands of ISDN within LATAs or within the networks of individual interexchange carriers. This is not much of an accomplishment for a nation that usually has been in the lead in the deployment of new service capabilities. The fact is that not one new nationwide, ubiquitous network service has been deployed since divestiture. New services offered by the BOCs and based on forwarding the calling number to the called customer have been limited to local LATAs for lack of CCS interconnection with interexchange carriers. For similar reasons, the new national network services offered by the interexchange carriers are limited to customers with direct interconnection to those carriers. More local and interexchange carriers now are offering and enhancing 800 service, but the basic service existed before divestiture.

A nationwide compatible version of ISDN, albeit with limited capabilities, is expected to have first deployment in 1992, more than a decade after the divestiture agreement was reached. Achieving this milestone has meant the time-consuming redesign of proprietary switching system software, digital access lines, and customer premises equipment to achieve compatibility with a common set of requirements.

The industry also appears to be coming together on the deployment of high-speed data services of the kind needed to interconnect local-area networks (LANs). A switched multimegabit data service (SMDS) defined by Bellcore seems to have gained the industry support necessary for implementation and nationwide deployment. A key ingredient of both ISDN and SMDS has been national and international agreement on the protocols and interfaces to be used. This is likely to be the pattern for the future deployment of services. The standards environment in which such agreements are reached is discussed in another section of this article.

Equipment Compatibility

Divestiture not only separated the BOCs from AT&T's long-distance network, but also from their principal telecommunications equipment supplier, the Western Electric Company. The intent of this provision of the divestiture agreement was to open the equipment market to free and equal competition. Compliance meant that the BOCs had to move as rapidly as possible toward the procurement of products to meet their network needs independent of AT&T and of each other.

As one might expect, similar products from different manufacturers are not exactly alike. For example, switching systems have different software features and transmission systems have different operating and maintenance features. It is quite natural for manufacturers to differentiate their products to gain competitive advantages, but those same product differences make it very difficult to mix and match the pieceparts of a network. The ability to mix and match products from multiple vendors is essential to achieve the full advantage of competition foreseen in the divestiture agreement.

The technical effort that made mixing and matching of products possible was the development of generic requirements and modular interfaces unrelated to any specific manufacturer's equipment. The responsibility for this work fell to Bellcore (Bell Communications Research), the research, systems engineering, and software development organization formed from parts of Bell Laboratories, AT&T, and Western Electric at the time of divestiture (11). Each regional company owns one-seventh of Bellcore and uses its output selectively in the conduct of its business. Bellcore's recommended requirements are usually, but not always, the basis for BOC procurement of the telecommunications equipment needed.

The process followed by Bellcore in the development of generic requirements begins with a BOC consensus view of what the priority needs are, usually related to high-technology, high-risk areas of the BOC networks. After soliciting relevant information from the industry at large, Bellcore issues requirements in draft form, called *Technical Advisories*. Frequently, but not always, these advisories are explained at Technology Requirements Industry Forums held under sponsorship of the BOCs (12). Further information received during a comment period leads to more mature requirements issued as *Technical References*. Even these are subject to change in subsequent issues if implementation by manufacturers turns up significant problems.

The real test of the process is whether it results in the availability of equipment that actually conforms to the generic requirements produced. Bellcore has a technical analysis activity that examines equipment for its conformity to requirements, and the results have been very positive. The principal problem has been the length of time that it takes to go from a new concept to its implementation and the actual deployment of equipment in the field. On the other hand, the free flow of requirements has stimulated a true multivendor environment with robust competition in features, price, and quality.

Quality has become an increasingly important element of competition and is worthy of further mention. Quality assurance on behalf of the BOCs has evolved over the past eight years from product inspection to manufacturing process auditing to supplier certification. Obviously, buyers favor those suppliers that provide the highest quality with the least oversight on the buyer's part. Divestiture created a competitive environment that provided an incentive for the manufacture of very good products from technical as well as economic standpoints. The quality and reliability of the pieceparts of networks has become absolutely essential as the networks become ever more complex.

Standards Environment

Divestiture changed the telecommunications standards environment in the United States drastically, and had a major impact on the rest of the world as well. Since AT&T had served 80% of the telephone customers in the United States, its practices became the de facto standards of others participating in the U.S. market. AT&T freely shared its technical knowledge with others, but had no strong motivation either to conform to international standards or to influence those standards strongly, except as they influenced the ability of AT&T to interwork with global networks. AT&T dominated the U.S. representation at the CCITT, the organization under the International Telecommunication Union (ITU) that sets telephone and telegraph standards.

The role AT&T played for over a century ended with divestiture. In the months preceding divestiture, the FCC published a Notice of Proposed Rulemaking in which it expressed concerns for network continuity and the need for a planning mechanism to provide the interconnection and interoperability standards necessary to assure the viability of a public telecommunications network made up of multiple networks. The notice evoked broad industry support for a new national telecommunications standards committee proposed by the newly formed Exchange Carriers Standards Association, a trade association of major telephone operating companies. Under the proposal, the standards committee would operate under the aegis of the American National Standards Institute (ANSI) and be open to participation by members of all interest groups (13).

This committee, known as ANSI/T1, became a reality in February 1984 (14). It soon became the largest of the ANSI-affiliated committees, with over 100 member organizations representing exchange carriers, interexchange carriers, resellers, manufacturers, user groups, government agencies and independent consultants. Because of the multinational character of the communications industry, some of the members of this U.S. committee have affiliations with companies headquartered in Canada, Japan, Sweden, Germany, France, and elsewhere in the world. This international participation has had the effect of increasing the influence of ANSI/T1 throughout the world. In fact, several initiatives begun in T1 have resulted in global recommendations from CCITT on relatively short time schedules (15).

The principal focus of the T1 committee's work has continued to be on interconnection, interoperability, and performance standards, which are major concerns for the U.S. network. The committee has done well in maintaining network viability while moving forward on new initiatives, many begun before divestiture. An important by-product has been contributions to U.S. positions at CCITT on such matters as ISDN and CCS. Acceptance of U.S. positions internationally is in the best interests of achieving seamless interconnection of the U.S. network to the global network and of opening world markets to U.S. manufacturers of communications products.

The Future

The general availability of new technology unleashed the political and economic forces that led to divestiture. In particular, low-cost, point-to-point microwave systems offered an alternative to the use of monopoly carrier facilities and began the breakup of the carrier monopoly. In the end, market forces prevailed in the application of technology to individual user needs, but the process was neither smooth nor easy.

This march of technology continues at what seems to be an accelerating pace. More capable integrated circuits, cheaper and faster computer memories, distributed computing, and higher-speed optical transmission systems all have an impact on both the use and implementation of communications networks. There is a strong incentive for individual competitors to be first in offering the benefits new technology allows, with the hope that compatible standards will follow. These offerings typically begin by supporting the needs for more sophisticated interconnection of computer applications within a single customer enterprise, with the expectation of somehow providing interconnection among different customers at a later time.

At present, T1 is the principal technical forum for moving the U.S. network forward to provide new, ubiquitous, national services in an orderly way. It has achieved some success in the ISDN area and in establishing a framework, known as SONET (Synchronous Optical Network), for the best utilization of optical fiber transmission systems (16). However, the process is flawed by the lack of an overall unifying vision of the future national communications network infrastructure all competitors can support as a goal of their standards participation.

Today, worldwide attention is focused on two major initiatives: broadband communications at speeds into the gigabit-per-second range and on personal communications networks with wireless access. Within the United States, the traditional carriers, the cellular radio carriers, the cable television carriers, and the alternative local-access carriers all expect to participate in these future markets. Under the present laws of the United States, there is no organization with the clear responsibility and authority to establish a vision for future communications encompassing all of these dimensions and to determine how these capabilities should evolve from the existing networks. Other countries are bringing their national views on an integrated network evolution to the CCITT for consideration in the setting of international recommendations. While the U.S. network still remains the best in the world, this lack of a forward-looking planning process could result in the loss of the U.S. leadership position.

Acknowledgments: TIRKS is a trademark of Bellcore. 4-ESS is a trademark of AT&T.

Glossary

AMERICAN NATIONAL STANDARDS INSTITUTE (ANSI). A coordinating body for working standards committees.

AMERICAN TELEPHONE AND TELEGRAPH COMPANY (AT&T). The principal defendant in the antitrust suit that led to the divestiture agreement.

AUTOMATIC NUMBER IDENTIFICATION (ANI). Identifies a calling number.

BELL COMMUNICATIONS RESEARCH (Bellcore). The research and development consortium owned equally by the seven Bell Operating Companies.

EXCHANGE CARRIERS STANDARDS ASSOCIATION (ECSA). A trade association of local-exchange carriers.

EXCHANGE NETWORK FACILITIES FOR INTERSTATE ACCESS (ENFIA). The subject of FCC hearings prior to divestiture.

INTEGRATED SERVICES DIGITAL NETWORK (ISDN). An international plan for deploying end-to-end digital network capabilities.

INTERNATIONAL TELECOMMUNICATION UNION (ITU). The parent of International Telegraph and Telephone Consultative Committee (CCITT).

LOCAL ACCESS AND TRANSPORT AREA (LATA). Defined by the terms of the divestiture agreement.

LOCAL-AREA NETWORK (LAN). Local network for the interconnection of computers and peripherals, typically at high data speeds.

NO. 4 ELECTRONIC SWITCHING SYSTEM (4-ESS). AT&T's switching system for large toll and tandem applications.

NUMBERING PLAN AREA (NPA). Numbering plan in the North American area.

NXX. Local central-office code, with 0 and 1 not allowed in the first-digit position.

SIGNAL TRANSFER POINT (STP). A packet-switching node in a common channel signaling network.

SWITCHED MULTIMEGABIT DATA SERVICE (SMDS). Service for the interconnection of LANs and similar high-speed data applications.

TRAFFIC SERVICE POSITION SYSTEM (TSPS). System for operator-assisted services.

References

1. Antitrust Suit, United States District Court for the District of Columbia, United States of America, Plaintiff, v. American Telephone and Telegraph Company; Western Electric Company, Inc.; and Bell Telephone Laboratories, Inc., Defendants; Civil Action No. 74-1698; November 20, 1974.

2. Plan of Reorganization, United States District Court for the District of Columbia, United States of America, Plaintiff, v. Western Electric Company, Inc., and American Telephone and Telegraph Company, Defendants; Civil Action No. 82-0192; December 16, 1982.

3. Mercer, R. A., What Equal Access Means to the Telcos, *Telephone Engineer and Management*, 99–101 (November 1, 1983).

4. Douglas, J. W., and Profili, G., Inside the Switch to Equal Access, *Bell Communications Research Exchange* (September–October 1985).

5. Pearce, A., Computer Inquiry I, II, and III — Computers and Communications: Convergence, Conflict, or Policy Chaos? In: *The Froehlich/Kent Encyclopedia of Telecommunications*, Vol. 4 (F. E. Froehlich and A. Kent, eds.), Marcel Dekker, New York, 1992, pp. 219–329.

6. Mardon, D. M., Plan for Access Tandem and Trunking, *Telephone Engineer and Management*, 140–142 (November 15, 1983).

7. Manseur, B., Merrill, H. S., Spang, T. C., and Vitella, M. E., Estimated Voice Transmission Performance of Equal Access Service, *ICC Conf. Rec.*, 1:347–353 (1985).

8. Cavanaugh, J. R., Hatch, R. W., and Sullivan, J. L., Transmission Rating Model for Use in Planning of Telephone Networks, *Globecom '83*, 2:683–688 (1983).

9. Goldberg, R. R., Common Channel Signaling. In: *The Froehlich/Kent Encyclopedia of Telecommunications,* Vol. 3 (F. E. Froehlich and A. Kent, eds.), Marcel Dekker, New York, 1992, pp. 163–181.

10. Fredericks, R. M., Design of Switched Exchange Access Service, *Globecom '85*, 1:31–35 (1985).

11. Bell Communications Research (Bellcore). In: *The Froehlich/Kent Encyclopedia*

of Telecommunications, Vol. 2 (F. E. Froehlich and A. Kent, eds.), Marcel Dekker, New York, 1991, pp. 57–64.

12. Hawley, G. T., O'Brien, S. K., and Benke, L. R., What in the World is a TRIF? *Telephony,* 100–106 (May 6, 1985).

13. Morgan, V., The American National Standards Institute. In: *The Froehlich/Kent Encyclopedia of Telecommunications,* Vol. 1 (F. E. Froehlich and A. Kent, eds.), Marcel Dekker, New York, 1990, pp. 137–141.

14. Lifchus, I. M., Standards Committee T1-Telecommunications, *IEEE Commun.,* 34–37 (January 1985).

15. Irmer, T., CCITT. In: *The Froehlich/Kent Encyclopedia of Telecommunications,* Vol. 2 (F. E. Froehlich and A. Kent, eds.), Marcel Dekker, New York, 1991, pp. 281–288.

16. Ballart, R., and Ching, Yau-Chau, SONET: Now It's the Standard Optical Network, *IEEE Commun.,* 8–15 (March 1989).

FREDERICK T. ANDREWS

Dolbear, Amos Emerson

In the last quarter of the 19th century, when the heady atmosphere of invention mixed with the bottom line of corporate capitalism, a battle of priority over the telephone was waged. Alexander Graham Bell, Elisha Gray, Thomas Edison, and Amos Emerson Dolbear, among others, formed armed camps behind the Bell Telephone Company and Western Union. Did Dolbear's self-professed claims and oath-bound testimony point toward a wronged inventor or a charlatan who tried to grab some of Bell's glory?

Perhaps Dolbear's childhood in industrializing New England was responsible for his technical bent. Born in Norwich, Connecticut, in November 1837, Dolbear was orphaned at a young age and began working on a farm in New Hampshire when he was 10 years old. At 16, he took a job at a pistol factory in Worcester, Massachusetts, giving it up 2 years later in favor of a schoolteacher's position in Missouri. But he was back in New England, working at the Springfield Armory and in a Taunton, Massachusetts, machine shop, by the time he was 20. Dolbear then gave up his machinist's career in favor of a college education. After his 1866 graduation from Ohio Wesleyan University, he taught at a number of universities, finally settling in as a professor of physics at Tufts College, near Boston, in 1874. There he stayed for the next 32 years.

Dolbear's interest in electrical engineering took hold during his undergraduate years with experiments with electromagnets and telegraphs. So, he may have been especially curious about Bell's telephone. In any case, Bell was doing all that he could to ensure that Boston's scientific and intellectual circles knew about his work. After the now-legendary, "Mr. Watson, come here, I want to see you," of March 1876, Bell hosted numerous public demonstrations and the telephone was a "hot" topic in both the professional and popular press. Dolbear, like many others, believed that he could improve upon Bell's instrument. "I began my experiments in Telephony in August 1876," he told the American Academy of Arts and Sciences (1, p. 77).

Dolbear had an assistant begin the construction of his telephone late in September 1876. His design incorporated two permanent magnets wrapped with wire and an iron armature. "About the first of October," Dolbear recalled, he changed his original scheme of using a rubber diaphragm for the transmitter, substituting instead a thin piece of sheet iron. The use of the permanent magnet and metal diaphragm differed from Bell's first telephone, which employed an electromagnet and a stretched membrane. Sometime during that same autumn, though, Bell also decided to switch to a permanent magnet and a thin steel disk for the transmitter membrane. The first clash of these two inventors was imminent.

Progress on Dolbear's telephone instruments was slow. His assistant only worked on Saturdays and did not complete the job before going away for several weeks. In addition, "I had not thought of the very great value of the invention [and] I was extremely busy with my college work," Dolbear later stated, "and also in getting a book through the press, and couldn't attend to it very well myself" (1, p. 79). But a mutual friend of his and Bell's, Percival Richards, thought that Dolbear might be onto something patentable. Richards told Gardiner Hubbard, Bell's backer and soon-to-be father-in-law, about Dolbear's work. Hubbard told Richards that Bell had begun using permanent magnets and metal diaphragms in mid-October and, supposedly also told Richards that Bell already had patented these ideas (Hubbard later denied this). With this news, Dolbear saw no need to hurry the completion of his telephones, finally finishing and operating them in February 1877. He then learned that Bell had not applied for a patent covering the permanent-magnet telephone until January 15, 1877.

Dolbear's opinion of Bell soured with this news. "I think that a mutual acquaintance informed [Bell] as to what I was doing and before I could get in working order he, by the aid of a skilled electrician and working nights and Sundays, completed his and got it patented instanter, so I suppose that I shall lose both the honor and the profit of the invention," he wrote in March (2, p. 265). Regardless of whether animosity or economics drove him, Dolbear began negotiations with Western Union around the time of the founding of the Bell Telephone Company in July 1877. In return for legal fees, one-third of any profits from his inventions, and a research stipend, Dolbear agreed to apply for and assign patents to Western Union's telephone subsidiary, the Gold and Stock Company. His first patent application under this agreement, filed on October 31, 1877, earned him a place in Western Union's and Bell Telephone's infringement and priority contests.

Alexander Graham Bell's telephone patents were the subject of over 600 lawsuits. Perhaps the most important of these was *Bell Telephone Company et al. v. Peter A. Dowd*. On September 12, 1878, Bell Telephone filed against Peter Dowd, an agent of the Gold and Stock Company, for infringement of Bell's fundamental patents, Numbers 174,465 and 186,787. The defense was taken up by Western Union, which, through its subsidiaries, the Gold and Stock Company and the American Speaking Telephone Company, controlled the telephone patents of Dolbear, Elisha Gray, and Thomas Edison. Western Union alleged that the inventions of these last three men had priority over Bell's claims and that Bell's first patent described an apparatus that had no hope of actually transmitting speech.

Western Union's lawyer however, urged a settlement, and a contract between the two companies was signed on November 10, 1879. In this agreement, Western Union conceded Bell's priority in the invention of the telephone. Furthermore, Western Union consented to leave the telephone business and to license all of its telephone patents to the Bell Company. In return, Bell would purchase all of the telephones and telephone exchanges owned by Western Union and pay the telegraph company 20% of Bell's telephone-rental income for the next 17 years. Dolbear's patents, then, along with Gray's and Edison's, became the property of the American Bell Telephone Company.

This did not end Dolbear's legal battles with Bell, however. In October 1880, Dolbear applied for a patent on a telephone receiver, and received that patent the following April. He already had incorporated the Dolbear Electric Telephone Company when, in October 1881, he again was charged with infringing on Bell's patents. Once more, the court upheld Bell's priority, stating that Dolbear's receiver was only a variation on Bell's apparatus, created "by using his general process or method" (3, p. 69). This time, the appeal went to the U.S. Supreme Court; the original ruling stood.

Amos Dolbear continued his scientific work in spite of the defeat of his telephone patents. In 1882, he received a patent for an invention in wireless communication, a field in which Heinrich Hertz and Guglielmo Marconi later would achieve fame. Dolbear also remained an active and apparently (from his students' nickname for him, "Dolly") well-liked professor of physics at Tufts. But he never gave up his claim of inventing the telephone. A former student recalled Dolbear, holding up a box for the class to see, asserting, "That is the first telephone that was ever invented. I invented it" (2, p. 278).

Bibliography

Bell, Alexander Graham, *The Bell Telephone*, American Bell Telephone Company, Boston, 1908.

Dolbear, Amos E., *The Telephone: An Account of the Phenomena of Electricity, Magnetism, and Sand, as Involved in Its Action*, Lee & Shepard, Boston, 1877.

Hounshell, David A., Elisha Gray and the Telephone: On the Disadvantages of Being an Expert, *Technology and Culture*, 16:133–161 (1975).

Prescott, George B., *Bell's Electric Speaking Telephone: Its Invention, Construction, Application, Modification and History*, D. Appleton, New York, 1884.

References

1. Dolbear, Amos E., Researches in Telephony, *Proc. American Academy of Arts and Sciences*, 14:77–91 (1879).
2. Bruce, Robert V., *Bell: Alexander Graham Bell and the Conquest of Solitude*, Little, Brown, Boston, 1973.
3. Rhodes, Frederick Leland, *Beginnings of Telephony*, Harper and Brothers, New York, 1929.

JOYCE E. BEDI

DS1 Services and Standards

Introduction

The purpose of this article is to provide a technical overview of DS1 (Digital Signal Level 1), including a description of services, applications, and relevant standards. Although the focus is on the current state of DS1 services and standards, the article puts them in the context of their evolution to date and provides an outlook toward the future.

The article begins by describing DS1 at the technical level (tutorial overview), including the characteristics of the digital signal, the position of DS1 in the digital hierarchy, differences among countries, and references to the key standards.

The article then describes services and applications made possible by DS1, including a brief historical review of their evolution. Fractional DS1 (FDS1) and Integrated Services Digital Network (ISDN) services also are addressed.

The relevant standards that make DS1 services possible are summarized to facilitate the identification of relevant standards in various parts of the world. The intent is to provide a global perspective including international and regional/national standards. Thus, the reader should be able to locate readily the standards, as required.

Finally, the article concludes with a summary and a look into the future. Since it is not possible to cover in this brief article all the details of DS1 services and standards, a comprehensive list of references (both technical papers and standards) is provided to allow the reader to pursue the subject further.

Digital Signal Level 1

Digital networks are based on standardized, hierarchical bit rates (1,2), built upon a primary level bit rate of either 1544 kilobits per second (kb/s) (e.g., United States, Canada, Japan, and Caribbean countries except Haiti) or 2048 kb/s (e.g., Europe, Africa, Australasia, Central and South America, Mexico, and Haiti). The 1544-kb/s primary level signal is referred to as the Digital Signal Level 1, or DS1. The term "DS1" is used mainly in North America. The 2048-kb/s digital signal and the corresponding systems/standards have been referred to by a variety of terms such as E-1 (3), CEPT-1 (4), DS1A (5), 2048 MHz (6), and D2048U/D2048S (7).

While DS1 refers to the signal itself, the equipment used to process and transmit/receive that signal usually is referred to as T1-carrier equipment. In the traditional sense, *T1-carrier* refers to copper-based (twisted-pair) line transmission systems. Other transmission media for DS1 may include microwave radio, communications satellite, coaxial cable, optical fiber, or even as part of community antenna television (CATV) applications (8).

DS1 and the Digital Hierarchies

The position of DS1 with respect to other bit rates used in digital networks, including the asynchronous and synchronous digital hierarchy levels, is summarized in Table 1 (1,2,5,9,10).

The left-most column in Table 1 indicates the asynchronous and synchronous International Telegraph and Telephone Consultative Committee (CCITT) digital hierarchy levels (1,2). The table also includes other bit rates used in North America, Japan, and Europe. The asynchronous bit rates are used in equipment that obtain the timing information (clock) from the signal itself and that can handle bit rates that are close but not necessarily identical. The synchronous (or isochronous) bit rates are used with equipment that have precise reference clocks.

It should be noted that the list in Table 1 of asynchronous bit rates based on 1544 kb/s contains two threads (United States and Japan). The asynchronous bit rates in North America and Japan are different at Level 3 and higher, as shown in the table.

The synchronous digital hierarchy (SDH), which starts with STM-1 at 155,520 kb/s, is common for both the 1544- and 2048-kb/s-based systems (11). The usable bit rate for information (payload) is shown in the table as multiples of 64 kb/s to indicate how the hierarchies are formed; however, other multiplexing alternatives exist and mappings of other tributary bit rates into the SDH levels are possible, giving rise to other payload capacities (e.g., Ref. 12). STS-N refers to the Synchronous Transport Signal Level N, for optical interfaces (9).

While all the bit rates in Table 1 are used in digital networks, the ones most commonly made available by telecommunications service providers to customers are DS0, DS1, and DS3.

Technical Characteristics of DS1

A summary of the most significant technical characteristics of DS1 is provided here. In particular, the DS1 format structures and line codes are described, following the American National Standards Institute (ANSI) standards. For further details, refer to the standards (e.g., Refs. 5, and 13–15) and the literature on the subject (e.g., Refs. 16–23). For a description of differences between DS1 and the 2048-kb/s-based systems refer also to Refs. 3, 4, and the applicable standards described below.

DS1 Frame Format

In digital telephony, the analog speech signal is sampled and digitized at the rate of 8000 samples per second and 8 bits per sample. This results in a digital

TABLE 1 Bit Rates and Digital Hierarchies

CCITT Digital Hierarchy Level	U.S. ANSI	Japan TTC	Based on 1544 kb/s		Based on 2048 kb/s	
			Usable ($n \times 64$ kb/s) $n =$	Total bit rate (kb/s)	Usable ($n \times 64$ kb/s) $n =$	Total bit rate (kb/s)
Asynchronous						
0	DS0	0	1	64	1	64
1	DS1	1	24	1,544	30	2,048
	DS1C		2×24	3,152		
2	DS2	2	4×24	6,312	4×30	8,448
3		3	$5 \times 4 \times 24$	32,064	$4 \times 4 \times 30$	34,368
3	DS3		$7 \times 4 \times 24$	44,736		
			$3 \times 5 \times 4 \times 24$	97,728	$4 \times 4 \times 4 \times 30$	139,264
4	DS4NA	4	$3 \times 7 \times 4 \times 24$	139,264		
	DS4		$6 \times 7 \times 4 \times 24$	274,176*		
5		5	$4 \times 3 \times 5 \times 4 \times 24$	397,200	$4 \times 4 \times 4 \times 4 \times 30$	564,992*
Synchronous						
	STS-1	0	$7 \times 4 \times 24$	51,840		
STM-1	STS-3	1	$3 \times 7 \times 4 \times 24$	155,520	$4 \times 4 \times 4 \times 30$	155,520
	STS-9		$3 \times 3 \times 7 \times 4 \times 24$	466,560		
STM-4	STS-12	4	$4 \times 3 \times 7 \times 4 \times 24$	622,080	$4 \times 4 \times 4 \times 4 \times 30$	622,080
	STS-18		$6 \times 3 \times 7 \times 4 \times 24$	933,120		
	STS-24		$8 \times 3 \times 7 \times 4 \times 24$	1,244,160		
	STS-36		$12 \times 3 \times 7 \times 4 \times 24$	1,866,240		
STM-16	STS-48	16	$16 \times 3 \times 7 \times 4 \times 24$	2,488,320	$16 \times 4 \times 4 \times 4 \times 4 \times 30$	2,488,320

*These bit rates are in use, but they have not been standardized in the digital hierarchies.

rate of 64 kb/s and a time interval per sample of 125 microseconds (µs). The sampling rate is based on the Nyquist rate of twice the bandwidth in the analog signal. Since the nominal band of telephone speech is 300 hertz (Hz) to 3400 Hz, there is a 900-Hz guard band when the band-limited analog signal is sampled at 8000 Hz. This must accommodate the transition band of the filters in the transmission system (24). This technique of transmitting analog information in a digital form is referred to as pulse code modulation (PCM) (16).

The basic timing parameters of the 64-kb/s signal must be maintained when several channels are multiplexed. The definition of DS1 is based on multiplexing 24 64-kb/s channels: 24 8-bit words in series (1 per channel) are preceded by 1 framing bit and are coded into 193 time slots in a 125-µs time interval, which is referred to as a *frame*. This establishes the first-level bit rate of 1.544 megabits per second (Mb/s) (DS1).

The framing bits identify the beginning of a new frame and multiple frames are associated to form a multiframe structure, referred to as the *DS1 frame structure*. The channel formed by the framing bits (one every 125 µs, i.e., 8 kb/s) is referred to as the *frame overhead*. The frame overhead can be encoded as a frame alignment signal and as maintenance channels to transmit information relating to alarms, signal quality, and other networks' operations data.

Over the years, DS1 has had different applications, in which the significance of certain bits and the coding have changed and the standards may offer options. Hence, implementations must ensure that each part of the equipment supports the same options in the standards.

The information payload may be structured into channels of 64 kb/s or other bit rates. Unchannelized structures are also possible, in which the total number of payload bits is available as one channel (1536 kb/s). The payload structure also may contain overhead channels carrying information necessary for a specific network use, depending on the application.

In telephony applications, such signaling information as setting up a connection and supervising its completion is included in the digital bit stream by robbing the least significant digit from each codeword in every sixth frame (see Fig. 1). This provides 24 bits every sixth frame (i.e., 32 kb/s) that can be used

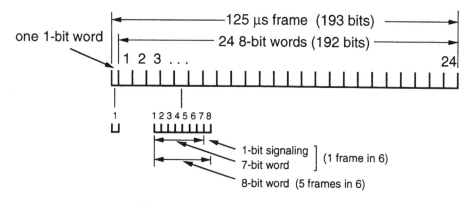

FIG. 1   DS1 frame format: Bit robbing.

for voice channel signaling information. This procedure causes every sixth frame to use 7-bit words to encode the voice channels. This technique is referred to as *bit robbing*. Although the effect of bit robbing on the overall quantization distortion of the voice is not significant, it prevents the use of the full 24 × 64 kb/s in data communications applications. Other schemes of bit robbing are possible. The signaling channel(s) created by the bit-robbing technique form part of the payload overhead.

As indicated above, multiple DS1 frames are associated to form a structure. Two DS1 frame structures have been standardized (13): superframe (SF) (12-frame multiframe) and extended superframe (ESF) (24-frame multiframe).

The SF structure consists of 12 DS1 frames (12 × 193 = 2316 bits). The 12 framing bits in each SF are used for frame and signaling phase alignment, depending on their position in the SF. An SF alignment channel is formed by the 6 framing bits contained in the odd-numbered frames of the 12 framing bits, referred to as S1, S2, S3, S4, S5, and S6. The other six framing bits (contained in the even-numbered frames) of the SF form a frame alignment channel and are referred to as F1, F2, F3, F4, F5, and F6. The SF alignment channel signal (S1 = 0, S2 = 0, S3 = 1, S4 = 1, S5 = 1, and S6 = 0) is used to locate all 12 frames. The frame alignment channel signal (F1 = 1, F2 = 0, F3 = 1, F4 = 0, F5 = 1, and F6 = 0) is used to locate all 24 payload channels.

The ESF structure consists of 24 DS1 frames (24 × 193 = 4632 bits). The 24 framing bits are shared among an ESF alignment signal, a cyclic redundancy check (CRC), and an operations channel data link (DL). The ESF format provides additional operations and maintenance capabilities in DS1 facilities. It was developed when advancements in technology permitted the establishment and maintenance of framing with fewer overhead bits than the 1 in 193 used in the 12-frame SF format. Hence, there was spare capacity in the frame overhead and the SF was extended to 24 frames with the 24 framing bits allocated as follows:

Framing alignment signal: 6 bits (Frames 4, 8, 12, 16, 20, and 24) (referred to as F1, F2, F3, F4, F5, F6), 2 kb/s
CRC of previous SF: 6 bits (Frames 2, 6, 10, 14, 18, 22) (referred to as C1, C2, C3, C4, C5, C6), 2 kb/s
Operations channel DL: 12 bits (odd-numbered frames) (referred to as M1, M2, M3, . . . , M12), 4 kb/s

In this configuration (i.e., ESF), each 3 bits provide a data rate of 1 kilobit per second (kb/s).

The standardized signals and functions for ESF include only the minimum essential subset of functionality needed for all applications. Additional functionality can be accommodated within the ESF for specific applications (25).

The 2048-Mb/s frame format is quite different (26). Each frame has 256 bits (32 8-bit words in series) and the frame alignment signal is carried in the first 8-bit word in alternate frames. The frames that do not contain the frame alignment signal carry operational messages in the first 8 bits of the frame. The remaining 31 8-bit words in each frame carry speech/data or signaling information. The 8-bit word in Time Slot 16 usually is reserved for signaling. Bit-robbing techniques are not used and 30 64-kb/s channels are available for payload. Frames are grouped into multiframes of 16 frames each.

DS1 Line Codes

In order to make efficient use of the transmission media, the line code for DS1 is a bipolar code known as alternate mark inversion (AMI). The zeros correspond to zero volts and successive ones are coded as pulses that alternate between positive and negative voltages (nominally ± 3 volts). This is done to remove very low frequency energy from the digital signal in order to allow transformer coupling of repeaters and cable pairs. The direct current (DC) powering for the line repeaters also is carried on the same pairs of wires as the digital signals. With AMI, two or more successive pulses of the same polarity are termed *bipolar violations*. Unless introduced intentionally by the coding system in a predetermined manner under certain conditions (see below), bipolar violations indicate an error in the signal.

The bipolar code requires that sufficient ones are transmitted so that the transmitter clock rate can be recovered effectively at each receiver. The pulse density requirements are that no more than 15 consecutive zeros are present and there are at least N ones in each and every time window of $8(N + 1)$ digit time slots (where N can equal 1 to 23); that is, the average density of ones has to be at least 1 in 8.

To ensure that DS1 can provide 64-kb/s clear channels for ISDN and other applications, modifications of the bipolar line code have been devised to ensure that long strings of zeros could be transported. DS1 standards offer two alternatives for line codes that support clear channels: bipolar 8-zero substitution (B8ZS) and zero-byte time-slot interchanger (ZBTSI). B8ZS is considered the long-term method of providing clear-channel capability in DS1, and ZBTSI is considered as one alternative interim method for providing it in the short term.

The B8ZS code uses an algorithm that replaces occurrences of eight consecutive zeros (00000000) with exactly defined bipolar violation pairs in the fourth- and seventh-bit position of the inserted code ($000VB0VB$) where B represents a normal bipolar pulse and V represents a pulse violating the bipolar rule, that is, V has the same polarity as the preceding pulse. The receiving equipment monitors the incoming DS1 signal for B8ZS codewords and replaces them with eight zeros. B8ZS is transparent to format and framing structures, treating all bits equally, and maintains the density requirements.

ZBTSI is a technique in which zero octets are replaced by an address chain that is decoded by the receiving terminal. It operates by buffering multiple frames prior to transmission, scanning the buffered data for all-zero patterns and replacing them with other bit patterns. The ZBTSI algorithm requires 2 kb/s of overhead information carried within the ESF DL. The ZBTSI overhead bits are referred to as Z-bits and each one is associated with the 96 octets that immediately follow it. The other frame bits between Z-bits also are present in the 96-octet group but they are not involved in the ZBTSI process in any way. The frame bits are buffered to experience the same delay as the information bits so that the original SF may be restored at the decoder. The ZBTSI algorithm is specified in T1.107 (13).

Other schemes of zero-code substitution are possible that maintain the pulse density requirements and do not introduce DC components. For example, the 2048-kb/s signal uses HDB3 (high-density bipolar of order 3) (27). Each block of four successive zeros (0000) is replaced by either $000V$ or $B00V$ so that the number of B pulses between consecutive V pulses is odd. That is, successive V pulses are of alternate polarity and no DC component is introduced.

DS1 Performance Aspects

Although the performance aspects of DS1 are outside the scope of this article, Refs. 22, 23, and 28–40 may be consulted. For network configurations, refer to Ref. 41.

Applications of DS1

Some of the different ways in which DS1 can be used include the following (13):

Unchannelized: The 1536-kb/s information payload is available to the user. One of the techniques available to provide a clear-channel capability (e.g., B8ZS or ZBTSI) may be used to allow for an unrestricted number of zeros in the payload.
Twenty-four channels at 64 kb/s: Each DS1 frame carries 24 octet-interleaved channel time slots.
Twenty-four channels with bit-robbing signaling channel: This is the traditional application with channel banks in digital telephony. Eight bits are available for payload in 5 out of 6 of the DS1 frames.
ISDN Primary Rate Access (PRA): The DS1 frame carries 24 octet-interleaved channel time slots. The channel time slot may be used for a variety of applications. If there is a D channel, it is assigned Time Slot 24.
Twenty-four channels with ISDN basic access signal.
Forty-four or 48 channels using 32-kb/s adaptive differential pulse code modulation (ADPCM): With ADPCM, the usual 24-channel capacity of DS1 is doubled. In the 44-channel format, 4 channels are designated for signaling, maintenance, and a DL. In the 48-channel format, signaling is provided either outside the DS1 signal or inside (e.g., using bit-robbing techniques).
Twenty-three channels at 64 kb/s with overhead channel: Channels 1–23 are used for payload and Channel 24 is used as a payload overhead channel.

DS1 facilities are built into systems to provide services that end users will use for their own applications. This is the subject of the next section.

DS1 Services

Evolution of DS1 Services

DS1 transmission systems were first introduced into telecommunications networks by the early 1960s (16,17). However, initially they were used exclusively by telephone companies for interconnections between digital channel banks, primary rate multiplexers, and switching centers. DS1 facilities were made available to commercial customers for the first time in the late 1970s on a special assembly basis for point-to-point private lines. A typical application was circuit consolidation by means of multiplexers for voice, data, and video applications. By the early 1980s, as a result of an increasing demand by customers to have access to these facilities, they eventually were tariffed and made available on a wider scale. Thus, DS1 traditionally has been used to implement private-network backbones and to provide access from customer premises equipment (CPE) to public networks. The evolution of DS1 service capabilities may be assessed by comparing the literature on the subject over the years (e.g., Refs. 42–57).

The introduction of digital cross-connect equipment, which initially was intended to provide greater operations flexibility to carriers, introduced new capabilities in the network that were made available to customers on a reservation basis. At the DS1/DS0 level, digital cross-connect systems allow the assignment and redistribution of 64-kb/s channels among the various DS1 lines connected to them (58).

Circuit-mode services at DS1 and FDS1 rates became available from major carriers as private-line or special-services networks (47,52). These services are "slow switched" by the very nature of their implementation using digital cross-connect systems. Typical applications for these services include the creation of private networks, access to public network services, and wide-area data networks. The trend has been an increasing demand by end users for more control over the management of circuit-mode DS1 services, thus reducing channel reservation requirements, increasing connectivity, and eventually leading to the introduction of fully dialable services at DS1 and FDS1 rates.

One of the limitations in the introduction of dialable services at high bit rates has been the characteristics of the existing switched network, which has evolved from the analog network by modifying portions of it for 56-kb/s (if bit robbing is used for signaling) and 64-kb/s switched services. Although switched digital services are not yet widely available, current networks are designed for switched services at 64 kb/s. Higher bit rates may be accommodated by using multiple 64-kb/s channels; however, the original digital network implementations made no attempt to maintain any timing relationship between any two or more 64-kb/s channels that could be subject to different delays when routed through the network. In order to maintain data integrity, it is necessary that the data are delivered by the network at the destination in exactly the same order as they were received from the originating end.

Technology advancements in the digital switched network have enabled this capability (52,53,59–63). Major carriers have announced dialable services at DS1 and FDS1 rates and service-definition standards are being finalized in

CCITT within the scope of ISDN. The primary rate user-network interface offers the required capabilities to provide access to these services.

While the emphasis here has been on circuit-mode services, DS1 also can be used to transport packet-mode services, particularly frame relay services (64–66).

International Telegraph and Telephone Consultative Committee Service-Definition Standards

CCITT has a three-stage methodology for describing telecommunications services (67). The first stage is the most important because it defines the characteristics of the service from the user's point of view. It consists of a service definition and description in prose form, a static description of the service by means of attributes, and a dynamic description of the service using graphical means. Based on the Stage 1 definition (which is the responsibility of CCITT Study Group I), then the subsequent stages specify a functional model, functional entities, protocols, and the signaling required to provide the service.

CCITT has published a series of Recommendations (I.231.x) that define circuit-mode, unrestricted information transfer bearer services at various rates, including 1536 kb/s and 1920 kb/s (68). All CCITT-defined circuit-mode services allow a variety of information transfer attributes including:

Establishment of the communication: demand, reserved, permanent
Symmetry: bidirectional symmetric, bidirectional asymmetric, unidirectional
Communication configuration: point-to-point, multipoint

The latest addition to this family of circuit-mode bearer services, still under study in CCITT, is Draft Recommendation I.231.10 (69), which defines a multirate circuit-mode bearer service for an ISDN interface that allows users to request from the ISDN on a demand basis the establishment and release of circuit-mode connections supporting unrestricted information transfer rates at integer multiples of 64 kb/s up to the maximum rate of the interface.

Examples of DS1 Services/Applications

A comprehensive table of possibilities for services and their applications has been produced by CCITT in preparation for the definition of broadband services for ISDN (70).

The following descriptions may refer to services (if offered as such by a service provider) or applications (if the user is accessing a more general-purpose service). No special distinction is made here between dedicated facilities and switched facilities since these services/applications may be implemented in either environment. Furthermore, while these services may be offered at various transport bit rates, depending on CODECs (coders/decoders) utilized and other factors, they all can be offered at DS1 and FDS1 rates.

Private-Network Backup and Rapid Provisioning

Large private networks are becoming more and more complex and require dynamic means of adding circuits for disaster recovery, time-of-day alternate routing, overflow/peak-period support, network reconfiguration, and so on (45,52, 53,71,72). These applications are ideal for switched DS1 services.

Local-Area Network Interconnection

The interconnection of local-area networks (LANs) using DS1 facilities can be implemented using bridges and routers with dedicated DS1 facilities (73) or using such packet-mode services over DS1 as frame relay (66).

Videoconferencing

Videoconferencing was one of the first applications for DS1 services at their full rate (51,53,74). While desk-to-desk and some conference room videoconferencing are possible using one or two 64-kb/s channels, the higher quality provided by video CODECs at 384 kb/s and above is better suited for use in conference rooms and studios. Videoconferencing can be in point-to-point or in multipoint configurations; hence, they will require switched services. One key standard for videoconferencing applications is the CCITT Recommendation H.261, also referred to as the $p \times 64$ (p times 64) standard for videoconferencing. Recommendation H.261 specifies the coding and decoding methods for the moving picture component of audiovisual services at rates of $p \times 64$ kb/s, with p in the range of 1 to 30 (75). Other relevant CCITT Recommendations are H.221, H.230, H.242, and H.320.

Recommendation H.221 defines a frame structure for audiovisual teleservices in single or multiple 64-kb/s channels, or a single primary rate channel, by making use of the characteristics and properties of CCITT audio and video encoding algorithms and transmission frame structures (76). Recommendation H.230 provides procedures, definition of symbols, and requirements for control and indication signals in transmission frame synchronous applications (77). Recommendation H.242 specifies a scheme in which a channel accommodates speech and, optionally, video or data at several rates up to 2 Mb/s, in a number of different modes (78). Recommendation H.320 provides the system requirements for visual telephone services with channel rates up to 1920 kb/s and covers such topics as system description, terminal requirements, intercommunications, maintenance and human factor aspects (79).

Desktop Multimedia Communications

One form of videoconferencing is for desk-to-desk communications. In this application, communications are more involved than for person-to-person interactions because multimedia information will be accessed at the same time (see Refs. 80–82). The need to access multimedia databases and shared visual spaces will increase the demand for DS1 services from desktop workstations. In the residential environment, multimedia terminals and personal computers also will be used for information retrieval, teleshopping, and teleconsultation (83).

Four scenarios are likely to develop for communications in this area:

1. 64 to 128 kb/s over ISDN basic rate access
2. Multimedia information over high-speed isochronous LANs
3. ISDN PRA with $n \times$ 64-kb/s circuits
4. Broadband ISDN (B-ISDN) at 155 Mb/s

Several standards activities are underway to facilitate multimedia services: audiovisual interactive services (service definition in CCITT Study Group I and protocols and terminal characteristics in Study Group VIII), multimedia and hypermedia information coding standards (International Organization for Standardization/International Electrotechnical Commission [ISO/IEC] Joint Technical Committee 1, Subcommittee 29 [JTC1/SC29]), and network capabilities to support multimedia services (CCITT Study Group XVIII).

Image Communications

There are professional groups and government departments (e.g., medical, research, earth resources, police) that have a need to communicate high-resolution images. Their requirements traditionally have been met using film-based systems that are slow, cumbersome, and expensive to handle. Possibly the most critical need for image communications is in the medical profession in which imaging is used extensively for medical diagnosis and records (84,85). Radiographs, thermographs, tomograms, ultrasound scans, and nuclear medicine images are typical examples of applications in the medical field. Depending on the application, DS1 may provide the ideal capability in terms of quality/cost tradeoffs. Image acquisition systems and image workstations distributed throughout hospitals and remote medical offices would communicate with each other and have access to databases of digital images and medical records using DS1 facilities and B-ISDN. The images in the image base would be accessed on demand from the visual workstations and desk-to-desk medical specialist consultations would take place. Digital images allow additional processing capabilities (e.g., Ref. 86).

Residential Video on Demand

Video on demand will enable residential users to access video sources and select video programs, via menus on the screens of their television sets, for entertainment, education, or other applications, offered by information providers (87–90). New coding technologies and standards now permit digital video coding at about 1.5 Mb/s with quality sufficient for many business and residential applications (89,91,92). This bit rate is well suited for compact optical disks, digital tape, magnetic disks, and DS1 facilities. Furthermore, new transport technologies permit the transmission at DS1 rates and higher over a single wire pair (93,94). Hence, a combination of these elements indicates that video on demand could be one of the mass applications for DS1 services.

High-Quality Music on Demand

Similar to on-demand video distribution, compact-disk-quality digital stereo music requires about 1.5 Mb/s, hence it is well suited for on-demand music via switched DS1 facilities. Other encoding techniques at lower bit rates (e.g., 384 kb/s) that provide equivalent quality also are possible.

Personal Computer Communications

A large range of personal computer applications is possible with switched DS1 and FDS1 services. These applications include computer-to-computer communications for the exchange of software and files, access to such remote peripheral devices as a CD-ROM (compact disk read-only memory), access to remote databases, and even to realize virtual personal computers (95). CD-ROM devices have interface bit rates of about 1.5 Mb/s, hence they are very suitable for remote access via DS1 services, referred to as *virtual CD-ROM*. A virtual CD-ROM is functionally identical to a real CD-ROM co-located with the personal computer.

DS1 Standards

While many standards are applicable to DS1 systems and services, some of which are mentioned in the previous sections, the emphasis in this section is on the digital hierarchy and the physical-layer standards for 1544- and 2048-kb/s transmission systems. The major standards organizations considered here are CCITT (96), ANSI (97), European Telecommunications Standards Institute (ETSI) (98,99), and Telecommunication Technology Committee (TTC) (48, 100).

The main standards for the DS1 physical layer are summarized in Table 2. More complete lists of references based on what is available currently in published or draft form are given in each section. Since standards organizations are updating and upgrading the standards constantly, the reader should consult the corresponding standards organizations to ascertain the status of any given standard or possible new standards.

Table 2 indicates how international, regional, and national standards ad-

TABLE 2 Relevant DS1 Standards

	International	Regional/National		
	CCITT Rec.	U.S. ANSI	Europe ETS	Japan TTC
Digital hierarchy				
Asynchronous	G.702	T1.102	300 166	JT-G702
Synchronous	G.707	T1.105	300 147	JT-G707
Physical/electrical	G.703	T1.102	300 166	JT-G703
interface characteristics		T1.403		
Frame structure	G.704	T1.107	300 167	JT-G704
Frame alignment	G.706	T1.107	300 167	JT-G706
Timing requirements	G.811	T1.101	(DI/BT-2006)	JT-G811
	G.812			
ISDN PRA Layer 1	I.431	T1.408	300 011	JT-I431

dressing similar aspects compare; however, that does not mean that there is a one-to-one correspondence between the standards, let alone equivalency, since there are significant differences in scope and content among them. Furthermore, many other standards are applicable (e.g., those pertaining to signaling, performance, management, electromagnetic compatibility [EMC] requirements, connectors, safety and protection, etc.) that are outside the scope of this article, but should not be difficult to locate with Table 2 and the list of references below as a basis.

International Telegraph and Telephone Consultative Committee

The CCITT is one of five permanent organs (another is the International Radio Consultative Committee [CCIR]) of the International Telecommunication Union (ITU). CCITT was formed in 1956 and its mandate is to study and issue recommendations on technical, operating, and tariff questions relating to worldwide standardization of telecommunications, except for radio communications. For background on CCITT, see Ref. 96.

The G-series recommendations contain the digital hierarchies (1,2), and the key DS1 standards at the physical layer (27,26,101–103). The ISDN PRA is specified in the I-series recommendations on ISDN (104).

Other relevant standards are the service-definition recommendations (64,67–70) and the video-coding and transmission recommendations (75–79).

American National Standards Institute

ANSI was founded in 1918 to coordinate voluntary standards activities in the United States. ANSI approves American National Standards (ANSs). For background on ANSI, see Ref. 97. For an overview of the development of DS1 standards in the United States, including evolution considerations in the light of divestiture considerations, refer to Ref. 19.

The relevant ANSI standards for DS1 include those for digital hierarchies (5,9), the physical layer (5,6,13–15,25,105), and ISDN PRA (106). The following are the key DS1 standards at the physical layer:

T1.102 describes the digital hierarchy in the United States and the electrical and physical characteristics for signals appearing at the appropriate level digital interface, including DS1 (5).

T1.107 describes the formats at each level of the digital hierarchy, including DS1, and identifies selected frame and payload structures, including channelization (13).

T1.403 describes the metallic interface between a carrier and customer installation, referred to as the network interface, at the DS1 rate (14).

Electronic Industries Association/Telecommunications Industry Association 547 (EIA/TIA-547) establishes performance and technical criteria for interfacing and connecting the network channel terminating equipment with various elements of the public telecommunications network (15).

T1.101 specifies the synchronization performance of references passed between networks and the synchronization performance of primary reference sources used to control synchronous networks to ensure that slip performance will be acceptable (6).

T1.408 establishes performance and technical criteria for interfacing and interconnecting the various functional groups in the ISDN primary rate functional reference configuration (106).

With regard to evolving technologies and standards, ANSI Committee T1 (Technical Subcommittee T1E1) is investigating the standardization of new digital subscriber-line technologies that will enable the transmission of DS1 on existing, nonrepeatered copper loops (19,93,94,107–109). Two such technologies are high-rate digital subscriber line (HDSL) and asymmetric digital subscriber line (ADSL).

HDSL is a symmetric capability using two copper pairs without repeaters and is limited to a range of 12,000 feet from the central office. ADSL is an asymmetric capability intended to transport a DS1 (or a DS3) bit stream from the telephone network toward the customer without repeaters, while simultaneously providing a 16-kb/s control channel from the customer to the network. ADSL has a design goal of 18,000 feet on one copper pair for DS1 and somewhat less for DS3 (94). ADSL can coexist with the basic telephone service (or basic rate ISDN) on the same copper pair because the two systems are independent and transparent to each other even though they use the same copper pair.

European Telecommunications Standards Institute

ETSI was founded in March 1988 by the member administrations of the CEPT with the mandate to develop technical specifications for European Telecommunication Standards (ETSs) (98,99). In some instances, ETSs replace CEPT recommendations that were previously in use.

In the area of DS1 standards, the relevant ETSs correspond to CCITT recommendations as indicated in Table 2. The ETSs in Table 2 specify what parts or options of the CCITT recommendations are normative, informative or not relevant, and may specify other modifications as appropriate (110–114).

ETSI has a comprehensive plan for the development of standards for Open Network Provision (ONP) of leased lines (7). ONP refers to the open and efficient access to public telecommunications networks and, where applicable, public telecommunications services and the efficient use of those networks and services. The first phase of the study assessed the situation and concluded that the currently available standards do not cover the requirements for ONP leased lines adequately. For example, work will be required to supplement ETS 300 166, which in itself does not provide a complete interface specification. Hence, a detailed plan for the development of the necessary standards has been developed (7).

Telecommunication Technology Committee (Japan)

In Japan, the TTC was established on October 25, 1985, with the mandate to develop standards related to the interconnection of telecommunications networks (100). In the area of DS1 standards, TTC standards correspond to CCITT recommendations as indicated in Table 2. The differences between them are summarized in Ref. 100.

Summary

This article provides a technical overview of DS1, including a description of services, applications, and relevant standards. DS1 is one of the two primary building blocks for digital hierarchies. In terms of bandwidth capabilities, DS1 is proving to be a stepping stone in the evolution of digital networks. Indeed, while it began for grouping 64-kb/s channels into a single transmission system, and that is how it was devised originally, it then became a suitable mode of transmission for high-quality audio, video, and multimedia information services in its own right. It is expected that DS1 services will stimulate the demand for broadband services within ISDN.

In terms of functionality, DS1 services and applications have evolved from dedicated lines, through private DS1 networks managed by end users, including FDS1, to fully dialable switched services providing clear-channel flexible bandwidth allocation between 64 kb/s and 1536 kb/s on demand.

It is expected that higher and higher digital access rates will be used as new applications and services are developed. The power of personal computers in general, and more specifically multimedia desktop communications in the workplace and intelligent video in the home, including music, compact disk interactive (CDI), video on demand, and so on, will see a growth of DS1 services and eventually even higher bit rates. Standards, both existing and under development, will be a key enabler for the orderly mass-market introduction of DS1 and subsequent developments.

It is now possible to deliver DS1 and FDS1 services to every desk and to every home. Even with B-ISDN, which will provide access capabilities up to 155 Mb/s, DS1 will continue to play an important role. The relative price/performance of the various technologies and the driving need for services from the user's point of view will determine the competitive technologies. In the meantime, DS1 provides at least a good stepping stone toward the higher bit rates. Eventually, B-ISDN will phase out DS1 services at some point in the future; however, only time will tell what will be the relative significance of DS1 in years to come and the role it will play in the introduction of broadband networks.

List of Acronyms

ADPCM	adaptive pulse code modulation
ADSL	asymmetric digital subscriber line
AMI	alternate mark inversion
ANS	American National Standard
ANSI	American National Standards Institute
B-ISDN	Broadband ISDN
B8ZS	bipolar 8-zero substitution
CATV	community antenna television
CCIR	International Radio Consultative Committee
CCITT	International Telegraph and Telephone Consultative Committee
CD-ROM	compact disk read-only memory
CDI	compact disk interactive
CEPT	European Conference of Postal and Telecommunications Administrations
CPE	customer premises equipment
CRC	cyclic redundancy check
DC	direct current
DL	data link
DS1	Digital Signal Level 1
EMC	electromagnetic compatibility
ESF	extended superframe
ETS	European Telecommunication Standard
ETSI	European Telecommunications Standards Institute
FDS1	fractional DS1
HDB3	high-density bipolar of order 3
HDSL	high-rate digital subscriber line
IEC	International Electrotechnical Commission
ISDN	Integrated Services Digital Network

ISO	International Organization for Standardization
ITU	International Telecommunication Union
JTC1	Joint Technical Committee 1
LAN	local-area network
NTT	Nippon Telegraph and Telephone
ONP	Open Network Provision
PCM	pulse code modulation
PRA	Primary Rate Access
SDH	synchronous digital hierarchy
SF	superframe
STS	Synchronous Transport Signal
TTC	Telecommunication Technology Committee
ZBTSI	zero-byte time-slot interchanger

References

1. International Telegraph and Telephone Consultative Committee, Digital Hierarchy Bit Rates, Recommendation G.702. In: *CCITT Blue Book*, Recommendation G.702, ITU, Geneva, 1989.

2. International Telegraph and Telephone Consultative Committee, *Synchronous Digital Hierarchy Bit Rates*, Recommendation G.707, CCITT, Geneva, 1991.

3. Rickard, N., Bridging between E-1 and T-1, *Telecommun.*, 24:66–68 (January 1990).

4. Reichard, D. W., Digital Cross-Connect Systems: CEPT Gateway Application, *Proc. ICC*, 1:331–335 (June 12–15, 1988).

5. American National Standards Institute, *American National Standard for Telecommunications—Digital Hierarchy—Electrical Interfaces*, ANSI T1.102-1987, ANSI, New York, 1987 (revision in progress, ref. T1X1.4/91-004).

6. American National Standards Institute, *American National Standard for Telecommunications—Synchronization Interface Standards for Digital Networks*, ANSI T1.101-1987, ANSI, New York, 1987 (revision in progress, ref. T1X1.3/90-026R1).

7. European Telecommunications Standards Institute, *Open Network Provision Technical Requirements Standardization Requirements for ONP Leased Lines* (Management Planning Information and Technical Aspects), ETSI Work Item No. BT-2012, ETSI, Valbonne, Cedex, France, December 13, 1991.

8. Pyle, K., Using T-1 in CATV Applications, *Commun. Engineering and Design*, 16:50–54 (May 1990).

9. American National Standards Institute, *American National Standard for Telecommunications—Digital Hierarchy—Optical Interface Rates and Formats Specifications*, ANSI T1.105-1988, ANSI, New York, 1988.

10. Ferguson, S. P., Plesiochronous High-Order Digital Multiplexing. In: *Transmission Systems* (J. E. Flood and P. Cochrane, eds.), Peter Peregrinus, London, 1991, pp. 210–242.

11. Pellegrini, G., and Wery, P.H.K., Synchronous Digital Hierarchy, *Telecommun. J.*, 58:815–824 (November 1991).

12. Cubbage, R. W., and Littlewood, P. A., DS1 Mapping Considerations for the Synchronous Optical Network, *Proc. SPIE: Fiber Optics in the Subscriber Loop*, 1363:163–171 (1990).

13. American National Standards Institute, *American National Standard for Telecommunications — Digital Hierarchy — Formats Specifications*, ANSI T1.107-1988, ANSI, New York, 1989.

14. American National Standards Institute, *American National Standard for Telecommunications — Carrier to Customer Installation DS1 Metallic Interface*, ANSI T1.403-1989, ANSI, New York, 1989.

15. Electronic Industries Association, *EIA/TIA Standard — Network Channel Terminal Equipment for DS1 Service*, ANSI/EIA/TIA-547-1989, EIA, Washington, DC, 1989.

16. Cattermole, K. W., *Principles of Pulse Code Modulation*, Iliffe Books, London, 1969.

17. Davis, C. G., An Experimental Pulse Code Modulation System for Short-Haul Trunks, *Bell Sys. Tech. J.*, 41:1–24 (January 1962).

18. Haggerty, C., An Overview of T-Carrier Concepts. In: *Management of Telecommunications*, MT20-615, Datapro, Delran, NJ, January 1991, pp. 101–128.

19. McNamara, W. J., Physical Layer Standards for DS1 Rate Access and Transport, *Proc. ICC*, 1:297–301 (June 23–26, 1991).

20. Ruffalo, D. R., Understanding T1 Basics: Primer Offers Picture of Networking Future, *Data Commun.*, 16:161–169 (March 1987).

21. Molloy, K. H., ABCs of Digital Formats, *Telephone Engineer and Management*, 93:70–75 (June 15, 1989).

22. Owen, F.F.E., *PCM and Digital Transmission Systems*, McGraw-Hill, New York, 1982.

23. Flood, J. E., and Cochrane, P. (eds.), *Transmission Systems*, Peter Peregrinus, London, 1991.

24. Wallace, A. D., Humphrey, L. D., and Sexton, M. J., Analogue/Digital Conversion and the PCM Primary Multiplex Group. In: *Transmission Systems* (J. E. Flood and P. Cochrane, eds.), Peter Peregrinus, London, 1991, pp. 150–181.

25. American National Standards Institute, *Application Guidelines for Use of the DS1 Extended Superframe Format Data Link*, ANSI TR No. 12, Committee T1 — Telecommunications Technical Report, ANSI, New York, November 1991.

26. International Telegraph and Telephone Consultative Committee, *Functional Characteristics of Interfaces Associated with Network Nodes*, Recommendation G.704, CCITT, Geneva, 1991.

27. International Telegraph and Telephone Consultative Committee, *Physical/Electrical Characteristics of Hierarchical Digital Interfaces*, Recommendation G.703, CCITT, Geneva, 1991.

28. Wright, M. A., Distortion and Error in DS1 Facilities, *Telephone Engineer and Management*, 92:68–70 (January 1, 1988).

29. Lesea, A., T1 Link Problem: Wander, *Telephone Engineer and Management*, 91:53–54 (November 1, 1987).

30. Lesea, A., T1 Link Problem: Jitter, *Telephone Engineer and Management*, 92:66–67 (January 1, 1988).

31. Stauffer, K., and Brajkovic, A., DS1 Extended Superframe Format and Related Performance Issues, *IEEE Commun.*, 27:19–23 (April 1989).

32. Eu, J. H., An Evaluation of Error Performance Estimation Schemes for DS1 Transmission Systems Carrying Live Traffic, *IEEE Trans. Commun.*, 38:384–391 (March 1990).

33. Rux, P. T., T1 testing and the Standard Answer, *Telephone Engineer and Management*, 94:50, 53, 57 (January 1, 1990).

34. Ingle, J. F., A 21-Tone Signal-to-Total-Distortion Testing Standard, *Proc. ICC*, 1:307–311 (June 23–26, 1991).

35. Duncan, B., The Hand-Off and the Electronic Cross-Connect: A Pivotal Pair, *Telephony*, 219:30–35 (August 20, 1990).

36. Walker, H., Ins and Outs of DS1 and DS3 Measurement, *Telephony*, 217:36–46 (September 11, 1989).

37. Mauk, G., DS1 Testing Sees the Light, *Telephony*, 219:41–42 (July 30, 1990).

38. Zinnikas, M. J., Cooper, T. R., Lindsay, P. D., and Heidt, R. C., DS1 Facility Performance: Digital Radio as Long-Haul Technology, *Proc. ICC*, 2:655–658 (June 1987).

39. Eu, J. H., and Rollins, W. W., A Performance Review of Out-of-Frame Detection Schemes for DS1 Signals, *IEEE Trans. Commun.*, 39:1004–1009 (June 1991).

40. Nilsson, A., Perry, M., and Sutton, M., Frame Synchronization Failure: Detection and Recovery, *IEEE Trans. Commun.*, 39:613–618 (April 1991).

41. May, J. E., Digital Channel Banks (Network Terminal Systems). In: *The Froehlich/Kent Encyclopedia of Telecommunications*, Vol. 5 (F. E. Froehlich and A. Kent, eds.), Marcel Dekker, New York, 1992, pp. 369–418.

42. Hutcheson, W. D., and Shaw, R. K., The Evolution of DS1 Digital Transmission Service Capabilities, *Proc. Globecom*, 1:422–425 (November 26–29, 1984).

43. Bobilin, R. T., Liu, A. Y., and Whitehead, P. C., Intelligent Transmission Networks Provide Higher Bit Rate Services, *Telephony*, 208:52–63 (January 28, 1985).

44. Gaz, R., and Relyea, W. E., Integrated Voice/Data for Large Business Customers Using DS1 Facilities, *Proc. ICC*, 3:1693–1699 (June 1987).

45. Anding, D. W., Service Offerings Utilizing Digital Cross-Connect Systems—A BellSouth Perspective, *Proc. ICC*, 1:336–339 (June 12–15, 1988).

46. Sharma, R., T1 Network Design and Planning Made Easier, *Data Commun.*, 17:199–207 (September 1988).

47. Mier, E. E., Fractional T1: Carriers Carve Out Bandwidth for Users, *Data Commun.*, 18:84–98 (November 1989).

48. Nakano, K., Inauguration of New Enhanced Super Digital Service, *NTT Rev.*, 2:36–41 (September 1990).

49. Sakamoto, M., and Tamaki, N., New High Speed Digital Leased Circuit System, *NTT Rev.*, 2:42–51 (September 1990).

50. Technology behind High-Speed Digital Service, *NTT Rev.*, 2:34–35 (November 1990).

51. ARCO: Fill'er Up with Bandwidth on Demand, *Commun. News*, 28:24–25 (February 1991).

52. Das, R., and Carbone, P., Dialable Services at FDS1 and DS1 Rates—The Services Network of the Early 1990s, *Proc. ICC*, 1:287–291 (June 23–26, 1991).

53. Rathmann, J. E., and Ballart, R., The Public Switched Fractional DS1 Service Capability, *Proc. ICC*, 1:292–296 (June 23–26, 1991).

54. Arredondo, G. A., and Floyd, P. H., The Future of Subrate Data and Private Line Networking, *IEEE Commun.*, 30:30–35 (March 1992).

55. Zerbiec, T. G., Considering the Past and Anticipating the Future for Private Data Networks, *IEEE Commun.*, 30:36–46 (March 1992).

56. Taylor, S. A., Frame Transport Systems, *IEEE Commun.*, 30:66–70 (March 1992).

57. Brosemer, J. J., and Enright, D. J., Virtual Networks: Past, Present and Future, *IEEE Commun.*, 30:80–85 (March 1992).

58. Struck, B. W., The Digital Cross-Connect: Cornerstone of Future Networks?, *Data Commun.*, 16:165–166, 169–170, 173–174, 177–178 (August 1987).

59. MacNeil, D. R., and Chahal, M. S., Wideband Circuit Switching Applications and Networking Implications, *Proc. Supercomm/ICC*, 4:1805–1809, Chicago,

(June 14–18, 1992).

60. Bellcore, *Generic Requirements for the Switched Fractional DS1 Service Capability from a Non-ISDN Interface (SWF-DS1/non-ISDN)*, TR-NWT-001068 (Issue 1), Bellcore, Morristown, NJ, November 1991.

61. Bellcore, *Generic Requirements for the Switched DS1/Switched Fractional DS1 Service Capability from an ISDN Interface (SWF-DS1/ISDN)*, TR-NWT-001203 (Issue 1), Bellcore, Morristown, NJ, December 1991.

62. Conference of European Postal and Telecommunications Administrations, *International Switched 2 Mbit/s Bearer Service*, Recommendation T/SF 69 E (Madrid 1991), CEPT, Paris, 1991.

63. American National Standards Institute, *American National Standard Supplement to ANSI T1.620 on Circuit-Mode Bearer Service Category Description (Multi-Rate Circuit-Mode Bearer Service for ISDN)*, ANSI T1.620a-1992, ANSI, New York, 1992.

64. International Telegraph and Telephone Consultative Committee, Frame Mode Bearer Services, Draft Recommendation I.233, CCITT, Geneva, 1992.

65. Schriftgiesser, D., and Levy, R., SMDS vs. Frame Relay: An Either/Or Decision?, *Business Commun. Rev.*, 21:59–63 (September 1991).

66. Smith, G., WAN Internetworking—Emerging Standards and Services, *Telecommun.*, 26:7–10, 22, (January 1992).

67. International Telegraph and Telephone Consultative Committee, Method for the Characterization of Telecommunication Services Supported by an ISDN and Network Capabilities of an ISDN. In: *CCITT Blue Book*, Recommendation I.130, ITU, Geneva, 1989.

68. International Telegraph and Telephone Consultative Committee, Circuit-Mode Bearer Service Categories. In: *CCITT Blue Book*, Recommendation I.231, ITU, Geneva, 1989.

69. International Telegraph and Telephone Consultative Committee, *Circuit-Mode Multirate Unrestricted, 8 kHz Structured Bearer Service Category*, Draft Recommendation I.231.10, CCITT Study Group I, 1992.

70. International Telegraph and Telephone Consultative Committee, *B-ISDN Service Aspects*, Recommendation I.211, CCITT, Geneva, 1991.

71. Dayanim, J., Disaster Recovery: Options for Public Networks, *Telecommun.*, 25:48–51 (December 1991).

72. Bodson, D., and Harris, E., When the Lines Go Down, *IEEE Spectrum*, 29:40–44 (March 1992).

73. Mier, E. E., Adding to Your Net Worth with T1-to-LAN Devices, *Data Commun.*, 18:103–118 (September 1989).

74. Thuston, F., Video Teleconferencing: The State of the Art, *Telecommun.*, 26:63–65 (January 1992).

75. International Telegraph and Telephone Consultative Committee, *Video Codec for Audiovisual Services at p × 64 kbit/s*, Recommendation H.261, CCITT, Geneva, 1990.

76. International Telegraph and Telephone Consultative Committee, *Frame Structure for a 64 to 1920 kbit/s Channel in Audiovisual Teleservices*, Recommendation H.221, CCITT, Geneva, 1990.

77. International Telegraph and Telephone Consultative Committee, *Frame-Synchronous Control and Indication Signals for Audiovisual Systems*, Recommendation H.230, CCITT, Geneva, 1990.

78. International Telegraph and Telephone Consultative Committee, *System for Establishing Communication between Audiovisual Terminals Using Digital Channels up to 2 Mbit/s*, Recommendation H.242, CCITT, Geneva, 1990.

79. International Telegraph and Telephone Consultative Committee, *Narrow-band Visual Telephone Systems and Terminal Equipment*, Recommendation H.320, CCITT, Geneva, 1990.

80. Tomita, Y., and Nakano, S., Multimedia Teleconference System Using ISDN Primary Rate Interface, *NTT Rev.*, 3:44–50 (July 1991).

81. Wilson, D., Wrestling with Multimedia Standards, *Computer Design*, 31:70–88 (January 1992).

82. Biggar, M. J., Emerging Digital Image and Video Services, *Australian Telecommunication Research*, 25:1–8 (1991).

83. Hase, M., Kawakubo, S., and Shoman, M., Advanced Videophone System Using Synchronized Video Filing Equipment—Telewindow Shopping System, *NTT Rev.*, 3:29–36 (July 1991).

84. Costa, J. M., Medical Image Communication Systems, *Proc. SPIE: Digital Radiography*, 314:380–388 (1981).

85. Georganas, N. D., Goldberg, M., Barbosa, L. O., and Mastronardi, J., A Multimedia Communications System for Medical Applications, *Proc. ICC*, 3:1496–1500 (1989).

86. Costa, J. M., Digital Tomographic Filters for Radiographs. In: *Multidimensional Systems: Techniques and Applications* (S. G. Tzafestas, ed.), Marcel Dekker, New York, 1986, pp. 585–623.

87. Judice, C. N., Addeo, E. J., Eiger, M. I., and Lemberg, H. L., Video on Demand: A Wideband Service or Myth? *Proc. ICC*, 3:1735–1739 (June 22–25, 1986).

88. Sawyer, W.D.M., Salladay, J. H., and Snider, M. W., The Gateway Role in an Open-Access, High-Capacity Network, *Proc. ISSLS*, 401–407 (April 1991).

89. Anderson, M. M., Video Services on Copper, *Proc. ICC*, 1:302–306 (June 23–26, 1991).

90. Lawrence, R. W., Switched Simplex High Bit Rate Services in Today's Residential Environment, *Proc. Supercomm/ICC*, 4:1799–1804, Chicago (June 14–18), 1992.

91. International Organization for Standardization/International Electrotechnical Commission, *Information Technology—Coding of Moving Pictures and Associated Audio—For Digital Storage Media Up to about 1.5 Mbit/s,* ISO/IEC Draft International Standard (DIS) 11172, ISO/IEC, Geneva, 1992.

92. Jurgen, R. K., Digital Video, *IEEE Spectrum*, 29:24–30 (March 1992).

93. Mordecai, B. A., Hsing, T. R., Waring, D. L., and Wilson, D. S., Repeaterless T-1 Technology Brings Copper and Fiber Together, *Telephony*, 219:28–30, 32–33 (July 2, 1990).

94. Manchester, E. E., New Uses for Residential Copper, *Telephony*, 220:27–28, 32 (June 10, 1991).

95. Sawyer, W.D.M., The Virtual Computer: A New Paradigm for Educational Computing, *Educational Technology*, 32:7–14 (January 1992).

96. Irmer, T., CCITT. In: *The Froehlich/Kent Encyclopedia of Telecommunications*, Vol. 2 (F. E. Froehlich and A. Kent, eds.), Marcel Dekker, New York, 1991, pp. 281–288.

97. Morgan, V., The American National Standards Institute. In: *The Froehlich/Kent Encyclopedia of Telecommunications*, Vol. 1 (F. E. Froehlich and A. Kent, eds.), Marcel Dekker, New York, 1990, pp. 137–141.

98. Besen, S. M., The European Telecommunications Standards Institute—A Preliminary Analysis, *Telecommun. Policy*, 14:521–530 (December 1990).

99. Temple, S., *A Revolution in European Telecommunications Standards Making*, Kingston Public Relations, Hull, United Kingdom, 1991.

100. The Telecommunication Technology Committee, *TTC Standards (Summary) 1990*, The Telecommunication Technology Committee, Tokyo, Japan, 1990.

101. International Telegraph and Telephone Consultative Committee, Frame Alignment and Cyclic Redundancy Check (CRC) Procedures Relating to Basic Frame Structures Defined in Recommendation G.704, Recommendation G.706, CCITT, Geneva, 1991.

102. International Telegraph and Telephone Consultative Committee, Timing Requirements at the Outputs of Primary Reference Clocks Suitable for Plesiochronous Operation of International Digital Links. In: *CCITT Blue Book*, Recommendation G.811, ITU, Geneva, 1989.

103. International Telegraph and Telephone Consultative Committee, Timing Requirements at the Outputs of Slave Clocks Suitable for Plesiochronous Operation of International Digital Links. In: *CCITT Blue Book*, Recommendation G.812, ITU, Geneva, 1989.

104. International Telegraph and Telephone Consultative Committee, Primary Rate User-Network Interface — Layer 1 Specification. In: *CCITT Blue Book*, Recommendation I.431, ITU, Geneva, 1989.

105. American National Standards Institute, *American National Standard for Telecommunications — Carrier to Customer Metallic Interface — Digital Data at 64 kbit/s and Subrates*, ANSI T1.410-1992, ANSI, New York, 1992.

106. American National Standards Institute, *American National Standard for Telecommunications — ISDN Primary Rate — Customer Installation Metallic Interfaces Layer I Specification*, ANSI T1.408-1990, ANSI, New York, 1990.

107. Walkoe, W., and Starr, T.J.J., High Bit Rate Digital Subscriber Line: A Copper Bridge to the Network of the Future, *IEEE J. Sel. Areas Commun.*, 9:765–768 (August 1991).

108. Lechleider, J. W., High Bit Rate Digital Subscriber Lines: A Review of HDSL Progress, *IEEE J. Sel. Areas Commun.*, 9:769–784 (August 1991).

109. Werner, J. J., The HDSL Environment, *IEEE J. Sel. Areas Commun.*, 9:785–800 (August 1991).

110. European Telecommunications Standards Institute, *Integrated Services Digital Network (ISDN): Primary Rate User-Network Interface — Layer 1 Specification and Test Principles* (DE/TM-3010 old, T/L 03-14), ETS 300 011, ETSI, Valbonne, Cedex, France, 1991.

111. European Telecommunications Standards Institute, *Transmission and Multiplexing: Synchronous Digital Hierarchy — Multiplexing Structure* (DE/TM-3001), ETS 300 147, ETSI, Valbonne, Cedex, France, 1991.

112. European Telecommunications Standards Institute, *Transmission and Multiplexing: Physical/Electrical Characteristics of Hierarchical Digital Interfaces for Equipment Using the 2048-kbit/s-based Plesiochronous or Synchronous Digital Hierarchies* (DE/TM-3002), ETS 300 166, ETSI, Valbonne, Cedex, France, 1991.

113. European Telecommunications Standards Institute, *Transmission and Multiplexing: Functional Characteristics of 2 Mbit/s Interfaces* (DE/TM-3006), ETS 300 167, ETSI, Valbonne, Cedex, France, 1991.

114. European Telecommunications Standards Institute, *Synchronization Strategies for Digital Private Telecommunications Networks (PTNs) and Timing Requirements for Digital Private Telecommunications Network Exchanges (PTNXs)*, draft proposed Interim ETS (DI/BT-2006), Version 8, ETSI, Valbonne, Cedex, France, September 1991.

JOSE M. COSTA

Duobinary Signaling   (see Correlative Line-Coding
[Partial-Response Signaling] Techniques)

Echo-Canceling Algorithms

Introduction

Echo is a phenomenon in which a delayed and distorted version of the original sound or signal is reflected back to the source. Echoes arise in various situations in the telecommunications network and impair communication quality.

An end-to-end connection of the current telephone network consists of both two-wire links, as in a subscriber loop, and four-wire links, as in a long-haul link. A hybrid transformer is used at the connection point, by which perfect conversion between bidirectional transmission using a two-wire link and two unidirectional transmissions using a four-wire link can be accomplished in an impedance-matched condition. An imperfect impedence matching causes echoes to be generated. Figure 1 shows an example of an echo canceler placed in a network for such a case.

Figure 2 illustrates a case in which acoustic echoes are generated. This type of echo results from the reflection of sound waves and acoustic coupling between the microphone and the loudspeaker. This echo disturbs natural dialogue and sometimes causes howling. Voice switches and directional microphones have been used frequently to solve these problems, but they often cause unnatural restrictions on the dialogue. Echo cancelers provide solutions that enable easier and more natural conversation. The echo canceler estimates the characteristics of the echo path, then generates a replica of the echo. This echo replica then is subtracted from the real echo, resulting in the improvement of the communication quality.

The theoretical basis for echo cancelers has been studied extensively for the past few decades, and practical adaptive echo-cancellation algorithms were developed by the mid-1960s (1,2). In the 1970s, several organizations evaluated the subjective transmission quality of the echo canceler (3,4). However, due to the intensive processing required, widespread implementation had to wait for advances in large-scale integration (LSI). The first practical very-large-scale integration (VLSI) echo canceler appeared in 1980 (5), and since that time rapid improvements in cost and size have been achieved.

Basic Principles

An echo canceler is a two-input, one-output system as shown in Fig. 1. One input is $x(j)$, called the *receive-in signal*, taken from the receive path, and the

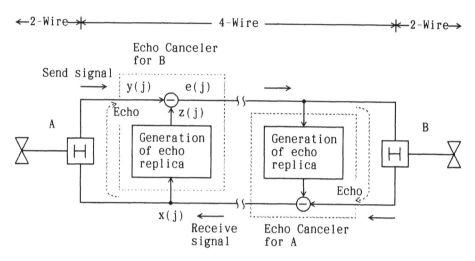

FIG. 1 Echo canceler for telephone networks.

other input is $y(j)$, called the *send-in signal*, taken from the send path. The echo canceler generates an echo replica $z(j)$ and the output $e(j)$ is obtained by subtracting $z(j)$ from $y(j)$. This output is the echo-canceled version of the send signal and, in the case of Party A being silent, this becomes the residual error signal.

To generate a replica of the echo, an adaptive digital filter (ADF) usually is used. Figure 3 shows an example of an ADF using a finite impulse-response (FIR) type of digital filter. The ADF is required to estimate and generate a

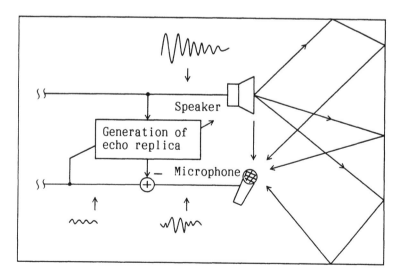

FIG. 2 Acoustic echo canceler.

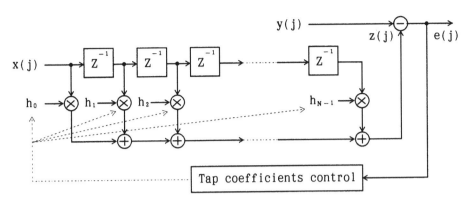

FIG. 3 Adaptive digital filter.

good echo replica since the echo path is usually unknown and time varying. Coefficients of the ADF are adjusted so that the error between the real echo and the echo replica is minimized. Because the send-in signal $y(j)$, in general, includes voice signal and noise in addition to the echo signal, considerations for the effects of these components are required in realizing adaptation algorithms. Algorithms to control tap coefficients depend on the error criteria functions and also on the ADF structures.

Performance of the echo canceler is evaluated by the echo return loss (ERL) or the echo return loss enhancement (ERLE). The ERL is defined as

$$ERL = 10 \log_{10} \frac{E[x^2(j)]}{E[e^2(j)]} \tag{1}$$

where $E[.]$ denotes the expectation. On the other hand, the ERLE is defined as

$$ERLE = 10 \log_{10} \frac{E[y^2(j)]}{E[e^2(j)]} \tag{2}$$

ERL includes the attenuation achieved by the hybrid transformer, whereas ERLE indicates the attenuation level achieved purely by the echo canceler.

Applications and Requirements

Table 1 lists the applications and requirements for echo cancelers. Requirements are quite different and depend on the applications. For example, in voice communication systems using a long-haul transmission line (such as satellite communication systems), the required ERLE is around 30 decibels (dB). When double-talk (a situation in which both parties talk on the line at the same time) is detected, the adaptation must be stopped to prevent erroneous adaptation.

TABLE 1 Applications and Requirements for Echo Cancelers

Echo Generation	Application	Requirement
Hybrid transformer impedance mismatch	Voice communication Long-haul transmission (e.g., satellite communication)	Specified in CCITT G.165 Echo cancellation of about 30 dB Double-talk control Center clipper Tone disabler
	Data communication Voiceband full-duplex data transmission (e.g., V.32 data modem)	Echo cancellation of over 70 dB Adaptation under double-talk condition Two echo cancelers for near-end echo and far-end echo
	Baseband full-duplex data transmission (e.g., ISDN subscriber loop)	Echo cancellation of over 70 dB Adaptation under double-talk condition (echo duration is short)
Sound wave reflection and acoustic coupling	Speaker microphone system (e.g., teleconferencing)	Simple algorithm Fast convergence

For full-duplex data modems, a considerably higher level of echo cancellation is required since the received signal can be about 40 dB below the transmitted signal level. The echoes must be suppressed to about 70 dB to maintain the necessary signal-to-noise ratio of about 30 dB. Also, because the full-duplex data-transmission system is inherently a double-talk system, an adaptation algorithm that can operate properly under a double-talk condition is needed. When implementing practical echo cancelers, we need to consider these different conditions according to the application. Examples of implementation for various applications in Table 1 are described in more detail in the section, "Examples of Implementation," toward the end of this article.

Basic Algorithms

As described above, the coefficients of the echo canceler are adjusted so that the output error of the echo canceler approaches the minimum value. There are various kinds of structures for echo cancelers, but for simplicity we start with the structure of the FIR-type ADF shown in Fig. 3.

In this case, the echo replica $z(j)$ is given by

$$z(j) = \sum_{i=0}^{N-1} h_i(j)x(j - i) \tag{3}$$

where $h_i(j)$, $i = 0, 1, \ldots, N - 1$ are the tap coefficients of ADF at time j. N denotes the tap length. Assuming a real-valued, time-invariant echo-path impulse response as g_i, $i = 0, 1, \ldots, \infty$, the send-in signal $y(j)$, is

$$y(j) = \sum_{i=0}^{\infty} g_i x(j - i) + n(j) \tag{4}$$

where $n(j)$ is the noise.

The residual echo or output error $e(j)$ is then

$$e(j) = y(j) - z(j)$$
$$= \sum_{i=0}^{N-1} [g_i - h_i(j)]x(j - i) + v(j) \tag{5}$$

where

$$v(j) = \sum_{i=N}^{\infty} g_i x(j - i) + n(j) \tag{6}$$

Using vector expressions, Eq. (5) can be expressed as follows:

$$e(j) = [G - H(j)]'X(j) + v(j) \tag{7}$$

where

$$G = (g_0, g_1, \ldots, g_{N-1})' \tag{8}$$
$$H(j) = [h_0(j), h_1(j), \ldots, h_{N-1}(j)]' \tag{9}$$
$$X(j) = [x(j), x(j - 1), \ldots, x(j - N + 1)]' \tag{10}$$

and $'$ denotes the transpose operation.

The tap coefficients $H(j)$ are controlled adaptively to minimize a criterion function $F\{e(j)\}$.

As for $F\{e(j)\}$, the expectation value of the squared error signal is adopted most widely, that is,

$$F\{e(j)\} = E[e^2(j)] \tag{11}$$

Assuming that $x(j)$ is statistically stationary during N samples, it follows from Eq. (7),

$$E[e^2(j)] = [G - H(j)]'R[G - H(j)] + 2[G - H(j)]'P + E[v^2(j)] \tag{12}$$

where

$$R = E[X(j)X'(j)] \tag{13}$$

and

$$P = E[X(j)v(j)] \tag{14}$$

R is the autocorrelation matrix of the receive-in signal $x(j)$, and P is the correlation vector between $x(j)$ and $v(j)$. The derivative of Eq. (12) with respect to $h_i(j)$ can be represented in a vector form as follows:

$$\frac{dE[e^2(j)]}{dH(j)} = -2R[G - H(j)] - 2P \tag{15}$$

By setting Eq. (15) to zero, the optimum tap coefficients can be obtained as

$$H_{opt}(j) = G + R^{-1}P \tag{16}$$

This is called the Wiener solution (6).

This is the optimum solution when the statistical characteristics of $x(j)$, $v(j)$, and the echo-path impulse response are time invariant. However, in practical systems, these values are generally time variant. Therefore, instead of the Wiener solution, two basic categories of adaptation algorithms generally are used. The first category is the gradient algorithm and the second is the recursive least squares (RLS) algorithm.

Gradient Algorithm

In the gradient algorithm, taps are adjusted according to the stochastic steepest descent algorithm.

Least-Mean-Square Algorithm

The criterion function of the least-mean-square (LMS) algorithm is the same as for Eq. (12). In practice, however, the instantaneous value of the squared error is used in place of Eq. (12), namely,

$$F\{e(j)\} = e^2(j) \tag{17}$$

The coefficients are controlled using its derivative with respect to the tap coefficients. From Eq. (5),

$$\frac{de^2(j)}{dh_i(j)} = -2e(j)x(j - i) \tag{18}$$

Therefore, the steepest descent algorithm leads to the following tap adaptation algorithm.

$$h_i(j + 1) = h_i(j) + 2\mu e(j)x(j - i), \qquad i = 0, 1, \ldots, N - 1$$

Or, using vector expressions of Eqs. (9) and (10),

$$H(j + 1) = H(j) + 2\mu e(j)X(j) \tag{19}$$

where μ is a constant that controls the stability and convergence rate.

The criterion function $e^2(j)$ is a function of tap coefficients $h_i(j)$ as shown in Fig. 4 for the case of two taps. The tap coefficients are adjusted in the opposite direction of the derivatives with regard to the tap coefficients as in Eq. (19) to approach the minimum point. The convergence rate of the LMS algorithm depends on the characteristic of the input signal. This can be understood intuitively from Fig. 4. The shape of $e^2(j)$ varies depending on the input signal. When the eigenvalues of the input signal autocorrelation matrix are equal or nearly equal, the gradient vector points to the bottom or close to the bottom of the mean-square error (MSE) surface because contour lines of $e^2(j)$ become closer to circles (see Fig. 4-*a*). Fast convergence thus can be achieved. However, when the eigenvalues are distributed in a wide range, such as in the case of speech signals or other strongly autocorrelated signals, the contour lines become ellipses. In this situation, the gradient vector (∇ in Fig. 4-*b*) does not point to the bottom of the MSE surface, thus leading to slower convergence (6).

The value of μ also plays an important role in determining the convergence speed, stability, and residual error after convergence. The convergence of the LMS algorithm is guaranteed if the following condition is satisfied (6).

$$0 < \mu < 1/\lambda_{max} \tag{20}$$

where λ_{max} is the maximum eigenvalue of the autocorrelation matrix of the input signal R. Because the value of λ_{max} depends on the characteristic of the input signal and cannot be predetermined in practical situations, μ must be set to a very small value to maintain stability even in the worst case. This is a disadvantage of the LMS algorithm.

Normalized Least-Mean-Square Algorithm

Normalizing the correction term in Eq. (19) by the input signal power, the next equation can be obtained.

$$h_i(j + 1) = h_i(j) + \alpha \, [e(j)x(j - i)]/ \sum_{k=0}^{N-1} x^2(j - k) \tag{21}$$

$$0 < \alpha < 2 \tag{22}$$

This method is called the learning identification (normalized LMS, or NLMS)

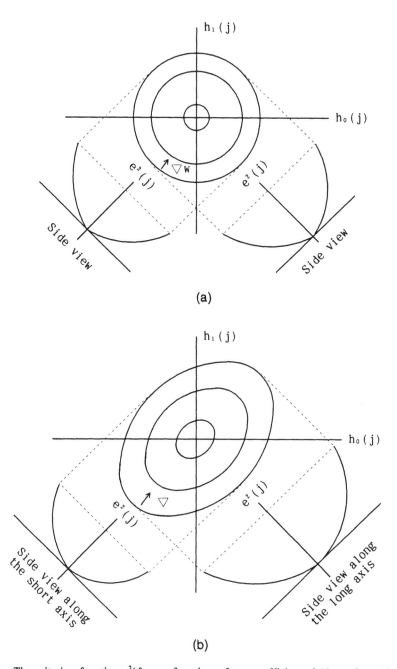

FIG. 4   The criterion function $e^2(j)$ as a function of tap coefficients $h_i(j)$: a, for white-noise input; b, for speech input.

method (7). This algorithm is used widely because constant α, as in Eq. (22), can be chosen independent of the input signal characteristics.

Using vector notation, the NLMS algorithm can be represented as

$$H(j + 1) = H(j) + \frac{\alpha e(j)}{||X(j)||^2} X(j) \tag{23}$$

where

$$||X(j)||^2 = \sum_{k=0}^{N-1} x^2(j - k) \tag{24}$$

By defining a modified ERLE as

$$\text{ERLE}_M = 10 \log_{10} \frac{E[\hat{y}^2(j)]}{E[\hat{e}^2(j)]} \tag{25}$$

where

$$\hat{y}(j) = y(j) - n(j) \tag{26}$$

$$\hat{e}(j) = e(j) - n(j) \tag{27}$$

and assuming no correlation between $h_i(j)$, $x(j)$, and $n(j)$, the following equation is obtained (8).

$$\text{ERLE}^* = \text{S/N} + 10 \log_{10}(2/\alpha - 1) \tag{28}$$

where

$$\text{S/N} = 10 \log_{10} E[\hat{y}^2(j)]/E[n^2(j)] \tag{29}$$

From Eq. (28), it can be seen that ERLE* is proportional to the S/N of the send-in signal. In addition, a large ERLE* is obtained by choosing a small value for α.

Although the NLMS algorithm has the advantage of requiring only modest computation, its major disadvantage is the dependence of convergence performance on the characteristics of the receive-in signal. Figure 5 shows an example of the convergence characteristics of the NLMS algorithm obtained by simulation. The number of taps is 150, and the assumed echo path is characterized to be $g_i = R_i\exp(-i/66.7)$, $i = 0, 1, \ldots, 149$, where R_i is a random number restricted between -1 and $+1$. Figure 5-a shows the dependence of convergence on the characteristic of the receive-in signal. As shown, when the receive-in signal is an actual speech signal, convergence becomes quite slow compared with a white-noise input. The use of 15th-order autoregressive (AR) process signals as a receive-in signal is also shown in Fig. 5-a.

Figure 5-b shows the difference of performance according to the value of α. Input signal is a white noise. When using a large value of α, convergence is

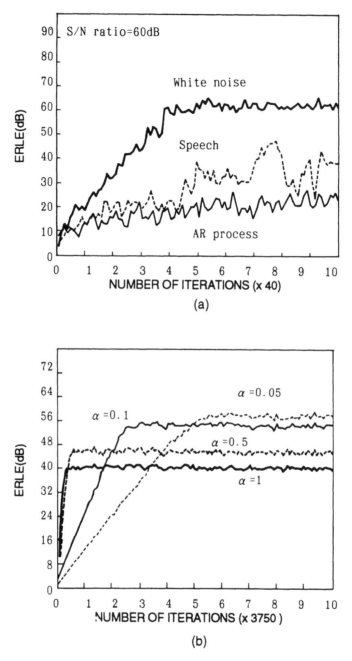

FIG. 5 Convergence performance of the normalized least-mean-square algorithm: *a*, influence of input signals; *b*, influence of different values of α.

fast, but the cancellation level is smaller. If a small value of α is used, a larger cancellation level is obtained even though convergence is slower. To improve convergence performance, several algorithms that control the gain α have been proposed (9,10).

Block Least-Mean-Square Algorithm

In the block LMS algorithm (11), the following criterion function is used.

$$F\{e(k)\} = \sum_{\ell=0}^{L-1} e^2(kL + \ell) \tag{30}$$

where k shows the index of the block that contains L samples, and tap coefficients are adjusted once per block. The derivative of Eq. (30) with respect to the tap coefficients becomes

$$\frac{dF\{e(k)\}}{dh_i} = -2 \sum_{\ell=0}^{L-1} e(kL + \ell)x(kL + \ell - i) \tag{31}$$

The tap adjustment following the stochastic steepest descent algorithm becomes

$$h_i(k + 1) = h_i(k) + 2\mu \sum_{\ell=0}^{L-1} e(kL + \ell)x(kL + \ell - i) \tag{32}$$

The convergence of the block LMS algorithm is guaranteed when

$$0 < \mu < 1/L\lambda_{\max} \tag{33}$$

The computational complexity of the block LMS algorithm can be reduced using fast Fourier transforms (FFTs) with the overlap-save method (12).

For the convolution of $X(kL)$ and $H(k)$, the circular convolution technique can be used since $h_i(k)$ is constant during L samples. Defining

$$H_{\mathrm{BLMS}}(k) = [h_0(k), h_1(k), \cdots, h_{N-1}(k), \underbrace{0, 0, \cdots 0}_{L}]' \tag{34}$$

$$X_{\mathrm{BLMS}}(k)$$
$$= [x(kL - N), \cdots, x(kL - 1), \underbrace{x(kL), \cdots, x(kL + L - 1)}_{L}]' \tag{35}$$

$$Z_{\mathrm{BLMS}}(k) = [z(kL), z(kL + 1), \cdots, z(kL + L - 1)]' \tag{36}$$

where $z(.)$ is the echo replica sequence as defined by Eq. (3). $Z_{\mathrm{BLMS}}(k)$ can be obtained as follows.

$$Z_{\text{BLMS}}(k) = \text{last } L \text{ samples of the inverse FFT (IFFT)}$$

$$\text{of (FFT of } H_{\text{BLMS}}(k)) \circledast (\text{FFT of } X_{\text{BLMS}}(k)) \tag{37}$$

where $\circledast$ denotes element by element multiplication.
Defining the following

$$E_{\text{BLMS}}(k) = [\underbrace{0, 0, \cdots 0}_{N}, e(kL), \cdots \qquad e(kL + L - 1)]' \tag{38}$$

$$c_i(k) = \sum_{\ell=0}^{L-1} e(kL + \ell)x(kL + \ell - i) \tag{39}$$

$$C(k) = [c_0(k), c_1(k), \ldots, c_{N-1}(k)]', \tag{40}$$

$C(k)$ can be calculated as

$$C(k) = \text{first } N \text{ elements of IFFT of (FFT of } E_{\text{BLMS}}(k))$$

$$\circledast (\text{FFT of } X_{\text{BLMS}}(k))^* \tag{41}$$

where $*$ denotes the complex conjugate. Because the tap adjustments in the time domain can be expressed as

$$H_{\text{BLMS}}(k + 1) = H_{\text{BLMS}}(k) + 2\mu[C(k)', \underbrace{0, \ldots, 0}_{L}]',$$

tap adjustments in the frequency domain can be obtained as

FFT of $H_{\text{BLMS}}(k + 1)$

$$= \text{FFT of } H_{\text{BLMS}}(k) + 2\mu \text{ FFT}[C(k)', \underbrace{0, \cdots 0}_{L}]' \tag{42}$$

Figure 6 is the structure of the block LMS algorithm implemented using FFTs.

Recursive Least Squares Algorithm

The RLS algorithm uses the past reference signals and the corresponding echoes to determine the coefficients that minimize the squared error summed over time. To fade out the past data gradually and enable finite dimension arithmetic, exponentially decreasing weights sometimes are assigned to the past errors.

Direct Method

The criterion function of the RLS algorithm is defined as

$$F[e(j)] = \sum_{i=-\infty}^{j} e^2(i)w(j - i) \tag{43}$$

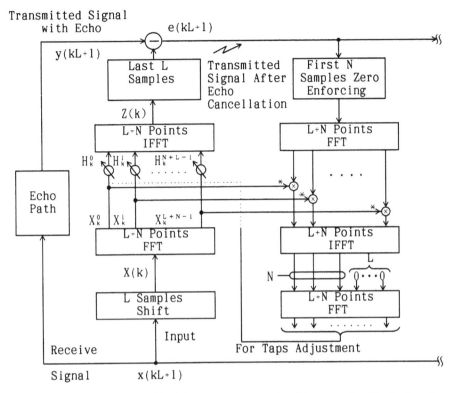

FIG. 6 Implementation of the block least-mean-square algorithm using fast Fourier transforms.

where $[w(0), w(1), \ldots]$ is a sequence of weights. For example, an exponential weighting sequence expressed as follows can be used.

$$w(i) = (1 - \lambda)^i, \qquad 0 < \lambda < 1 \tag{44}$$

Through manipulations similar to the LMS case, the following equations are obtained.

$$e(j) = y(j) - X'(j)H(j) \tag{45}$$

$$H(j + 1) = H(j) + \lambda R^{-1}(j)X(j)e(j) \tag{46}$$

$$R(j) = (1 - \lambda)R(j - 1) + \lambda X(j)X'(j) \tag{47}$$

Comparing Eq. (46), the tap adaptation equation, with Eq. (19), the correction term is multiplied by $R^{-1}(j)$. This has the effect of "whitening" the signal $X(j)$. Therefore, the convergence becomes independent of the signal characteristics. This is the main advantage of the RLS algorithm.

The direct implementation of Eqs. (45) through (47) is very complex because the computation of $R^{-1}(j)X(j)$ requires in the order of $N^3/6$ multiplications per iteration. However, instead of updating and inverting the $R(j)$ matrix, it is possible to update $Q(j) = R^{-1}(j)$. By using this feature, the following equations can be derived (13).

$$H(j + 1) = H(j) + \lambda P(j)Q(j - 1)X(j)e(j) \qquad (48)$$

where

$$P(j) = \frac{1}{(1 - \lambda)/\lambda + X'(j)Q(j - 1)X(j)} \qquad (49)$$

$$Q(j) = 1/(1 - \lambda)[1 - P(j)Q(j - 1)X(j)X'(j)]Q(j - 1) \qquad (50)$$

In this case, the required number of operations is $O(2N^2)$ multiplications per iteration.

Fast Implementation of Recursive Least Squares Algorithm

Considering that $X(j + 1)$ is only a "pushed down" version of $X(j)$, updating the gain vector $R^{-1}(j)X(j)$ can be done more efficiently, providing a fast implementation of the RLS algorithm (14). In this case, the required number of operations becomes $O(10N)$, which is only about five times the number required in the NLMS algorithm.

By writing the cross-correlation vector between the send-in signal $y(j)$ and the receive-in signal $x(j)$ as $Cr(j) = E[x(j)y(j)]$, it can be shown that the right side of Eq. (46) becomes $R^{-1}(j)Cr(j)$. This fact shows that when there is no correlation between the receive-in signal $x(j)$ and the noise $n(j)$, performance of the echo cancellation is not restricted by noise. This is the other advantage of the RLS algorithm.

Structures and Algorithms

There are several different types of structures for an ADF other than the transversal FIR type. These ADF structures and related adaptation algorithms are described next.

Infinite Impulse-Response Filter Echo Canceler

The infinite impulse-response (IIR) filter may be more suitable if the echo path can be modeled better by a combination of zeros and poles. The key points here are to guarantee stability by confining the poles within the unit circle and to obtain an unbiased estimate of the coefficients that provide the optimum global performance.

There are two formulations of IIR adaptive echo cancelers. In the first configuration, called the *serial-parallel structure* (Fig. 7-*a*), the pole adaptation is achieved essentially in the all-zero domain by adding a parallel adaptive FIR filter. Thus, the well-known adaptation procedures for FIR as described above can be applied. The coefficients of the added parallel filter then are copied to the all-pole filter in the serial path. Monitoring the zeros of $1 - B(z)$ is required to keep the system stable. As seen from the structure in Fig. 7-*a*, coefficients may not converge to the optimum value if there is an additive noise in the send-in signal. This is one of the disadvantages of the serial-parallel structure.

The structure in Fig. 7-*b* is a more direct formulation. In the figure,

$$z(j) = \sum_{m=1}^{N-1} b_m(j)z(j - m) + \sum_{m=0}^{M-1} a_m(j)x(j - m) \qquad (51)$$

This equation can be represented compactly as

$$z(j) = K'(j)I(j) \qquad (52)$$

where

$$K(j) = [b_1(j), \ldots, b_{N-1}(j), a_0(j), \ldots, a_{M-1}(j)]' \qquad (53)$$

$$I(j) = [z(j - 1), \ldots, z(j - N + 1), x(j), \ldots, x(j - M + 1)]' \qquad (54)$$

Using these notations, the adaptation algorithm can be written as (15)

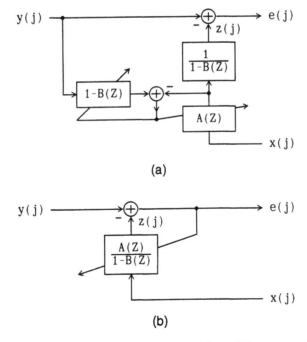

(a)

(b)

FIG. 7 Infinite impulse-response echo canceler: *a*, serial-parallel structure; *b*, parallel (direct) structure.

$$K(j + 1) = K(j) + \alpha R^{-1}(j + 1)F(j)I(j)G(j)e(j) \qquad (55)$$

$$R(j + 1) = (1 - \alpha)R(j) + \alpha F(j)I(j)\{F(j)I(j)\}' \qquad (56)$$

where $F(j)$ and $G(j)$ are functions defined in several ways to reduce the computational complexity, to improve the steady-state performance, and so on (15). When $R = I$, Eq. (55) becomes the gradient-search-based method.

Assuming that $K(j)$ varies slowly during N samples, the recursive prediction error (RPE) algorithm is obtained:

$$K(j + 1) = K(j) + \alpha R^{-1}(j + 1)[b(j) * I(j)]e(j) \qquad (57)$$

where $b(j)$ is an impulse response of $1/(1 - B(z))$.

The RPE algorithm can be simplified by using an approximation of $F(j) = G(j) = 1$, then Eq. (55) becomes

$$K(j + 1) = K(j) + \alpha R^{-1}(j + 1)I(j)e(j) \qquad (58)$$

This algorithm is called a *pseudolinear regression (PLR) algorithm.*

In the direct structure of Fig. 7-b, the output error is a nonlinear function of the coefficients. This means that the MSE surface can have multiple local minima. Adaptive algorithms converge very slowly because of the assumption made in introducing Eqs. (57) and (58), and also it might converge to a local minimum. These are the drawbacks of the algorithm. In addition, stability monitoring is required in this structure. Further research is needed to clarify the convergence properties of this structure.

Transform-Domain Echo Canceler

There are two types of transform-domain echo cancelers (Fig. 8). The purpose of transform-domain structures is to improve convergence performance and/or to reduce the computational complexity.

Lattice Structure

Figure 9 is a lattice filter that belongs to the category shown in Fig. 8-*a*. The input signal is whitened by the lattice filter, and the resulting signal is fed into the FIR filter. The gradient algorithm now can be applied and, because the lattice structure "orthogonalizes" the updates of the coefficient vector and eliminates the interaction among the adaptation of the filter coefficients, the convergence rate is faster, and is similar to that obtained when a white sequence is used as an input of a transversal filter as in Fig. 5-*a*. In this structure, the correlation of the input signal is in effect "stored" in the coefficients of the transversal filter. From this feature, it is seen that the structure is effective in data-transmission applications because the characteristic of the data signal is stationary. On the other hand, since the correlation of a speech signal varies

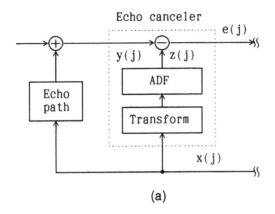

(a)

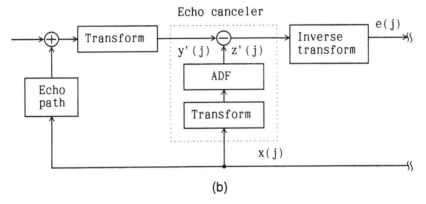

(b)

FIG. 8 Transform-domain echo cancelers: *a*, Type 1; *b*, Type 2.

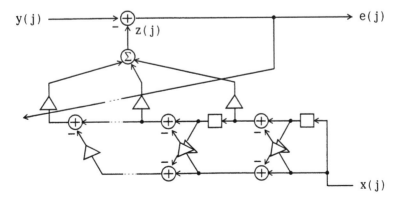

FIG. 9 Lattice structure echo canceler.

quite rapidly, the lattice structure might not be able to track the correlation variation for voice applications.

Frequency-Domain Structure

Figure 10-*a* shows a structure, of the category shown in Fig. 8-*b*, based on short-time frequency analysis. In this example, the input and reference signals are applied to one-quarter overlapped Hamming windows and FFTs. After that, a gradient-based algorithm is applied for each frequency component. Because the echo canceler for each component operates at the down-sampled rate, the required number of operations is reduced inversely proportional to the number of components. The performance of this structure is limited by the stop-band attenuation of the spectrum of the Hamming window. Figure 10-*b* is an example of the convergence characteristic of this structure.

By dividing the input and reference signals into several subbands and applying a gradient algorithm to each subband, a subband echo canceler is obtained. This canceler also belongs to the category shown in Fig. 8-*b*.

Figure 11 shows the general structure of the subband echo canceler. This structure contains band-pass filter banks, which can be implemented efficiently using polyphase digital filters and FFTs (18). To maintain high performance, quadrature mirror filter (QMF) (19) banks are used with the oversampling technique, in which a decimation factor smaller than the number of subbands is used.

Subband echo cancelers based on the gradient-adaptation algorithm are effective not only in reducing the computational complexity, but also in improving the convergence as compared with the full-band approach. For speech, because the signal in each subband becomes "more white" than the full-band signal and the normalization can be applied to each subband independently, the convergence rate is improved dramatically. Figure 12 shows an example of the performance of the subband echo canceler. The convergence rate shown is almost independent of the characteristics of the input signals. This result is obtained when the number of subbands is larger than 32 (the decimation factor larger than 16) (20).

A block LMS algorithm also can be implemented using FFTs as shown in Fig. 6. This structure sometimes is called a frequency-domain structure because the processing is realized after FFTs.

Memory-Type Echo Canceler

In the memory-type echo canceler, echo replicas corresponding to each sequence of the transmitted data are stored in memory (Fig. 13). Therefore, it is not necessary to generate the echo replica as in most ADF-type echo cancelers; it simply can be read from the memory using the data sequence as the address. To track the change of echo-path characteristics, the memory contents are updated according to the gradient-based algorithm.

The structure is suitable when the echo duration is short. This structure also

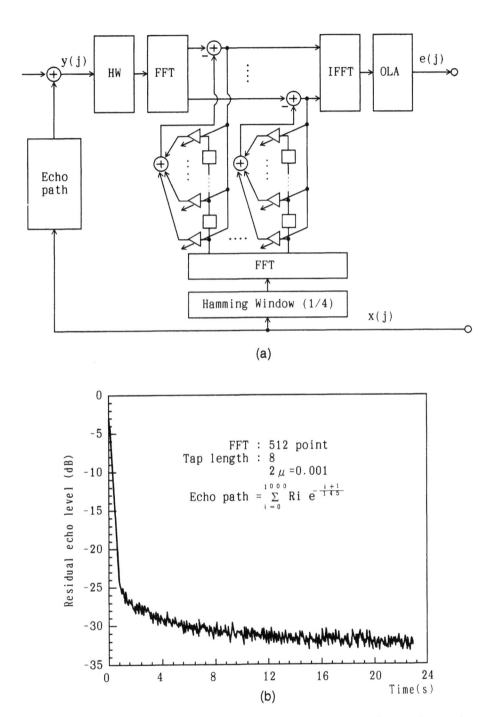

(a)

(b)

FIG. 10 Echo canceler based on short-time spectrum analysis (FFT = fast Fourier transform; IFFT = inverse fast Fourier transform; HW = Hamming window; OLA = overlap add): *a*, structure; *b*, example of performance.

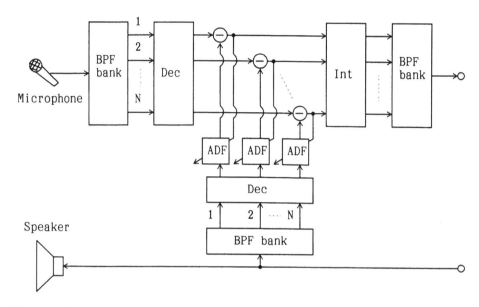

FIG. 11   Subband echo canceler (BPF = band-pass filter; Dec = decimation; Int = Interpolation; ADF = adaptive digital filter).

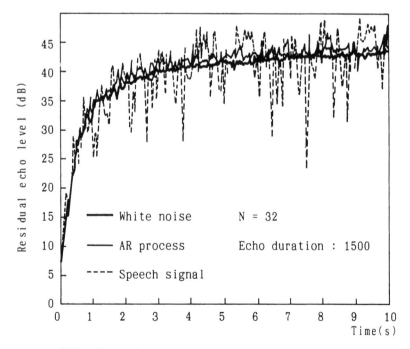

FIG. 12   Performance example of subband echo canceler.

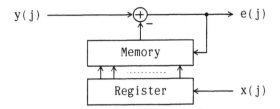

FIG. 13 Memory-type echo canceler.

has an advantage in that nonlinear effects in the echo path also can be included since table lookup is not restricted to linear functions.

Double-Talk Detection

For voice communication systems, double-talk control and detection is necessary to prevent erroneous adaptation when both parties on the line talk at the same time. This is because the other party's voice can be misinterpreted as an echo.

There are two methods for controlling double-talk. The first is to improve the adaptation algorithms and the second is to stop the adaptation during double-talk periods. Although the first method may provide better performance, it usually results in intensive computational load. Therefore, the second approach is used more often in practical applications. In this case, tap values are frozen during the double-talk periods to the values obtained just before the double-talk occurrence.

In the hybrid transformer of telephone circuits, the echo-path loss is larger than 6 dB even in the worst condition. By exploiting this feature, a method can be used in which the sending signal level is compared with the threshold level set at 6 dB below the receive-in signal level. The double-talk status is detected when the value of the send-in signal becomes larger than this threshold. Since the signal level measurement causes delay in the double-talk detection, degradation in echo cancellation performance results. To mitigate this problem, judging the double-talk status using instantaneous sample value is sometimes used (21).

For speaker/microphone systems, the methods described above cannot be used because the loss of the signal power in the echo path is not guaranteed. One method of detection is to place a second echo canceler for double-talk detection in parallel with the normal echo canceler. The second echo canceler is adapted even during double-talk periods, and thus the ERL of this echo canceler decreases. Using this feature, the double-talk condition can be detected. The convergence of this echo canceler must be fast enough, however, to prevent erroneous detection of double-talk in case the echo-path change occurs.

Examples of Implementation

Echo Canceler for Telephone Circuits

FIR implementation is used widely for telephone circuit application since the echo duration is rather short, on the order of 10–60 milliseconds (ms). For a sampling rate of 8 kilohertz (kHz), the required number of taps is below 500 and the NLMS algorithm can be applied conveniently for the adaptation.

The required computation for each tap is one multiplication and one addition for the convolution, and one multiplication and one addition for the recursive coefficient update of Eq. (23). Assuming a tap length of 512, the required multiplication rate becomes

$$512 \times 2 \times 8 \, \text{kHz} = 8.2 \times 10^6 \, \text{multiplications/second}$$

The required memory amounts to 1024 words. The processing requirements can be met easily with a single VLSI chip that has a multiplying time of 100 nanoseconds (ns).

To develop a practical echo canceler, many peripheral functions need to be implemented. Double-talk control is necessary to prevent erroneous adaptation. A center clipper is used to remove the low-level signal caused by line noise, CODEC (coder/decoder) quantization error, and so on that cannot be canceled by the echo canceler. This function is disabled when there is a send-in signal. A tone disabler also must be provided to facilitate full-duplex data transmission over voice channels. This disconnects the canceler by a tone indication signal sent prior to data transmission. These miscellaneous functions can be integrated on the same chip with the main canceler (22,23), or can be implemented separately with a general-purpose signal processor (24). A hardware configuration example for using a signal processor is shown in Fig. 14. Echo cancelers for three channels are commonly controlled by a single digital signal processor (DSP).

Echo Canceler for Teleconferencing

In teleconferencing applications, there is a long echo duration and a straightforward FIR implementation requires more than 4000 taps. This results in complex

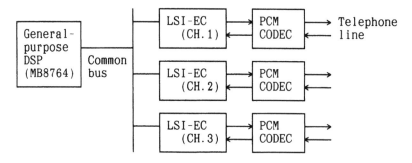

FIG. 14 Echo canceler for three-channel processing.

hardware and slow convergence. One approach to handle these problems is to use the subband echo canceler described in structures and algorithms section. An example of a two-band implementation in which band-pass filters have sharp cutoffs and the filter characteristics of each subband do not overlap is shown in Fig. 15 (25,26). This system covers an echo duration of 250 ms, with a speech bandwidth of 7 kHz. The total system requires 22 DSP chips. If the structure shown in Fig. 11 with QMF filter banks is used, more compact hardware can be implemented.

Echo Canceler for Full-Duplex Data Modem

Full-duplex data modems, such as the one specified in International Telegraph and Telephone Consultative Committee (CCITT) Recommendation V.32, require echo cancelers to be placed at the line interface where hybrids connect the modems to two-wire loops (27). A considerably higher level of echo cancellation is required compared to telephone applications. Since the received signal can be about 40 dB below the transmitted signal level, the echoes must be suppressed to about 70 dB to obtain the necessary signal-to-noise ratio of 30 dB. Therefore, high-precision (usually using more than 24 bits word length) is needed.

Far-end echo due to the far-end hybrid also must be taken into account. In this case, the echo delay depends on the entire length of the connection, which in some cases includes a satellite link. Therefore, two echo cancelers for the near-end and far-end echoes are required. Careful design is needed to prevent

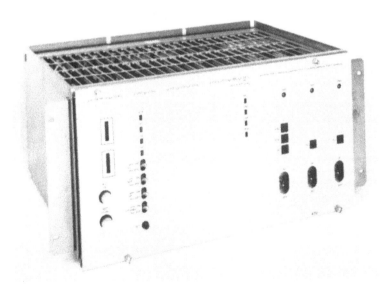

FIG. 15 Acoustic echo canceler using two-band split structure.

interference with automatic equalizers. Figure 16 shows an example of a modem configuration with full-duplex transmission.

For the adaptation algorithm, the block LMS algorithm is used. Echo cancelers of both the local and remote sides are adapted first by using training data. In the training phase, adaptations in a half-duplex mode are carried out first, followed by the adaptation in the full-duplex mode. After the training period, normal adaptation begins.

To achieve the adaptation correctly under the double-talk condition, the constant in Eq. (32) is set to a very small value and thus high-precision arithmetic is required. To make this condition easier to meet, the block LMS algorithm is used.

The echo cancelers for full-duplex data modem applications normally are implemented with high-precision DSPs together with the various data modulation-demodulation functions.

Echo Canceler for Integrated Services Digital Network Subscriber Transmission

Integrated Services Digital Network (ISDN) access requires 144-kilobits-per-second (Kbps) full-duplex data transmission over two-wire subscriber loops. The American National Standards Institute (ANSI) T1 committee has adopted a four-level 2B1Q code with a baud rate of 80 kHz for standard baseband transmission, and echo cancelers that operate with this scheme are required.

This application is characterized by its much higher data rate in comparison

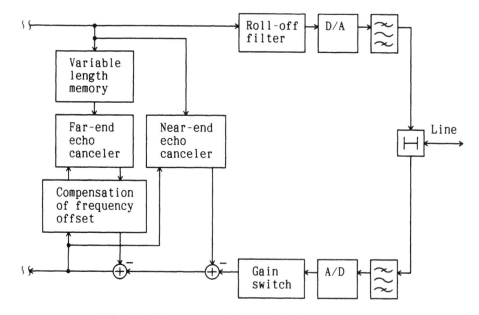

FIG. 16 Echo cancelers for full-duplex data transmission.

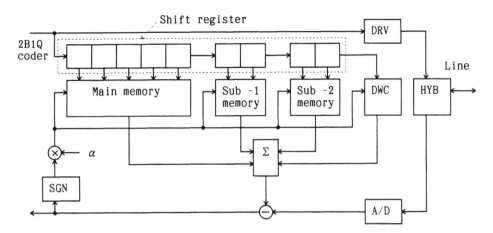

FIG. 17 Echo canceler for Integrated Services Digital Network subscriber transmission (DRV = driver; DWC = DC wandering compensator; HYB = hybrid; SGN = sign).

to voiceband applications. However, since the length of the subscriber loop is usually less than 7 kilometers (km), the echo duration is limited to the order of 100 microseconds (μs), or equivalent to about an eight-symbol length. Therefore, either FIR or memory-type echo cancelers are being adopted. The echo-cancellation requirement is similar to voiceband data applications and is about 70 dB. An example of the implementation of memory-type echo cancelers is shown in Fig. 17 (28).

Conclusions

The principle, structures, and applications of echo cancelers have been examined. The RLS algorithm provides the best performance, but the requirement for computational complexity is its main disadvantage. The LMS algorithm, on the other hand, can be realized with relatively low complexity and thus is useful in many practical applications. There are various kinds of structures for the adaptive filter; for the IIR structures, further research is required. Requirements for the algorithm are quite different according to the applications, and it is necessary to choose the optimum structure and algorithms.

References

1. Sondhi, M. M., and Presti, A. J., A Self-Adaptive Echo Canceler, *Bell Sys. Tech. J.*, 45:1851–1854 (1966).

2. Becker, F. K., and Rudin, H., Application of Automatic Transversal Filters to the Problem of Echo Suppression, *Bell Sys. Tech. J.*, 45:1847–1850 (1966).

3. Suyderhoud H. G., et al., Results and Analysis of an Adaptive Impulse Echo Canceler, *COMSAT Tech. Rev.*, 15(2):253 (1972).

4. Demytko, N., and English, K. S., Echo Cancelation on Time Variant Circuits, *Proc. IEEE*, 65:444–453 (March 1977).

5. Duttweiler, D. L., and Chen, Y. S., A Single-Chip VLSI Echo Canceler, *Bell Sys. Tech. J.*, 59(2):149–160 (1980).

6. Widrow, B., et al., Stationary and Nonstationary Learning Characteristics of LMS Adaptive Filters, *Proc. IEEE*, 64(8):1151–1162 (August 1976).

7. Nagumo, J., and Noda, A., A Learning Method for System Identification, *IEEE Trans.*, AC-12(3):282–287 (June 1967).

8. Itakura, H., and Nishikawa, Y., On Some Characteristics of An Echo Canceler Using a Learning Identification Algorithm, *Trans. IEICE*, J60-A(11):1015–1022 (November 1977).

9. Hoge, H., Analysis of an Adaptive Echo Canceler with Optimized Gradient Gain, *Siemens Res. Dev. Rep.*, 3:127–131 (1975).

10. Yamamoto, S., and Kitayama, S., An Adaptive Echo Canceler with Variable Step Gain Method, *Trans. IEICE*, E-65:1–8 (January 1982).

11. Clark, G. A., Mitra, S. K., and Parker, S. P., Block Implementation of Adaptive Digital Filters, *IEEE Trans.*, ASSP-29(3) (June 1981).

12. Ferrara, E. R., Fast Implementation of LMS Adaptive Filters, *IEEE Trans.*, AS-SP-28(4):474–475 (1980).

13. Gritton, C.W.K., and Lin, D. W., Echo Cancelation Algorithms, *IEEE ASSP Magazine*, 30–38 (April 1984).

14. Falconer, D., and Ljung, L., Application of Fast Kalman Estimation to Adaptive Equalization, *IEEE Trans.*, COM-26(10):1439–1446 (October 1978).

15. Shynk, J. J., Adaptive IIR Filtering, *IEEE ASSP Magazine*, 4–21 (April 1989).

16. Messerschmitt, D. G., Echo Cancelation in Speech and Data Transmission, *IEEE Trans.*, JSAC-2(2):283–297 (1984).

17. Friedlander, B., Lattice Filters for Adaptive Processing, *Proc. IEEE*, 70(8):829–867 (1982).

18. Bellanger, M., Bonnerot, G., and Coudreuse, M., Digital Filtering by Polyphase Network: Application to Sample-Rate Alteration and Filter Banks, *IEEE Trans. ASSP*, ASSP-24(2):109–114 (April 1976).

19. Vaidyanathan, P. P., Quadrature Mirror Filter Banks, M-Band Extensions and Perfect-Reconstruction Techniques, *IEEE ASSP Magazine*, 4–20 (July 1987).

20. Perez, H., Amano, F., and Unagami, S., On the New Structures of Sub-band Type Echo Canceler, *Proc. IEICE*, SP89-119 (February 1990).

21. Duttweiler, D. L., A Twelve-Channel Digital Echo Canceler, *IEEE Trans.*, COM-26(5):647–653 (May 1978).

22. Furuya N., et al., Custom VLSI Echo Canceler, *Proc. Nat'l. Conv. IEICE Japan*, 600 (October 1984).

23. Takahashi, K., Sakamoto, T., and Tokizawa, I., A Cascadable Echo Canceler, *IEEE ICC '82*, 5F3.1–3.5 (June 1982).

24. Kobayashi, N., et al., An Adaptive Transversal Filter VLSI, *Proc. ICASSP '86*, 1529–1532 (April 1986).

25. Sakai, Y., et al., Hardware Implementation of Echo Canceler by Using MB86232, *Proc. IEICE Symposium on DSP*, A4-4:225–230 (December 1988).

26. Yasukawa, H., Furukawa, I., and Ishiyama, Y., Characteristics of Acoustic Echo Canceler Using Sub-band Sampling and Decorrelation Methods, *Elect. Lett.*, 24(16):1039–1040 (August 1988).

27. International Telegraph and Telephone Consultative Committee, *A Family of*

Two-Wire, Duplex Modems Operating at Data Signaling Rates of Up to 9600 bps for Use on the General Switched Telephone Network and on Leased Telephone-Type Circuits, CCITT Recommendation V.32, ITU, Geneva, 1989.

28. Fukuda, M., et al., An Approach to LSI Implementation of A 2B1Q Coded Echo Canceler for ISDN Subscriber Loop Transmission, *Globecom '88* (November 1988).

KAZUO MURANO
FUMIO AMANO

Echo Suppressors

Echo is a phenomenon that can be pleasing or irritating, depending on the circumstances in which it is experienced. We have all had the pleasure of shouting and having our voice reflected and returned to us a short time later. Few of us, however, have experienced the difficulty of holding a conversation in the presence of echo; those who have, found it hard to communicate and, what is more, the level of difficulty seemed to increase as the delay time of the echo increased. Echo time is a function of round-trip distance traveled and the medium in which the signal is propagated. Sound travels at approximately 750 miles per hour in air and is a function of the temperature of the air. Sound in the form of electrical current travels much faster as it traverses the electrical conductors. It travels still faster when converted into light waves and is sent over fibers to its destination. The fact that the speed is so much faster when the sound is converted to electrical current or light impulses means that the signal must travel much further to be of consequence with respect to echo.

As sound energy, echo is caused by the reflection of the sound waves striking a nonabsorbing surface and bouncing back; electrical echo is caused by the electrical current striking an impedance mismatch (i.e., the electrical equivalent to the reflecting surface of sound). The telephone network has many of these impedance irregularities and the industry devotes large amounts of resources to reduce the conditions for echo as well as compensating for echo when it is present.

Echo only exists where there is an opportunity for the electrical signal to travel in both directions. Although the typical long-distance telephone circuits do not possess this property since they are unidirectional, at some point in the network there is a conversion from unidirectional to bidirectional conductors. Since it takes two wires to complete a circuit path, then for both directions in the typical long-distance circuit, four wires are required (Fig. 1-*a*). In a bidirectional circuit, the signals are permitted to travel in both directions and only require two wires (Fig. 1-*b*), thus achieving cost and space savings. For this reason, the terms *four-wire circuit* and *two-wire circuit* are used to describe a configuration quality of the network.

To convert a four-wire circuit to a two-wire circuit, a transformer-type device called a *hybrid* is employed. The hybrid couples the two paths of the four-wire circuit to the bidirectional path of the two-wire circuit (Fig. 1-*c*). The hybrid was designed to make this transition with a minimum of impedance mismatch, or unbalancing. A technique is used to balance the hybrid for the purpose of minimizing echo. This is difficult to achieve because the balance is dependent on the impedance of the circuits connected to the hybrid. During use, these circuits are changed as the network, together with the telephone switch and other elements, are reconfigured to complete a customer connection.

The use of the telephone network is not limited to voice communication. Data transmission has been a significant component of the information movement. However, there are two characteristics of data transmission that make it immune to effects of echo conditions: the data is transmitted in only one direction at a time, or the data is transmitted at specific different frequencies in each direction. In fact, as we will see, certain properties of echo suppressors reduce

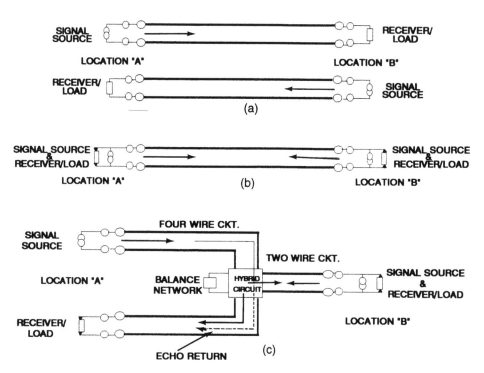

FIG. 1   Circuit paths: *a*, four-wire circuit; *b*, two-wire circuit; *c*, two-wire/four-wire conversion.

the accuracy of data transmission and therefore they are "disabled" for data communication.

The tolerance for echo depends on how loud it is as well as the echo time. The term *delay* is used to describe the time for a signal to reach a distant point. *Round-trip delay* is used to describe the elapsed time for a signal to travel to a point of reflection and then back; this is the *echo time*. Round-trip delay that produces echoes that are approximately the time between syllables is the point at which most people find it difficult to hold a conversation.

Echo times can range from zero seconds to six-tenths of a second on a single satellite. There is a point at which the round-trip delay, even without echo, causes difficulty in holding a conversation. At any instant, because of delay time, it cannot be determined if the person on the other end is talking. On a transmission circuit with unusually long delay, it is not uncommon for someone to start talking and then stop when the voice from the other end reaches them. At the other end, a similar event is taking place, resulting in the effect that neither party knows when to start or stop talking. It takes on the illusion of a battle of interruption when, in fact, it is an attempt for both parties to be polite and let the other person speak.

Solving the Echo Problem

There are several alternatives to reduce the echo on the network. One method is to design the network with a configuration that will not allow two-way transmission on a facility path. A second is to minimize the delay time on the circuit path, and a third is to compensate for the echo generated. All three of these merit serious considerations, but each has limitations. In practice, each method is used to some extent. Building a network configuration that is totally four wire in which signals are not allowed to travel in both directions would be very expensive. Yet, that design is used over the longest of circuits. This arrangement minimizes the possibilities of multiple echoes and permits other means to be employed to improve the quality of the connection further. The idea of reducing the delay time seems unlikely since the signals are traveling nearly at the speed of light. However, we can reduce the delay time by selecting the path and facility type the signal must take. In this regard, the use of satellite circuits is designed to attempt to employ only one satellite hop per connection. The reason for this is not so much to control echo, but to minimize the effect in which the delay time alone makes it difficult to hold a conversation. However, it would be harder to achieve a compensation effect using echo suppressors if the delay time was greater than that produced by one satellite. The third alternative, compensation of the echo, is achieved by reducing the level of the echo after it has been generated. This is the purpose of the echo suppressor.

An additional method to achieve compensation is by use of the echo canceler. This is the new and current technology. For much of the world's telephone networks, it has replaced the echo suppressor. The method of compensation using the echo canceler is to insert into the return path, energy that is equal and opposite to the incoming signal at just the correct level to "cancel" the echo.

The echo suppressor, on the other hand, achieves its compensation by introducing loss into the echo signal path with accurate control of the time the loss is inserted. Subjective testing was used to determine the amount of loss required to satisfy most people. This loss varied between 30 to 50 decibels (dB). The loss value was dependent on the amount of round-trip delay time. For shorter delays, up to about 100 milliseconds (ms), 30 dB was satisfactory. Beyond the 100-ms level and up to the 600-ms level (the round-trip delay time of a synchronous satellite), 50 dB was required. This variation in loss requirements led to different methods of compensation. When the round-trip delay was relatively short, the introduction of continuous loss was possible since the amount of loss required to achieve acceptable echo reduction did not reduce the principal signal significantly. As the delay time increased, continuous loss was not possible so a means of introducing loss at select times and conditions was a principle of the echo suppressor design.

The basic operation of echo suppressors is to detect a voice signal from one end of the transmission circuit and at that time place significant loss in the other direction of that circuit. Therefore, by the nature of their operation, it can be seen that echo suppressors only can be deployed in four-wire systems, otherwise the echo suppressor could not distinguish from which end the voice energy was produced. By this single action, the echo suppressor in effect reduces the transmission path to a single direction. This operation alone is not sufficient to satisfy all the conditions that are required for maintaining a normal conversation. For instance, a listener would have to wait for the talker on the other end to finish speaking before he or she could talk back. It would be an operation common in radio communication; the person talking and holding down the microphone button in the "ON" position signifies he or she is finished by saying the word "over."

To correct for this less desirable operation, a feature was designed that permitted the listener to interrupt by overcoming the suppressor action and thereby establishing a talking path to the talker. This feature is known as *break-in*. The breakin action is initiated when the level of the listener's voice is relatively higher than the talker's voice. A differential comparator within the echo suppressor compares the signals in both transmission path directions and develops a reference voltage that maintains the suppression action. Speech energy from an interrupting talker easily can exceed the reference voltage and cause breakin to occur. For a short instant, the transmission paths in both directions are open. This characteristic of the breakin feature is called *double-talking*. This condition is permitted to exist for only a short interval without suppression returning; otherwise, unwanted echo would be present. If the listener continued to talk, then the suppression or loss would be inserted in the opposite direction so that the returning echo will not reach the original listener, who has interrupted and now is talking.

In each of the operations discussed, suppression and breakin, there are timing restraints critical for the purpose of establishing a talking environment that simulates natural conversation. In each case, there are periods designed into the echo suppressor that must pass before the operation takes place. The terms *pickup time* and *holdover time* are used to describe these parameters. *Pickup time* is defined as the time it takes for the operation to take place after the echo suppressor first detects sufficient energy on the line. The main purpose of the pickup time is to reduce the incidence of spurious energy on the line, operating the echo suppressor. *Holdover time* is the time it takes to release the operation after the echo suppressor detects that energy has been removed from the line. The main purpose of the holdover time is to hold the echo suppressor in an operated condition during short breaks in the speech pattern of the talker as well as to compensate for the delay of the energy reaching the echo suppressor.

These echo suppressors work reasonably well, but there are occasions when speech may be "clipped" as a result of the echo suppressor action. This most likely occurs during the initial syllables spoken by an interrupting listener. The extent of clipping will worsen as the round-trip delay of the circuit increases.

At one time, echo suppressors were installed at only one end of the circuit, at which point suppression was placed selectively, determined by the direction of the speech. This required the round-trip delay to be considered in setting the operating parameters of the echo suppressor. To remedy this, "split" echo suppressors were developed. A split echo suppressor is installed at each end of the circuit, and is able to insert suppression only in the transmit side of the circuit. This effectively eliminates the circuit delay time as a concern since suppression is placed at the location where the echo is generated.

To differentiate between the two types of echo suppressors, the word *full* was applied to the original type of echo suppressor, in which suppression was placed in either side from a single end (see Figs. 2-*a* and 2-*b*). The full echo suppressors generally are used on circuits of less than 1500 miles, while the split suppressors are used for longer-delay systems and satellite facilities.

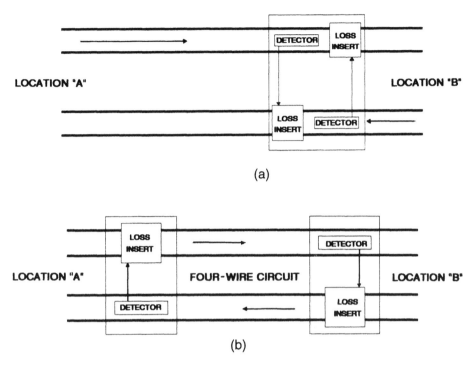

FIG. 2 Echo suppressors: *a*, full echo suppressor; *b*, split echo suppressor.

Disabling, mentioned above, is an additional operation of the echo suppressor. It places the echo suppressor into a mode that does not permit it to apply suppression. As discussed above, this is used specifically during the time data are to be transmitted. The usefulness of this function is to prevent clipping of the data content, which can occur if operating normally. The echo suppressor is disabled by applying a specific frequency tone, approximately 2100 hertz (Hz), in the absence of any other frequency signal. After a pickup time has expired, the echo suppressor will go into the disabling mode and remain there as long as any signal remains on the line.

As with suppression and breakin, the disabling function also has a pickup and holdover time. The pickup time guards against spurious tones causing the echo suppressor to go into the disabling mode, while the holdover time permits breaks in the transmission of tone on the line without dropping the disabling operation. When the data stream does drop for a period greater than the holdover time, the echo suppressor will return to its normal operating mode.

Echo Suppressor Performance

Echo suppressors have been used in the network for over 30 years and during that period they have performed very well. The tolerances of the requirements for the echo suppressor operation are not stringent. For this reason, maintaining proper echo suppressor operation has not been difficult from that perspective. When they work properly, there are no distractions during a conversation. When they do not, there are many indications that could be experienced individually or in combination. Some of the common offenders are excessive loss, speech clipping, and, of course, echo.

For a number of years, echo suppressors were tested in two ways. The principal method was to make a call over the circuit with the suspected faulty echo suppressor and hold a conversation with someone at the other end. This method was used when there were customer complaints. The problem with this procedure was that the complaints could not be associated with any specific telephone long-distance circuit since the circuits are selected as the calls are placed. Someone in California calling New Jersey, for instance, may experience echo on the line. When they call to make the complaint, they have dropped the faulty circuit and all that is known to the repair facility is the locations of the two terminating ends. The call could have been routed over many different possible combinations of circuit facilities. When many complaints were received between two general locations, a tedious procedure was used to identify the circuit or circuits with the faulty echo suppressor operation.

The second method used to maintain echo suppressors was to isolate them routinely from the circuit and run tests on their operation. These tests took a relatively long time and the results generally did not uncover faulty echo suppressors. This is reasonable since most echo suppressors worked as expected. A second problem with this procedure for testing was that it did not test the echo suppressor in the environment of the circuit in which it was installed. The operation of the echo suppressor was influenced greatly by the circuit parameters. For instance, excessive noise or gain on one side (one direction) of the circuit could bias the echo suppressor operation.

A system to test echo suppressors quickly in their circuit environment was developed. This allowed the testing to associate the circuit parameters and the echo suppressor parameters and thus achieve a more realistic testing capability. The result of the introduction of this test system was to uncover many faulty echo suppressors and thus improve the overall quality of the network.

Echo suppressors are being replaced slowly by echo cancelers. In the U.S. domestic network, echo suppressors all have been replaced. In Europe, there is now a predominance of echo cancelers. In other parts of the world, echo suppressors are still the only means of echo compensation. As new technology replaces old, and augmentation of the current network capacity takes place, the usefulness of echo suppressors will continue to diminish.

Bibliography

Rugo, J. J., Checking Echo Suppressors, *Bell Lab. Record*, 80–84 (March 1970).
Unrue, J. R., Jr., Controlling Echo in the Bell System, *Bell Lab. Record*, 233–238 (August 1969).

JOHN J. RUGO

Economic Theories of Regulation in Telecommunications

General Theories

In most countries of the world, telecommunications systems are owned and operated by a monopolistic government agency. In North America (and increasingly elsewhere), networks are being provided by private companies operating subject to close government regulation. Indeed, few industries are supervised as closely by government as telecommunications, raising questions as to why such control is being exercised.

There are two broad categories of explanations for a governmental regulation of markets. The first, known as *public-interest theory*, argues that regulation is required to protect the public from various forms of harm: monopolies that overcharge, unsafe products, unqualified professionals, chemicals that are pollutants, securities that are capable of being manipulated, unsound banks, and so forth. This view of regulation reconciles a market economy—private economic entities freely engaging in economic activities—with governmental control and intervention. A number of causes for "market failures" are offered, among them natural monopoly (i.e., the inability of any competing provider to survive), collusion among rival firms, such negative neighborhood effects as pollution, or informational asymmetries. Public-interest theorists believe that the function of regulation is to restore or protect the market. For example, in the financial securities field, the provision of accurate information is to be assured by regulation-imposed disclosure requirements. Regulation's function, in that view, is largely to ensure results that match Pareto efficiency, a condition in which no one could be made better off without someone being made worse off.

Proponents of the public-interest view of regulation were typically political centrists or liberals. In the intellectual field, they included the empiricist Mitchell (1), the reformer Commons (who introduced public utility regulation, the nation's first, to Wisconsin's governor La Follette) (2), and the theorist of social overhead costs, J. M. Clark (3). In the political sphere, their ideas translated into the creation of regulatory structures and institutions. Where problems persisted despite regulation, it was viewed as a challenge to devise and administer regulatory policies more effectively. Regulation became an area of specialization, particularly in the field of public utilities (4,5), with a body of increasingly sophisticated economic analyses such as Kahn's (6).

Advocates of public-interest theory recognized the existence of political expediency in the actual exercise of regulation. As a consequence, they usually advocated independent expert regulatory agencies with judicial, legislative, and executive powers. The classic institutional model was the Interstate Commerce Commission, established in 1887 on the federal level to regulate railroads. Similar commissions were created to regulate such vital infrastructure services as transportation and public utilities, including telecommunications, under the dual impetus of prairie populism and business efforts to avoid European-style nationalizations. In the United States, an intermediate path—privately-owned but regulated utilities—was followed. But even such moderate policy came under severe political challenge as an interference in property rights, and was sanctioned by the courts with reluctance in such landmark cases as *Munn v. Illinois* (regulatory powers over private businesses "affected with the public interest") (7) and *Nebbia v. New York* ("the state may regulate a business in any of its aspects") (8).

Key elements of utility regulation included the limitation of monopolistic pricing behavior by the imposition of price restrictions or the control of its profitability through rate-of-return regulation. This inevitably also led to regulation of service quality, investments, expenditures, and rate structure, since all of these are interrelated. Perhaps most importantly, utility regulation led to restriction of entry by the requirement of licenses and franchises, which tended to establish and protect monopolies and divide markets.

In time, the expectation that regulatory mechanisms would restore efficiency and protect the public proved disappointing to many; some agencies were inefficient, others misguided, and still others biased. Their independence often was illusory (9). From this, public-interest theorists drew the conclusion that better and more scientific regulation was necessary. Others attacked this view from various directions. The left, viewing government as a tool of business interests, saw regulation as part of the support structure of the capitalist order, for example, by stabilizing markets otherwise subject to ruinous competition, by deflecting popular discontent through the illusion of control, or by protecting monopolies and oligopolies from competitive entry (10,11). In that view, even in those rare cases in which regulation might have been initiated with the intention of protecting the public, it was destined to be captured soon by the powerful subjects of the regulation.

Free-market advocates mounted an intellectual and political challenge, arguing that regulation was costly in operation, distortive in terms of resource allocations, and usually captured by various industry and political-interest groups that used the process to obtain protection and redistribution in their favor. They took up the criticism of the progressive Horace Gray that monopoly was the creation of public policy (12), and charged that the public-interest approach naively assumed the political-regulatory process to be a black box into which good intentions were put and the public interest emerged.

Instead, free-market advocates offered, as a second major approach to the explanation of regulation, an analysis that factored in the motives of politicians and regulators and the multiple constituencies with which they must deal. This position was advanced by such members of the Chicago school as Simons (13), Friedman (14), Stigler (15), Peltzman (16), and Becker (17), with antecedents provided by the Austrian classic liberals. The essence of this approach is that regulators and politicians weigh the benefits and costs of various courses of action in a political framework in which the attainment of a voting majority determines success. In its broadest interpretation, this approach emphasizes a balancing of interest-group strengths at the margin, with the outcome determined by the stake that the various groups have in the outcome and the efficiency with which they can influence the regulatory process. Other interpretations were offered by McCraw (18), Olson (19), Owen and Braeutigam (20), Noll (21), and Wilson (22).

An illustration for this analysis is the Coase theorem (23). It postulates that free transactions between individuals will result in the economically most efficient type of activity, regardless of the regulatory rule. An example is a railroad that emits sparks that burn grazing land adjoining the tracks, and where the sparks can be prevented by appropriate guard technology, though at a cost. What should be the regulatory rule? A simplistic public-interest view of regulation would argue that the railroad, as a business, should be prevented from harming farmers, the people. A more sophisticated public-interest approach would take into account the relative cost of the protective technology and its impact on transportation costs, and weigh them against the probability and severity of fires and their harmful effects on things such as food production.

The determination of this question, involving issues of fact, probability, and technology, would be lodged in a regulatory body, which in time would come up with appropriate rules and enforcement mechanisms. But, according to Coase, that regulatory effort would be immaterial to the actual outcome (protective guards or the periodic burning of the grazing land). Where the railroad is liable for damages, if it is cheaper for it to keep causing fires and pay off the farmers' damages rather than install the guard technology, it will do so. Conversely, even if the railroad is under no legal obligation to install the guard, it would do so anyway if the harm to the farmers is such that they would pay the railroad for the installation. Thus, underlying economies prevail over whatever the regulatory rule is, provided the parties can transact among themselves.

This is not to say that regulatory rules have no impact. However, the impact is not on the outcome (e.g., spark guards) but on the distribution of wealth. In one case, the railroad either is free to emit sparks or is being paid not to do so. In the other, the farmers either can graze or are being paid not to do so. This simple analysis provided one of the foundations for the view that the function of regulation is not so much to affect outcomes but to distribute wealth. This theme was developed further by Posner in related literature on law and economics (24), and by Tullock with his analysis of interest-group monopoly "rent seeking" (25). Coase discussed how the regulatory rule should be set. He concluded that it should be imposed on the lowest-cost avoider because this would reduce subsequent transaction costs. In that, he is close to another strand of literature, that of Williamson's transaction costs and organizations (26). But for all of Coase's intellectual elegance, for which he received a Nobel Prize, essentially his theory is applicable only to situations in which the various parties can organize and transact easily, and there is no free riding. As soon as these conditions are not met, such as in the case of air pollution, group representation shifts to the political process and thus to government regulation. Here, an inherent practical as well as theoretical problem with regulation is that government must pursue numerous objectives with only a limited number of policy variables available, thus leading to various compromises and contradictions—a problem formalized by Kenneth Arrow's impossibility theorem, for which he, too, received a Nobel Prize (27).

Application to Telecommunications

Regulatory developments followed the ascendancy of various schools of economic thought, from the progressives to the institutionalists to the free-market liberals of the Chicago school. In the American telecommunications industry, the trust-busting public-interest sentiment of the Progressive Era led to the 1913 Kingsbury Commitment, a deal between AT&T and the government to contain AT&T's expansion (28). Institutionalist economists and lawyers provided the basis for state utility regulation, as well as for the Communications Act of 1934. As part of the public-interest orientation of regulation, cross-subsidies were built into the system to help achieve universal service and political acceptance (29,30). But at the same time, AT&T's monopoly was being protected from competitors.

Following the 1934 Communication Act, concerns were raised in a 1939 Federal Communications Commission (FCC) study over profit shifting between equipment and services in the regulation of AT&T's rates. After World War II, an antitrust suit based on classic public-interest principles was brought, ending in 1956 with a consent decree that kept AT&T intact and contained. But the seeds of instability were sowed at the same time by technology, entrepreneurism, and economic thinking. The *Above 890* decision allowed microwave competition from private-line services (31). This led to such new general microwave services as those of MCI (32) and eventual public switched voice offerings (33). At about the same time, the *Carterfone* case opened the door to equipment competition (34).

In the early 1970s, disenchantment with the performance of monopolies led to a convergence of schools of thought (35). The procompetitive Chicago approach (exemplified by antitrust chief Baxter) (36), joined with the antimonopoly approach of the public-interest advocates (exemplified by Judge Greene) (37) to provide the theory behind a government antitrust suit against AT&T (38). A competing body of economic theory was developed at its Bell Laboratories and at Princeton University to defend the legitimacy of its monopoly. It led to a general reappraisal of industrial organization analysis. One such theory, that of contestability, stated that the threat of entry by a new competitor in a monopolized market would create the same efficiencies as actual competition (39). It also analyzed the sustainability of multiproduct monopolists.

The emergence of competition did not spell an end to regulation, but formed a complex system of telecommunications (40). It became part competitive, part monopolistic, and more complex to administer than the simpler one-company system. Hence, the near future of telecommunications regulation appears to be a complicated web of partial regulations. This has led to efforts to provide theories for partially regulated firms. Whereas the old regulatory analyses emphasized the control of rates and profits, with theories to deal with these issues (41), the new policy agenda is likely to concentrate on the problems of interconnection in a network of networks and technical standards (42), content flows, quality of service, and privacy protection. New theories, such as Noam's (43), no doubt will surface to explain this new orientation of government regulation.

Regulatory Theories in Other Countries

Germany

In Germany, the dominant view on government control was rooted in a variant of the public-interest or social economy theory (*Gemeinwirtschaftlehre*), which sought to imbue private enterprises oriented toward the fulfillment of public tasks. Schmoller argued for embedding economic analysis in its social setting (44). Wagner took an ethical approach to social economy (45) that saw state intervention as morally superior to the market described by Smith (46), and postulated a law of inevitable growth in the social economic sector. Sax, in contrast, advocated private but regulated infrastructure enterprises that stressed marginal utility and presaged much current economic theory (47). Marx and Engels earlier had developed broad theories of the inevitability of socializing the means of production (48). In the political arena, their views were softened later by social democratic "revisionists," but those, too, advocated publicly owned infrastructure monopolies.

On the other hand, classic liberals, especially those of the Austrian school such as Hayek and Mises, argued for competition as an autonomous, decentralized, and efficient discovery procedure. Another Austrian economist, Schumpeter, pointed out the "creative destruction" process of capitalist development, which led to monopolistic state enterprises (49).

Broad political and intellectual support was given to the state telecommunications monopoly, which remained stable for a century. Eventually, reform proposals began to surface in Germany from market-oriented authors (50–52). In time, these notions led to a restructuring of the state monopoly into a semi-autonomous operating company, publicly owned but regulated.

Great Britain

Adam Smith synthesized the beliefs of French physiocrats and English classicists to develop the science of economic inquiry, strongly based on free-market principles, though conceding the need for controls against collusion. The market analysis was taken further into the realm of political economy by Ricardo (53), Mill (54), and Malthus (55). They were followed by utopian socialist and reformist Fabian movements, including Wells (56), and the Webbs (57). They advocated the state ownership of infrastructure for natural monopolies, and influenced the thinking of the Labour Party for a long time.

British telegraph companies were nationalized in 1868, as were telephone companies by 1911. For most of the century, the British Post Office Department set telecommunications policy and goals according to the demands of the government in power. After a long period of stability, institutional stagnation, and cautious technological progress, in the early 1980s various voices argued for a market-oriented approach to telecommunications regulation (58,59). Soon, the conservative Thatcher government, pursuing a program of privatization, turned to telecommunications and privatized British Telecom. It also led to permitting the entry of the rival long-distance carrier, Mercury, and the establishment of value-added services networks.

France

French thinking on the role of the state in guiding the economy always was torn between classically liberal and statist traditions, with the latter usually dominant. Classical liberals included, in the 18th century, Turgot (who opposed the earlier mercantilism of Cardinals Richelieu and Mazarin) (60), Say (61), Du Pont de Nemours (62) and Quesnay (63) (both physiocrats who emphasized natural law and property rights), Condillac (64) (who pioneered the applications of utility theory in France), Sismondi (65), and Cournot (66) (whose mathematical models for pricing under monopoly and duopoly conditions still are used today). The physiocrats discarded the mercantilist belief that wealth and its increase were due to exchange. By valuing accumulation instead, they laid the foundations for the modern French regulationists. In contrast, advocates of a strong state involvement in the 19th century were Navier (who presented early cost-benefit analyses of public goods), Minard, and Depuit (and later utopian socialists and Marxists), as well as economic nationalists, many of whom were graduates of the prestigious state engineering schools that long dominated the discipline of economics in France.

The strong tradition of state industrial policy and government ownership yielded a close involvement between operational and regulatory functions. This led to the development, beginning in the 1960s, of the French (or Paris) school of economic regulation (67), which sought to explain accumulation in terms of the role of the state in supporting industry. The regulationists saw a structure or network of capitalist institutions forming to respond to crises in national economies. Their theories were countered by the "new conservatives" such as Glucksmann, Henri-Levy, and Raymond Aron. The technocratic position was taken by the statists Nora and Minc, who advocated a state-led computerization of French society under the banner of *telematique* (68).

With the advent of a socialist government in 1981, the French telecommunications equipment and computer industries were nationalized. In time, some segments of the market were privatized and opened to competition, but the monopoly remained fairly secure. However, it, like similar systems, was being challenged by the European Commission in Brussels, which represented another powerful trend, that of European integration, in which there was no long-term room for national monopolies.

Japan

After the 1860s, the Mejii oligarchy, and later the militarists, controlled economic development in Japan. Their economic interventionist views were shared by the National Socialists, a group of prewar bureaucrats led by Kishi. After the war, Marxist traditions were in vogue following Katayama (69). These soon faded in the debate between industrial policy advocated by Arima and the

Ricardian comparative advantage arguments of Ichimada. The former theory won out, leading to a stronger mandate for governmental economic development policies.

For a long time, these political, economic, and intellectual forces supported the monopoly system. But in the 1980s, under the leadership of the MITI economics ministry and segments of the private sector, and against the opposition of telecommunications traditionalists, Japanese telecommunications were moved from a rigid monopoly to an open, competitive market (70). This approach combined a free-market orientation with the industrial policy goal of strengthening Japan's international competitiveness.

Outlook

As similar evolutions began to take place in other industrialized countries, it was recognized that they were following broader economic and technological forces (71). The growth of technological and operational alternatives, in conjunction with the economics of network growth and the merging of technologies, undercut the economies of scale and scope once offered by the centralized network. As governments expanded networks toward universal service, technological developments created opportunities for large users to exit the telecommunications network to obtain specialized services. A phenomenal growth in user demand for services resulted, and many commercial groups began interlinking through telecommunications, a force that in turn was based on the shift toward a service-based economy. These forces led to a transformation of institutions that had been stable through most of the century. And while regulation sometimes took a leading role in advancing change, the opposite was more often the case. Typically, traditional regulatory institutions were protective of the traditional status quo. In many instances, the rhetoric of public-interest theory was used to rationalize actions more compatible with the less-idealistic free-market view of regulatory reality.

Yet, ironically, regulatory institutions emerged from the changes of deregulation to become more important than ever before. They now hold a more genuine role in refereeing among contesting forces, as opposed to the past, when they were merely an appendix to the giant operating monopolies. In consequence, one can expect the need for regulatory analysis and theory to be more important than ever, and in need of exploration as well as leading a sector in rapid transformation.

References

1. Mitchell, Wesley C., *The Backward Art of Spending Money*, McGraw-Hill, New York, 1937.
2. Commons, John R., *Distribution of Wealth*, Macmillan, New York, 1893.

3. Clark, John M., *Competition as a Dynamic Process*, Brookings Institute, Washington, DC, 1961.

4. Bonbright, James, Danielsen, Albert, and Kamerschen, David, *Principles of Public Utility Rates*, 2d ed., Public Utilities Reports, Arlington, VA, 1988.

5. Trebing, Harry, Regulation of Industry: An Institutionalist Approach, *J. Economic Issues*, 21(4):1707–1737 (December 1987).

6. Kahn, Alfred, *The Economics of Regulation*, Vols. 1 and 2, MIT Press, Cambridge, MA, 1970.

7. *Munn v. Illinois*, U.S. Supreme Court, 94 U.S. 113 (1877).

8. *Nebbia v. New York*, U.S. Supreme Court, 291 U.S. 502 (1934).

9. Bernstein, Marver, *Regulating Business by Independent Commission*, Princeton University Press, Princeton, NJ, 1955.

10. Kolko, Gabriel, *The Triumph of Conservatism: A Reinterpretation of American History 1900–1916*, Free Press, New York, 1963.

11. Baran, Paul, and Sweezy, Paul, *Monopoly Capital*, Modern Reader Press, New York, 1966.

12. Gray, Horace M., The Passing of the Public Utility Concept, reprinted in the American Economic Association, *Readings in the Social Control of Industry*, Blakiston Company, Philadelphia, 1942.

13. Simons, Henry C., *Economic Policy for a Free Society*, University of Chicago Press, Chicago, 1948.

14. Friedman, Milton, *A Program for Monetary Stability*, New York, 1960.

15. Stigler, George, The Theory of Economic Regulation, *Bell J. Economics and Management Science*, 2:3–21 (1971).

16. Peltzman, S., Toward a More General Theory of Regulation, *J. Law and Economics*, 19:211–240 (August 1976).

17. Becker, G., A Theory of Competition among Pressure Groups, *Quarterly J. Economics*, 96:371–400 (1983).

18. McCraw, Thomas, *Prophets of Regulation*, Belknap, Cambridge, MA, 1984.

19. Olson, Mancur, *The Logic of Collective Action*, Harvard University Press, Cambridge, MA, 1965.

20. Owen, Bruce, and Braeutigam, Ronald, *The Regulation Game: Strategic Use of the Administrative Process*, Ballinger, Cambridge, MA, 1978.

21. Noll, Roger, *Reforming Regulation*, Brookings Institution, Washington, DC, 1971.

22. Wilson, James Q., *The Politics of Regulation*, Harper and Row, New York, 1983.

23. Coase, R. H., The Problem of Social Cost, *J. Law and Economics*, (1960).

24. Posner, Richard A., Theories of Economic Regulation, *Bell J. Economics and Management Science*, 5:337–352 (Autumn 1974).

25. Tullock, Gordon. In: *Toward a Theory of the Rent-Seeking Society*, (James M. Buchanan, Robert Tollison, and Gordon Tullock, eds.), Texas A&M, College Station, TX, 1980.

26. Williamson, Oliver, *Markets and Hierarchies, Analysis and Antitrust Implications: A Study in the Economics of Internal Organization*, Free Press, New York, 1975.

27. Arrow, Kenneth, *Social Choice and Individual Values*, John Wiley and Sons, New York, 1951.

28. Brock, G. W., *The Telecommunications Industry*, Harvard University Press, Cambridge, MA, 1981.

29. Wenders, John T., and Egan, Bruce L., The Implications of Economic Efficiency for U.S. Telecommunications Policy, *Telecommunications Policy*, 10:33–40 (March 1986).

30. Wenders, John T., The Economic Theory of Regulation and the U.S. Telecommunications Industry, *Telecommunications Policy*, 12:16–26 (March 1988).
31. FCC, Report and Order, FCC Docket 11866, Above 890 Mc., July 29, 1959, 27 FCC 359 at 403-13, recon. denied, 29 FCC, 825 (1960).
32. FCC, First Report and Order, FCC Docket 18920, Specialized Common Carriers, June 3, 1971, 29 FCC 2d 870, aff'd sub nom. Wash. Util. & Trans. Commission v. FCC, 513 F. 2d 1142 (9th Cir. 1975), cert. denied, 423 U.S. 836 (1975).
33. FCC, Decision, FCC Docket 20640, June 30, 1976, 60 FCC 2d 25 at 42-3. Judge Skelly Wright, Decision, July 28, 1977, MCI v. FCC, No. 75-1635, 561 F. 2d 365 at 374, (D.C. Cir.), cert. denied 434 U.S. 1040 (1978).
34. FCC, Decision, FCC Docket 16942, Carterfone, adopted June 26, 1968, 13 FCC, 2d 420.
35. Temin, Peter, and Galambos, L., *The Fall of the Bell System*, Cambridge University Press, Cambridge, England, 1987.
36. *Telecommunications Competition and Deregulation Act of 1981* (S. 898).
37. *United States v. AT&T, Civil Action 74-1698*, 1974.
38. Wenders, John T., On Modifying the MFJ, *Telecommunications Policy*, 11:243–246 (September 1987).
39. Baumol, William, Panzar, John, and Willig, Robert, *Contestable Markets and the Theory of Industry Structure*, Harcourt Brace Jovanovich, New York, 1982.
40. Huber, Peter W., *The Geodesic Network, 1987 Report on Competition in the Telephone Industry*, Antitrust Division, U.S. Department of Justice, Washington, DC, 1987.
41. Averch, Harvey, and Johnson, Leland, Behavior of the Firm under Regulatory Constraint, *Amer. Econ. Rev.*, 52:1052–1069 (1962).
42. Besen, Stanley M., and Saloner, Garth, The Economics of Telecommunications Standards. In: *Changing the Rules: Technological Change, International Competition and Regulation in Communications*, (Robert W. Crandall and Kenneth Flamm, eds.), The Brookings Institution, Washington, DC, 1989.
43. Noam, Eli M., *Telecommunications in Europe*, Oxford University Press, New York, 1991.
44. Schmoller, Gustav Von, *Grundriss der allgemeinen Volkswirtschaftslehre*, Leipzig, Duncker, and Humblot, Erster-[Zweiter] Teil, 1904.
45. Wagner, Adolf, Finanzwissenschaft und Staatssozialismus, *Zeitschrift für die Gesamte Staatswissenschaft*, 43:37–122, 675–746 (1887).
46. Smith, Adam, *The Wealth of Nations*, Methuen and Company, London, 1904.
47. Sax, E., *Die Verkehrsmittel in Volks-und Staatswirtschaft*, Grundlegung der theoretischen Staatswirtschaft, Vienna, 1887.
48. Marx, Karl, and Engels, Frederich, *The Communist Manifesto*, Penguin, Harmondsworth, 1848.
49. Schumpeter, Joseph A., *The Theory of Economic Development*, Harvard University Press, Cambridge, MA, 1912.
50. Mestmäcker, Ernst-Joachim (ed.), Kommunikation ohne Monopole, *Über Legitimation und Grenzen des Fernmeldemonopols*, Nomos Verlag, Baden-Baden, 1980.
51. von Weizsäcker, Carl Christian, Wettbewerb im Endgerätebereich, *Jahrbuch der Deutschen Bundespost*, 25:578-587 (1984).
52. Witte, Eberhard, *Neuordung der Telekommunikation*, R. V. Decker's Verlag, Heidelberg, 1987.
53. Ricardo, David, *Principles of Political Economy and Taxation*, J. Murray, London, 1817.
54. Mill, John Stuart, *Principles of Political Economy*, Little, Brown, Boston, 1848.

55. Malthus, Thomas R., *Definition of Political Economy*, J. Murray, London, 1827.
56. Wells, Herbert G., in Shaw, George B., *The Fabian Essays*, Humboldt, London, 1891.
57. Passfield, Sidney, *Problems of Modern Industry*, Longmans, Green, and Co., London, 1898.
58. Littlechild, Stephen C., Deregulation of UK Telecommunications: Some Economic Aspects, *Econ. Rev.*, 1(2):29 (1983).
59. Beesley, Michael E., *Liberalization of the Use of British Telecommunications Network Report to the Secretary of State*, Her Majesty's Stationery Office, London, 1981.
60. Turgot, Anne, *Reflections on the Foundation and the Distribution of Riches*, Macmillan, New York, 1898.
61. Say, Jean Baptiste, *Treatise on Political Economy*, C. R. Prinsep, Boston, 1821.
62. Du Pont de Nemours, Pierre S., *Physiocratie, ou constitution essentielle du gouvernement le plus avantageux au genre humain*, Recveil Publishers, Paris, 1768.
63. Quesnay, Francois, *Tableau Oeconomique*, Macmillan, New York, 1894.
64. de Condillac, Etienne B., *Le Commerce et le gouvernement consideres l'un a l'autre*, Paris, 1776.
65. Sismondi, Jean Charles, *New Principles of Political Economy, or of Wealth and Its Relation to Population*, Chez Delaunay, Paris, 1879.
66. Cournot, Augustin, *The Mathematical Principles of the Theory of Wealth*, L. Hachette, Paris, 1838.
67. Aglietta, Michel, *A Theory of Capitalist Regulation*, NLR, London, 1978.
68. Nora, Simon, and Minc, Alain, *The Computerization of Society*, MIT Press, Cambridge, MA, 1980.
69. Katayama, Sen, *The Labor Movement in Japan*, Kerr, Chicago, 1918.
70. Komatsuzaki, Seisuke, An Economic Impact of Informationization, *Keio Communication Rev.*, 7:13–23 (March 1986).
71. Noam, Eli M., The Public Telecommunications Network: A Concept in Transition, *J. Commun.*, 30–48 (Winter 1987).

ELI M. NOAM
JOHN T. WENDERS

EIA-232-D Standard. *See* **Binary Serial Data Interchange:**
EIA-232-D Standard *Volume 2, pages 113–123*

Electrical Filters

Introduction

History

An electrical filter is a device that transforms a given input signal into a desired output signal. The transformation or filtering may be carried out in the frequency or time domain by a variety of physical means (electrical, mechanical, acoustical, etc.) depending on the frequency range of the signals and on the application in question. The most commonly used electrical filters traditionally have been wave or frequency filters; this explains why most literature on electrical filters is based on the theory and practice of frequency filters.

Electrical frequency filters are frequency-selective networks designed to "pass" or transmit sinusoidal waves in one or more continuous frequency bands (pass band) and to "stop," reject, or attenuate sinusoidal waves in the complementary bands ("stop band"). Filters with single pass bands typically are classified as low-pass, high-pass, and band-pass, depending on the bands of frequencies passed. For example, the pass band of the band-pass filter in Fig. 1 extends from frequency ω_1 to ω_2 (where ω is the angular frequency expressed in radians per second and is equal to the frequency f in cycles per second, or hertz (Hz), multiplied by 2π). There are other filter types, such as all-pass, frequency-emphasizing, and frequency-rejection filters, all of which are discussed in more detail below.

Filters are among the most important components of present-day communications systems. For example, in the common radio receiver, filters are used to smooth out the pulsating direct current (DC) from the rectifier to give a steady output of the power supply. In superheterodyne receivers, filters in the intermediate frequency stages provide the desired selectivity. Telephone carrier circuits are possible only because of filters used to separate the various channels and to route the signals to the proper paths. In fact, while filters generally are used very widely, the telephone systems traditionally have been among the biggest users. However, with the trend toward digital integrated circuits and the development of the Integrated Services Digital Networks (ISDNs), frequency-selective filters largely are being replaced by filters operating in the time domain, that is, digital filters and digital signal processors (DSPs). Nevertheless, also in modern digital communication systems, frequency-selective filters still have their place (e.g., antialiasing filters, smoothing filters), albeit to a lesser extent than in the classical frequency-based (frequency-carrier) communications systems. Other applications include noise suppression, the elimination of parasitic, signaling, or pilot frequencies from a desired signal, and auxiliary operations associated with modulation, sampling, and modern digital signal processing techniques.

The invention of electrical filters is attributed to K. W. Wagner and G. A. Campbell, who proposed the principles independently in 1915, and O. J. Zobel, who first disclosed a practical synthesis method in 1923. The first contributions toward practical and exact two-port synthesis were made by R. M. Foster in

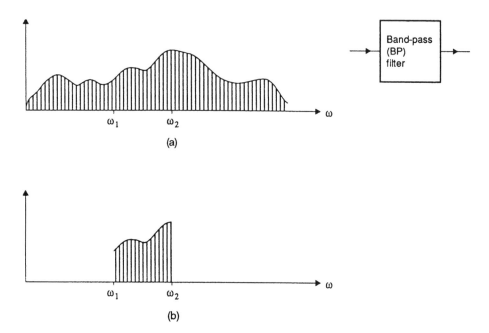

FIG. 1   Typical input and output frequency spectrum of band-pass filter: a, frequency spectrum of the input signal; b, frequency spectrum of the output filter.

1924 and W. Cauer in 1926 for the synthesis of the lossless one-port, and by O. Brune in 1931 for the synthesis of the resistor-inductor-capacitor (RLC) one-port. The synthesis of lossless symmetrical two-ports was described in 1927 by A. C. Bartlett, Brune, and Cauer. The general synthesis of the lossless two-port was achieved independently by S. Darlington, Cauer, and H. Piloty between 1938 and 1941. The problems raised by the insertion of piezoelectric resonators were solved by W. P. Mason between 1934 and 1937.

The practical design of filters by exact methods was encumbered first by numerical difficulties. Not until the advent of the digital computer could the (so-called effective) parameter design method replace the previously used image parameter theory of Zobel. The first table compiled by computers was published in 1963 by R. Saal. However, present-day synthesis programs still are limited to relatively low-degree filters. Practically, the maximum tractable degree for a filter is the same as the number of significant digits in the computer.

In addition to the synthesis problem, there is the important problem of finding a transfer function that will approximate best the ideal filter. The Butterworth characteristic was found in 1930 and Cauer demonstrated in 1931 that Chebyshev polynomials and elliptic functions can be used to define optimal approximations to the ideal attenuation characteristic of a filter. The optimal approximation of a constant delay was solved by W. E. Thomson in 1949 for the maximally flat criterion, and by T. A. Abele, E. Ulbricht, and Piloty in 1960 for the Chebyshev criterion. Although the Chebyshev criterion for the equiripple approximation of most specifications generally is considered optimum, B. Rakovich in 1972 showed that this is not true for the constant-delay approximation.

As early as 1938, H. H. Scott proposed the first active RC (resistor-capacitor) filter, built with an RC twin-T two-port and an amplifier, but the cost and power consumption of vacuum tubes made this circuit impractical. The discovery of the transistor and the development of integrated circuits (ICs) subsequently enlarged the choice of components and technologies and, in 1953, J. G. Linvill proposed a filter with a single negative impedance converter, resistors, and capacitors. However, this never was implemented successfully due to its high sensitivity to component values.

In 1955, P. R. Sallen and E. L. Key introduced the concept of the RC active filter as a cascade of biquadratic sections (biquads). Yet the sensitivity of these filters still remained higher than that of conventional LC (inductor-capacitor) filters. The sensitivity issue was addressed seriously first by H. J. Orchard in 1968 when he explained the low sensitivity of classical LC filters, and by G. S. Moschytz in 1971 when he introduced the concept of the gain-sensitivity product and emphasized the advantages of stable RC components using hybrid (thin/thick film) integrated networks. As a result, many parallel attempts were undertaken to design cascades of active RC biquads and to simulate classical LC filters by means of RC active filters incorporating gyrators. In 1970, another solution, based on feedback loops in a cascade of biquadratic sections, was proposed by F. E. J. Girling and E. F. Good. This solution was generalized under the acronym LF (leapfrog) in 1973 by G. Szentirmai. Similarly, in 1969 L. T. Bruton proposed the FDNC/FDNR (frequency-dependent negative conductor/frequency dependent negative resistor) concept. Finally, in 1974, K. R. Laker and M. S. Ghausi presented a family of structures called FLF (follow-the-leader feedback) that also rely on feedback loops without explicit reference to an LC filter simulation.

In the quest for integrated silicon filter chips, the switched-capacitor filter consisting of op amps (operational amplifiers), switches, and capacitors was introduced by D. L. Fried in 1972. A precursor to this approach was introduced by K. W. Cattermole in 1958. A filter including only capacitors, switches, and op amps was described by R. Boite and J. R. V. Thiran in 1968, but the first metal-oxide-semiconductor (MOS) integrated filter chip was introduced by B. J. Hosticka, R. W. Brodersen, and P. R. Gray in 1977. A further step, in which a signal is quantized not only in time but also in amplitude, resulted in the digital filter. One of the initiators of this approach was J. F. Kaiser in 1966. A. Fettweis introduced the wave digital filter in 1971, which actually simulates an LC filter and therefore enjoys the same insensitivity properties pointed out by Orchard.

This brief survey mentions only a few of the many contributors to the vast field of electrical filters. Moreover, it is restricted to lumped-element filters. Distributed component filters are, for example, mechanical filters, monolithic filters, transmission line filters, and surface wave filters, some of which are mentioned below.

Classifications and Definitions

Filter functions generally are characterized by their pass- and stop- or attenuation-frequency bands. *Low-pass filters* allow electrical signals having frequency components from DC to a specified frequency (the cutoff frequency $\omega_c = 2\pi \cdot f_c$) to pass with little or no attenuation, and cause electrical signals having frequencies beyond this to be rejected by at least 6 decibels (dB) per octave attenuation, or multiples thereof. *High-pass filters* accept signals above the cutoff frequency ω_c and reject them below this frequency. *Band-pass filters* accept signals within a defined band or spectrum and reject them outside the band. *Band-rejection filters* reject signals within a defined band and accept or pass them outside the band.

Figure 2 shows typical frequency responses of the four filter types. The notion of signal acceptance or rejection is one of degree only, and the "transition zone" of the filter between signal acceptance and rejection exists in the region of the cutoff frequency (see Fig. 3). The degree of signal acceptance, signal rejection, and the frequency rate of crossover are three of several properties that define the characteristics of filter performance. The frequency rate of crossover is a measure for the slope of the filter amplitude characteristic, in dB/octave or dB/decade, as it crosses a given gain or attenuation level (e.g., zero dB, or -60 dB).

The principal characteristics commonly used to specify the performance of an electrical filter are insertion loss, stop-band rejection, cutoff frequency, bandwidth, Q factor, shape factor, impedance level, and power-handling capacity (see Fig. 3).

Insertion loss is the ratio of the amplitude of the desired signal before filter insertion to that value at the filter output terminals after insertion, also referred to as the *insertion loss in the filter pass band*. Ordinarily, it is desired that this value be held at a minimum, especially for filters with a main signal of interest that carries significant power, since a substantial amount of heat dissipation or undesired loss may be involved. For many low-power communication filters, conducting signal power of less than about 1 watt (W), an insertion loss of a decibel or two is generally acceptable. For filters passing high power in the acceptance band, such as those used for 60- or 400-Hz AC (alternating current) power mains, or radar or communications transmitters, it often is desirable to keep insertion losses to about 0.1 dB or less.

Stop-band rejection is the ratio of the amplitude of unwanted frequency components before filter insertion to the amplitude existing after filter insertion, also referred to as the *insertion loss in the filter stop band*. This is the principal property of a filter in the sense that it is a measure of the extent to which it rejects unwanted signals. Typical values may range from 20 dB for certain harmonic-rejecting, high-power transmitter filters to more than 100 dB for some preselector or tunable filters used to protect the front end of sensitive receivers. The degree of rejection will vary over the stop-band frequency, and beyond some frequency the rejection may degrade or fail altogether.

Cutoff frequency is the frequency between the signal acceptance and rejection bands corresponding to an attenuation of 3 dB below the pass-band insertion loss. Although this characteristic often is specified by the application engineer, it sometimes is better to allow the designer to choose the cutoff frequency, as long as the insertion-loss and rejection-band requirements are satisfied properly.

Bandwidth is the frequency acceptance window, or band, measured between the 3-dB cutoff frequencies in a band-pass filter. In a low-pass filter, it is the

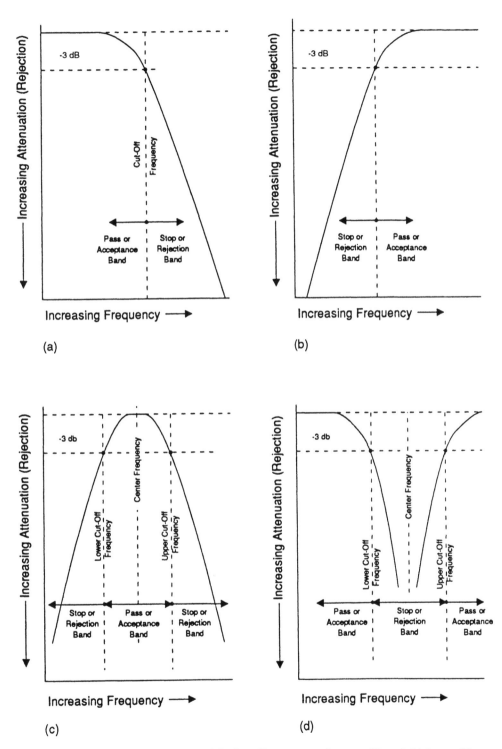

FIG. 2 Typical frequency response of the four filter types: *a*, low-pass filter; *b*, high-pass filter; *c*, band-pass filter; *d*, band-rejection filter.

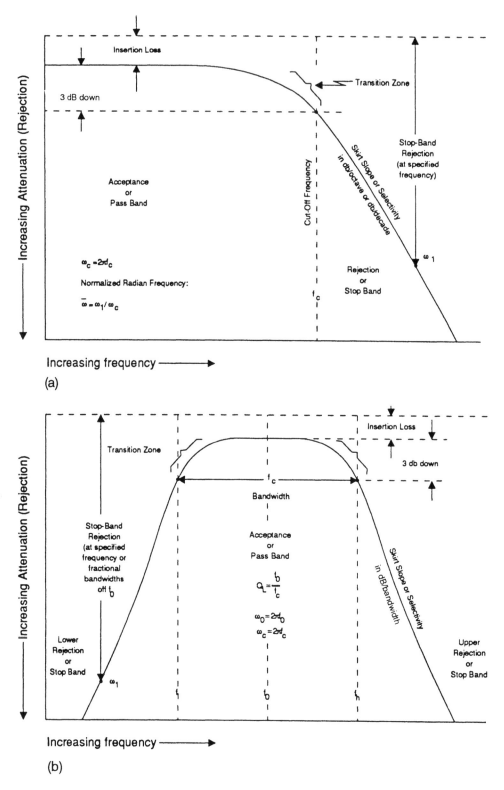

FIG. 3 Terminology used to describe filter characteristics: *a*, low-pass filter terms; *b*, band-pass filter terms.

band between zero frequency and the cutoff frequency or, simply, the cutoff frequency.

The Q *factor* is the ratio of the center frequency to the bandwidth in a band-pass filter. The center frequency is the geometric mean frequency between the 3-dB cutoff frequencies and may be approximated by the arithmetic mean when the Q factor is higher than about 10; thus:

$$Q = \frac{f_0}{f_c} = \frac{\sqrt{f_h f_L}}{f_h - f_L} \approx \frac{1}{2} \frac{f_h + f_L}{f_h - f_L} \tag{1}$$

where
  f_0 = the band-pass filter center frequency
  f_c = the 3-dB filter bandwidth (see Fig. 3)
  f_h = the upper 3-dB cutoff frequency
  f_L = the lower 3-dB cutoff frequency

The term Q-factor, used in specifying band-pass filters, often is called the *loaded Q-factor,* or Q_L, inasmuch as the driving and terminating impedance loads are connected to the filter when inserted in a network. This is to be distinguished from the term *unloaded Q-factor,* or Q_u, which is a measure of the performance of the individual components making up a filter. The unloaded Q-factor is not needed in specifying filter characteristics since the term is implicit in the other characteristics. However, Q_u is very important when specifying the quality of filter components required for a given filter design.

The *shape factor* for band-pass filters is the ratio of the bandwidth at the 60-dB points below insertion loss to the bandwidth at the 6-dB points below insertion loss; that is,

$$\text{shape factor (60/6 dB)} = \frac{f_{60\,dB}}{f_{6\,dB}} \text{ (conventional definition).} \tag{2}$$

Since the 3-dB bandwidth is specified more commonly, the shape factor also can be specified as

$$\text{shape factor (60/3 dB)} = \frac{f_{60\,dB}}{f_{3\,dB}} = \frac{f_{60\,dB}}{f_c} \text{ (practical definition).} \tag{3}$$

The shape factor is useful for certain applications, as when specifying intermediate-frequency amplifier filters. In general, however, the terms *stop-band rejection* and *bandwidth* are used more commonly.

Impedance level is the value, specified in ohms, of both the filter source (driving or input) impedance and the terminating (load or output) impedance. Generally, the input and output impedance levels specified are the same, especially for communication filters in which 50-, 75-, and 300-ohm transmission lines are used frequently. On the other hand, power filters, especially those used in 50- and 400-Hz generator lines to reject harmonics, rarely have equal input and output impedances since the internal voltage drop at the generator input should be small.

Power-handling capacity is the rated average power in watts beyond which the performance of the filter may degrade or fail altogether due to burn out. Occasionally, peak power is used to specify power-handling capacity, especially when a breakdown of components or a gas inside a hollow transmission line is involved. This specification generally becomes important for filters handling more than about 1 W.

Another way of classifying filters is by their physical realization. The most important types are lumped-element electrical LC filters, distributed-element or transmission-line filters, lumped–distributed-element filters, mechanical filters, monolithic crystal filters, active RC filters, switched-capacitor (SC) filters, charge-coupled device (CCD) filters, surface acoustic wave (SAW) filters, and digital filters.

1. *Lumped-element electrical LC filters*, consisting of inductors (L) and capacitors (C), are usable from DC to approximately 500 megahertz (MHz). Parasitic capacitance and lead length inductance make higher-frequency applications impractical since the component performance behaves very differently and often unpredictably. Below about 100 Hz, inductance and capacitance values of tuned circuits become so large that they become impractical, except when size and weight are not an issue.

Substantial progress has been made in the development of ferrite materials used in inductors and dielectrics used in capacitors. This permits the realization of inductor Q_u factors of the order of 300 at low frequencies and in comparatively small physical sizes, generally of the toroidal shape. Capacitor Q_u factors are significantly higher than their inductor counterparts, with a ratio of capacitor to inductor Q_u factors of 5 to 10 being typical. An important advantage of LC filters is that they are virtually noise free.

2. *Distributed-element or transmission-line filters* are useful from about 200 MHz to 100 gigahertz (GHz) or higher. The filters in these lines contain a wide variety of types such as open and shorted sections of transmission lines, cavities, ring filters, and direct and quarter-wave coupled chambers. Slow-wave structures such as helical and serrated lines used in traveling-wave tubes provide additional design flexibility. Gyromagnetic resonance and gaseous absorption effects sometimes are used to realize certain filter characteristics. In addition, mode generators and suppressors are useful, respectively, to transfer energy to a waveguide or coaxial mode providing better filtering properties or to eliminate undesired energy in certain modes.

Hollow pipe and ridge waveguides, coaxial and strip line, and open dual-conductor lines are the more common transmission media. Their sizes may vary in cross-section from as big as 200 square inches in P-band (300-MHz waveguides), to perhaps 0.05 square inches for X-band (10-GHz) strip line, or 0.01 square inches for 70-GHz waveguides. Unloaded Q_u factors in the order of 30,000 are obtainable with L-band (1-GHz) waveguides; Q_u reduces to about 8000 at X-band. Coaxial or air-dielectric strip lines of about 1 square inch in cross-section will have Q_u factors of about 600 at 100 MHz and 1500 at L-band.

3. *Lumped–distributed-element filters* are regarded as hybrid types employing the advantages of both lumped and distributed elements, particularly in the "awkward" frequency spectrum between about 100 MHz and 1 GHz. In this range, lumped-element parasitics become objectionable and the wavelength is

comparatively long. Hollow waveguide pipes, operated well beyond cutoff frequency, for example, may use magnetic coupling between a series of axially supported, resonant metallic loops. Another hybrid version uses stub-supported transmission lines to realize the equivalent of individual LC elements in a filter in which the remaining elements are of the lumped type. A third hybrid may use lumped capacitors in a coaxial or strip line as the series interface elements of direct-coupled chambers.

4. *Mechanical filters* generally consist of coupled mechanical resonators, and rely on the electrostrictive, magnetostrictive, or piezoelectric effect of separate electromechanical transducers for the interaction between electrical and mechanical energy. They may be used over the frequency spectrum from a fraction of a hertz to approximately 1 MHz. At low frequencies, they occur as mechanical resonators. Diaphragm resonators are useful at the lowest frequencies, below 1 Hz to about 1 kilohertz (kHz). Resonators in the form of vibrating reeds, tuning forks, or bender-type crystals afford high Q factors of 100 to 10,000 and a frequency coverage from approximately 30 Hz to 20 kHz. From a few hundred hertz to about 300 kHz, rod-type longitudinal resonators are used; they reduce to the form of plates at the higher frequencies. The plate structure exists primarily in quartz crystals, which are used from about 10 kHz to 10 MHz (up to 200 MHz on overtones). Q factors in the order of several thousand are realizable (as previously mentioned, crystals frequently are used with LC elements to obtain high-loaded Q factors because of the practical upper limit on the Q factor of inductors).

Mechanical filters consisting of coupled resonators cover the higher-frequency spectrum and exist in three major forms combining rod-, disk-, and plate-type resonators. Loaded Q factors of 1000 are easily realizable. Other than the crystal filter, all mechanical resonators require the electromechanical transducers for electrical applications. A disadvantage is that the resulting insertion losses are typically of the order of 10 to 30 dB.

5. *Monolithic crystal filters* use coupled mechanical vibrations in a piezoelectric (generally quartz) material to provide narrowband band-pass filters. A filter consists of a crystalline quartz wafer onto which pairs of metal electrodes are deposited. The piezoelectrical coupling factor of the quartz material causes a coupling of the mechanically resonating crystal to the electrodes. These in turn transform the resonator to an equivalent LC circuit exhibiting two closely spaced resonances (one serial and one parallel). The Q factors of these resonances are very high and can reach 100,000 and more; hence, very narrow filters are achievable. In bulk resonator designs, the bandwidth can be widened by adding discrete inductors. Bulk resonator filters can be designed in the range of 50 kHz to several hundred kHz with a relative bandwidth of about 5%. Overtone designs go to higher frequencies; however, parasitic capacitors are likely to limit the bandwidth considerably. At higher frequencies, from several MHz to tens of MHz, the monolithic designs are used. Temperature stability is excellent due to the inherent temperature stability of quartz resonators. At moderate signal levels, nonlinear distortions are small and noise virtually is absent.

6. *Active RC filters* combine resistors (R), capacitors (C), and gain devices (generally, op amps or operational transconductance amplifiers [OTAs]) to provide the frequency filtering function. They can be designed for use from several

hundred Hz to approximately 1 MHz. Their main constraints are related to the gain rolloff of the active devices used in the design. The rolloff limits the gain and increases the phase in the feedback path that, in turn, is necessary to generate complex pole pairs. This not only may limit the applications of active filters at higher frequencies, it also may render the feedback loop unstable if the Qs required for steeper filters are too high. The achievable temperature stability of active filters depends on the temperature coefficients of the resistors and capacitors, as well as on the temperature stability of the active devices.

To improve the temperature stability, RC components often are realized using thick-film or thin-film hybrid IC technology with compensating temperature coefficients. Active filters do not require matching terminations since the output of the amp can be used as a voltage source, and the input port generally has a high impedance and is driven by a voltage source. This makes the cascade connection of second-order sections (biquads) relatively easy; transfer functions of higher-order filters then can be realized by a cascade of second- or first-order sections without the individual sections influencing each other. The choice of pole-zero pairing has an impact on the dynamic range of the resulting cascaded filter. LC ladder filters also can be simulated by active RC circuits using various types of active impedance inverters (e.g., gyrators), converters (e.g., negative impedance converters), and FDNRs. Such LC filter simulations have the sensitivity advantages of the doubly terminated ladder structure (Orchard). The active devices in active RC filters introduce distortion at high signal levels; low signal levels are limited by the noise introduced by the active devices. Thus, the inherent active-device noise and nonlinear distortion determine the practical dynamic range of active filters.

7. *Switched-capacitor filters* consist of FET-type (field-effect-transistor-type) switches, capacitors, and amplifiers. They were developed for implementation using complementary-metal-oxide-semiconductor very-large-scale-integrated (CMOS VLSI) technology, and provide the long-awaited "filter on a chip." Switched-capacitor combinations replace resistors, the latter having too wide a tolerance to be feasible in VLSI technology. SC filters are "sampled-data" filters, quantized in time but not in amplitude. Thus, the input signal must be band limited to prevent aliasing. Also, SC filters are subject to constraints similar to those of active filters. Additional limitations arise in accordance with the MOS technology used. As the switching times reach the time constant of the FET switches (capacitance times switch on resistance), performance deterioration sets in due to incomplete charge transfer. Accordingly, sampling rates are limited to several hundred kHz for standard CMOS VLSI designs. The accuracy of the frequency response depends on the accuracy of the capacitor ratios and on the stability of the sampling rate. The former depends on the attainable photolithography precision, the latter on the clock oscillator stability. The dynamic range is determined by linear distortion and noise of the CMOS amplifiers.

8. *Charge-coupled device filters* are based on charge transfer in CMOS structures. They differ fundamentally from SC filters in that they are nonrecursive, making use of only transfer function zeros to shape the frequency response. By contrast, SC filters are recursive, making use of both zeros and poles. Thus, CCD filters mainly are used as transversal or FIR (finite impulse-response) filters, such as when linear phase is required or for matched filtering when sharp separations between stop- and pass-band regions are not required. Best performance is obtained from so-called split electrode configurations in which the electrodes of a CCD delay line are split into two or three parts. CCD transversal filters are used frequently in correlators in which the charge leakage and nonlinear effects in CMOS devices are not critical.

9. *Surface acoustic wave filters* are passive, noiseless devices that in principle resemble FIR or transversal filter designs. Utilizing the piezoelectric effect, a wave is induced on the surface of a crystal and is propagated along the filter structure until it finally is reconverted into electrical signals. SAW filters provide the same stability as crystal filters; however, as FIR filters, they require a large number of taps for narrowband, as well as wideband filters.

Photolithographic technology allows the implementation of several hundred taps for filters in the range of 50 MHz to several GHz. SAW band-pass filters are used for such varied purposes as television intermediate frequency (IF) filtering; channel selection in satellite transmission systems; pulse compression in chirp radar systems; paging and mobile telephone systems; timing recovery in optical-fiber digital transmission systems; frequency control of high-frequency oscillators; broadband signal processing; matched filtering; and many other passive and dynamic filtering applications. For relatively low signal levels, nonlinear effects are negligible. As passive devices, these filters have to be terminated properly at the input and output.

10. *Digital filters* process signals that are quantized in both time and amplitude. As such, these filters are in a class by themselves. They are closer to digital signal processors (DSPs) than to electrical filters (although, in analogy, the latter sometimes are referred to as analog signal processors). The performance of digital filters depends on the number of bits representing the signal and coefficients. The more bits used for the signal representation, the larger the dynamic range of the signal will be. With the advent of DSPs based on floating-point arithmetic, the signal range within the device virtually is unlimited. The number of bits in the coefficients determines the precision with which a desired frequency response can be obtained. A sharp and narrow filter requires more coefficient bits than a filter with a modest rise between pass band and stop band. Digital filters have virtually no limitations in their functional performance as long as the logic circuitry is expanded sufficiently in order to implement the necessary binary arithmetic. However, with DSPs realized as VLSI chips using CMOS technology, the required logic circuitry is becoming an insignificant constraining factor. Thus, the conditions under which the digital filter can work is dependent solely on the choice of the technology for the logic circuitry. On the other hand, a natural limit for the number of bits that can be processed remains the number of bits in the analog-to-digital (A/D) and digital-to-analog (D/A) conversion. Depending on sampling rates, below 100 kHz 16 bits may not be unreasonable, whereas at hundreds of MHz, 4–6 bits more likely will be the limit.

Frequency Response and Transfer Function

Most filter networks belong to the family of linear lumped-parameter finite (LLF) networks. Those not falling into this category typically are nonlinear, distributed, nonfinite, or any combination of these.

The output signal of an nth-order LLF network can be found in terms of the input signal by solving a linear nth-order differential equation of the form

$$a_n \frac{d^n y}{dt^n} + a_{n-1} \frac{d^{n-1} y}{dt^{n-1}} + \cdots + a_1 \frac{dy}{dt} + a_o y$$

$$= b_m \frac{d^m x}{dt^m} + b_{m-1} \frac{d^{m-1} x}{dt^{m-1}} + \cdots + b_1 \frac{dx}{dt} + b_o x \quad (4)$$

where

$x(t)$ = input signal
$y(t)$ = output signal
$n \geq m$

Applying the Laplace transform to this equation, we obtain the transfer function $T(s) = Y(s)/X(s)$ as the ratio of two polynomials $N(s)$ and $D(s)$, namely,

$$T(s) = \frac{N(s)}{D(s)} = \frac{b_m s^m + b_{m-1} s^{m-1} + \cdots + b_1 s + b_o}{a_n s^n + a_{n-1} s^{n-1} + \cdots + a_1 s + a_o} \quad (5)$$

where $s = \sigma + j\omega$ is the complex frequency and $N(s)$ and $D(s)$ are polynomials in s with real coefficients a_i and b_j. Expressing $N(s)$ and $D(s)$ in their factored form, we obtain the poles and zeroes of the transfer function:

$$T(s) = K \frac{(s - z_1)(s - z_2) \ldots (s - z_m)}{(s - p_1)(s - p_2) \ldots (s - p_n)} = K \frac{\prod_{i=1}^{m} (s - z_i)}{\prod_{j=1}^{n} (s - p_j)}. \quad (6)$$

The poles p_j and zeros z_i may be either real or complex conjugates. Combining a complex conjugate zero pair with a complex conjugate pole pair, we obtain the special case of a second-order transfer function:

$$T(s) = K \frac{(s - z)(s - z^*)}{(s - p)(s - p^*)} = K \frac{s^2 + (\omega_z/q_z)s + \omega_z^2}{s^2 + (\omega_p/q_p)s + \omega_p^2} \quad (7)$$

where

$$z, z^* = -\sigma_z \pm j\tilde{\omega}_z$$

$$p, p^* = -\sigma_p \pm j\tilde{\omega}_p. \quad (8)$$

The poles and zeros can be displayed in the complex frequency or s-plane as shown in Fig. 4. Note that

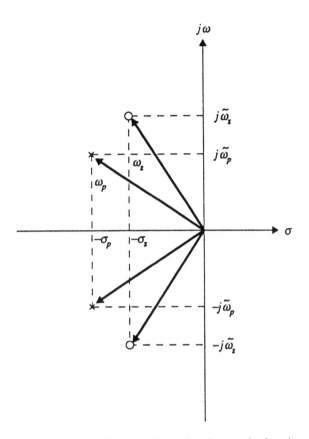

FIG. 4 Pole-zero diagram of second-order transfer function.

$$\omega_z^2 = \sigma_z^2 + \tilde{\omega}_z^2; \qquad \omega_p^2 = \sigma_p^2 + \tilde{\omega}_p^2 \tag{9}$$

and

$$q_z = \frac{\omega_z}{2\sigma_z}; \qquad q_p = \frac{\omega_p}{2\sigma_p}. \tag{10}$$

To obtain the frequency response of the network described by Eq. (4), we assume a sinusoidal input signal and, because the network is linear, obtain a sinusoidal response. The response is obtained by letting $s = j\omega$ in Eq. (6); thus,

$$T(j\omega) = K \frac{\displaystyle\prod_{i=1}^{m} (j\omega - z_i)}{\displaystyle\prod_{j=1}^{n} (j\omega - p_j)} = |T(j\omega)| e^{j\phi(\omega)}. \tag{11}$$

Taking the logarithm of Eq. (11), we obtain

$$\ln T(j\omega) = \ln|T(j\omega)| + j \arg T(j\omega) \tag{12}$$

$$= \alpha(\omega) + j\phi(\omega)$$

where $\alpha(\omega)$ and $\phi(\omega)$ are the gain given in nepers and phase response given in degrees, respectively. To obtain the gain response in decibels, we have

$$\alpha_{\text{dB}}(\omega) = 20 \log|T(j\omega)| = \frac{20}{\ln 10}\alpha(\omega) = 8.686\alpha(\omega) \tag{13}$$

and to obtain the group delay

$$\tau_g(\omega) = -\frac{d\phi(\omega)}{d\omega}. \tag{14}$$

The amplitude and phase response associated with a transfer function $T(s)$ can be obtained graphically from the pole-zero plot of $T(s)$. If $T(s)$ is given in terms of its poles and zeros as in Eq. (6), then, for any frequency ω_o, the corresponding amplitude and phase can be obtained from the expression

$$T(j\omega_o) = |T(j\omega_o)|e^{j\phi(\omega_o)} = \frac{(j\omega_o - z_1)(j\omega_o - z_2) \ldots (j\omega_o - z_m)}{(j\omega_o - p_1)(j\omega_o - p_2) \ldots (j\omega_o - p_n)}. \tag{15}$$

Each factor in Eq. (15) represents a complex phasor that can be displayed graphically in the s-plane. With

$$(j\omega_o - z_i) = A_{zi}(\omega_o)\exp j\theta_{zi}(\omega_o) \tag{16}$$

$$(j\omega_o - p_j) = A_{pj}(\omega_o)\exp j\theta_{pj}(\omega_o) \tag{17}$$

we obtain, from Eq. (15),

$$|T(j\omega_o)| = \frac{A_{z1}A_{z2} \ldots A_{zm}}{A_{p1}A_{p2} \ldots A_{pn}} \tag{18}$$

and

$$\phi(\omega_o) = \theta_{z1} + \theta_{z2} + \cdots + \theta_{zm} - \theta_{p1} - \theta_{p2} - \cdots \theta_{pn}. \tag{19}$$

Example. Consider the transfer function

$$T(s) = K\frac{s^2 + \omega_z^2}{(s^2 + 2\sigma_p s + \omega_p^2)(s + \alpha)}. \tag{20}$$

The corresponding poles and zeros in the s-plane are shown in Fig. 5. With Eq. (13), the gain at ω_o is

$$\alpha_{\text{dB}}(\omega_o) = 20 \log \frac{A_{z1}(\omega_o)A_{z2}(\omega_o)}{A_{p1}(\omega_o)A_{p2}(\omega_o)A_{p3}(\omega_o)} \tag{21}$$

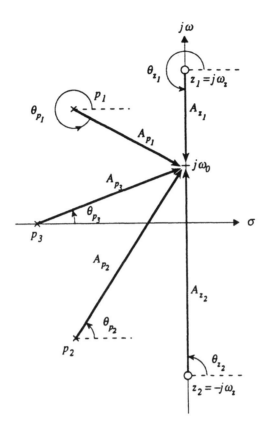

FIG. 5 Pole-zero diagram for transfer function of Eq. (20).

and the phase is

$$\phi(\omega_o) = \theta_{z1}(\omega_o) + \theta_{z2}(\omega_o) - \theta_{p1}(\omega_o) - \theta_{p2}(\omega_o) - \theta_{p3}(\omega_o). \qquad (22)$$

Note that $A_{z1}(\omega_z) = 0$ and therefore $|T(j\omega_z)| = 0$, meaning that the transmission zeros, which are located on the imaginary axis at $\pm j\omega_z$, cause the amplitude to be zero at the frequency ω_z.

Example. Consider the following fifth-order transfer function:

$$T(s) = \frac{K}{[s^2 + (\omega_{p1}/q_{p1})s + \omega_{p1}^2][s^2 + (\omega_{p2}/q_{p2})s + \omega_{p2}^2](s + \alpha)} \qquad (23)$$

Note that $T(s)$ has only finite poles since the numerator is a constant (the five zeros are said to be at infinity). The locations of the poles are shown in Fig. 6 and the gain, phase, and group-delay responses in Fig. 7.

The transfer function $T(s)$ provides the gain response, that is, the output/input ratio of a network, whereas filter specifications very often are expressed

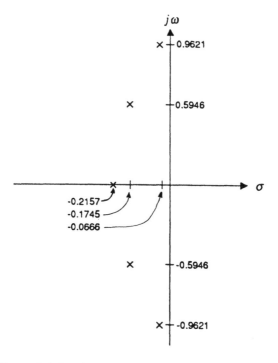

FIG. 6   Pole location for transfer function given by Eq. (23).

in terms of the attenuation response, or the input/output ratio. Except for a correction term that relates to the input and output termination of a general filter, the attenuation response is basically the inverse of the gain function, that is, it can be obtained from the inverse of the transfer function as given by Eq. (5). For active filters, this inverse relationship between gain and attenuation is an exact one because the active filter generally is driven from a voltage source (zero source impedance) and the output signal taken from the output of an op amp (voltage source). This corresponds to the operating conditions given in Fig. 8, for which the output-to-input voltage ratio is

$$T(s) = \frac{V_{out}}{V_{in}}(s). \tag{24}$$

The attenuation function $H(s)$ then is defined as

$$H(s) = \frac{V_{in}}{V_{out}}(s) = \frac{1}{T(s)}. \tag{25}$$

For the sinusoidal input case, we define the attenuation and phase response as

$$\ln H(j\omega) = \ln|H(j\omega)| + j \arg H(j\omega)$$
$$= A(\omega) + jB(\omega) \tag{26}$$

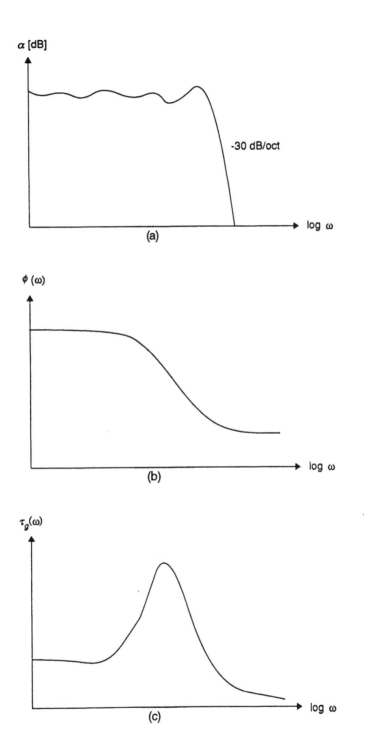

FIG. 7 Responses for transfer function given by Eq. (23): *a*, gain; *b*, phase; *c*, group delay.

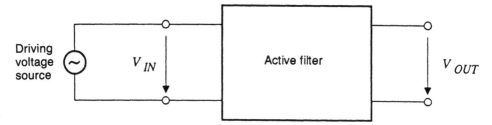

FIG. 8   Typical operating conditions for active filter.

and the attenuation response in decibels

$$A_{dB}(\omega) = 20 \log |H(j\omega)|. \tag{27}$$

The attenuation response of the network function of Eq. (23) is as shown in Fig. 9. A_{dB} corresponds to the insertion loss defined in the section, "Classifications and Definitions," and provides the link to conventional LC filter tables and other classical analytical tools. A_{max} and the cutoff frequency ω_c of the fifth-order low-pass filter characterize the passband, the frequency span $\omega_s - \omega_c$ determines the selectivity, and A_{min} is the minimum loss in the stop band.

Basic Filter Types

One of the most basic problems in filter design is the so-called approximation problem. It entails finding a rational nth-order transfer function of the form given by Eq. (5) such that, for $s = j\omega$,

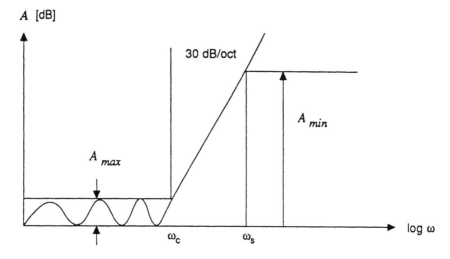

FIG. 9   Attenuation response corresponding to the transfer function given by Eq. (23).

$$\alpha(\omega) = \text{Re}\{\ln T(j\omega)\} = \ln|T(j\omega)| \tag{28}$$

and

$$\phi(\omega) = \text{Im}\{\ln T(j\omega)\} = \arg T(j\omega) \tag{29}$$

correspond to a specified amplitude and phase response, respectively, where Re stands for the real past of the complex function ln $T(jw)$ and Im for the corresponding imaginary past. An alternative is to find directly a filter structure based on the technology at hand (e.g., LC passive, active RC, crystal, mechanical, etc.), with a transfer function that satisfies the given specifications. This approximation problem has been solved for a variety of basic filter types with characteristics that have been related to commonly occurring attenuation specifications. The results are tabulated and programmed for computer use for both alternatives, that is, relating the given specifications either to a transfer function (poles and zeros) or to a given filter structure, generally a passive LC ladder network. These basic filter types generally are presented in terms of the characteristics of a low-pass filter. The other filter types (e.g., band pass, high pass, etc.) then can be derived from the low-pass filter by simple frequency transformations. Thus, having solved such an approximation problem as the derivation of a rational function $T(s)$ to satisfy a specified amplitude and phase response in the low-pass domain, it is relatively simple to obtain the design parameters corresponding to such other filter types as high pass, band pass, and so on.

The amplitude response of an ideal low-pass filter is shown in Fig. 10-*a*, and the corresponding insertion loss in decibels in Fig. 10-*b*. The ideal low-pass filter is characterized by (1) zero loss and ripple in the pass band, (2) an infinite attenuation slope at the cutoff frequency f_c (i.e., a zero transition region), and (3) infinite attenuation in the stop band. The ideal low-pass filter with this kind of response often is referred to as a "brick-wall" filter. In general, a linear phase response also is assumed. The ideal low-pass filter is distinguished by the fact that no rational transfer function $T(s)$ exists to describe it accurately. Thus, any analytical description of the ideal low-pass filter, at best, can be an approxima-

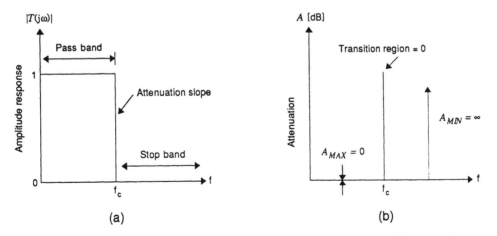

FIG. 10 Characteristics of an ideal low-pass filter: *a*, amplitude response; *b*, insertion loss.

tion. Many such approximations have been derived in the field of classical network theory. The best and most commonly used ones can be grouped into basic types, the characteristics of which are shown qualitatively in Fig. 11. Briefly, they can be described as follows.

Maximally flat or Butterworth filters. The maximally flat or Butterworth approximation of an ideal low-pass filter is shown in Fig. 11-*a*. It has a maximally flat amplitude response in the pass band, but this is achieved at the expense of phase linearity and steepness of attenuation slope. However, the attenuation slope of the Butterworth filter is quite good, it has a reasonably good impulse response, and, because it provides an excellent general-purpose approximation to the ideal filter response, it is one of the most commonly used filter types.

Equiripple or Chebyshev filters. If steepness of attenuation slope, especially in the region of cutoff, is more important than pass-band flatness or phase linearity, the Chebyshev response (see Fig. 11-*b*) often is applicable. The Chebyshev filter exhibits increased overshoot when driven by a step function, and is designed with a prescribed ripple (i.e., equiripple) in the pass band (e.g., 0.01-dB ripple up to a possible 3-dB ripple). In return for the lack of smooth response in the pass band, there are advantages in a very much higher rate of cutoff around the edge of the pass band. The amplitude response curve at frequencies beyond the cutoff region runs parallel, but nearer, to the pass band than that of the equivalent-order Butterworth filter.

Both the Butterworth and the Chebyshev low-pass filters achieve infinite attenuation only at infinite frequency, that is, all the zeros of transmission occur at infinite frequency. At any other frequency, some signals will pass through the filter (i.e., also in the stop band). If infinite attenuation at particular frequencies in the stop band is required, the inverse Chebyshev response (see Fig. 11-*c*), may be used. There is no ripple in the pass band, but ripple does exist in the stop band and attenuation is infinite at certain frequencies (so-called attenuation poles).

Elliptic or Chebyshev–Cauer filters. Elliptic or Chebyshev–Cauer filters (also sometimes called complete Chebyshev, double Chebyshev, Darlington, or Zolotarev filters) have ripple in the pass band and stop band, descend rapidly to a prescribed attenuation outside the pass band, and maintain a specified minimum attenuation to undesirable frequencies outside the pass band. Like the inverse Chebyshev filters, at certain finite frequencies in the stop band they have infinite attenuation, that is, attenuation poles. A typical amplitude response is shown in Fig. 11-*d*. Elliptic filters are probably the most efficient filters in terms of component count for approximating the amplitude response of an ideal filter. For a given filter order, it is possible to produce filters more economically with either a very sharp cutoff or a very high attenuation in the stop band. On the other hand, the attenuation does not drop off smoothly to minus infinity (in dB) outside the pass band, but is maintained at a predetermined level. Note that the Chebyshev and the inverse Chebyshev filters are special cases of the more general Chebyshev–Cauer filters.

Optimum-Monotonic or Legendre filters. A Butterworth filter has a maximally flat pass-band response, and the Chebyshev family of filters provides a good attenuation slope. In some applications, the attenuation slope of a Butterworth filter is inadequate and the ripple of a Chebyshev filter is not tolerable.

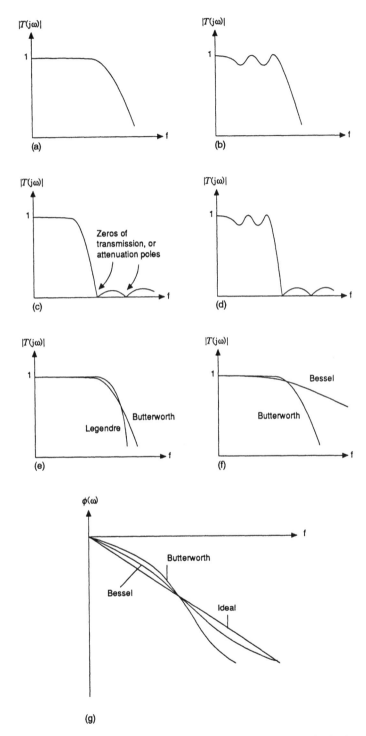

FIG. 11 Responses of basic filter types: *a*, Butterworth (amplitude); *b*, Chebyshev (amplitude); *c*, inverse Chebyshev (amplitude); *d*, elliptic or Chebyshev–Cauer (amplitude); *e*, Legendre (amplitude); *f*, Bessel (amplitude); *g*, ideal, Butterworth, and Bessel phase responses.

Designing a Chebyshev filter with a very small or zero ripple does not help because Chebyshev and Butterworth filters are of the same family; a Chebyshev filter with zero ripple is a Butterworth filter. A solution in this instance may be to use a Legendre or optimum-monotonic filter. The amplitude response for such a filter is shown in Fig. 11-*e*. For purposes of comparison, the amplitude response of a Butterworth filter also is shown. Note that the Legendre response is not as flat as that of the maximally flat Butterworth response in the pass band, but that the attenuation slope of the Legendre response is steeper. A typical property of Butterworth and Legendre filters is their monotonic character, that is, for any value of gain there is a unique frequency. This is in contrast to the Chebyshev filters, in which a particular value of gain will occur at several frequencies because of ripple. The Legendre characteristic attempts to combine the best characteristics of the Butterworth and Chebyshev characteristics. Here the attenuation slope is made as steep as possible with the restriction that the amplitude response remains monotonic.

Linear-phase or Bessel filters. The four filter types discussed above are characterized mainly in terms of their amplitude responses, which are plots of gain (or attenuation) versus frequency. However, these plots do not describe the complete transmission properties of a filter; for example, the phase characteristic of a network is one of the most important parameters of a filter designed for the transmission of square-wave or pulse-shaped signals. When a rectangular pulse is passed through a Butterworth, Chebyshev, or Legendre filter, overshoot or ringing will appear on the pulse at the output. If this is undesirable, one of the members of the so-called Gaussian family of filters can be used. The most common is called a Bessel filter, because Bessel polynomials occur in the denominator of the transfer functions. Bessel filters also are sometimes called Thomson filters after the originator, in 1949, of the design method, W. E. Thomson.

If ringing or overshoot must be avoided when pulses are filtered, the phase shift between the input and output of a filter must be a linear function of frequency; stated differently, the rate of change of the phase with respect to frequency or the group delay must be constant. The net effect of a constant group delay in a filter is that all frequency components of a signal transmitted through it are delayed by the same amount, that is, there is no dispersion of signals passing through the filter. Accordingly, since a pulse contains signals of different frequencies, no dispersion takes place (its shape will be maintained) when it is filtered by a network that has a linear-phase response or constant group delay. Just as the Butterworth filter is the best approximation to the ideal of perfect flatness of the amplitude response in the filter pass band, so the Bessel filter provides the best approximation to the ideal of perfect flatness of the group delay in the pass band because it has a maximally flat group-delay response. However, this applies only to low-pass filters because high-pass and band-pass Bessel filters do not have the linear-phase property.

Figure 11-*f* compares the amplitude response of a Bessel filter with that of a Butterworth filter of the same order. Note that the Bessel filter is a poorer approximation to the ideal, both in flatness in the pass band and in steepness of attenuation. There is no point in the pass band at which the loss becomes zero; it falls off very gradually toward the cutoff (or 3-dB) frequency and then continues to fall just as gradually toward the eventual cutoff slope determined by the order of the filter. This attenuation slope will ultimately (i.e., asymptotically) run parallel with the Butterworth and Chebyshev curves, but will be further from the pass band.

Figure 11-*g* compares the phase response of an ideal low-pass filter with those of Butterworth and Bessel filters. For an ideal filter, the phase shift is linear with frequency; its group delay is constant at all frequencies. The Butterworth filter group delay is not constant because the plot of the phase angle versus the frequency is nonlinear. In contrast, the Bessel filter has a reasonably linear phase-angle-versus-frequency response in the pass band; it therefore provides a good approximation to constant group delay.

Typical responses of various filters to a square-wave input are illustrated qualitatively in Fig. 12. The ringing in the Butterworth and Chebyshev filters, which is the result of their nonlinear-phase characteristics, is evident; the absence of ringing in the Bessel filter shows how well this type of filter approximates the desired linear-phase response. As expected, the response of the Chebyshev filter is inferior to the other two since its phase response is even less linear than that of the Butterworth filter.

Transitional filters exist that have compromise characteristics trading off the best properties of two types of filters. One of the most common is the Butterworth–Thomson filter that attempts to combine the maximally flat ampli-

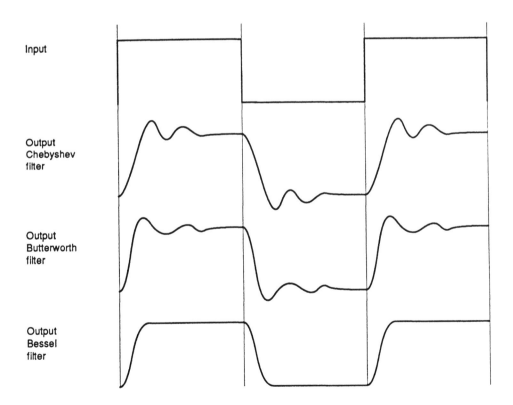

FIG. 12 Square-wave response of various filters.

tude of the Butterworth with the maximally flat group delay of the Bessel, or Thomson, filter.

Passive Filters

Filters generally are called *passive* if they contain no active components (such as amplifiers or other gain elements). There are other definitions of activity and passivity, relating to the absorbed or produced energy within a filter, but these are more of a theoretical than practical nature. Thus, passive filters generally are realized with inductors, capacitors, and resistors (LCRs), but other passive components such as mechanical resonators, transmissions lines, and so on having equivalent LCR filter characteristics are also possible, as discussed below. Passive filters are analog devices, that is, the electrical value of the continuous-time input and output signal directly represents a real value (e.g., number) at all times.

Lumped-Element Inductor-Capacitor-Resistor Filters

The lumped-element LCR filters are the oldest and, up until the widespread use of semiconductor ICs, the most commonly used type of filters. Classical filter theory is based on the properties of LCR filters, and most modern filter realizations (active, digital, switched-capacitor, etc.) either can be derived from, or their properties can be related to, those of equivalent LCR filters.

Lumped-element electrical LCR filters consist of inductors and capacitors in a ladder structure with resistors as terminations. They are usable from DC to approximately 500 MHz. Parasitic capacitance and lead-length inductance make higher frequency applications impractical since the individual components behave very differently, and often unpredictably, beyond these frequencies. Below about 100 Hz, inductance and capacitance values of tuned circuits become so large that they become impractical except where size and weight are not issues.

Substantial progress in the development of ferrite materials for inductors permits the realization of inductor Q_u factors (unloaded Q-factor) of the order of 300 at low frequencies and in comparatively small physical sizes, generally of the toroidal shape. Capacitor Q_u factors are significantly higher, with a typical ratio of capacitor to inductor Q_u factors of 5 to 10. An important advantage of LCR filters is that they are virtually noise free.

Elements of Classical Filter Theory

The theory and classification of passive LC filters terminated at both ends by a resistor is based on the two-port configuration shown in Fig. 13. The maximum available power is obtained in the load resistor with the filter removed or short-circuited and with $R_1 = R_2$, thus

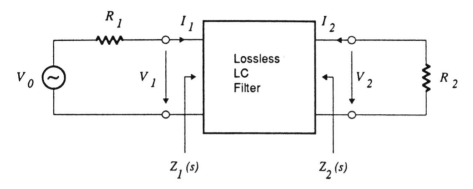

FIG. 13 Typical operating conditions for inductor-capacitor filter.

$$P_{max} = \frac{V_o^2}{4 R_1}. \tag{30}$$

Inserting the filter between source and load, the output power is

$$P_2 = \frac{|V_2|^2}{R_2}. \tag{31}$$

The transmission coefficient defined as the ratio of the power P_2 actually absorbed by the load R_2 to the maximum (real) power P_{max} available from the source is defined as

$$|t(s = j\omega)|^2 = |t(j\omega)|^2 = \frac{P_2}{P_{max}} = \frac{4 R_1}{R_2} |T(j\omega)|^2 \leq 1 \tag{32}$$

where $T(s) = V_2/V_o$ is the voltage transfer function with respect to the source voltage V_o. The ratio P_2/P_{max}, or its inverse, provides a well-defined measure of the power-transmission efficiency of a terminated two-port. For $P_2/P_{max} = 1$, all the available source power is transmitted to the load; for $P_2/P_{max} = 0$, none of it is transmitted. It is common to define the transducer function $H(s)$, which is the inverse of the transmission coefficient $t(s)$, as

$$H(s) = \frac{1}{t(s)} = \frac{1}{2} \sqrt{\frac{R_2}{R_1}} \frac{V_o}{V_2(s)}. \tag{33}$$

Finally, with the input impedances $Z_1(s)$ and $Z_2(s)$ of the doubly loaded reactance two-port in Fig. 13, we define a reflection coefficient

$$\rho(s) = \frac{Z_1(s) - R_1}{Z_1(s) + R_1} \tag{34}$$

which describes the matching of the filter input (driving-point) impedance to its termination R_1 at the input terminals of the two-port. An equivalent expression, involving $Z_2(s)$ and R_2, can be defined for the matching at the output (load) terminals. With these expressions, and in analogy to transmission-line theory, it is now illustrative to interpret the power distribution from source to load hypothetically as follows: the source is assumed to supply its maximum power P_{max} to the primary port of the lossless reactive filter two-port at all times. However, because of the mismatch between $Z_1(j\omega)$ and R_1, a part of this power P_{max}, namely, the reflected power P_r, is reflected from the two-port back to the source, where it is reabsorbed. Classical, or insertion loss filter synthesis, now can be interpreted as the problem of synthesizing a reactive LC two-port, with the input impedance $Z_1(j\omega)$ such as to reflect into the source by mismatch with R_1 as much power as possible in the stop band and, by optimum matching with R_1, to transmit into the load as much power as possible in the pass band. It can be shown readily that the reflected power P_r can be expressed by P_{max} and $\rho(j\omega)$ as follows:

$$P_r = \left| \frac{Z_1(j\omega) - R_1}{Z_1(j\omega) + R_1} \right|^2 P_{max} = |\rho|^2 \cdot P_{max} \tag{35}$$

With the interpretation above, the following power balance holds:

$$P_2 + P_r = P_{max} \tag{36}$$

Thus, with Eqs. (35) and (36), it follows that

$$|t(j\omega)|^2 + |\rho(j\omega)|^2 = 1 \tag{37}$$

or

$$t(s) \cdot t(-s) + \rho(s) \cdot \rho(-s) = 1. \tag{38}$$

This is known as the *Feldtkeller equation* after the network theorist R. Feldtkeller. Defining the so-called characteristic function $K(s)$ by

$$|H(s)|^2 = \frac{P_{max}}{P_2} = 1 + |K(s)|^2 \qquad s = j\omega \tag{39}$$

we have from Eqs. (35) and (36)

$$|\rho(s)|^2 P_{max} = |K(s)|^2 \cdot P_2 \qquad s = j\omega \tag{40}$$

or with Eq. (34)

$$|\rho(j\omega)|^2 = \frac{|K(j\omega)|^2}{|H(j\omega)|^2} = |K(j\omega)|^2 \cdot |t(j\omega)|^2. \tag{41}$$

The characteristic function therefore also can be defined by

$$K(s) = \rho(s) \cdot H(s) \tag{42}$$

and the Feldtkeller equation written as

$$H(s)H(-s) + K(s)K(-s) = 1. \tag{43}$$

Note that the reactive two-port is assumed lossless, therefore $P_1 = P_2$, that is, the power flow from the input to the output is lossless. If we let $\rho(s)$ in Eq. (34) be $\rho_1(s)$ and introduce the equivalent $\rho_2(s)$ for the output terminals, it therefore follows that $\rho_1(s) = \rho_2(s) = \rho(s)$.

Rather than use the linear measures $t(s)$ and $\rho(s)$ for the forward and reflected power flow, respectively, it is useful, in a practical context, to introduce corresponding logarithmic measures as follows. The transducer constant Γ_m is defined as the natural logarithm of $H(j\omega)$, thus

$$\Gamma_m = \ln H(j\omega) = A_m(\omega) + jB_m(\omega) \tag{44}$$

and the *transducer loss* (also called *insertion loss, effective attenuation,* or *transfer loss*)

$$A_m(\omega)[\text{nepers}] = \ln|H(j\omega)| \tag{45}$$

and the *phase lag* (output voltage V_2 lagging behind the source voltage V_o)

$$B_m(\omega)[\text{radians}] = \arg H(j\omega). \tag{46}$$

Thus

$$e^{2A_m} = \frac{P_{\max}}{P_2} = |H(s)|^2 \qquad s = j\omega \tag{47}$$

and, similarly,

$$e^{2A_r} = \frac{P_{\max}}{P_r} = \frac{1}{|\rho(s)|^2} \qquad s = j\omega \tag{48}$$

where $A_r(\omega)$ [nepers] is the return loss or the echo attenuation in a transmission system. To obtain A_m and A_r in decibels, the quantities in nepers must be multiplied by $20/\ln 10 \approx 8.686$ as in Eq. (13). With the logarithmic quantities A_m and A_r, the Feldtkeller equation becomes

$$e^{-2A_m} + e^{-2A_r} = 1. \tag{49}$$

Note that the linear quantities $t(s)$ and $\rho(s)$, and the corresponding logarithmic quantities A_m and A_r, respectively, have some practical (and also historical) significance. The transmission coefficient $t(s)$ is used for the classification of filters in tables. The transducer loss A_m is useful in defining filter specifications and in table classification, and the return loss A_r is a useful measure in communications, particularly in telephone systems. Naturally, only one quantity need be specified; the other three then can be computed by a combination of the Feldtkeller equation and a logarithmic transform.

The Classification of Normalized Low-pass Filters

Filters generally are tabulated in the form of normalized low-pass filters. If other filters are required (e.g., band pass, high pass, etc.) they can be obtained by applying a frequency transformation to the low-pass transfer function or to its individual components. The actual filter frequencies and impedance levels are obtained from the normalized low-pass filters by frequency and impedance scaling. This is achieved by denormalization using frequencies and impedance levels (i.e., resistive terminations) determined by the desired filter specifications.

The normalized low-pass filters are tabulated by types according to the approximation method used. Thus, typically, Butterworth, Chebyshev, Chebyshev–Cauer or elliptic, Bessel, and Legendre filters may be listed in catalog form. The more parameters required to characterize such a type, the more extensive must be the corresponding tables. Thus, for a Butterworth low-pass filter, having established the necessary filter order, it is sufficient to list the coefficients of the denominator polynomial because the numerator is a constant. It is a so-called all-pole filter; the zeros (with a number that corresponds to the order of the denominator polynomial) are said to be "at infinity" (i.e., at infinite frequency). A filter that features both a denominator and a numerator polynomial (i.e., poles and finite zeros), such as a Chebyshev–Cauer filter, requires more parameters to characterize it in catalog form. In fact, Chebyshev–Cauer filters, being the most efficient in terms of component count (resulting in a correspondingly low filter order) for a given filter selectivity, require the most parameters for their representation in catalog form.

Referring to the normalized low-pass specifications for a Chebyshev–Cauer filter as shown in Fig. 14-*a*, we require the four parameters below to locate an appropriate normalized Chebyshev–Cauer low-pass filter in a typical filter catalog. The catalog entry may be either in the form of a transfer function (poles and zeros) or as an LCR filter structure.

1. *The order* n *of the filter*. This corresponds to the order of the denominator polynomial, or to the number of poles of the transfer function. It is also a measure for the complexity or number of reactive elements required in the filter. The order n is determined by the maximum permissible ripple in the pass band (A_{max}), the minimum permissible attenuation in the stop band (A_{min}), and the attenuation slope between pass band and stop band (transition zone) given by the frequency ratio.

$$\Omega_s = \frac{\omega_s}{\omega_c} \tag{50}$$

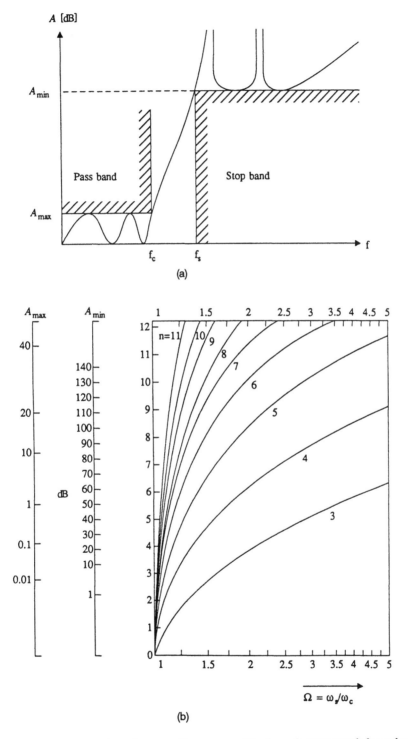

FIG. 14 Chebyshev–Cauer low-pass filters: *a*, specifications; *b*, nomograph for order *n*.

Because the relationship among these three parameters is complicated, particularly for Chebyshev–Cauer filters, it is useful to represent it in nomograph form (see Fig. 14-*b*).

2. *The reflection coefficient ρ.* This quantity is proportional to the ripple A_{max} in the pass band. However, A_{max} is unsuitable for tabulation purposes because it theoretically can vary continuously from zero to relatively large values. A set of concise tables requires a measure that is limited between well-defined values and, preferably, can be given in linear, integer quantities. The reflection coefficient, which is related to A_{max} as

$$\rho = \sqrt{|1 - 10^{-0.1A_{max}}|} \tag{51}$$

serves this purpose. The qualitative plot of ρ versus A_{max} in Fig. 15 shows this clearly. As A_{max}, plotted in decibels, increases indefinitely, ρ, plotted in percentages, increases from 0 to 100.

Most tables are ordered according to integer numbers of ρ in percent, namely, ρ = 1%, 2%, 3%, 4%, 5%, 8%, 10%, 15%, 20%, 25%, and 50%. The corresponding values of A_{max} are listed in Table 1. Thus, an in-band ripple of 0.3 dB corresponds approximately to ρ = 25%. Notice that the integer values of ρ permit only relatively few values of A_{max} to be obtained exactly. To be on the safe side of a design, the value of A_{max} therefore must be rounded down to the next lower value of ρ. Thus, for example, for A_{max} = 0.1 dB, a value of ρ = 15% should be used.

3. *The modular angle θ.* The modular angle θ is related to Ω_s, which is the

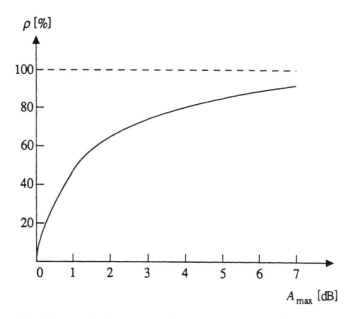

FIG. 15 Qualitative plot of ρ (in percent) versus A_{max} (in decibels).

TABLE 1 Relationship between the Reflection Coefficient ρ and the In-Band Ripple A_{max}

$|\rho|$ [%]	A_{max} [dB]
0	0
1	0.00043
2	0.0017
3	0.0039
4	0.007
5	0.011
8	0.028
10	0.044
15	0.099
20	0.18
25	0.28
35	0.57
50	1.25

measure for the attenuation slope between pass band and stop band, given by Eq. (50). As is the case with A_{max}, Ω_s is also an unsuitable quantity for the purpose of tabulation. Typically, Ω_s, which must be larger than unity (since brick-wall filters are not physically realizable), will have values between unity and, say, five. Theoretically, however, Ω_s can be arbitrarily large. The question is how to transform Ω_s into a measure that permits a classification in linear terms while giving emphasis to the low values of Ω_s ranging just beyond unity. The measure used, namely, the modular angle θ, accomplishes this objective. It is related to Ω_s as follows:

$$\theta = \sin^{-1}\left(\frac{1}{\Omega_s}\right) \qquad (52)$$

The plot of θ versus Ω_s, shown schematically in Fig. 16, demonstrates that θ is limited between 0° and 90° and that the most linear and spread-out portion of the curve covers the values of Ω_s just beyond unity. Thus, for example, $\Omega_s = 1.5$ corresponds to a modular angle $\theta = 42°$. Again, to be on the safe side in any given design, the linear distribution of θ in tables requires a rounding off of Ω_s.

4. *The symmetry factor K.* This value indicates the required equal or unequal termination resistances of a filter. It is relevant only for passive LC filters, or for other filter types (e.g., active, digital, switched capacitor) if they are based on a simulated LC filter.

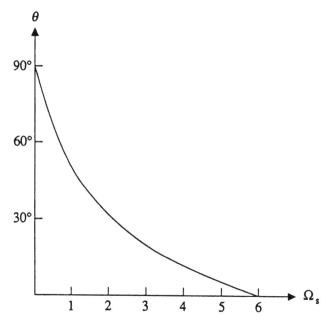

FIG. 16   Qualitative plot of θ (in degrees) versus Ω_s.

Example.   As an illustrative example, we now shall find the Chebyshev–Cauer, or CC, transfer function corresponding to the filter specifications given in Fig. 17. From the nomograph in Fig. 14-b, we find the order $n = 5$. From Table 1, we find $\rho = 25\%$, and, from Eq. (52), we compute $\theta = 49°$. In terms of a typical filter handbook, the desired filter therefore would be classified as

$$\text{CC} \qquad 05 \qquad 25 \qquad 49 \tag{53}$$

Chebyshev–Cauer   Polynomial order n Reflection coeffi-   Modular angle $\theta[°]$
                                              cient $\rho[\%]$

From one of the numerous filter catalogs listed in the bibliography at the end of this article, we obtain the normalized poles and zeros of the transfer function:

$$
\begin{aligned}
p_o &= -\sigma_o = -0.54010 \\
p_1, p_1^* &= \sigma_1 \pm j\Omega_1 = -0.08058 \pm j1.0277 \\
p_3, p_3^* &= \sigma_3 \pm j\Omega_3 = -0.32410 \pm j0.7617 \\
z_2, z_2^* &= \pm j\Omega_2 = \pm j1.9881 \\
z_4, z_4^* &= \pm j\Omega_4 = \pm j1.3693
\end{aligned}
\tag{54}
$$

and the corresponding normalized transfer function

$$T(s) = K \frac{(s^2 + 3.9525)(s^2 + 1.8750)}{(s + 0.5401)(s^2 + 0.1612s + 1.0627)(s^2 + 0.6482s + 0.68523)}. \tag{55}$$

This transfer function is normalized with respect to the transmission level and the cutoff frequency ω_c. For a given cutoff frequency ω_c, the transfer function must be scaled

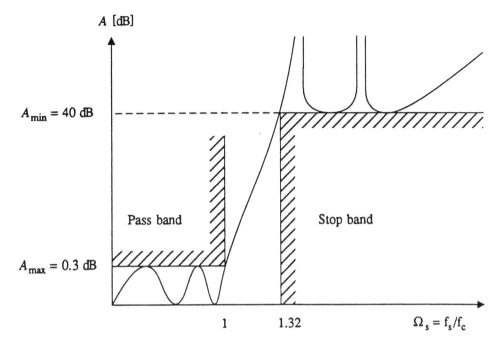

FIG. 17 Example of filter specifications for Chebyshev–Cauer filter.

accordingly; each normalized frequency term must be multiplied by ω_c such that the resulting $T(s)$ remains dimensionless. Similarly, to obtain a given signal level, say at $\omega = 0$, K must be selected accordingly. For $|T(0)| = 1$, K must be selected such that

$$K \frac{3.9525 \times 1.8750}{0.5401 \times 1.0627 \times 0.68523} = 1 \tag{56}$$

or $K = 0.053$. Beside the poles and zeros, a typical filter handbook also supplies a normalized LC filter structure for each pair of n and ρ values. Since each LC network has an LC dual, both networks are given. One is a minimum L structure, in which the minimum possible number of inductors for a given filter order occur, the other is the corresponding minimum C structure. For passive LC filters, the minimum L structure is advantageous because of the size and cost of inductors; for some active filter realizations the minimum C structure is preferred. For the specifications of our example, as given in Fig. 17, we obtain the two normalized dual networks shown in Figs. 18-*a* and 18-*b*. They both are designated by the classification according to Eq. (53), have exactly the same transfer function and transfer characteristics, but differ significantly physically, and thus differ from a practical point of view. The normalized L' and C' values given in Figs. 18-*a* and 18-*b* correspond to a symmetrically terminated network (i.e., $R_1 = R_2$ in Fig. 13). Another set of values is available in most tables for an unloaded output: $K^2 = \infty$. As is shown in the next section, in order for the excellent sensitivity properties (e.g., low sensitivity to element tolerances) of an LC ladder filter to apply, the doubly and symmetrically terminated version must be used (i.e., $R_1 = R_2$, or $K^2 = 1$).

For each normalized (dimensionless) inductance value L_i' and capacitance value C_j', the corresponding denormalized inductance L_i and capacitance C_j is obtained in terms

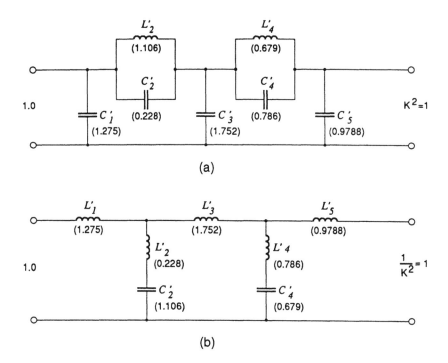

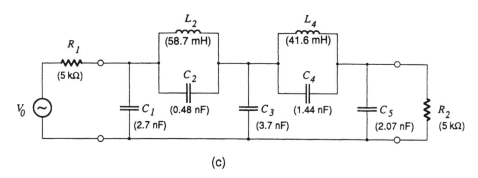

FIG. 18 CC-05-25-49 low-pass filter: *a*, normalized minimum L version; *b*, normalized minimum C version; *c*, denormalized minimum L version.

of the specified cutoff radian frequency ω_c and the impedance level R (corresponding to the termination resistance) as

$$L_i = L_i' \cdot \frac{R}{\omega_c}\,; \qquad C_j = \frac{C_j'}{\omega_c R} \qquad (57)$$

To obtain other than low pass filters (e.g., high pass, band pass, band stop), appropriate frequency transformations must be applied either to the filter transfer function (e.g., Eq. 55) or to each reactive component of the filter network. For this, the reader is referred to the filter handbooks listed at the end of this article.

For a doubly-terminated filter of impedance level 5 $k\Omega$ ($R_1 = R_2 = 5$ $k\Omega$) and a cutoff frequency $\omega_c = 2\pi \cdot 15$ kHz, the denormalized minimum L version of the CC-05-25-49 filter is shown in Fig. 18-c. The corresponding minimum C filter is readily obtained from Fig. 18-b.

The Reactive Ladder Filter and Its Properties

In the introduction to this article, special reference is made to the extraordinarily low sensitivity of reactive LC ladder networks to variations of individual element values. Thus, for example, if properly designed and resistively terminated at both ends, the tolerances of the individual components of an LC ladder filter may be orders of magnitude larger than the tolerance of the resulting frequency response in the pass band. In other words, a 0.1-dB ripple can be guaranteed easily with components having no more than 1% accuracy if the order of the filter is sufficiently high. This remarkable feature of doubly (resistively) terminated LC ladder filters was pointed out first by H. J. Orchard in 1968, decades after LC filters first had come into widespread use. It also is responsible for the fact that subsequent approaches to filter design based on modern technology (e.g., active RC, SC, digital) attempted to recapture, often successfully, this excellent low-sensitivity property of the ladder network. This was accomplished by simulating the behavior and the properties of an LC ladder structure, even though the actual components were entirely different. This low-sensitivity property, sometimes called the Orchard theorem after the originator of its formulation, is responsible for the fact that the doubly terminated LC ladder structure (real or simulated) plays a central role in filter theory and design, no matter what the technology used for the actual filter realization. Because of its importance for filter theory as a whole, the Orchard theorem therefore is outlined in more detail below.

The relative sensitivity of a quantity or function F to relative variations of a component x, can be defined by the following equivalent expressions:

$$S_x^F = \frac{dF/F}{dx/x} = \frac{d[\ln F]}{d[\ln x]} = \frac{dF}{dx} \cdot \frac{x}{F}. \tag{58}$$

Thus, the relative variation of F caused by a relative change in the component x becomes

$$\frac{\Delta F}{F} = S_x^F \frac{\Delta x}{x} \tag{59}$$

where the difference values ΔF, Δx, rather than the differential values dF, dx, are used to denote physical, or measured, changes. If, now, we wish to find the sensitivity of the transmission coefficient $t(x,\omega)$ of a doubly terminated reactive ladder network with respect to a component x, we can write

$$\frac{\Delta |t(x,\omega)|}{|t(x,\omega)|} = S_x^{|t(x,\omega)|} \frac{\Delta x}{x}. \tag{60}$$

The relative sensitivity term in Eq. (60) becomes, with Eq. (37) and the sensitivity definition in Eq. (58)

$$S_x^{|t(x,\omega)|} = S_x^{\sqrt{1-|\rho(x,\omega)|^2}} = -\frac{|\rho(x,\omega)|^2}{|t(x,\omega)|^2} \cdot S_x^{|\rho(x,\omega)|} \tag{61}$$

From Eqs. (32) and (35), we have, in the filter pass band,

$$\frac{|\rho(x,\omega)|^2}{|t(x,\omega)|^2} = \frac{P_r}{P_2} \ll 1. \tag{62}$$

This becomes clear immediately if we look at the typical $|t(\omega)|^2$ and $|\rho(\omega)|^2$ frequency response for a low-pass filter, shown quantitatively in Fig. 19. In the pass band, $|t(\omega)|^2 \approx 1$ and $|\rho(\omega)|^2 \ll 1$. In the stop band, the situation is reversed. Thus, in the pass band, the ratio $|\rho(\omega)|^2/|t(\omega)|^2$, which multiplies the sensitivity $S_x^{|\rho|}$ in Eq. (61), will be smaller as the filter order increases and the more frequently $|\rho(\omega)|$ is zero (so-called reflexion zeros in the pass band). Accordingly, at the reflexion zeros (which occur only in the pass band since at these frequencies $|t(\omega)| = 1$), the transmission sensitivity $S_x^{|t|}$ is actually identically equal to zero. It follows from Fig. 19 that the smaller the specified ripple of $|t(\omega)|$ within the pass band and, therefore, the higher the filter order (and with it the number of reflexion zeros in the pass band), the closer $|\rho(\omega)|$ will be to zero in-between the reflexion zeros. Thus, the ratio $|\rho(\omega)|^2/|t(\omega)|^2$, which is part of the product making up the transmission sensitivity on the right-hand side of Eq. (61), will be much smaller than unity within the pass band. What about the other multiplicand in Eq. (61), that is $S_x^{|\rho(\omega,x)|}$? Since

$$S_x^{|\rho(\omega,x)|} = \frac{d|\rho(\omega,x)|}{dx} \cdot \frac{x}{|\rho(\omega,x)|} \tag{63}$$

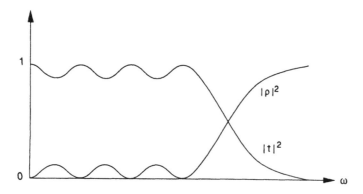

FIG. 19   Typical frequency plot of $|t(\omega)|^2$ and $|\rho(\omega)|^2$.

the question is whether anything can be said about the derivative $d|\rho|/dx$ in Eq. (63). To answer this question, we refer to Orchard's brief intuitive argument as it was first published:

> If one designs a flat-passband reactance ladder filter to operate from a resistive source into a resistive load, and arranges that, at the frequencies of minimum loss over the passband, the source delivers its maximum available power into the load, one finds, to a first order of approximation, that, at every frequency in the pass band and for every component, the sensitivity of the loss to component tolerances is zero. This is easily checked by noting that, when one has zero loss in a reactance network, a component change, either up or down, can only cause the loss to increase; in the neighbourhood of the correct value, the curve relating loss to any component value must therefore be quadratic, and, consequently, *d(loss)/d(component) must be zero!* (Italics added. 1, pp. 224–225)

In this text, "loss" is synonymous with the reflexion function $|\rho(\omega,x)|$.

To summarize, Orchard's theorem says that any doubly terminated passive LC ladder network will have a transmission sensitivity with respect to variations of its components that will be smaller in the pass band as the pass-band ripple decreases and the filter order increases. This is in contrast to most other filter structures (and, indeed, linear systems), which become all the more prone to instability and to possessing a high component sensitivity, as the filter complexity and order increases. This explains why, in whatever technology (active RC, digital, SC), the ladder structure simulating a doubly terminated LC ladder network is the preferred structure when high performance is required. Here, performance pertains both to a high selectivity and order of the filter as well as to a low tolerance of the pass-band characteristics and to a low sensitivity to component variations in the pass band.

Note that Orchard's theorem does not make any claims about insensitivity to component variations in the stop band. Indeed, it can be shown that such other structures as cascades of second-order (biquadratic) filters (or *biquads*) well may display a lower sensitivity to component variations in the stop band. Since the specifications in the pass band generally have a higher priority than those in the stop band, Orchard's theorem retains its importance. Nevertheless, since ladder networks are more difficult to manufacture (as LC structures or in simulated form) economical considerations often dictate the production of biquad cascades or variations thereof (see the sections, "Active Resistor-Capacitor Filters" and "Switched-Capacitor Filters"). In fact, depending on the application and the overall requirements, either one of the two basic filter types may constitute the more appropriate realization.

Electromechanical Filters

A designer of frequency-selective networks or filters intended for microcircuit implementation or for low-frequency applications will delete, a priori, all electromagnetic items (transformers, inductors, etc.) from the list of possible components. This eliminates all components characterized by two of Maxwell's four

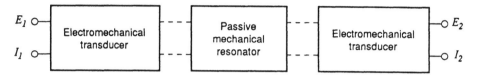

FIG. 20 Generalized electromechanical filter.

equations* governing electromagnetic phenomena. Instead, it necessitates the inclusion of other components with analogous resonance properties that, with the help of suitable transducers, can be made to interact with electric signals and networks. Certain electromechanical devices that combine mechanical resonators with energy transducers based on the piezoelectric, the magnetostrictive, or the electrostrictive effect come closest to meeting this requirement.

The frequency-sensitive element of the generalized electromechanical filter shown in Fig. 20 is a mechanical transmission device in which mass or moment of inertia, and elastic compliance or stiffness, interact in resonance at a particular frequency. A direct analogy to electromagnetic resonance can be found in that there is a correspondence of mass to inductance, of stiffness to capacitance, and of mechanical resonance to electrical resonance. The conversion of the energy within the mechanical resonator into electrical energy and the coupling of the converted electrical signals to an electric network are obtained by means of an electromechanical transducer, which generally is based on the piezoelectric, magnetostrictive, or electrostrictive effect.

Broadly speaking, there are two basic categories of electromechanical filters: those that provide mechanical resonance properties and the capability for energy conversion in one device, and those that combine two separate materials or devices to perform these functions. The most important groups in the former category are monolithic crystal and ceramic filters; in the latter category, the group generally called mechanical filters is most important.

Monolithic Crystal Filters

Monolithic crystal filters use coupled mechanical vibrations in a piezoelectric material to provide band-pass-type filter functions. A filter consists of a crystalline quartz wafer onto which pairs of metal electrodes are deposited. The operation of the filter is made possible by two factors:

1. The crystal is piezoelectric. It can transfer electrical energy into a mechanical form, specifically, into a transverse shear wave (see Fig. 21) and back

*James Clerk Maxwell (b. November 13, 1831, Edinburgh; d. November 5, 1879, Cambridge, England) ranks next to Newton for his fundamental formulation of electromagnetic field theory as well as numerous other scientific contributions (color vision, mechanics, kinetic theory of gases, etc.). His electromagnetic theory culminated in the formulation of four fundamental equations relating electrical and magnetic field phenomena. Two of the equations pertain to electric field strength and flux density, two to magnetic field strength and flux density. Capacitors and inductors, as passive frequency-dependent filter components, are governed by the former and latter pair of equations, respectively.

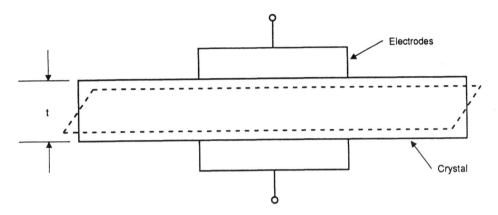

FIG. 21   Transverse thickness shear wave of AT-cut crystal slice.

again. Therefore, in addition to serving as an interresonator coupling medium, it performs the transducing functions.

2. The metal electrodes lower the resonance frequency of the transverse shear wave in the plated region as compared with unplated quartz. As a result, the resonance created in the plated region does not propagate into the areas without electrodes, but remains trapped under the electrodes, which are thin metal films.

In Fig. 22-*a*, a monolithic filter is shown in its simplest form. It consists of an input and output resonator formed by a pair of electrodes deposited onto opposite faces of a quartz crystalline wafer. Vibration, induced by resonance between the plates, decays very rapidly outside the plated region; therefore, it essentially is trapped under the electrodes. Although the vibration is confined to the electrode area, the displacement decays exponentially in the surrounding region and this, mechanically or acoustically, couples adjacent resonators. The dimensions and separation of the resonators determine the coupling, bandwidth, transmission characteristics, and terminal impedance. The equivalent electrical network is shown in Fig. 22-*b*.

The amount of coupling between adjacent resonators may be controlled within limits since it depends upon the dimension of the resonators, the thickness of the metal electrodes, and the spacing between resonant regions. In AT-cut crystals, for example, the range of coupling coefficients permits the realization of pass bands ranging from about 0.001% to about 0.2% of the center frequency if the fundamental mode of the thickness shear vibration is used. The techniques available to shape the wafer permit center frequencies from about 5 MHz to approximately 150 MHz.

In general, the important factors in the monolithic filter are the number of resonant elements, the coefficient of coupling between adjacent resonant elements, and the impedance of the resonance elements. More resonant elements result in a higher rate of cutoff, that is, a more rapid increase of attenuation as a function of frequency in the frequency range adjacent to the flat pass band. This is shown qualitatively in Fig. 23 for a four-resonator filter. A cross-section

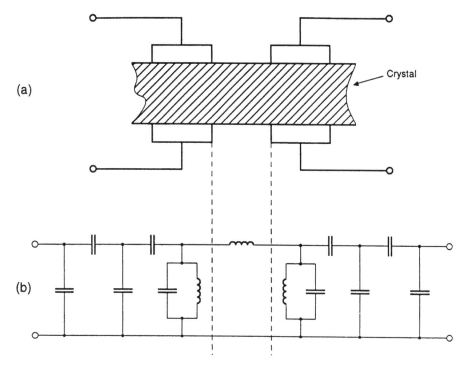

FIG. 22   Monolithic crystal filter: *a*, simplest form; *b*, equivalent circuit.

of the filter with its electrode arrangement is shown in Fig. 23-*a* and the equivalent circuit diagram in Fig. 23-*b*. The frequency-response curves in Fig. 23-*c* show the variation of the current with frequency as the input voltage is held at 1 volt (V). The improvement in selectivity as the current passes through the resonators is apparent from the increased steepness of the slope at the band edges. The "bumps" inside the pass-band region indicate reflections in the filter. This reflective interaction eventually leads to the desired flat response in the band by the time the current passes the last resonator.

Table 2 presents a summary of the salient features of monolithic crystal filters. The advantages of these filters from the standpoint of Q capability and frequency stability are apparent. In fact, there is no comparable technique to match this performance. However, the frequencies are limited to the MHz range. One possible method of using monolithic crystal filters for low-frequency applications is to heterodyne the low-frequency signal into the crystal frequency range and then to demodulate the filtered signal back to the low frequencies.

Functionally, the versatility of monolithic crystal filters is limited. They are inherently all-pole (more specifically, band-pass) filters with the number of transmission poles corresponding to the number of electrode pairs. As such, they permit the realization of any standard band-pass function (Chebyshev, Butterworth, etc.). Other filter functions (band-elimination, transmission zeros, low pass, high pass, etc.) can be realized only with additional components, although a degree of functional variability can be obtained by splitting electrodes.

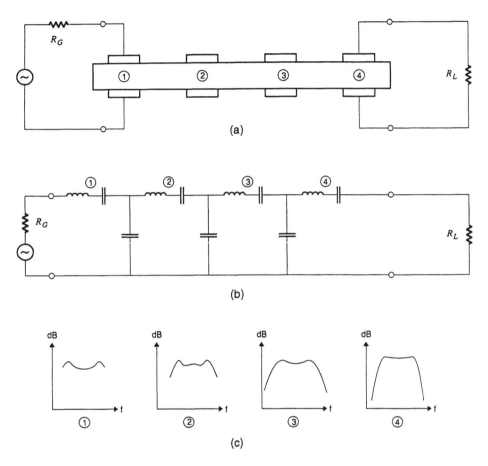

FIG. 23 Four-resonator filter: *a*, cross-section; *b*, equivalent circuit diagram; *c*, frequency-response curves after successive resonators.

Initial filter adjustment and tuning can be carried out very accurately (e.g., 0.1-dB ripple in the pass band of, say, an eight-pole band-pass filter can be reproduced consistently). However, because of unwanted vibrations within the quartz plate, invariably there are additional pass bands at higher frequencies. The amount of interference with the required frequency response will depend on how close these spurious resonant modes are to the frequency band of interest. The magnitude of the unwanted modes may be reduced to some extent by adjusting the areas of the various resonators appropriately. However, this in turn may degrade the Q of the useful main shear mode and effect the desired out-of-band frequency rejection of the filter.

The mechanical or acoustical coupling, which determines the bandwidth of the filter, depends principally on the electrode separation and, to a lesser extent, on the electrode mass and length. The bandwidth is controlled by electrode separation, mass, and area. It can be adjusted by placing stripes between the electrodes and either adding to these stripes by evaporation or removing them by laser. The center frequency, determined principally by the plate thickness, is

TABLE 2   Characteristics of Electromechanical Filters*

Type	Frequency Range	Pole Q	$\frac{\Delta f^\dagger}{f}$ ppm/°C	$\frac{\Delta Q^\dagger}{Q}$ %	Functional Versatility	Functional Accuracy	Tunability	Signal Dynamic Range, dB
Monolithic crystal	5-150 MHz	1000-250,000	± 1	0.1	Fair (band-pass and frequency rejection networks)	Good (with exception of spurious modes)	Initial tuning good; system adjustment poor	40-80 (depending on proximity to spurious tones)
Ceramic	0.1-10 MHz	30-1,500	± 100	0.2	Fair (requires additional components for most functions)	Good (with exception of spurious modes)	Initial tuning good; system adjustment poor	40-80 (depending on proximity to spurious tones)
Mechanical	0.5kHz-1MHz	50-30,000	± 50	0.1	Fair (band-pass and frequency rejection; numerous nonfilter functions possible)	Good	Good	60-80

* The characteristics listed may be mutually exclusive.
† Frequency range over temperature range from 0° to 60° C and aging.

fine tuned by the mass of electrodes. These operations can be controlled tightly in manufacture. However, once the filter is assembled into a system, it is very difficult to make any additional adjustments. Furthermore, since the entire resonator structure is coupled, it is impossible to measure the uncoupled frequency of any one resonator. Adjustments of resonators, therefore, must take this coupling into account with the help of theoretical expressions derived for this purpose.

Compared with their conventional counterparts, monolithic crystal filters permit considerable size reduction since no transformers, inductors, or other discrete components are needed. Typical applications for monolithic crystal filters traditionally have been in telephone carrier systems (frequency range, 2.6–19 MHz); point-to-point wire and radio transmission systems with single-sideband, double-sideband, or narrowband frequency modulation; and broadband telephone multiplex systems. All of these systems, in addition to being in the MHz frequency range, have stringent requirements for Q and frequency stability that are hard to meet economically by other methods.

Ceramic Filters

Two major limitations of monolithic crystal filters are their restriction to frequencies in the MHz range and their narrow bandwidths (i.e., high Qs). Both of these limitations can be overcome partially by ceramic (e.g., lead-titanate-zirconate [Pb-Ti-Zr] composition) filters. This is so because of the difference in material properties (principally, the piezoelectric coupling coefficient) between ceramic and crystal filters, rather than because of any basic difference in concept. Thus, ceramic filters combine the functions of the mechanical resonator and the electromechanical transducer shown in Fig. 20, due to their combined mechanical resonance and piezoelectric properties, in the same way that monolithic crystal filters do.

In contrast to monolithic crystal filters, ceramic filters are used more in combinations of individual two- or three-electrode resonators of the kind shown in Figs. 24 and 25 than as monolithic structures. Two-electrode resonators may be interconnected with capacitors and/or amplifiers to provide high-order filter configurations. A building-block approach to ceramic filter design in which identical two-electrode resonators are combined with capacitors in a ladder configuration is shown in Fig. 26. By selecting appropriate capacitor combinations instead of a variety of different resonators, a wide range of frequency responses can be obtained. Composite filters using three-electrode resonators also behave essentially like ladder structures consisting of cascades of resonators, coupled either directly or by series or shunt capacitors. Figure 27 shows a two-resonator filter network.

As with crystal filters, efforts are made, when additional components are necessary to realize a desired network function, to avoid inductors and to use active auxiliary networks instead. In this way, ceramic filters can be incorporated into or combined with hybrid integrated circuits. An example of a resona-

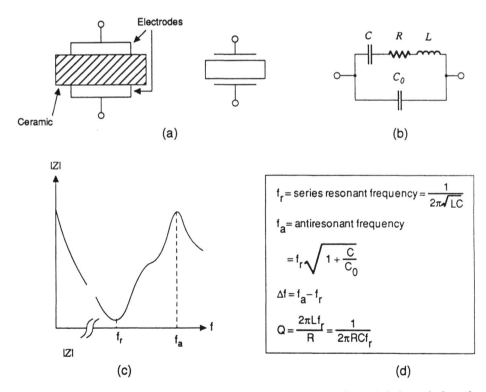

FIG. 24 Basic two-electrode resonator: *a*, cross-section and circuit symbol; *b*, equivalent circuit; *c*, impedance frequency characteristics; *d*, design equations.

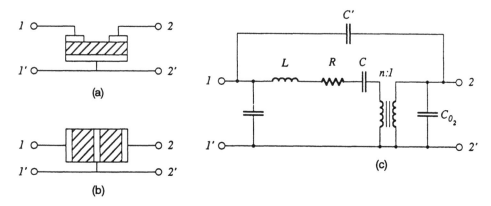

FIG. 25   Three-electrode resonators: a, basic form; b, composite three-electrode resonator obtained by bonding together two single resonators; c, equivalent circuit.

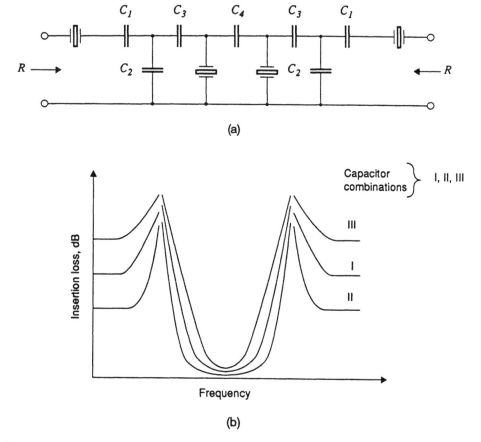

FIG. 26   Four-resonator network with identical resonators and four different capacitors: a, circuit diagram; b, frequency response.

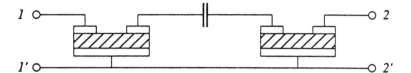

FIG. 27 Two-resonator version of a composite filter using three-electrode resonators.

tor-op amp (with differential output) combination is shown in Fig. 28-*a*. The transfer function of this circuit is given by

$$T(s) = \frac{e_{\text{out}}}{e_{\text{in}}} = -\frac{R_F}{R_G} \frac{Z_b - Z_a}{0.75[R^2 + R(Z_a + Z_b) + Z_a Z_b]^{1/2}} \tag{64}$$

This function corresponds to a hybrid-lattice filter of the kind shown in Fig. 28-*b*, except that it is realized with an op amp instead of a transformer. The frequency response that corresponds to the selectivity of a critically coupled double-tuned LC circuit is shown in Fig. 28-*c*.

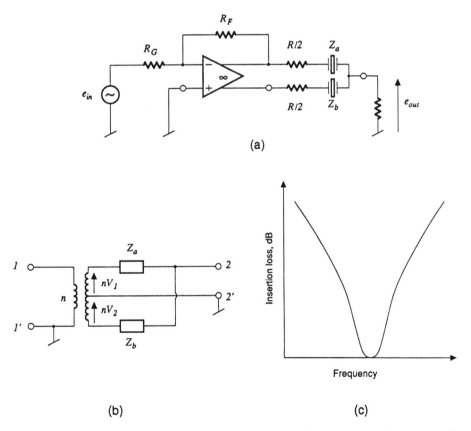

(a)

(b)   (c)

FIG. 28 Hybrid lattice filter using ceramic resonators: *a*, active realization; *b*, passive realization; *c*, frequency response.

The characteristics of ceramic filters, as determined by the ceramic resonators, are summarized in Table 2. The lower frequency range and wider bandwidths (lower Q values) are apparent. The electrical resonator parameters may be varied over a range, depending upon the ceramic material, processing, and resonator geometry. The quantities $\Delta f/f_r$ and C_o (see Fig. 24) may be adjusted over a wide range by the manufacturing process. This makes the ceramic resonator a more versatile electric circuit element than, for example, the quartz resonator (the equivalent circuit of which is the same as that of Fig. 24 but with a $\Delta f/f_r$ that is a material constant of small and fixed value).

All equivalent circuits of the piezoelectric resonator are valid only in the vicinity of the operating frequency. For radial resonators, the circuit equivalent of Fig. 24-*b* is accurate for frequencies up to approximately $1.5f_r$. At higher frequencies, other resonances (spurious modes) occur due to overtones of the radial mode and to other vibrational modes.

Spurious responses within the individual resonator may be suppressed by optimizing both the processing and the geometry of the resonator. However, some of these approaches are not compatible with economy. For instance, one known method completely eliminates all radial overtones, but results in a more costly resonator and a bulkier, less rugged package. On the other hand, complete elimination of one radial overtone, or partial suppression of two neighboring radial overtones, may be achieved by controlling the resonator geometry without sacrificing economy, size, or ruggedness.

Mechanical Filters

Mechanical filters generally consist of a mechanical resonator, and rely on the electrostrictive, magnetostrictive, or piezoelectric effect of a separate transducer for the interaction between electrical and mechanical energy, particularly for a direct relation between electrical and mechanical resonances. The H-shaped resonator shown in Fig. 29 is an example. It consists of two balanced masses (i.e., the bars of the H) connected by a flexible web. As the piezoelectric input transducer contracts and expands, the web flexes and the bars oscillate in opposite directions. The resonance signal can be picked off by a coil near a bar or by another piezoelectric transducer underneath the web. Since the bars rotate counter to each other at their nodal points (i.e., points of least deflection), there is virtually no net transmission of energy through the common plane of the base. Thus, in contrast with most other mechanical resonators, the resonant energy loss that stems from vibrations transmitted by the resonator to its mounting base is avoided. Since the Q of a mechanical resonator is given by the ratio of conserved energy to dissipated energy, very high Q values can be obtained.

Another common problem of low-frequency mechanical resonators—their susceptibility to external shock and vibrations—also is reduced greatly. In the same way that transmission of vibrations from the resonator to the mount is prevented by the symmetrical push-pull configuration of the H-shaped resonator, external shocks are prevented from traveling from the mount to the resonator. This process is similar to the rejection of a common-mode signal at the

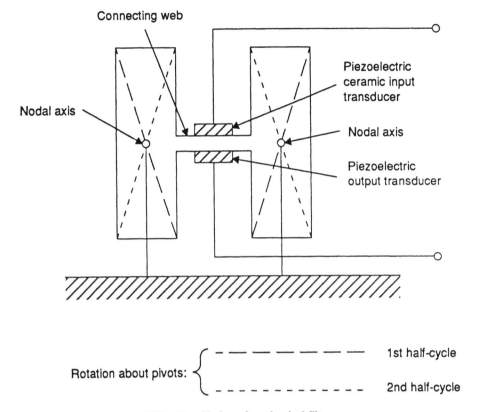

FIG. 29 H-shaped mechanical filter.

input to a differential amplifier in that any vibrations originating at the base are common to both bars of the resonator, and therefore cancel out.

The main characteristics of typical mechanical filters are listed in Table 2. The resonance frequencies range from a few hundred Hz to close to 1 MHz; Q values can vary anywhere between 50 and 5000. The response of a resonator with a Q of 4000 at 318 Hz is shown in Fig. 30. If appropriate materials are used, dimensional variations due to temperature changes have little or no effect on the performance of mechanical filters, and the frequency stability can be controlled very tightly. For the H-resonator, the resonance frequency can be shifted 20% to 30% by balanced pairs of threaded tuning slugs inserted in threaded holes running the entire length of each bar. Low values of Q are obtained by setting the slugs at differing distances from the nodes of the bars. Consequently, this particular mechanical resonator initially can be tuned accurately as well as readjusted after incorporation into a system. The H-resonator is unconventional as far as mechanical filters go in that it does not use electrostrictive or magnetostrictive transducers; instead, it uses a piezoelectric transducer. This is possible because of its geometrical configuration.

The functional versatility of mechanical filters, just as with other electromechanical filters, is quite limited. Inherently, these filters are resonators providing band-pass or frequency-rejection characteristics. Thus, to obtain, say, low-pass,

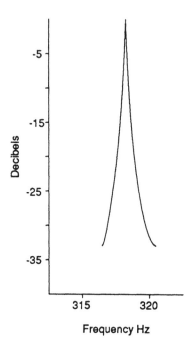

FIG. 30   Frequency response of mechanical H-resonator (f_c = 318.36 Hz; bandwidth = 0.081 Hz at −3 dB, 1 Hz at −20 dB; Q = 4000; insertion loss = −18 dB at 5 V).

high-pass, or elliptic filters with finite transmission zeros, auxiliary circuits, which may be active RC in order to avoid inductors, must be used.

Although the H-resonator and other mechanical filters may be considerably smaller than equivalent LC circuits, particularly at low frequencies, they still are far from directly compatible with IC techniques. However, work has gone on in the last few years to overcome this limitation, and reports of mechanical resonators fabricated by batch processes compatible with ICs have been reported.

Active Resistor-Capacitor Filters

One of the main problems with passive filters, both LC and electromechanically based, is their incompatibility with modern integrated circuit technology. Compared to VLSI circuits, LC filters in particular are bulky and defy the high-volume batch-processing techniques that have brought down the cost of manufacturing IC chips so dramatically. With regard to LC filters, these disadvantages outweigh their many significant advantages (e.g., unconditional stability, negligible power dissipation, low component sensitivity, low noise, offset-free DC connections, and, if desirable, total isolation using transformers), and have motivated the search for alternatives. Since most of the disadvantages of LC filters are related to the electromagnetic devices (i.e., the inductors), any viable alternatives essentially must be inductorless filters. The oldest of these are active RC filters, in which the absence of inductors is compensated by the presence of active gain elements (i.e., generally voltage- or current-gain devices). Although such filters go back to the vacuum-tube days of the late 1930s, it was only in the transistor age (which began commercially in the mid-1950s) that active RC filters became feasible.

Basic Active Network Elements

In the following treatment of active RC filters, we take a network-theoretical approach to their description, that is, we assume ideal components and ignore the circuit design considerations that, ultimately, make a practical realization possible. In the context of this survey, we have no other choice; for further details, the interested reader is referred to the suggested reading list at the end of this article.

The most important basic concepts and idealized network elements necessary for the description and analysis of active RC filters are the two-port; the nullator, norator, and nullor; the impedance converter; the impedance inverter; and the frequency-dependent negative resistor.

The Two-Port

Resorting to one of the most elementary concepts of network theory, the two-port, we refer to Fig. 31, in which a loaded (load impedance Z_L) is shown. The two-port can be described by the equations

$$V_1 = AV_2 - BI_2 \tag{65}$$

$$I_1 = CV_2 - DI_2 \tag{66}$$

and

$$Z_L = -\frac{V_2}{I_2} \tag{67}$$

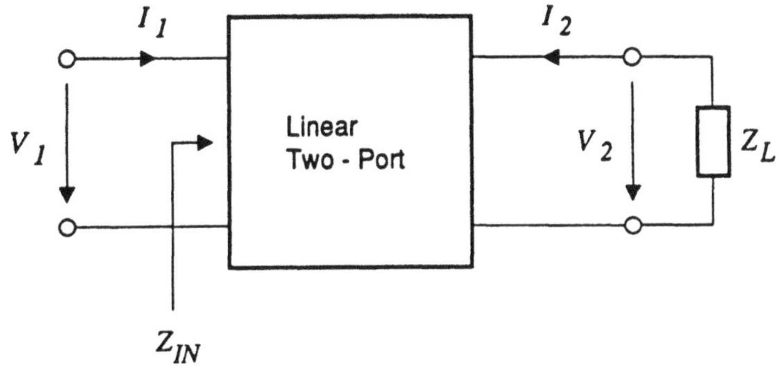

FIG. 31 Terminated linear two-port network.

The two-port equations are given in terms of the elements of the [$ABCD$] or transmission matrix. The corresponding matrix equation is then

$$\begin{bmatrix} V_1 \\ I_1 \end{bmatrix} = \begin{bmatrix} A & B \\ C & D \end{bmatrix} \begin{bmatrix} V_2 \\ -I_2 \end{bmatrix} \tag{68}$$

To obtain the input impedance Z_{IN} in terms of the [$ABCD$] parameters, a simple calculation involving Eqs. (65) to (67) results in

$$Z_{\text{IN}} = \frac{V_1}{I_1} = \frac{A\,Z_L + B}{C\,Z_L + D} \tag{69}$$

This expression provides the basis for some of the ideal network elements described below.

The Nullator, Norator, and Nullor

The nullator and norator belong to a physically nonrealizable, idealized class of network elements that have no conventional matrix representation. Nevertheless, they are very useful in the analysis and synthesis of idealized network elements. Naturally, the idealized network, at some point, must give way to a practical network with physically realizable components. The description of the network with nullators and norators then represents a sort of upper bound, with an idealized performance that, due to nonideal effects, only can be approached but never actually reached. The nullator (Fig. 32-a) is defined by the condition

$$I = V = 0 \tag{70}$$

the norator (Fig. 32-b) by

$$V = k_1, \qquad I = k_2 \tag{71}$$

where k_1 and k_2 are arbitrary. When a norator is used in a circuit, V and I take on the values needed to satisfy Kirchhoff's current and voltage laws. An ideal-

FIG. 32   Idealized network elements: a, nullator; b, norator.

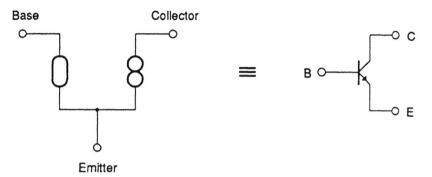

FIG. 33 Nullor equivalent of a transistor.

ized transistor and op amp can be represented by a nullator-norator combination (see Figs. 33 and 34, respectively). Nullators and norators always occur in nullator-norator pairs, also called *nullors*. The bridge from an abstract, idealized, and physically nonrealizable nullor circuit to a nonideal, physically realizable circuit is crossed by replacing each nullor either by a transistor (*Transistorization*) or by an op amp. Transistorization requires every nullor to have a common terminal (the *emitter*).

The Impedance Converter

The impedance converter is a two-port with an [*ABCD*] matrix given by

$$[ABCD] = \begin{bmatrix} A & 0 \\ 0 & D \end{bmatrix} \tag{72}$$

If A and D are frequency dependent, such that $A = A(s)$, $D = D(s)$, then we have a *general impedance converter* (GIC). For the case that A and D are constants, but of opposite polarity, that is,

$$A = \mp k_1, \qquad D = \pm \frac{1}{k_2} \tag{73}$$

FIG. 34 Nullor equivalent of an operational amplifier.

then the impedance converter loaded by Z_L has, according to Eq. (69), the input impedance

$$Z_{IN} = -k_1 k_2 Z_L \tag{74}$$

We then speak of a negative impedance converter (NIC) because, for $k_1 > 0$ and $k_2 > 0$, and for a realizable (i.e., positive) Z_L, the input impedance of the loaded two-port is negative. Thus, for $k_1 = k_2 = 1$, we have $Z_{IN} = -Z_L$.

A nullor realization of an NIC is shown in Fig. 35. Straightforward application of the defining expressions of the nullator and norator, Eqs. (70) and (71), respectively, gives the input impedance as

$$Z_{IN} = -\frac{Z_1}{Z_2} Z_L \tag{75}$$

Note that in Eq. (73) the pair $A = -k_1$, $D = +k_2^{-1}$ corresponds to an NIC with voltage reversal (VNIC), and the pair $A = +k_1$, $D = -k_2^{-1}$ corresponds to an NIC with current reversal (CNIC). The nullor configuration in Fig. 35 corresponds to a CNIC; the nullor/dual circuit (i.e., nullator and norator exchanged) results in a VNIC. Note, furthermore, that the configuration in Fig. 35 excludes a one-transistor realization of an NIC because the nullor has no common terminal. Using nullator-norator identities to extend the circuit, a two-transistor realization can be obtained. Such nullator-norator extensions are based on the fact that a nullator and norator in series are equivalent to an open circuit; in parallel, they are equivalent to a short circuit.

The Impedance Inverter

An impedance inverter is defined as a two-port with an [ABCD] matrix given by

$$[ABCD] = \begin{bmatrix} 0 & B \\ C & 0 \end{bmatrix} \tag{76}$$

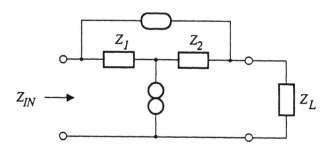

FIG. 35 Nullor realization of a loaded negative impedance converter.

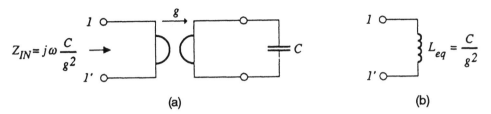

FIG. 36 Simulation of a grounded inductor: *a*, gyrator-capacitor combination; *b*, equivalent inductor.

For the special case that

$$B = 1/g_1, \quad C = g_2 \tag{77}$$

the input impedance to the loaded two-port becomes

$$Z_{\text{IN}} = \frac{1}{g_1 g_2 Z_L} \tag{78}$$

This is the defining expression for the so-called gyrator; its symbolic representation is shown in Fig. 36. The gyrator constants g_1 and g_2 have the dimension ohm^{-1}, and a capacitively loaded gyrator (see Fig. 36) has an inductive input impedance

$$Z_{\text{IN}} = j\omega \frac{C}{g_1 g_2} = j\omega L_{\text{eq}} \tag{79}$$

where the equivalent inductance L_{eq} is given by

$$L_{\text{eq}} = \frac{C}{g_1 g_2} \bigg|_{g_1 = g_2 = g} = \frac{C}{g^2} \tag{80}$$

The assumption that $g_1 = g_2 = g$ very often holds in practice. To simulate a floating inductance, we require two cascaded gyrators with a grounded capacitor between them (see Fig. 37). The equivalent inductance again is given by Eq. (79). An LC band-stop filter and its gyrator-R-C simulation is shown in Fig. 38. From Eq. (80), each gyrator constant g_i is given by

$$g_i = \left(\frac{C_{L_i}}{L_i}\right)^{1/2} \tag{81}$$

The nullor-R realization of an impedance inverter with gyrator characteristics is given in Fig. 39. It can be shown that a gyrator-type impedance inverter cannot be realized with less than three nullors. For the nullor gyrator of Fig. 39, we have

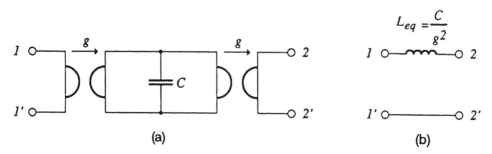

FIG. 37 Simulation of a floating inductor: a, gyrator-capacitor combination; b, equivalent inductor.

$$[ABCD] = \begin{bmatrix} 0 & R_2 \\ 1/R_1 & 0 \end{bmatrix} \tag{82}$$

and thus, with the load $Z_L = (s\, C_L)^{-1}$, we have, with Eq. (80),

$$L_{eq} = R_1 R_2 C_L \tag{83}$$

Note that the nullators in Fig. 39 are designated n_i, the norators N_i; they all are identical, however, and defined by Eqs. (70) and (71), respectively.

The Frequency-Dependent Negative Resistor

The FDNR appears as the result of an impedance-scaling procedure of an LCR network that is aimed at the substitution of resistors for all the inductors in the LCR network. Impedance scaling means multiplying all the impedances of a network by a dimensionless constant k or frequency-dependent factor $k(s)$ whereby, significantly, the transfer function of the network remains unchanged. Letting

$$k(s) = \frac{\omega_o}{s} \tag{84}$$

and scaling each L_i, C_j, and R_v by $k(s)$, the resulting scaled impedances are

$$Z'_i = \omega_o L_i = R'_i \tag{85}$$

$$Z'_j = \frac{\omega_o}{s^2 C_j} = \frac{1}{s^2 D_j} \tag{86}$$

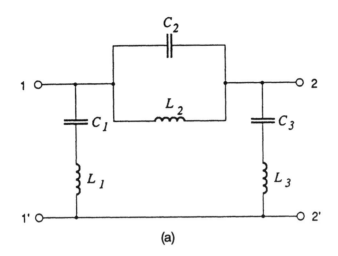

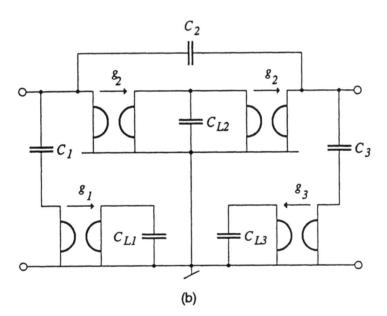

FIG. 38 Inductor simulation of an LC band-stop filter using gyrators: a, LC band-stop filter; b, gyrator-RC (resistor-capacitor) equivalent filter in which $g_1 = (C_{L1}/L_1)^{1/2}$, $g_2 = (C_{L2}/L_2)^{1/2}$, and $g_3 = (C_{L3}/L_3)^{1/2}$.

and

$$Z_v = \frac{\omega_o}{s} R_v = \frac{1}{s C_v'} \tag{87}$$

Note that every inductor L_i becomes a resistor $R_i' = \omega_o L_i$, every resistor R_v becomes a capacitor $C_v' = (\omega_o R_v)^{-1}$, and every capacitor C_j becomes a new

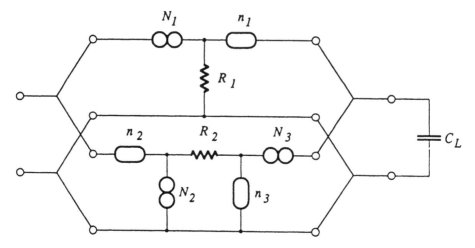

FIG. 39 Nullor-R realization of a gyrator with load capacitor C_L.

element $D_j = C_j/\omega_o$, known as an FDNR. The reason for this designation is that for $s = j\omega$ the impedance Z'_j is equal to $-(\omega^2 Dj)^{-1}$, where this negative, frequency-dependent quantity is real and has the dimension ohms. The network symbol for the FDNR is shown in Fig. 40-a. The FDNR transformation is most useful for low-pass-type filters comprising numerous floating inductors and grounded capacitors (see Figs. 40-b and 40-c). Since each LCR network has a dual, one of which is a minimum L and the other a minimum C network, the

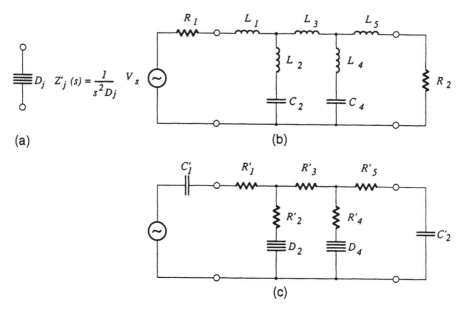

FIG. 40 Frequency-dependent negative resistor (FDNR) network: a, FDNR symbol; b, fifth-order elliptical LCR low-pass filter; c, FDNR-transformed equivalent circuit.

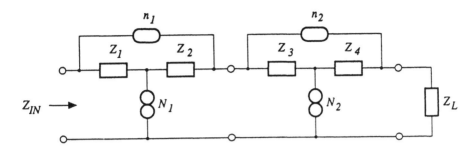

FIG. 41 Two-nullor-based negative impedance converters in cascade with loading impedance Z_L.

FDNR transformation preferably is applied to the minimum C version. The fact that the resistive terminations become capacitive may be a problem with the FDNR transformation; however, it generally can be solved by means ranging from a large-valued resistor in parallel with the terminating capacitors, to so-called GIC embedding, the details of which can be found in references suggested in the bibliography of this article.

Figure 41 shows the nullor realization of an FDNR obtained by cascading two NICs of the kind shown in Fig. 35. The input impedance of this configuration is given by

$$Z_{IN} = \frac{Z_1 Z_3}{Z_2 Z_4} Z_L \tag{88}$$

Depending on which of the five impedances in this expression are resistors and which capacitors, either an FDNR or a simulated inductor (equivalent to a gyrator-C combination) can be obtained. Thus, if one of the three pairs $Z_1 Z_3$, $Z_1 Z_L$, or $Z_3 Z_L$ is capacitive and the other three impedances are resistive an FDNR is obtained. Similarly, if either Z_2 or Z_4 is capacitive and the other four impedances resistive, a simulated inductor results. Using two op amps to replace the two nullors in the configuration of Fig. 41, an operative circuit is obtained. Again, there are various ways of combining the nullors (e.g., n_1, N_1 and n_2, N_2, or n_1, N_2 and n_2, N_1) and for each combination more than one method of connecting the op amps (remember, in contrast to a transistor, a nullor being replaced by an op amp requires no common terminal). Fig. 42 shows the FDNR-R-C fifth-order low-pass filter of Fig. 40-c with a component distribution and op amp connection for the FDNRs that has proved itself well in practice. Notice that in the realization of the LCR filter the source resistor R_1 was assumed equal to zero.

Cascaded Biquad Design

Cascaded biquad design is one of the oldest, and has proved to be one of the most enduring, of active RC filter design methods. This is due mainly to (1) the simplicity of design (second- or third-order filter sections, or *biquads*, are cascaded to provide any *n*th-order filter function), and (2) the excellent properties

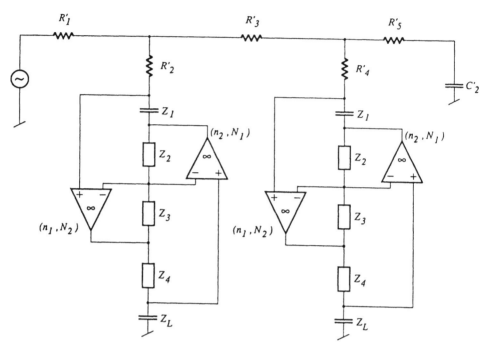

FIG. 42 Frequency-dependent negative resistor-based operational amplifier–resistor-capacitor realization of the low-pass filter in Figs. 40-*b* and 40-*c*.

and low cost of, first, bipolar, and then CMOS integrated op amps. The op amp is ideally a differential input, single-ended output amplifier with infinite gain, bandwidth, and input impedance and zero output impedance and offset voltage. In practice, the gain may be anywhere between 60 and 100 dB, the bandwidth several MHz, and the input impedance several megohms. The offset voltage may be several to tens of millivolts. These features may be mutually exclusive; an op amp will be designed to optimize any one or more of these features and, in addition, numerous others, such as dissipated power, noise, common mode and power supply rejection ratio, slew rate, and so on. As a network element, the idealized op amp can be considered to be a voltage-controlled voltage source with infinite gain. When used with negative feedback, it has a virtual ground at the input. More recently, current amplifiers also are being used in the form of OTAs. To a large extent these amplifiers are duals of the voltage-gain-based op amps. They have certain technology-related advantages when built in CMOS chips, and often are advantageous with respect to high-frequency operation. In the discussion that follows, the conventional op amp is assumed as the basic gain element; the change to OTAs has very little consequence with respect to the theoretical design aspects.

From Transfer Function to Active Filter

Referring to the section, "Frequency Response and Transfer Function" above, we recall that the transfer function of an *n*th-order filter network has the form

$$T(s) = \frac{N(s)}{D(s)} = \frac{b_m s^m + b_{m-1} s^{m-1} + \cdots + b_1 s + b_o}{a_n s^n + a_{n-1} s^{n-1} + \cdots + a_1 s + a_o} \qquad (89)$$

where $s = \sigma + j\tilde{\omega}$.

This is a rational function in the complex frequency variable s. The order of $T(s)$ is determined by the order of the denominator polynomial $D(s)$. $T(s)$ can be factored into a product of second- or third-order functions $T_i(s)$, depending on whether n is even or odd. Thus, for n even,

$$T(s) = \prod_{i=1}^{n/2} T_i(s) \qquad (90)$$

and for n odd,

$$T(s) = \frac{1}{s + \alpha} \cdot \prod_{i=1}^{(n-1)/2} T_i(s) \qquad (91)$$

where α is a negative real pole and the individual biquadratic functions $T_i(s)$ have the general form

$$T_i(s) = K_i \frac{s^2 + \frac{\omega_{z_i}}{q_{z_i}} s + \omega_{z_i}^2}{s^2 + \frac{\omega_{p_i}}{q_{p_i}} s + \omega_{p_i}^2} = K_i \frac{(s - z_i)(s - z_i^*)}{(s - p_i)(s - p_i^*)}. \qquad (92)$$

In general, and for any of the basic filter types outlined in this article, the zeros and poles of each biquad function $T_i(s)$ will be complex conjugates (see Fig. 43-a). The closer the poles are to the $j\omega$-axis, the higher will be the selectivity of the filter; the closer the zeros are, the larger will be the filter attenuation at those frequencies. Referring to Eqs. 9 and 10, the closeness to the $j\omega$-axis of the poles and zeros is indicated by the quantities q_p and q_z, respectively. If the critical frequencies (i.e., poles and zeros) are on the negative real axis, that is, far from the $j\omega$-axis, the corresponding q values will be less than 0.5; if they are complex conjugates, the q values will be larger than 0.5, and, in the limit (i.e., on the $j\omega$-axis), equal to infinity.

It can be shown that the poles of a passive RC network must be single and on the negative real axis (see Fig. 43-b). Note that there is no such limitation on the location of the zeros. In terms of q_p, this means that for a second-order RC network, that is, a pole pair on the negative real axis

$$(q_p)_{\text{RC}} = \hat{q} < 0.5 \qquad (93)$$

where we use the $\hat{}$ on the q and on any other pertinent quantity to indicate that it is associated with an RC network. On the other hand, we have seen that for

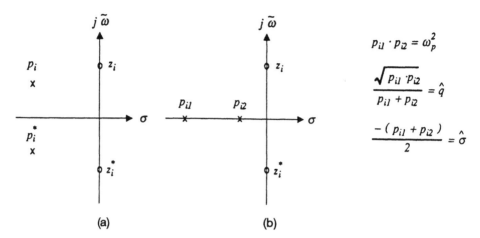

FIG. 43   Pole-zero pair of a biquad: a, general; b, passive resistor-capacitor.

any practical application, the selectivity specifications of the filter will require complex conjugate poles:

$$(q_p)_{\text{LCR}} = q_p > 0.5. \tag{94}$$

As indicated in this expression, this condition is satisfied readily by conventional LCR networks. Thus, the basic difference between a passive RC and a passive LCR network is the location of the poles: a passive RC network has negative real poles and poor selectivity, an LCR network, theoretically, can have poles arbitrarily close to, but not on, the $j\omega$-axis, and therefore high selectivity. Thus, in an active RC network, the purpose of a gain element, say β, is essentially to increase the $\hat{q}$ of each pole pair to a value larger than 0.5. This can be done in a number of different ways.

Q Multiplication (Negative Feedback)

Starting out with the biquadratic transfer function $T(s)$ of an RC second-order network

$$\hat{T}(s) = \frac{N(s)}{s^2 + \dfrac{\omega_p}{\hat{q}} s + \omega_p^2}. \tag{95}$$

The poles of this function are, by definition, negative real. Multiplying $\hat{q}$ by a quantity

$$\mu = 1 + \beta, \qquad \text{where } \beta > 0 \tag{96}$$

we obtain the same function as Eq. (95), but with complex conjugate poles, thus,

$$T(s) = \frac{N(s)}{s^2 + \dfrac{\omega_p}{q_p} s + \omega_p^2} \tag{97}$$

where

$$q_p = \hat{q}\,(1 + \beta) > 0.5. \tag{98}$$

After some manipulation, and with the assumption that $\beta \gg 1$, Eq. (97) with Eq. (98) can be written as

$$T(s) = \hat{T}(s) \cdot \frac{\beta}{1 + \beta \hat{t}_1(s)} \tag{99}$$

where

$$\hat{t}_1(s) = \frac{s^2 + \omega_p^2}{s^2 + \dfrac{\omega_p}{\hat{q}} s + \omega_p^2}. \tag{100}$$

Eqs. (99) and (100) correspond to an active RC network consisting of $\hat{T}(s)$, $\hat{t}_1(s)$ and the amplifier β in a negative feedback loop as shown in the block diagram of Fig. 44. A typical biquad with band-pass characteristics, which is based on Q multiplication, is shown in Fig. 45.

Sigma Reduction (Positive Feedback)

Instead of expressing the biquad transfer function in terms of the pole q as in Eq. (95), we can do so in terms of the negative real quantity $\hat{\sigma}$:

$$\hat{T}(s) = \frac{N(s)}{s^2 + 2\hat{\sigma}s + \omega_p^2} \tag{101}$$

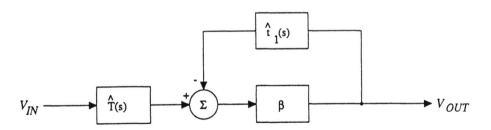

FIG. 44   Biquad function with complex-conjugate poles based on negative feedback.

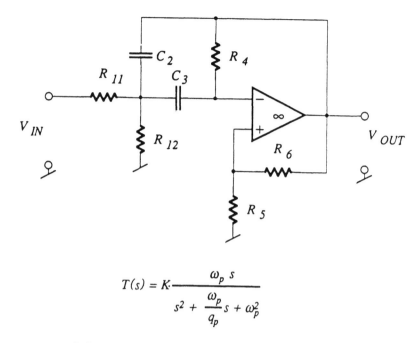

$$T(s) = K \cdot \frac{\omega_p \, s}{s^2 + \dfrac{\omega_p}{q_p} s + \omega_p^2}$$

FIG. 45   Band-pass biquad based on negative feedback.

where

$$\hat{\sigma} > \omega_p. \tag{102}$$

The inequality Eq. (102) indicates the fact that the two poles of $\hat{T}(s)$ are negative real. They can be made complex conjugates by decreasing the quantity $2\hat{\sigma}$ by some value κ, thus

$$T(s) = \frac{N(s)}{s^2 + (2\hat{\sigma} - \kappa)s + \omega_p^2} \tag{103}$$

where

$$0 < \kappa < 2\hat{\sigma}. \tag{104}$$

Equation (103) corresponds to Eq. (97) since

$$q_p = \frac{\omega_p}{2\hat{\sigma} - \kappa} > 0.5. \tag{105}$$

$T(s)$ in Eq. (103) can be rewritten as

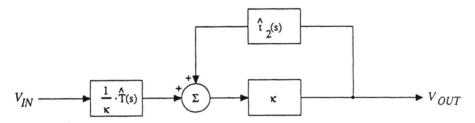

FIG. 46 Biquad function with complex-conjugate poles based on positive feedback.

$$T(s) = \frac{1}{\kappa} \cdot \hat{T}(s) \cdot \frac{\kappa}{1 - \kappa \hat{t}_2(s)} \qquad (106)$$

where

$$\hat{t}_2(s) = \frac{s}{s^2 + 2\hat{\sigma}s + \omega_p^2}. \qquad (107)$$

This expression corresponds to an active RC network consisting of $\hat{T}(s)$ and $\hat{t}_2(s)$ in a positive feedback configuration with gain κ. This is shown in the block diagram of Fig. 46. A low-pass biquad based on sigma reduction (i.e., positive feedback) is given in Fig. 47.

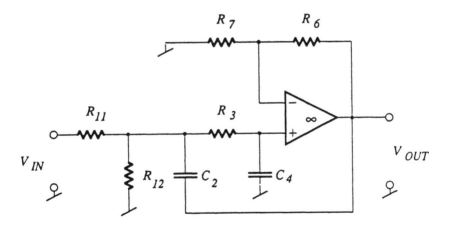

$$T(s) = K \cdot \frac{\omega_p^2}{s^2 + \dfrac{\omega_p}{q_p}s + \omega_p^2}$$

FIG. 47 Low-pass biquad based on positive feedback.

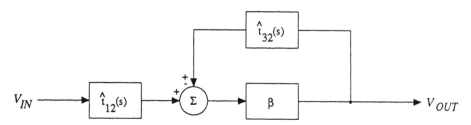

FIG. 48   Feedback configuration as basis for active RC biquads.

Single-Amplifier Biquads with Positive and Negative Feedback

As shown in the previous two sections, complex conjugate poles can be generated by Q multiplication or sigma reduction. The former method is based on negative feedback, the latter on positive feedback. Both can be combined in the general feedback structure shown in Fig. 48. Here, the RC network in the forward path $\hat{t}_{12}(s)$ determines the filter type (i.e., low pass, high pass, band pass, etc.). The RC network in the feedback path $\hat{t}_{32}(s)$ determines the necessary feedback polarity and the actual path on the root locus with respect to β, along which the initially negative real poles of $\hat{t}_{32}(s)$ travel, to become the complex conjugate poles of $T(s)$. It can be shown that there are essentially three basic feedback functions $\hat{t}_{32}(s)$ providing complex conjugate poles with negative feedback, and one providing them with positive feedback. This is the basis for the classification of single-amplifier biquads presented in Table 3. Which of these

TABLE 3   Classification of Single-Amplifier Biquads

Negative Feedback		
Class	$\hat{t}_{32}(s)$	Possible Filter Functions $T(s)$
1	lowpass	lowpass bandpass
2	highpass	highpass bandpass
3	bandstop	lowpass highpass bandpass
Positive Feedback		
4	bandpass	lowpass highpass bandpass bandstop

biquads is best for a particular filter type depends on the application. Further information on this and other practical design questions can be found in the references listed in the bibliography at the end of this article.

Example. Consider the transfer function of a 10th-order band-pass filter given by the general form

$$T(s) = K \frac{s^8 + a_7 s^7 + \cdots + a_2 s^2}{s^{10} + a_9 s^9 + \cdots + a_1 s + a_o}. \tag{108}$$

This transfer function can be factored into a product of five biquadratic functions $T_i(s)$, $i = 1, \ldots, 5$, which corresponds to a cascade of five biquad filters as shown in Fig. 49-*a*. We assume now that the five biquadratic functions are given as follows:

$$T_1(s) = 0.85 \frac{s^2 + 0.1039 \cdot 10^{12}}{s^2 + 0.5522 \cdot 10^5 s + 0.8844 \cdot 10^{11}}$$

$$T_2(s) = 0.75 \frac{s^2 + 0.135 \cdot 10^{12}}{s^2 + 0.06296 \cdot 10^6 s + 9.911 \cdot 10^{10}}$$

$$T_3(s) = \frac{s^2 + 0.06296 \cdot 10^6 s + 9.911 \cdot 10^{10}}{s^2 + 1.7412 \cdot 10^4 s + 9.911 \cdot 10^{10}}$$

$$T_4(s) = \frac{3.277 \cdot 10^{10}}{s^2 + 1.8668 \cdot 10^5 s + 4.357 \cdot 10^{10}}$$

$$T_5(s) = \frac{0.834 s^2}{s^2 + 5.056 \cdot 10^4 s + 4.594 \cdot 10^9} \tag{109}$$

The functions $T_1(s)$ and $T_2(s)$ correspond to so-called frequency-rejection networks (FRNs), $T_3(s)$ to a frequency-emphasizing network (FEN), $T_4(s)$ to a low-pass network, and $T_5(s)$ to a high-pass network. Note that the poles of $T_2(s)$ cancel the zeros of $T_3(s)$. The resulting "phantom pair" is introduced with a view to the filter realization (FEN cascaded with FRN), which has certain practical advantages. These practical considerations are responsible for an increase of the filter order from the actually required 8, to the practically preferable 10. The response of the five biquad filters and of the overall 10th-order cascade is shown in Fig. 49-*b*. Typically, the amplitude ripple may be specified at 0.1 dB with a ±0.5-dB tolerance. Such specifications are quite demanding, but also quite feasible with high-quality passive and active components.

Direct-Form Design Using Integrators

Just as the cascaded biquad design method described in the section above starts out with the desired rational transfer function $T(s)$ from which the active filter structure can be derived, the same is true for the so-called direct-form integrator design. However, instead of factoring $T(s)$ into individual biquadratic terms, each of which is designed individually, here the whole function is dealt with

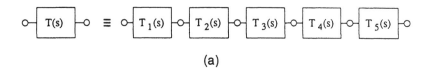

(a)

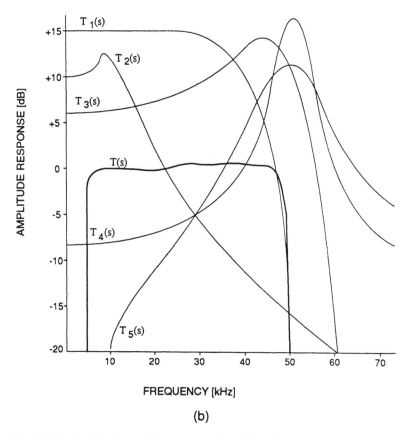

(b)

FIG. 49 Tenth-order band-pass filter: *a*, cascade of five biquads; *b*, amplitude response of individual biquads and of overall cascade.

directly, and realized as one wholly interconnected structure. Consider a given transfer function

$$T(s) = \frac{N(s)}{D(s)} = \frac{b_n s^n + b_{n-1} s^{n-1} + \cdots + b_o}{s^n + a_{n-1} s^{n-1} + \cdots + a_o}. \tag{110}$$

Compared to Eq. (89) in the previous section, Eq. (110) differs in two respects: (1) the degree of the numerator is assumed to be equal to that of the denominator and (2) a_n, the leading term in $D(s)$, is assumed to be equal to unity. With a little thought it will be apparent that these two assumptions in no way limit the generality of the method. Dividing both $N(s)$ and $D(s)$ by s^n, we obtain

$$T(s) = \frac{b_n + b_{n-1}\left(\frac{1}{s}\right) + \cdots + b_o\left(\frac{1}{s}\right)^n}{1 + a_{n-1}\left(\frac{1}{s}\right) + \cdots + a_o\left(\frac{1}{s}\right)^n}. \tag{111}$$

In order to relate this expression to a physical structure, it is useful to invoke the concepts of signal-flow graph (SFG) theory. SFGs use graphical means to represent systems of equations or rational functions. They serve to provide a bridge between an analytical description of a system and its physical architecture. Most important in this regard is Mason's rule, which provides the transfer function of a SFG in terms of the transmission paths and feedback loops of that graph. This, in turn, describes a system or a network. According to Mason, the transfer function of a general SFG is given by

$$T(s) = \frac{\Sigma P_k \Delta_k}{\Delta} \tag{112}$$

where

Δ = Graph determinant = $1 - \Sigma L_i + \Sigma L_i L_j - \ldots$ and $\Sigma L_i L_j$, $\Sigma L_i L_j L_v$ ... involves only nontouching loops L_i, L_j, L_v, and so on.

P_k = Transmission of kth path from input to output

Δ_k = Cofactor of kth path, that is, determinant of that part of the SFG that does not touch the kth path.

If we make the following two assumptions:

(i) every forward path P_k touches every loop, thus $\Delta_k = 1$ for all k
(ii) all loops touch each other, thus $\Sigma L_i L_j = \Sigma L_i L_j L_v = \ldots = 0$

then Eq. (111) can be equated to the following simplified version of Eq. (112):

$$T(s) = \frac{\Sigma P_k}{1 - \Sigma L_i}. \tag{113}$$

This permits the two direct representations (Direct Form I and Direct Form II) of $T(s)$ in the form of SFGs to be made as shown in Figs. 50-a and 50-b. The physical realization of these graphs follows directly, since every $1/s$ term corresponds to an integrator and every node with paths entering it to a summing (positive or negative) amplifier with gains directly related to the path transmission coefficients. Since the latter correspond to the coefficients of the transfer function, the graphs aptly are called *direct forms*. Note that the number of integrators is equal to the order of $T(s)$, that is, to the degree of $D(s)$.

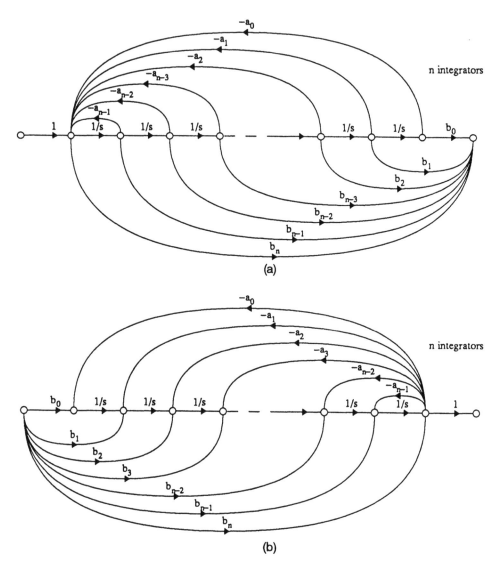

FIG. 50 Signal-flow graph representation of $T(s)$ given by Eq. (111): a, Direct Form I; b, Direct Form II.

Example. Consider the general biquadratic function $T(s)$:

$$T(s) = K\frac{s^2 \pm B_1 s + B_o}{s^2 + A_1 s + A_o} = K\frac{s^2 \pm \dfrac{\omega_z}{q_z}s + \omega_z^2}{s^2 + \dfrac{\omega_p}{q_p}s + \omega_p^2} \tag{114}$$

Dividing by s^2, we obtain

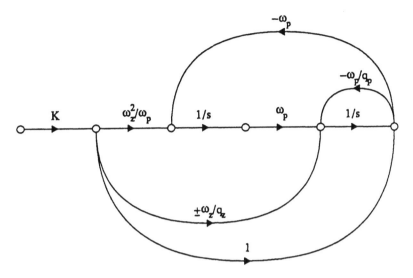

FIG. 51 Direct Form II signal-flow graph of general biquad.

$$T(s) = K \frac{1 \pm B_1 \left(\dfrac{1}{s}\right) + B_o \left(\dfrac{1}{s}\right)^2}{1 + A_1 \left(\dfrac{1}{s}\right) + A_o \left(\dfrac{1}{s}\right)^2} = K \frac{1 \pm \dfrac{\omega_z}{q_z}\left(\dfrac{1}{s}\right) + \omega_z^2 \left(\dfrac{1}{s}\right)^2}{1 + \dfrac{\omega_p}{q_p}\left(\dfrac{1}{s}\right) + \omega_p^2 \left(\dfrac{1}{s}\right)^2} \qquad (115)$$

The Direct Form II SFG corresponding to Eq. (115) is shown in Fig. 51. Using the op amp realizations of integrator, lossy integrator, and summer (see Fig. 52), we obtain the

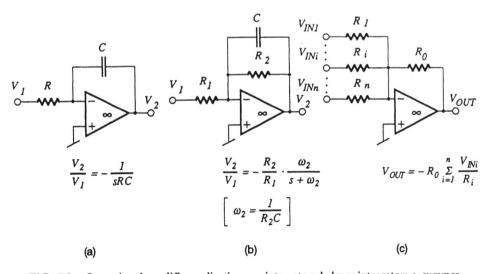

FIG. 52 Operational amplifier realizations: a, integrator; b, lossy integrator; c, summer.

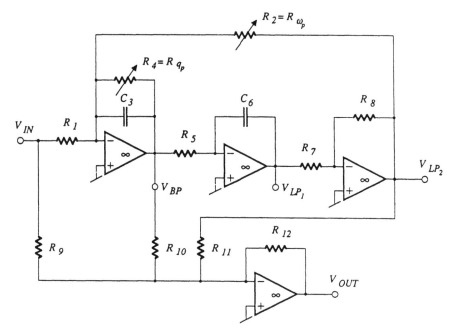

FIG. 53   General-purpose biquad.

active op amp version of a general biquad as shown in Fig. 53. The transfer function for this circuit is

$$T(s) = \frac{V_{OUT}}{V_{IN}} = \frac{R_2}{R_9} \cdot \frac{s^2 + s\left[\frac{\omega_p}{q_p}\left(1 - \frac{R_9}{R_{10}}K_b\right)\right] + \left[\omega_p^2\left(\frac{R_9}{R_{11}}K_l - 1\right)\right]}{s^2 + \frac{\omega_p}{q_p}s + \omega_p^2} \tag{116}$$

where
$$K_b = R_4/R_1$$
$$K_l = R_2/R_1$$
$$R_7 = R_8$$

and

$$K = R_{12}/R_9 \tag{117}$$

$$\omega_p^2 = \frac{R_8}{R_2 R_5 R_7 C_3 C_6} \tag{118}$$

$$q_p = R_4 C_3 \omega_p \tag{119}$$

$$\omega_z^2 = \omega_p^2\left(\frac{R_9}{R_{11}}K_l - 1\right) \tag{120}$$

$$\omega_z/q_z = \frac{\omega_p}{q_p} \left(1 - \frac{R_9}{R_{10}} K_b \right) \qquad (121)$$

With a general-purpose biquad as shown in Fig. 53, any arbitrary biquadratic filter function can be obtained. However, four op amps are required, which is often unacceptable from the point of view of both cost and necessary power dissipation. In the case of high-selectivity (high-pole Q) and high-precision applications, the LC filter simulation method discussed next, in which inductors either are replaced by gyrator-capacitor combinations, or eliminated by the FDNR transformation, therefore are preferable.

Inductor-Capacitor Filter Simulation

In the section, "The Reactive Ladder Filter and Its Properties," the remarkable property of low sensitivity to component variations in the pass band (i.e., Orchard's theorem) of LC ladder-filter structures is discussed. It is this property that has motivated the simulation of LC ladder networks when inductors may not be used. This is so in all IC realizations, both analog and digital. In the section, "Basic Active Network Elements," the two most important active elements used to build active RC simulated LC filter networks are described. They are (1) the gyrator for inductor simulation and (2) the FDNR as it occurs in conjunction with the FDNR transformation of LCR filter networks. Which of these two approaches is used depends on the LCR filter that is to be rendered inductorless.

Consider, for example, the fifth-order elliptic low-pass filter shown in Fig. 54. This is the minimum L version of the CC-05-25-49 filter already shown in Fig. 18a. Simulating the inductors by gyrator-C combinations (which is why the minimum L version is used), we obtain the active RC simulated inductor circuit in Fig. 55. Note that each floating inductor (i.e., L_2 and L_4) requires two gyrators, and each gyrator at least two op amps, for its realization. Thus the gyrator-C replacement of L_2 and L_4 will require eight op amps, which, for a fifth-order filter, is quite uneconomical, with regard to both component cost and dissipated power.

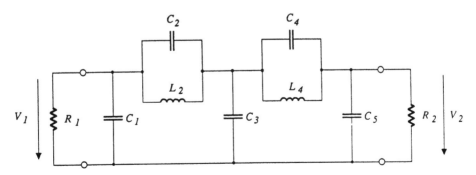

FIG. 54 Minimum L version of fifth-order inductor-capacitor low-pass-filter-type CC-05-25-49.

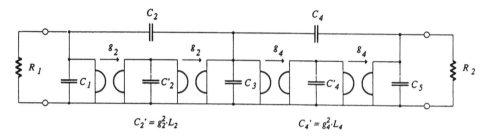

FIG. 55   Inductorless gyrator-C version of CC-05-25-49 filter in Fig. 54.

Floating inductors are typical for a low-pass filter because a true low-pass characteristic also must transmit DC. The floating inductors, of course, could be eliminated by an FDNR transformation. Carrying this transformation out on the LCR filter of Fig. 54, however, would be counterproductive. L_2 and L_4, indeed, would be transformed into resistors, but each of the five capacitors would be transformed into an FDNR. A grounded FDNR can be realized by two op amps, a floating FDNR requires more. Thus, an FDNR transformation of the minimum L filter in Fig. 54 requires well over 10 op amps, which is still more extravagant in op amp count than the gyrator-C version. Typically, the situation looks quite different for the minimum C version of a filter. For the CC-05-25-49 filter, this is shown in Fig. 56. Scaling each impedance in the filter by ω_o/s, where ω_o is selected as the filter cutoff frequency ω_c, we have the situation shown in Fig. 57. Here, each impedance-scaled inductor L_i becomes a resistor $R_i' = \omega_o L_i$, each resistor R_j a capacitor $C_j' = (\omega_o R_j)^{-1}$, and each capacitor C_v an FDNR $Z_v' = (s^2 D_v)^{-1}$, where $D_v = C_v/\omega_o$. The resulting DCR filter is shown in Fig. 58. Most important, there are only two active elements in the filter: the two impedance transformed capacitors, each of which becomes a grounded FDNR. With two op amps required for each FDNR (see the section on basic active network elements), the overall circuit now comprises four op amps and numerous resistors and capacitors (see Fig. 59). Thus, compared to the simulated inductor version of the filter (Fig. 55), the number of op amps now has been halved.

Using single-ended biquads (see the section, "Cascaded Biquad Design,"

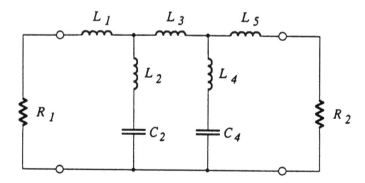

FIG. 56   Minimum C version of fifth-order inductor-capacitor low-pass-filter-type CC-05-25-49.

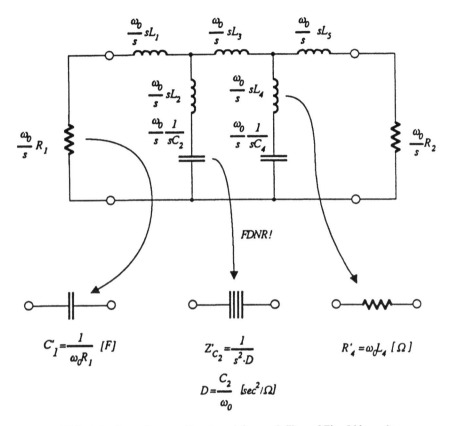

FIG. 57   Impedance scaling the minimum C filter of Fig. 56 by ω_0/s.

above) to realize the CC-05-25-49 low-pass filter discussed above, we obtain the circuit configuration shown in Fig. 60. Here the op amp count has been halved again, but this reduction will come at a price. Because the sensitivity to component variations is higher with the biquads than with the simulated LCR ladder filter, the performance in the pass band of the biquad cascade when subjugated to ambient changes involving temperature, humidity, or aging will be signif-

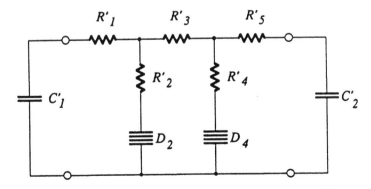

FIG. 58   DCR version of CC-05-25-49 fifth-order low-pass filter.

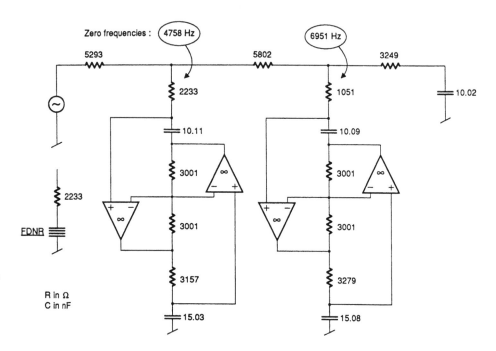

FIG. 59 FDNR realization of CC-05-25-49 fifth-order low-pass filter.

icantly worse for the biquads than for the simulated LCR ladder filter. Nominally, however, the filter response (see Fig. 61) of the two filters will be identical. Note that the specifications given in Fig. 17 are well satisfied by this filter.

The FDNR transformation generally provides the most efficient (in terms of op amp count) inductorless active RC circuit for low-pass specifications because low pass filters contain floating coils in order to guarantee a DC path. Floating inductors require two gyrators per inductor for inductor simulation and, as discussed above, this may result in twice as many op amps as a corresponding FDNR realization. This preference will not hold for a general filter application, however, in which a combination of gyrator-C, FDNR, and GIC embedding more typically will provide the most economical inductorless filter.

A GIC is introduced briefly in the discussion of basic active network elements above. It is a two-port with a transmission, or *ABCD*, matrix given by

$$[ABCD] = \begin{bmatrix} A(s) & 0 \\ 0 & D(s) \end{bmatrix} \tag{122}$$

Loading a GIC with an impedance Z_L at the output terminals, the input impedance given by Eq. (69) will be

$$Z_{IN} = \frac{A(s)}{D(s)} Z_L = k(s) \cdot Z_L \tag{123}$$

where $A(s)/D(s)$ is a dimensionless but frequency-dependent quantity designated $k(s)$.

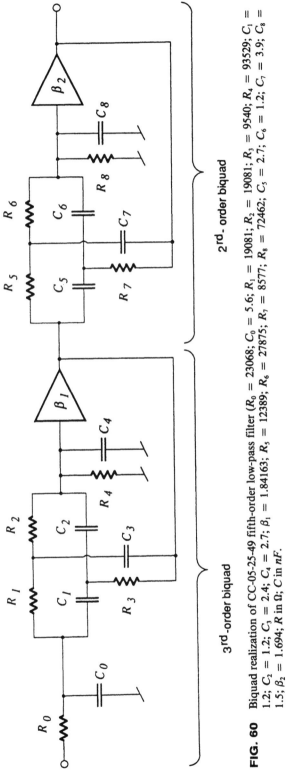

FIG. 60 Biquad realization of CC-05-25-49 fifth-order low-pass filter ($R_0 = 23068$; $C_0 = 5.6$; $R_1 = 19081$; $R_2 = 19081$; $R_3 = 9540$; $R_4 = 93529$; $C_1 = 1.2$; $C_2 = 1.2$; $C_3 = 2.4$; $C_4 = 2.7$; $\beta_1 = 1.84163$; $R_5 = 12389$; $R_6 = 72462$; $C_5 = 2.7$; $C_6 = 1.2$; $C_7 = 3.9$; $C_8 = 1.5$; $\beta_2 = 1.694$; R in Ω; C in nF.

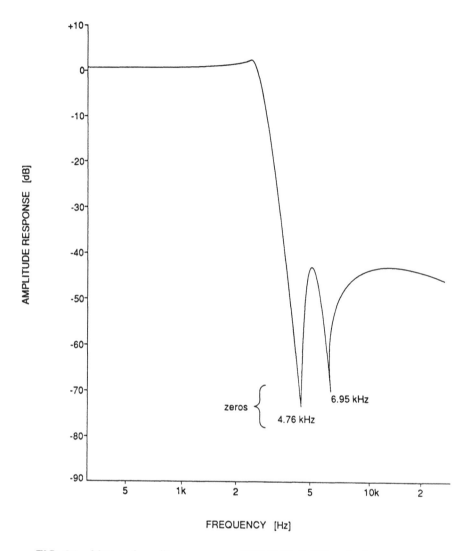

FIG. 61   Measured amplitude response of CC-05-25-49 fifth-order low-pass filter.

GIC embedding is based on the fact that a two-port, which is characterized by its *ABCD* matrix and "embedded" between two GICs (see Fig. 62), has the overall transmission matrix

$$[ABCD]_{\text{Tot}} = \begin{bmatrix} 1 & 0 \\ 00 & 1/k_1(s) \end{bmatrix} \begin{bmatrix} A & B \\ C & D \end{bmatrix} \begin{bmatrix} 1 & 0 \\ 0 & 1/k_2(s) \end{bmatrix}$$

$$= \begin{bmatrix} A & B/k_2(s) \\ C/k_1(s) & D/k_1(s) \cdot k_2(2) \end{bmatrix} \tag{124}$$

For $k_1(s) = k(s) = k_2^{-1}(s)$, this simplifies to

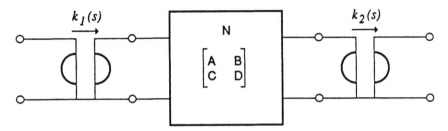

FIG. 62 Embedding the network N between two general impedance converters.

$$\begin{bmatrix} A & B/k(s) \\ C/k(s) & D \end{bmatrix} \tag{125}$$

Since the dimensionless transfer parameters A and D remain unaltered, and the impedance and admittance parameters are scaled by $k(s)$ and $1/k(s)$, respectively, embedding between two GICs is identical with impedance scaling of the embedded network by $k(s)$. If $k(s) = \omega_o/s$, then the embedded network undergoes an FDNR impedance transformation. Thus, for example, a resistive network embedded between two GICs with $k_1(s) = s/\omega_o$ and $k_2(s) = \omega_o/s$ appears as an all-inductive network (see Fig. 63). In practice, GIC embedding and gyrator-C substitution of inductors will be applied within the same network, depending on the configuration. This is shown in Fig. 64, in which GIC embedding and gyrator-C inductor simulation can be alternated to provide the most efficient

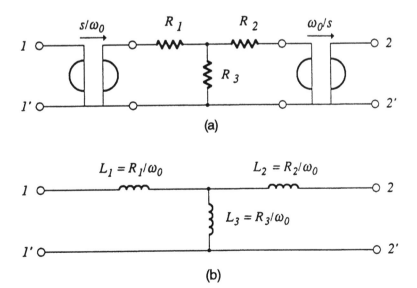

FIG. 63 General impedance converter embedding: a, resistive network embedded between two general impedance converters; b, equivalent inductive network.

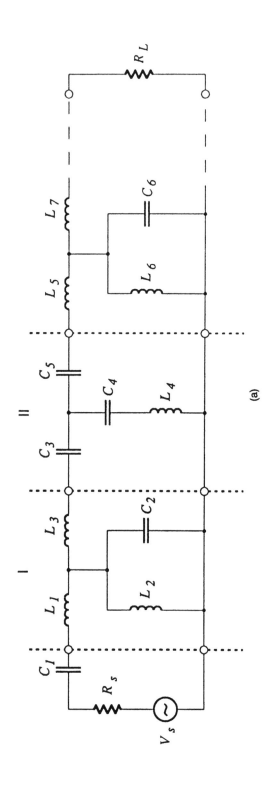

(a)

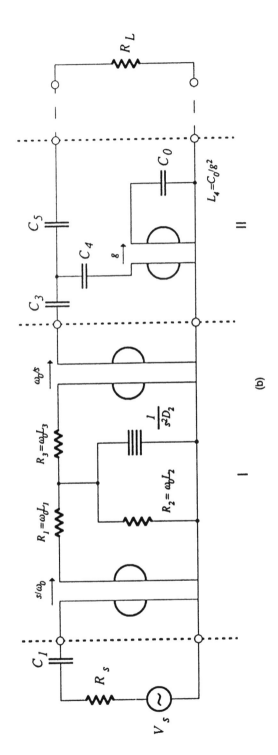

(b)

FIG. 64 Inductor-capacitor filter simulation using general impedance converter embedding and gyrator-C substitution of inductors: *a*, LCR filter network; *b*, inductorless equivalent circuit using GIC embedding and gyrator-C substitution.

simulated active RC ladder filter. Note that the embedding is introduced at the terminal end to leave the terminating resistor R (and also the first capacitor C_1) intact. This is the most elegant way of avoiding the R-to-C transformation of the terminal resistors that otherwise would occur in a straightforward FDNR transformation.

Switched-Capacitor Filters

Electrical filters, like most other circuits used in information-processing systems since the mid 1960s, have been influenced strongly by the rapid progress in semiconductor technology. In particular, the trend to VLSI circuits has resulted in ever-larger electronic systems being integrated on silicon IC chips. However, in contrast to most other circuits, lumped-element LCR filters (see section on this topic above) cannot be considered for IC realization because inductor coils and ferrite cores cannot be integrated on a silicon chip in any form. Being inherently three-dimensional devices, it can be shown that their quality factor Q, which is related directly to their frequency-selectivity capabilities, decreases with decreasing size.

Active RC filters (discussed above) can be integrated, albeit most easily as hybrid IC. This means that they are not easily manufacturable as monolithic silicon chips (although it is possible to do so), but as combinations of thick- or thin-film passive RC components combined with silicon IC amplifier chips. The main obstacles to an active RC IC chip are (1) the overly wide tolerances of the IC resistors and (2) there is no way of adjusting or "tuning" the filter to meet specifications after manufacture. Thus, due to the first obstacle, in general only very wide-tolerance IC filters can be realized. On the other hand, in the case of hybrid integrated active RC filters, the film resistors can be adjusted accurately by laser trimming after manufacture. This corresponds to an accurate tuning of the frequency response of the final filter to meet the required specifications. Thus, for many years, it was believed that only digital filters could be incorporated into a VLSI chip, whereas an analog and continuous-time "filter on a chip" would remain a practical impossibility.

This notion was dispelled in the late 1970s when the first publication of an integrated filter chip based on MOS IC technology appeared. The filter overcame the problem of wide-tolerance IC silicon resistors by eliminating them altogether. Instead, combinations of analog MOS switches and capacitors took over the resistive role. The resulting SC filter was analog, but not continuous time. In fact, SC filters fall into the category of analog discrete time, or sampled-data filters.

Sampled-data filters are related to digital filters (see Fig. 65). Note that the sampled-data filter operates directly on an analog input signal, producing signals that are analog in amplitude but quantized in time. Only after amplitude quantization with an A/D converter is the signal ready to be processed by a digital filter. The sampled-data filter processes the discrete-time analog-amplitude signals directly. Thus, one of the big advantages of SC filters is the

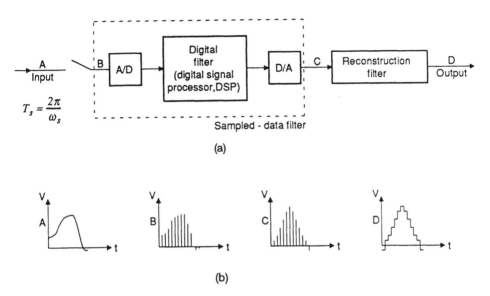

FIG. 65 Basic diagram of generalized sampled-data and digital filters: *a*, block diagram including input sampler and output reconstruction filter; *b*, typical waveforms at Points A through D.

fact that they do not require A/D converters. Another is their simplicity of overall structure, which permits processing speeds significantly higher than those exhibited by digital filters. Finally, it can be shown that for a given signal-processing task they very often require far less DC power than a corresponding digital filter or digital signal processor (DSP). It is because of these and other advantages that SC circuit techniques are being used in many applications other than filtering. Thus, SC circuits can be found alone or in combination with digital circuits in numerous VLSI chips used, for example, in telecommunications systems, measuring equipment, and instrumentation. They are particularly common in applications in which power is at a premium, such as in hand-held battery-powered equipment.

The Components of Switched-Capacitor Filters

The characteristics of the components used in SC filters depend directly on the technology used to manufacture them. Naturally, with the rapid progress in semiconductor circuit processing, these characteristics are subject to change in a relatively short time. Thus, the characteristics briefly outlined in what follows are typical of a particular technology (CMOS technology) at a particular time (early to mid-1990s). Rather than attempting any prognosis of how technology and components will change in the decade or so following this time, it seems wiser to regard only the present, and to leave any extrapolations into the future up to the reader.

The basic elements of an SC filter are capacitors, switches, and op amps or

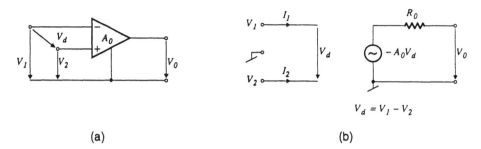

(a)   (b)

FIG. 66 Operational amplifier: *a*, basic diagram; *b*, equivalent voltage-controlled voltage source.

OTAs. The difference between an op amp and an OTA is shown in Figs. 66 and 67 (see also Fig. 34). The op amp can be represented by a voltage-controlled voltage source (VVS). The ideal op amp is defined as follows:

(1)	$A_o \to \infty$	(infinite open loop gain)
(2)	$I_1 = I_2 = 0$	(infinite input impedance)
(3)	$R_o = 0$	(zero output impedance)
(4)	$V_d \mid\ = 0$ $\qquad V_o = 0$	(zero offset voltage)
(5)	$V_o \mid\ = 0$ $\qquad V_1 = V_2$	(zero common-mode gain)
(6)	$A_o = $ constant	(infinite bandwidth)

Naturally, in practice, the real op amp will satisfy none of these characteristics exactly. It will depend on their importance in a given application which should be approximated as closely as possible.

The OTA (Fig. 67) is best represented by a voltage-controlled current source (VCS). In the ideal case, its characteristics are similar to those of the op amp except that they refer to a current source. Thus

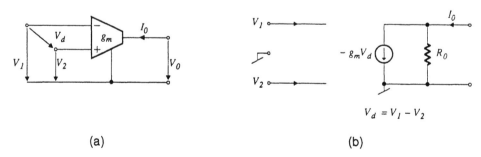

(a)   (b)

FIG. 67 Operational transconductance amplifier: *a*, basic diagram; *b*, equivalent voltage-controlled current source.

(1) the transconductance g_m is a well-defined circuit parameter determined by the circuit

(2) $I_1 = I_2 = 0$ (infinite input impedance)

(3) $R_o \rightarrow \infty$ (infinite output impedance)

(4) $V_d \mid_{I_o = 0} = 0$ (zero offset voltage)

(5) $I_o \mid_{V_1 = V_2} = 0$ (zero common-mode gain)

(6) $g_m = \text{constant}$ (infinite bandwidth)

As with the op amp, the real OTA will satisfy none of these ideal characteristics exactly; the application will determine which should come closest to the ideal. Where the op amp can be better approximated by a bipolar transistor circuit, the OTA will model more closely an amplifier designed with MOS FET circuits. In fact, depending on the application and the technology, the characteristics of a practical amplifier will be somewhere between those of an ideal op amp and an OTA. In what follows, we assume op amp characteristics since a majority of analog and SC circuits initially were designed with these in mind.

The most important characteristics of the three basic components of an SC network (i.e., capacitors, switches, and amplifiers) are given in Table 4. The performance of an SC filter is determined by the characteristics of each of these components, each in its own way. The coefficients of the filter transfer function, Eq. (89), are determined by ratios of capacitor values, and not by absolute values. This is a key feature of SC networks, because ratios, in contrast to absolute values, can be realized very accurately and conveniently using MOS IC technology. The switches are close to ideal in the off state; with an on resistance of up to 10 kΩ (kiloohms) they appear far from ideal in the on state. This conclusion is incorrect, however, because the impedance level of the whole SC circuit is very high. With the low capacitance values, which range from 0.1 picofarads (pF) to approximately 30 pF, the on resistance becomes noticeable only above several MHz.

TABLE 4   The Components of Switched-Capacitor Networks

Capacitors	Range approximately $0.1 \div 30$ pF. Absolute tolerance $\approx 10 \div 20\%$ Accuracy of capacitor ratios $< 0.1\%$
Switches	On resistance ≤ 10 kΩ Off resistance > 10 MΩ
Op amps	Gain, bandwidth, noise performance, power consumption, and so on depend on design tradeoffs. Optimization for given features (e.g. wide bandwidth or low power consumption)
Typical	Voltage gain > 60 dB Bandwidth > 1 MHz Dynamic range > 80 dB Power consumption $= 1 \div 3$ megawatt (mW)/op amp

The performance of the op amp is a result of design by tradeoff. Some of the important features, such as wide bandwidth and low power, are mutually exclusive. Here, again, the available technology plays a critical role. It is particularly with respect to higher frequencies that progress in technology has been, and is likely to continue to be, most pronounced. This is true for MOS technology, but even more so for the more recently emerging gallium arsenide (GaAs) technology. At the present time, MOS devices already have been fabricated that can be used to implement SC filters with clock frequencies up to 30 MHz; upper limits of up to 150 MHz are predicted. GaAs devices can be used with clock frequencies up to 250 MHz, and possibly beyond 500 MHz. Such technological advances will permit the creation of a new class of analog, high-speed signal-processing systems. They will be based on SC and, more recently, so-called switched-current devices, both of which use similar sampled-data processing techniques.

Another critical parameter, because it directly effects the chip cost, is chip size. Scaling down in line width, and consequently chip area, is one of the main thrusts of technological advances. Although this has fewer detrimental side effects for digital than for analog VLSI circuits, much already has been achieved for the latter. A high-performance op amp presently occupies a chip area of about $5 \cdot 10^4$ μm^2 (square micrometers), a switch area of about 50 μm^2, and a 1-pF capacitor area of about 2500 μm^2. Thus, an SC filter requiring 100 op amps, a total capacitance of 300 pF (e.g., 300 capacitors of 1 pF) and, say, 500 switches, will consume an area of $5 \cdot 10^7$ μm^2. Assuming one op amp per filter order (which is typical), this corresponds to a 100-pole filter, or any combination thereof (e.g., 10 filters of 10th order). Thus, present MOS technology already permits highly sophisticated filters to be fabricated on a relatively small chip area. Notice that most of the chip area is consumed by the capacitors; minimizing the total required capacitance area therefore takes high priority in any filter synthesis and optimization procedure.

The Basic Principles of Switched-Capacitor Filters

It is stated above that in SC networks combinations of analog MOS switches and capacitors take over the resistive role. It is, of course, not immediately obvious how this can be done. Rather than present a detailed and exact analysis of SC networks in which this statement becomes clear, it is more appropriate in the present context to give a simplified, somewhat intuitive, explanation for the workings of an SC network.

Consider a capacitor C with a current surge injected into it as shown in Fig. 68. The resulting incremental charge $q(\tau)$ stored on the capacitor plates is then

$$q(\tau) = \int_0^\tau i(t)dt. \tag{126}$$

Assume that the resulting voltage across C, after the surge is over at time τ, is $v(\tau)$, and that the initial voltage, before the surge began, is $v(0)$. Then Eq. (126), expressed in terms of these two voltages, becomes

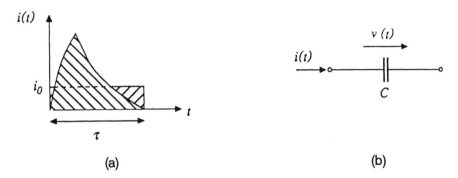

FIG. 68 Capacitor with current surge $i(t)$ injected into it.

$$q(\tau) = Cv(\tau) - Cv(0). \qquad (127)$$

The current surge in Fig. 68 lasts τ seconds (which, as we see below, is the minimum duration that a corresponding switch is closed). Expressing the incremental charge $q(\tau)$ by an average current i_o (see Fig. 68), such that Eq. (126) also can be expressed as

$$q(\tau) = i_o \cdot \tau \qquad (128)$$

and setting this equal to Eq. (127), we obtain for the average current during the time τ

$$i_o(\tau) = \frac{C}{\tau} v(\tau) - \frac{C}{\tau} v(0). \qquad (129)$$

We now compare this expression with that of a resistor R or a conductance $G = 1/R$ (see Fig. 69). The current at any time, $t = \tau$, is given by

$$i(\tau) = G \cdot v(\tau). \qquad (130)$$

Comparing Eqs. (129) and (130), the essential difference between the two expressions is in the second term in Eq. (129) that accounts for the storage or memory term associated with the capacitor C. If, in some way, we can eliminate this term by removing the initial charge $C \cdot v(0)$, then, at least at the time $t = \tau$, the capacitor will behave like a resistor.

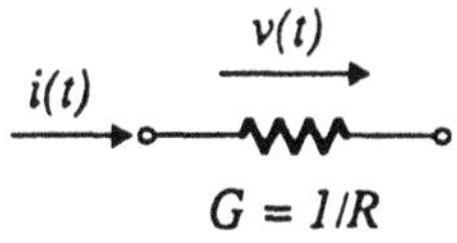

$v(t)$

$i(t)$

$G = 1/R$

FIG. 69 Resistor with current source $i(t)$ causing voltage $v(t)$.

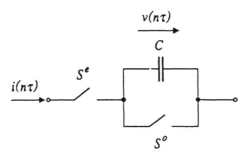

FIG. 70   Switched-capacitor combination.

Consider now the SC combination shown in Fig. 70. S^e and S^o are switches closed for τ seconds during even and odd times, respectively, as shown in Fig. 71. Incidentally, it is important that the "closed" times of the two switches do not overlap. For any even time interval $n\tau$ (i.e., n is even), the equation corresponding to Eq. (129) can be written as

$$i(n\tau) = \frac{C}{\tau} v(n\tau) - \frac{C}{\tau} v((n-1)\tau) \tag{131}$$

or, using the superscript e and o for even and odd, respectively,

$$i^e(n) = \frac{C}{\tau} v^e(n) - \frac{C}{\tau} v^o(n-1). \tag{132}$$

Note that, for convenience, we have omitted the τ in the time units (i.e., n instead of $n\tau$). Since the odd switch will short out the capacitor during odd time

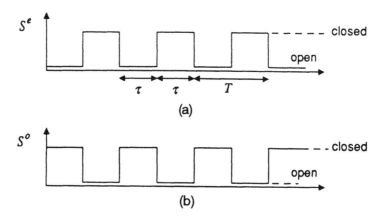

FIG. 71   Switches closed for τ seconds: a, S^e, closed for even times; b, S^o, closed for odd times.

intervals $n^o\tau$ (thereby eliminating the charge on the capacitor during odd times), we have

$$i^e(n) \;=\; \frac{C}{\tau}\, v^e(n). \tag{133}$$

Thus, comparing with Eq. (130), we see that the SC combination shown in Fig. 70 behaves like a resistor during even time intervals $n^e\tau$. This shorting out of charge during alternate time intervals, using nonoverlapping switches with a 50% duty cycle, is the method by which the storage property of capacitors in SC networks is eliminated; resistive behavior in the complementary time intervals thereby is obtained. Furthermore, by comparison of Eq. (130) with Eq. (133), the resistance is given by

$$R \;=\; \frac{\tau}{C}. \tag{134}$$

The SC realization of a resistor in Fig. 70 is, strictly speaking, a time-variant circuit: at even times $n^e\tau$ it behaves like a resistor; at odd times it is not accessible (because the even switch is open), but "internally" it behaves like a short circuit. This time variance can be represented spatially by looking at the circuit separately at even and odd times and "linking" the resulting time-invariant "even" and "odd" circuits (or "paths") by a so-called link two-port (LTP), which represents the capacitor. It is precisely because of its storage property that the capacitor links the even with the odd path (see Fig. 72). Notice that a closed switch

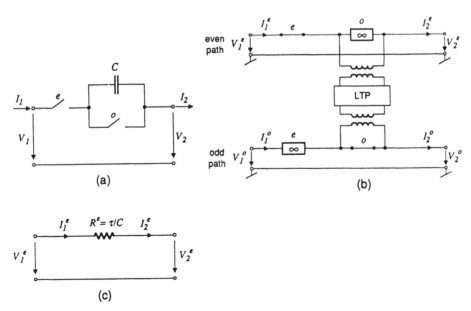

FIG. 72 Switched-capacitor realization of a series resistor $R^e = \tau/C$: a, switched-capacitor combination; b, four-port equivalent circuit with link two-port; c, equivalent series resistor.

appears as a short circuit in the corresponding path; an open switch appears as an infinite impedance. Furthermore, in the case of a series capacitance, the LTP is embedded in the circuit between two ideal one-to-one transformers.

The SC combination realizing a series resistor shown in Figs. 70 and 72 is not the only way of obtaining a resistor at even (or odd) times. There are various other switching combinations, all of which, however, are based on eliminating the storage property of a capacitor at the complementary times τ. One of these, together with the LTP equivalent circuit, is shown in Fig. 73. Notice that the shorting of the capacitor at odd time intervals corresponds to shorting the odd terminals of the LTP in the odd circuit path.

Having introduced the concept of the LTP and the four-port equivalent circuit of an SC circuit, we now can ask how a capacitor in an SC circuit behaves if its stored charge is not shorted out in alternating time intervals. This corresponds to the capacitor in Fig. 70 without the odd switch S^o across it. To answer this question, we must take a closer look at the LTP.

The Link Two-Port

The LTP is a four-port representation of a capacitor in a biphase sampled data system. (It has not been mentioned here before, but the concept of SC networks can be extended readily to multiphase sampled-data circuits. For m-phase switches, with individual closing times of $\tau = T/m$ where T is the overall clock period, the link two-port then becomes a link m-port (LMP). All other considerations relating to the biphase then can be extended to the m-phase case. In the present context, it must suffice to limit our discussion to the biphase case.) To

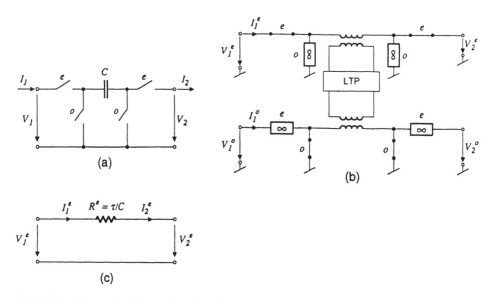

FIG. 73 Alternative SC combination of series resistor $R^e = \tau/C$: a, SC circuit; b, four-port equivalent circuit; c, equivalent series resistor.

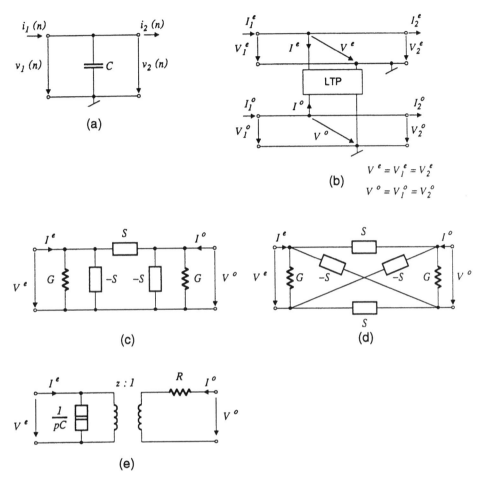

FIG. 74 Two-port with grounded capacitor: a, basic diagram; b, equivalent four-port; c, link two-port; d, balanced lattice version of link two-port; e, useful equivalent circuit for link two-port.

see this, consider the simple grounded-capacitor two-port shown in Fig. 74-a. Assuming discrete-time current surges where $i(t)$ at $t = n\tau$ is the value of the current surge at the end of the time interval $n\tau$, we can write the following nodal charge equations, which are based on the principle of charge conservation:

$$v_1(n) = v_2(n) \tag{135}$$

$$\frac{C}{\tau} v_1(n) = i_1(n) - i_2(n) + \frac{C}{\tau} v_1(n - 1) \tag{136}$$

Applying the z-transform to these difference equations, we obtain

$$V_1(z) = V_2(z) \tag{137}$$

$$\frac{C}{\tau} V_1(z) = I_1(z) - I_2(z) + \frac{C}{\tau} z^{-1} V_1(z) \qquad (138)$$

where $z = e^{s\tau}$. The equations above describe the simple capacitor two-port of Fig. 74-a for all time intervals $n\tau$. Thus, whether in the time domain, as in Eqs. (135) and (136), or in the frequency domain as in Eqs. (137) and (138), we can assume a series of voltage and current samples at times $n\tau$. In order to obtain a spatial representation of our two-port in which one circuit path represents even times $n^e\tau$ and the other odd times $n^o\tau$, we can break up our sequence of samples into even samples and odd samples, that is, in the time domain:

$$v(n) = v^e(n) + v^o(n) \qquad (139)$$

$$i(n) = i^e(n) + i^o(n) \qquad (140)$$

or, in the z-domain:

$$V(z) = V^e(z) + V^o(z) \qquad (141)$$

$$I(z) = I^e(z) + I^o(z) \qquad (142)$$

Remaining in the z-domain, we now partition the terms in Eqs. (137) and (138) into even and odd terms and group them into even-time equations and odd-time equations, where we take note of the fact that terms of the form $z^{-1} V^o(z)$ belong in an even-time equation, and terms of the form $z^{-1} V^e(z)$ in an odd-time equation:

$$V_1^e(z) = V_2^e(z) \qquad (143)$$

$$V_1^o(z) = V_2^o(z) \qquad (144)$$

$$I_1^e(z) = G V_2^e(z) - S V_2^o(z) + I_2^e(z) \qquad (145)$$

$$I_1^o(z) = G V_2^o(z) - S V_2^e(z) + I_2^o(z) \qquad (146)$$

where G is a conductance of value C/τ and S is a so-called storistor (resistor with storage property) of value $z^{-1} C/\tau$. Rewriting Eqs. (145) and (146) and letting $I_1^e(z) - I_2^e(z) = I^e(z)$ and $I_1^o(z) - I_2^o(z) = I^o(z)$, we obtain

$$V_1^e(z) = V_2^e(z) \qquad (147)$$

$$V_1^o(z) = V_2^o(z) \qquad (148)$$

$$I^e(z) = G V_2^e(z) - S V_2^o(z) \qquad (149)$$

$$I^o(z) = -S V_2^e(z) + G V_2^o(z) \qquad (150)$$

These four equations define the link two-port, linking the even and odd circuit paths as shown in Fig. 74-*b*. From these equations, the link two-port emerges as a well-defined two-port (Fig. 74-*c*) with a short-circuit admittance matrix equation given by

$$\begin{bmatrix} I^e \\ I^o \end{bmatrix} = \begin{bmatrix} G & -S \\ -S & G \end{bmatrix} \begin{bmatrix} V^e \\ V^o \end{bmatrix} \qquad (151)$$

Note that both a positive and negative storistor occurs in the equivalent circuit. A balanced lattice version of the LTP is shown in Fig. 74-*d*. Equation (151) can be rearranged to provide the transmission, or *ABCD* matrix of the LTP, namely,

$$\begin{bmatrix} V^e \\ I^e \end{bmatrix} = \begin{bmatrix} z & zR \\ z\,pC & z \end{bmatrix} \begin{bmatrix} V^o \\ -I^o \end{bmatrix} \qquad (152)$$

where
$$R = \tau/C$$
$$p = (1 - z^{-2})/\tau.$$

Finally, from the transmission matrix given by Eq. (152), an equivalent circuit for the LTP can be derived that often is useful in providing insight into the physical behavior of an SC network (see Fig. 74-*e*). The symbol used for the impedance $(pC)^{-1}$ is meant to indicate that this is a kind of *p*-transformed capacitor or a capacitor as it occurs in SC networks, as we explain below. The series resistor R equals τ/C, and the ideal "transformer" can be interpreted as a general impedance converter as defined in Eq. (122), with $A(s) = z$ and $D(s) = z^{-1}$. Although this equivalent circuit for the LTP appears to be nonreciprocal and unsymmetrical, it is neither, as a simple calculation will verify. In the series capacitor case, in which the corresponding LTP occurs floating, for example, as in Fig. 72-*b*, any one of the equivalent circuits for the LTP can be considered embedded between two ideal transformers with a one-to-one transformation ratio.

With the definition of the LTP, either as an equivalent circuit or with its two-port matrix parameters, it is easy to see that an LTP that is short-circuited by a closed switch at one of its ports will appear as a resistor at the other. This, as we may recall, is how we interpreted the circuit of Fig. 72-*a*, resulting in the series resistance of Fig. 72-*c*. A simple test for this is to use the *ABCD* matrix of the LTP given by Eq. (152) in the expression for the input impedance of a two-port loaded by Z_L, as used in Eq. (69). With a short circuit at the output terminals, we have $Z_L = 0$, thus, with Eq. (152),

$$Z_{\text{IN}} = \left. \frac{A Z_L + B}{C Z_L + D} \right|_{Z_{L=0}} = \frac{B}{D} = R. \qquad (153)$$

It now is easy to answer the question posed at the end of the section above, how does a capacitor in an SC network behave if its stored charge is not shorted out in alternating time intervals? With the LTP, the answer to this question is very simple. A capacitor with a storage property that is maintained cannot have a shorting switch across it at alternating times, that is, it corresponds to the circuit in Fig. 72-*a* without the switch S^o. Instead of a short circuit at the output (i.e., odd) terminals of the LTP, we now have an open circuit or $Z_L = \infty$. Using the same expression as in Eq. (153), we then obtain with Eq. (152)

$$Z_{\text{IN}} = \frac{A Z_L + B}{C Z_L + D} \Bigg|_{Z_{L=\infty}} = \frac{A}{C} = \frac{1}{pC} \tag{154}$$

where $p = (1 - z^{-2})/\tau$. The unswitched capacitor in an SC circuit therefore appears as a *p*-transformed capacitor as already anticipated above. It also is verifiable readily from the LTP equivalent diagram in Fig. 74-*e* when the output (i.e., odd) terminals of the two-port are left open.

The Integrator in Switched Capacitor Networks

In the section that deals with active RC filters above, it is shown that an arbitrary rational filter transfer function $T(s)$ can be converted into an expression involving only integrators, as in Eqs. (110) and (111). Using SFGs, Fig. 50 then shows how a network topology consisting only of integrators, multipliers (i.e., gain elements), and summing circuits, can be derived directly from the converted transfer function. Finally, in Fig. 52, integrators comprising op amps and RC combinations are shown, and these then are used for the synthesis of active RC filters, either in the form of biquads or filters of higher order. In SC filter design, a very similar procedure can be followed. Being a sampled-data network, one of the main differences will be that the continuous-time integration will have to be replaced by a numerical integration, or, in other words, the corresponding describing differential equation replaced by a difference equation.

With the LTP concept discussed in the previous section, an SC integrator can be derived directly from an active RC integrator. Consider the active RC integrator shown in Fig. 75-*a*. Its voltage transfer function is given by

$$T(s) = \frac{V_2}{V_1} = -\frac{1}{s R_1 C_0}. \tag{155}$$

From the discussion in the previous sections, and referring to Fig. 72, it is easy to see that an SC version of this integrator can be obtained with the circuit in Fig. 75-*b*. The corresponding four-port equivalent circuit using the LTP concept is shown in Fig. 75-*c*. From this and previous four-port equivalent circuits, it follows that, theoretically, four basic transfer functions can be defined for a biphase SC network, namely,

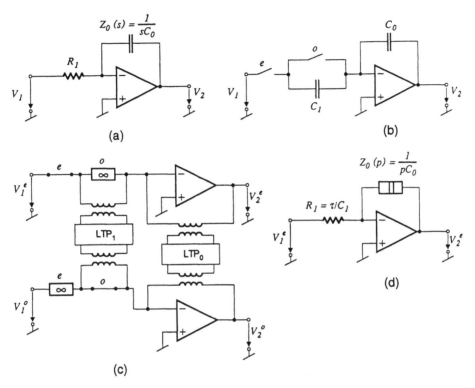

FIG. 75 Derivation of a switched-capacitor integrator from an active resistor-capacitor integrator: *a*, active RC integrator; *b*, SC integrator; *c*, equivalent four-port network; *d*, simplified equivalent circuit for $H^{ee}(z)$.

$$H^{ee}(z) = \left. \frac{V_2^e}{V_1^e} \right|_{V_1^o = 0} \tag{156}$$

$$H^{oe}(z) = \left. \frac{V_2^e}{V_1^o} \right|_{V_1^e = 0} \tag{157}$$

$$H^{eo}(z) = \left. \frac{V_2^o}{V_1^e} \right|_{V_1^o = 0} \tag{158}$$

$$H^{oo}(z) = \left. \frac{V_2^o}{V_1^o} \right|_{V_1^e = 0} \tag{159}$$

thus

$$V_2(z) = V_2^e(z) + V_2^o(z)$$

$$= V_1^e(z) [H^{ee}(z) + H^{eo}(z)]$$

$$+ V_1^o(z) [H^{oe}(z) + H^{oo}(z)] \tag{160}$$

In practice, it will depend on the phase of the input and output switch which of these four transfer functions exist. By inspection of Fig. 75-c, it should be clear that only two transfer functions exist, namely, $H^{ee}(z)$ and $H^{eo}(z)$. The first, $H^{ee}(z)$, results immediately if we recognize that in the even path LTP$_1$ is a short-circuited (in the odd path) LTP, and LTP$_0$ is an open-circuited (in the odd path) LTP. The circuit therefore can be simplified as in Fig. 75-d. Comparing with the RC integrator in Fig. 75-a, we obtain by inspection

$$H^{ee}(z) = \frac{V_2^e}{V_1^e} = -\frac{1}{p R_1 C_0}. \tag{161}$$

Since $R_1 = \tau/C_1$ and $p = (1 - z^{-2})/\tau$, this becomes

$$H^{ee}(z) = -\frac{C_1}{C_0} \cdot \frac{1}{1 - z^{-2}}. \tag{162}$$

Note that (1) the integrator constant is determined by a capacitor ratio, and (2) that a comparison between Eqs. (155) and (162) shows that an s-to-z transformation is implied here, namely,

$$s \to p = \frac{1 - z^{-2}}{\tau}. \tag{163}$$

This, in turn, corresponds to a difference equation defining a numerical integration known as the backward Euler integration. To obtain $H^{eo}(z)$ for the four-port equivalent circuit of Fig. 75-c, we can redraw the circuit of Fig. 75-c as shown in Fig. 76-a. This, in turn, can be simplified as in Fig. 76-b. Straightforward analysis then provides the transfer function

$$H^{eo}(z) = -\frac{C_1}{C_0} \frac{z^{-1}}{1 - z^{-2}} \tag{164}$$

This corresponds to a so-called lossless discrete integration (LDI) integrator with the s-to-z transformation

$$s \to \frac{1}{\tau} \frac{1 - z^{-2}}{z^{-1}} \tag{165}$$

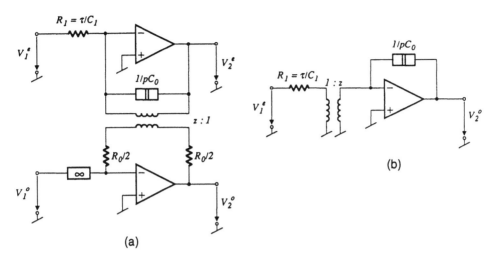

FIG. 76 Four-port equivalent circuit of SC integrator in Fig. 75-*b*: *a*, redrawn and simplified version; *b*, further simplified circuit for H^{eo}.

Note that merely by changing the phase of the output terminal switch of the SC integrator in Fig. 75-*b* from even to odd, a different numerical integration is obtained.

There are various ways of performing a numerical integration, the most important of which are shown in Fig. 77. The corresponding difference equations and *s*-to-*z* transformations are listed in Table 5. (It should be pointed out here that, because of our definition of $z = e^{s\tau}$ where $\tau = T/2$, that is, τ is half the clock frequency interval T [see Fig. 71], the complex variable z occurs either as a linear or a squared term. In the literature of digital filtering and also much of the SC literature, the definition $z = e^{sT}$ is also common. In this case, z^{-2} must be substituted by z^{-1}, and z^{-1} by $z^{-1/2}$ in the expressions given here.)

From Figs. 77 and Table 5, it follows that there are numerous ways of performing a numerical integration, each of which has a corresponding *s*-to-*z* transformation. Interestingly enough, for each numerical integration, there are more than one SC circuit capable of performing it. An assortment of these SC integrators and the type of integration they perform is shown in Table 6. Theoretically, all of them are acceptable, just as every form of numerical integration shown in Fig. 77 has its own merits. In practice, however, the situation is very different; their suitability as SC circuits varies considerably. There are two main reasons for this: (1) the "quality" of mapping from the *s*- to the *z*-plane is not the same and, more importantly, (2) the degree of insensitivity to parasitic capacitance of the corresponding SC integrators, when realized on an IC chip, is critically different.

To obtain some idea of the "quality" of mapping, the *s*-to-*z* mapping corresponding to the four types of integration in Table 5 are shown in Fig. 78. Ideally, the most desirable kind of mapping is obtained when the *jω*-axis is

(*text continues on p. 526*)

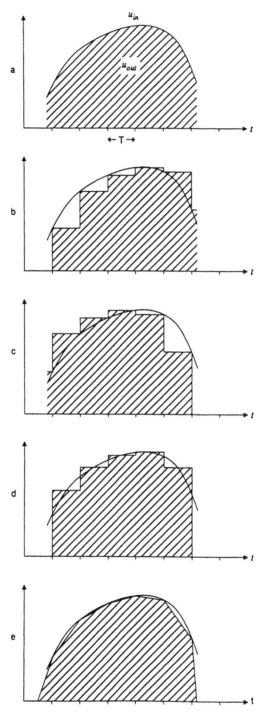

FIG. 77 Types of numerical (discrete-time) integration: *a*, continuous; *b*, forward difference; *c*, backward difference (*p*-transform); *d*, lossless discrete integration (intermediate rectangular); *e*, bilinear integration (trapezoidal).

TABLE 5 Types of Discrete-Time Integrators Used in Switched-Capacitor Networks

Types of Integrators	Discrete-Time Approximation of the Continuous-Time Integration: $\int_{(k-2)\tau}^{k\tau} x(t)dt = y(k\tau) - y((k-2)\tau)$, $T=2\tau$		$s \to z$ Transformation
	Differential / Difference Equation	s - / z - plane	
Continuous-Time, exact Integration	$y(k\tau) - y((k-2)\tau) = \int_{(k-2)\tau}^{k\tau} x(t)dt$	$Y(s) = X(s) \cdot \frac{1}{s}$	—
Forward Difference fd	$y[k] - y[k-2] = x[k-2] \cdot T$	$Y(z) = X(z) \cdot T \cdot \frac{z^{-2}}{1-z^{-2}}$	$s \to \frac{1}{T} \cdot \frac{1-z^{-2}}{z^{-2}}$
Backward Difference bd	$y[k] - y[k-2] = x[k] \cdot T$	$Y(z) = X(z) \cdot T \cdot \frac{1}{1-z^{-2}}$	$s \to \frac{1}{T} \cdot (1-z^{-2})$
Lossless Discrete ldl *	$y[k] - y[k-2] = x[k-1] \cdot T$	$Y(z) = X(z) \cdot T \cdot \frac{z^{-1}}{1-z^{-2}}$	$s \to \frac{1}{T} \cdot \frac{1-z^{-2}}{z^{-1}}$
Bilinear Integration bl **	$y[k] - y[k-2] = \frac{x[k]+x[k-2]}{2} \cdot T$	$Y(z) = X(z) \cdot \frac{T}{2} \cdot \frac{1+z^{-2}}{1-z^{-2}}$	$s \to \frac{2}{T} \cdot \frac{1-z^{-2}}{1+z^{-2}}$

*intermediate rectangular   **trapezoidal

TABLE 6 Switched-Capacitor Realization of Discrete-Time Integration

	SC Integrator	Transfer Function	Transform	Amplitude Error	Phase Error
1		$H^{eo} = -c \cdot \dfrac{z^{-1}}{1-z^{-2}}$ $H^{oe} = -c \cdot \dfrac{z^{-2}}{1-z^{-2}}$	Idi fd	$\varepsilon = \omega\tau/\sin\omega\tau - 1$ $\varepsilon = \omega\tau\, ctg\,\omega\tau - 1$	$\theta = 0$ $\theta = -\omega\tau$
2		$H^{oe} = -c \cdot \dfrac{1}{1-z^{-2}}$ $H^{eo} = -c \cdot \dfrac{z^{-1}}{1-z^{-2}}$	bd Idi	$\varepsilon = \omega\tau\, ctg\,\omega\tau - 1$ as 1	$\theta = +\omega\tau$ as 1
3		$H_{tot} = -c \cdot \dfrac{1+z^{-1}}{1-z^{-1}}$	bi	$\varepsilon = \dfrac{\omega\tau}{2} ctg \dfrac{\omega\tau}{2} - 1$	$\theta = 0$
4		$H_{tot} = -c \cdot \dfrac{1+z^{-2}}{1-z^{-2}}$	bi	$\varepsilon = \omega\tau\, ctg\,\omega\tau - 1$	$\theta = 0$
5		$H^{eo} = -c \cdot \dfrac{1+z^{-2}}{1-z^{-2}}$ $H_{tot} = -c \cdot \dfrac{1+z^{-1}}{1-z^{-1}}$	bi	as 4 as 3	as 4 as 3
6		$H_{tot} = -c \cdot \dfrac{1+z^{-2}}{1-z^{-2}}$	bi	as 4	as 4
7		$H^{oe} = -c \cdot \dfrac{1}{1-z^{-2}}$ $H^{eo} = -c \cdot \dfrac{z^{-1}}{1-z^{-2}}$	bd Idi	as 2	as 2
8		$H^{eo} = +c \cdot \dfrac{z^{-1}}{1-z^{-2}}$ $H^{oe} = +c \cdot \dfrac{z^{-2}}{1-z^{-2}}$	Idi fd	as 1	as 1
9		$H^{eo} = -c \cdot \dfrac{z^{-1}}{1-z^{-2}}$ $H^{oe} = -c \cdot \dfrac{z^{-2}}{1-z^{-2}}$	Idi fd	as 1	as 1
10		$H^{eo} = +c \cdot \dfrac{z^{-1}}{1-z^{-2}}$ $H^{oe} = +c \cdot \dfrac{z^{-2}}{1-z^{-2}}$	Idi fd	as 1	as 1

bd : backward difference   bi : bilinear
fd : forward difference   Idi : lossless discrete

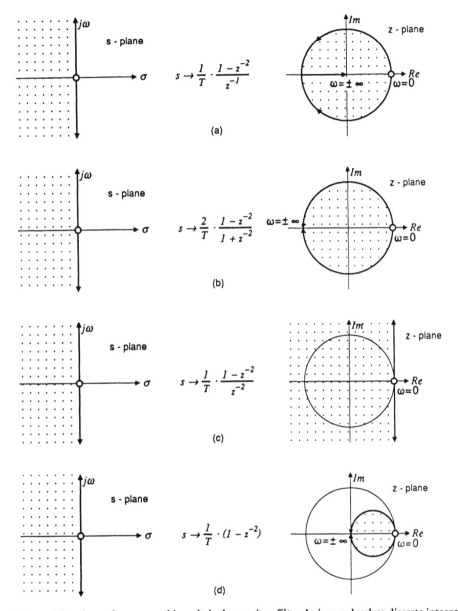

FIG. 78 Mappings of $s \to z$ used in switched-capacitor filter design: a, lossless discrete integration; b, bilinear integration; c, forward difference; d, backward difference (note that prewarping for the backward case may mean initially moving some poles outside the unit circle; when sampling, these poles then move inside the unit circle).

mapped once onto the unit circle in the z-plane. From Fig. 78, it is clear that this happens exactly only for the bilinear transform and, to a large extent, for the LDI transform. This is why, in digital filtering, the bilinear transform is the preferred method of s-to-z mapping, followed in preference by the LDI transform. All four methods of mapping require a degree of prewarping (see Table 7), but this is not unusual and has become a standard procedure in most sampled-data and digital filtering. Whereas signal aliasing must be avoided in all discrete-time and sampled-data filters, and this imposes limits on the lowest possible switching frequency, filter-response aliasing actually is avoided with the bilinear transform, and can be dealt with relatively easily using the LDI transform. As it happens, however, none of these considerations are really very important when selecting an SC integrator from the circuits in Table 6 because, in one way or another, they can be dealt with satisfactorily in a practical design situation. Far more serious, and ultimately decisive for the choice of an SC integrator, is its insensitivity to parasitic capacitance.

TABLE 7 Prewarping Conditions for the Four $s \rightarrow z$ Transformations Used in Switched-Capacitor Networks

LDI:
$$\boxed{s \rightarrow \frac{1}{T}(z - z^{-1})}$$

Sampling frequency $\omega_s = \frac{2\pi}{T}$

$z = e^{st}$ $\tau = \frac{T}{2}$ $s = j\omega$:

$$\frac{1}{T}(e^{st} - e^{-st})\Big|_{s=j\omega} = \frac{1}{\tau} \cdot \frac{e^{j\omega\tau} - e^{-j\omega\tau}}{2} = j\frac{\sin \omega\tau}{\tau} \quad :$$

$$\boxed{\omega_c = \frac{2}{T} \sin\left(\omega_d \frac{T}{2}\right) = \frac{\omega_s}{\pi} \sin\left(\pi \frac{\omega_d}{\omega_s}\right)}$$

Similarly:

BI:
$$\boxed{s \rightarrow \frac{2}{T} \cdot \frac{1 - z^{-2}}{1 + z^{-2}}}$$

$$\boxed{\omega_c = \frac{2}{T} \tan\left(\omega_d \frac{T}{2}\right) = \frac{\omega_s}{\pi} \tan\left(\pi \frac{\omega_d}{\omega_s}\right)}$$

FD:
$$\boxed{s \rightarrow \frac{1}{T} \cdot (z^2 - 1)}$$

$$\boxed{\omega_c = \frac{2}{T} \sin\left(\omega_d \frac{T}{2}\right) e^{j\omega_d T/2} = \frac{1}{T} \sin(\omega_d T) + j\frac{2}{T} \sin^2\left(\omega_d \frac{T}{2}\right)}$$

BD:
$$\boxed{s \rightarrow \frac{1}{T} \cdot (1 - z^{-2})}$$

$$\boxed{\omega_c = \frac{2}{T} \sin\left(\omega_d \frac{T}{2}\right) e^{-j\omega_d T/2} = \frac{1}{T} \sin(\omega_d T) - j\frac{2}{T} \sin^2\left(\omega_d \frac{T}{2}\right)}$$

By prewarping the filter specifications given in terms of the "digital" frequency ω_d according to the expressions given above, an undistorted frequency response results.

Parasitic Insensitive Switched-Capacitor Networks

Although there are many technological parameters and peculiarities that must be considered when designing CMOS VLSI circuits, none has been more critical for the success or failure of CMOS filter chips than their insensitivity to parasitic capacitance. This topic therefore merits separate treatment in our survey of SC filters. Such other nonideal effects as offset, bandwidth, slew rate, settling time, noise, clock feed-through, capacitor spread and total capacitance, power dissipation, and so on, to name only a few, may be serious in a given situation, but generally can be coped with to a more or less satisfactory degree. To learn more about them, the reader is referred to the professional literature and to the bibliography at the end of this article. As will become clear in what follows, parasitic insensitivity, as it has come to be known, is more basic in nature and tends to "make or break" a given application.

Consider the schematic of an MOS capacitor as shown in Fig. 79-a. The desired capacitor C_0 is lodged between two electrodes (the top one also may be a metal layer). Unfortunately, the two plates each have a parasitic capacitance to the substrate, so that instead of simply a capacitor C_0, we have the capacitive π-network shown in Fig. 79-b. Note that the top plate of the capacitor may have a parasitic capacitance ranging from 0.1% to 1% of the desired capacitance value C_0, the bottom plate parasitic capacitance may be as high as 20% of C_0. As a consequence, an SC filter chip, the accuracy of which usually depends on the accuracy of the capacitance ratios of its integrators, may be entirely unusable. The stray capacitance due to the parasitics will change the coefficients of the corresponding filter transfer function to the point at which, in a precision filter, the specified tolerance limits cannot be satisfied. Consider, for example, the first integrator in Table 6, which is shown again in Fig. 80-a. The parasitic capacitors of C_1 are designated C_{1_t} and C_{1_b} (top and $bottom$ plates, respectively), those of C_2 are designated C_{2_t} and C_{2_b}. Clearly, C_{1_b} can be ignored; so can C_{2_t} because it is in parallel with the output voltage source of the op amp. The same is true for C_{2_b}, which is in parallel to a virtual ground, that is, there (ideally) is no voltage across it. That leaves accounting for C_{1_t}. Redrawing the integrator in terms of the equivalent LTP circuit of Fig. 74-e, we obtain the equivalent circuit shown in Fig. 80-b. As indicated in the figure, the p-transformed capacitor parallel to the input voltage source can be ignored. The remaining transformer and the resistor R_1, together with the op amp and feedback impedance $(p\,C_2)^{-1}$, then gives (see Eq. [161]):

$$H^{eo}(z) = \frac{V_2^o}{V_1^e} = -\frac{z^{-1}}{p\,R_1\,C_2} = -\frac{C_1 + C_{1_t}}{C_2} \cdot \frac{z^{-1}}{1 - z^{-2}}. \qquad (166)$$

Note that the integrator constant is in error by the amount C_{1_t}/C_2, which may be up to 1% of the desired value C_1/C_2. In most precision filters, this error will be unacceptable, considering that subsequent fine-tuning is not possible.

Consider now Integrator 7 from Table 6 as shown in Fig. 81-a. Using the LTP equivalent circuit of Fig. 74-e, and taking into account the fact that an

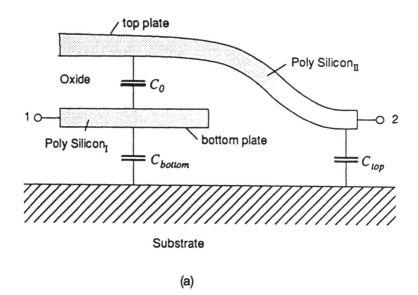

(a)

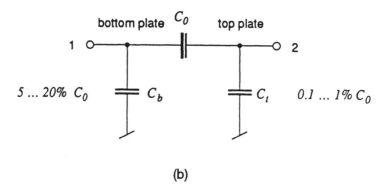

(b)

FIG. 79 Metal-oxide-semiconductor capacitor C_0: a, basic schematic; b, equivalent diagram
with parasitic capacitors C_t and C_b.

LTP that is short-circuited at the load (e.g., odd) side corresponds to a resistor
at the input (e.g., even) side, we obtain the equivalent circuit shown in Fig.
81-b. Note that besides the parasitic capacitors C_{2_b} and C_{2_t}, which can be neglected for the reasons given above, the parasitic resistors R_{1_b} and R_{1_t} also can
be neglected, the first because it is in parallel with the driving voltage source,
the second because it is across virtual ground. Thus, this integrator is parasitic
insensitive and its transfer function

$$H^{ee}(z) = \frac{V_2^e}{V_1^e} = -\frac{1}{p R_1 C_2} = -\frac{C_1}{C_2}\frac{1}{1 - z^{-2}}. \tag{167}$$

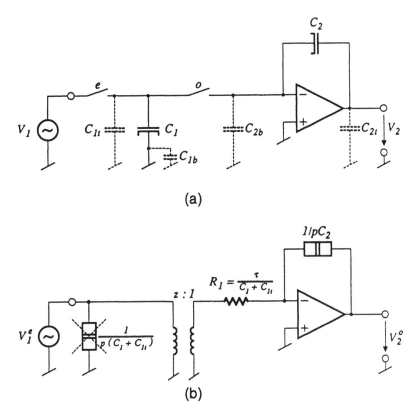

FIG. 80 Integrator 1 from Table 6: a, basic diagram; b, equivalent circuit using link two-port equivalent circuit given in Fig. 74-e.

Note that for the purposes of parasitic insensitivity, it is irrelevant which type of numerical integration is carried out. More to the point, the requirement for parasitic insensitivity is so important that it overrides any requirements made on the resulting s-to-z transform. On the other hand, by utilizing the degree of freedom provided by the phasing of the input and output switches, a limited variety of s-to-z transforms is obtainable. Thus, for example, the transform in Eq. 167 corresponds to a backward difference (BD) integration; by changing the output phase of the circuit from even to odd, the resulting transfer function $H^{eo}(z)$ provides an LDI transformation (see Table 6).

If we consider the SC integrators listed in Table 6, it is easy to show that only two of them, Integrators 7 and 8, are parasitic insensitive. (In the parlance of SC networks, such circuits also are called stray insensitive.) Note that neither of these integrators perform the well-known and, in digital filtering, most desirable bilinear transform. Integrators performing the latter are either stray sensitive or too complicated to be used in SC filter design. Nevertheless, Integrators 7 and 8, also referred to as inverting and noninverting integrators, respectively, do provide the LDI transform and the forward and backward difference integration. Of these, the LDI is the most useful, as a glance at Fig. 78 will show. In

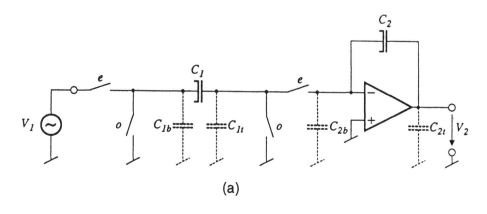

(a)

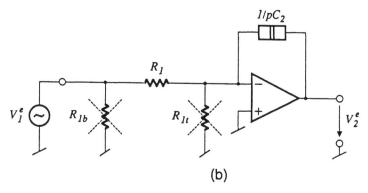

(b)

FIG. 81   Integrator 7 from Table 6: *a*, basic diagram; *b*, equivalent circuit using link two-port equivalent circuit given in Fig. 74-*e*.

practice, SC filters are designed by transforming the given continuous-time filter specifications into the discrete-time domain using the bilinear transform, and subsequently realizing the resulting transfer function in z, that is, $H(z)$, using LDI-type stray-insensitive integrators of the inverting and noninverting type as required. This is discussed in the next section.

Designing Switched-Capacitor Filters

Inductor-Capacitor Filter Simulation and Direct-Form Design

The design of SC filters can be carried out in much the same way as active RC continuous-time filters (discussed in the section of that title above). Considering LC filter simulation and direct-form design first, this can be carried out in five steps.

1.   Start out with the filter specifications as given in the sampled-data frequency domain. An example of a low-pass filter specification is shown in

Fig. 82. Note that the frequency axis is designated ω_d, where d stands for discrete time. This means that the frequency specifications will be periodic, that is, they will be repeated symmetrically about all integer multiples of the sampling frequency ω_s, as described in the literature on sampled-data and digital filters (see bibliography at the end of this article). Since the sampling frequency is assumed to be much larger than the corner frequency of the stop band ω_{sB} in Fig. 82, this periodicity is not shown; it merely is implied by the designation ω_d of the frequency axis.

2. Predistort the filter specifications (i.e., the discrete-time frequency axis) according to the inverse bilinear transform. For this, a sampling frequency ω_s must be selected such that it is at least twice as large as the highest significant frequency associated with the specifications (e.g., $\omega_s > 2\omega_{sB}$). This requirement is related to the sampling theorem and prevents aliasing of the frequency response. The sampling frequency ω_s is related to the sampling period T by $\omega_s = 2\pi/T$. The predistortion rule, which provides the corner frequencies of the filter specifications in the continuous-time frequency domain, is

$$\omega_c = \frac{2}{T} \tan\left(\frac{T}{2}\omega_d\right) = \frac{\omega_s}{\pi} \tan\left(\pi\frac{\omega_d}{\omega_s}\right). \tag{168}$$

This relationship also is given in Table 7. The subscript c of ω_c stands for continuous time. The new (i.e., predistorted) continuous-time specifications are depicted in Fig. 83.

3. Solve the approximation problem (see "Basic Filter Types" and "Lumped Element Inductor-Capacitor-Resistor Filters") in the predistorted continuous-time domain. Find a suitable filter transfer function $T(s)$, or an LC filter network, satisfying the continuous-time specifications.

4. Having obtained $T(s)$ in Step 3, there are various ways of continuing the design process. Three useful procedures follow.

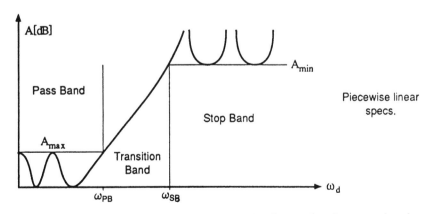

FIG. 82 Low-pass filter specifications given in the discrete-time frequency domain.

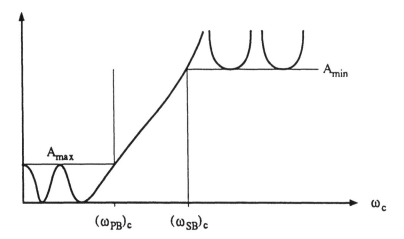

FIG. 83   Prewarped low-pass filter specifications given in the continuous-time frequency domain.

a.   Apply the bilinear transform to $T(s)$, namely,

$$T(s)\Bigg| = H(z) \qquad (169)$$

$$s = \frac{2}{T}\frac{1 - z^{-2}}{1 + z^{2}}$$

where $z = e^{s\tau}$ and $T = 2\tau$.

Having found $H(z)$, it is advisable to use such an SC filter synthesis program as SCSYN (Institute for Signal and Information Processing, ETH, Zurich) (2), which provides an SC filter using the LDI stray-insensitive integrator described in the previous section.

b.   Expand $T(s)$ as in the section, "Direct-Form Design Using Integrators," namely,

$$T(s) = \frac{N(s)}{D(s)} = \frac{b_n s^n + b_{n-1} s^{n-1} + \cdots + b_o}{s^n + a_{n-1} s^{n-1} + \cdots + a_o} \qquad (170)$$

where, without loss of generality, the polynomials $N(s)$ and $D(s)$ are assumed to have the same degree n. Dividing by s^n, we obtain

$$T(s) = \frac{b_n + b_{n-1}\left(\frac{1}{s}\right) + \cdots + b_o \left(\frac{1}{s}\right)^n}{1 + a_{n-1}\left(\frac{1}{s}\right) + \cdots + a_o \left(\frac{1}{s}\right)^n} \qquad (171)$$

As in the above-mentioned section, Eq. (171) can be depicted as an SFG comprising integrators and the coefficients of the numerator in the forward paths, and the coefficients of the denominator as the transmissions of the feedback paths (see Fig. 50). This is the so-called direct form corresponding to an nth-order transfer function. Instead of using RC active integrators (see Fig. 52), stray-insensitive SC integrators (Table 6, Integrators 7 and 8) now are used to obtain the corresponding SC filters.

c. If, in Step 3, an LC ladder structure is found as the solution to the approximation problem, the LC filter also can be converted into an SFG with integrators, inverters, and summers, as shown in Fig. 84. Here again, the integrators are realized by stray-insensitive SC circuits and combined with SC inverters and summing circuits. Special care must be taken for the SC realization of the terminating resistors, the details of which go beyond the scope of this discussion.

5. Having obtained the SFG corresponding to a set of desired filter specifications, it is a simple matter to scale the SFG for maximum dynamic range, minimum capacitance spread, and even minimum total capacitance. After that, it is the task of the IC designers, using such software layout tools as the above-mentioned SCSYN, to transform the SFG into a viable and functioning SC CMOS IC chip.

Example. Consider the band-pass filter specifications shown in Fig. 85-a. From filter tables, the corresponding sixth-order LC ladder filter shown in Fig. 85-b can be obtained. The corner frequencies are the geometric mean of the center frequency f_o. Using the procedure outlined in Step 4c above, the SFG comprising integrators, inverters, and summing points depicted in Fig. 85-c is shown. Using the stray-insensitive SC integrators 7 and 8 in Table 6, the SC filter network shown in Fig. 85-d is obtained. The corresponding amplitude response is shown in Fig. 85-e.

Cascaded Biquad Design

The procedure to be followed when designing SC filters by cascading individual SC biquads (i.e., subnetworks with transfer functions with numerator and denominator polynomials that are, at most, second order) is much the same as that described for active RC biquad design. The difference is that here we start out with a transfer function in the z-domain that can be considered as the Laplace-transformed domain of discrete-time transfer functions. The general nth-order transfer function $H(z)$ will have the form

$$H(z) = K \frac{1 + B_1 z^{-1} + \cdots + B_n z^{-n}}{1 + A_1 z^{-1} + \cdots + A_n z^{-n}} \tag{172}$$

where, without loss of generality, we have assumed that the order of the numerator is the same as that of the denominator, although as in Eq. (170) it, if anything, may be of lower order. $H(z)$ may be obtained by following the first

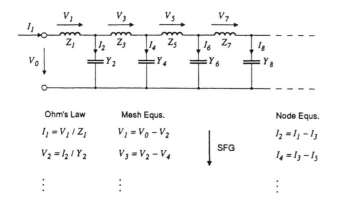

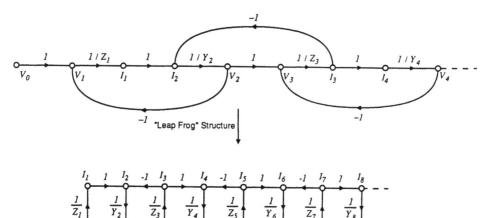

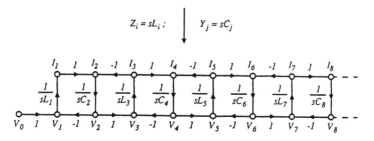

FIG. 84   Signal-flow graph derivation of inductor-capacitor ladder low-pass filter.

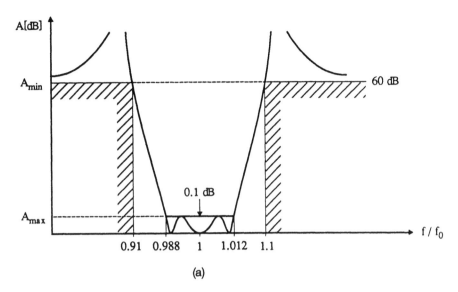

(a)

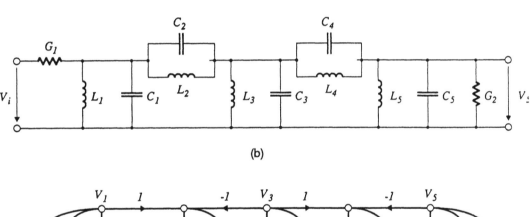

(b)

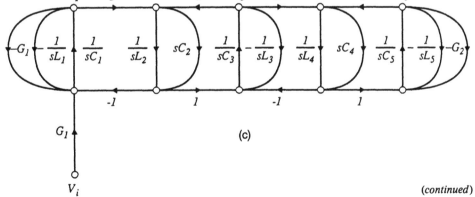

(c)

V_i

(continued)

FIG. 85 Design of a sixth-order band-pass filter: *a*, filter specifications; *b*, corresponding sixth-order inductor-capacitor band-pass filter; *c*, signal-flow graph representation; *d*, switched-capacitor realization; *e*, amplitude response.

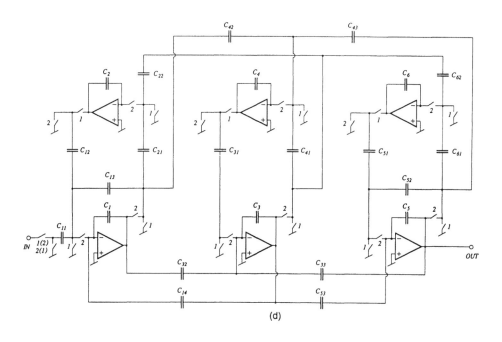

(d)

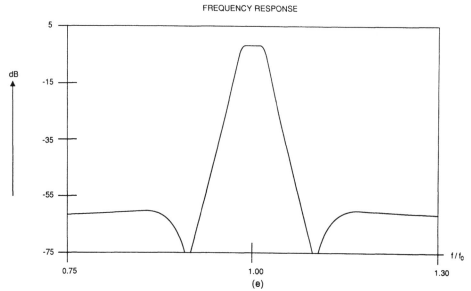

(e)

FIG. 85 Continued.

four steps outlined above, up to Step 4a, in which $H(z)$ is obtained using the bilinear transform. Having $H(z)$ as in Eq. (172), it can be expanded into the product of biquadratic terms (second-order in numerator and denominator) as follows:

$$H(z) = K \frac{(1 - z_1 z^{-1})(1 - z_2 z^{-1}) \cdots (1 - z_n z^{-1})}{(1 - p_1 z^{-1})(1 - p_2 z^{-1}) \cdots (1 - p_n z^{-1})} \qquad (173)$$

where z_i and $p_i (i = 1, 2, \ldots, n)$ are the zeros and poles, respectively, of $H(z)$.

For n even, we can pair the complex conjugate poles and zeros and obtain

$$H(z) = \prod_{i=1}^{n/2} H_i(z) \qquad (174)$$

where

$$H_i(z) = K_i \frac{1 + B_1 z^{-1} + B_2 z^{-2}}{1 + A_1 z^{-1} + A_2 z^{-2}}. \qquad (175)$$

If n is odd, there is an additional real-pole and zero term in the product of Eq. (174). As with active RC cascade design, each of the biquadratic functions $H_i(z)$, as in Eq. (175), is realized by an individual building block, commonly called a *biquad*. The resulting biquads then are cascaded, thereby realizing the nth-order transfer function $H(z)$.

A very useful general-purpose SC biquad circuit is shown in Fig. 86. The transfer function is given by

$$H(z) = - k_6 \frac{1 - z^{-1}[2 - k_5(k_1 + k_2)/k_6] + z^{-2}(1 - k_2 k_5/k_6)}{1 - z^{-1}[2 - k_5(k_3 + k_4)] + z^{-2}(1 - k_3 k_5)}. \qquad (176)$$

The k_i refer to the capacitor values given in Fig. 86.

Multiplying numerator and denominator of Eq. (175) by z^2, the general biquad can be factored as follows (where, for convenience, we drop the subscript i):

$$\begin{aligned}
H(z) &= K \frac{(z - z_1)(z - z_2)}{(z - p_1)(z - p_2)} \\
&= K \frac{z^2 - (z_1 + z_2)z + z_1 z_2}{z^2 - (p_1 + p_2)z + p_1 + p_2} \\
&= K \frac{z^2 - 2r_z \cos \theta_z + r_z^2}{z^2 - 2r_p \cos \theta_p + r_p^2}
\end{aligned} \qquad (177)$$

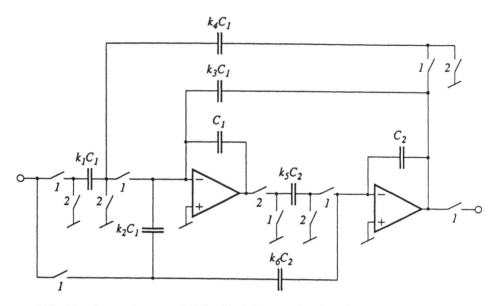

FIG. 86   A general-purpose building block for cascade switched-capacitor filter design.

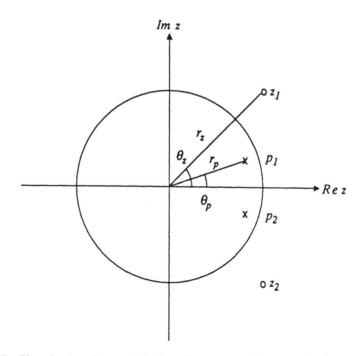

FIG. 87   Biquad pole and zero pair in the z-plane expressed in terms of polar coordinates.

The last term in Eq. (177) expresses $H(z)$ in terms of the polar coordinates as shown in Fig. 87. Comparing the coefficients with those of Eq. (176), we find that

$$1 - k_3 k_5 = r_p^2 \tag{178}$$

and

$$k_4 k_5 = 1 + r_p^2 - 2r_p \cos \theta_p \tag{179}$$

$$H(z) = -k_6 \frac{1 - z^{-1}(2 - k_5(k_1 + k_2)/k_6) + z^{-2}(1 - k_2 k_5/k_6)}{1 - z^{-1}(2 - k_5(k_3 + k_4)) + z^{-2}(1 - k_3 k_5)}$$

•Notch Filter:

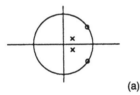

$k_2 = 0$

$$\overset{r_z^2}{\underset{\parallel}{}}$$

$$H(z) = \frac{-k_6 [z^2 - z(2 - k_1 k_5/k_6) + 1]}{z^2 - z(2 - k_5(k_3 + k_4)) + 1 - k_3 k_5}$$

(a)

•Low-pass Filter:

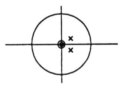

$k_2 = k_6 = 0$

$$H(z) = \frac{k_1 k_5 z}{z^2 - z(2 - k_5(k_3 + k_4)) + 1 - k_3 k_5}$$

(b)

•High-pass Filter:

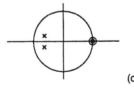

$k_1 = k_2 = 0$

$$H(z) = \frac{-k_6 [\overset{(z-1)^2}{\overbrace{z^2 - 2z + 1}}]}{z^2 - z(2 - k_5(k_3 + k_4)) + 1 - k_3 k_5}$$

(c)

•Band-pass Filter:

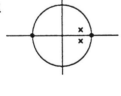

$k_1 = 0 \qquad k_5 k_2/k_6 = 2$

$$H(z) = \frac{-k_6 [\overset{(z-1)(z+1)}{\overbrace{z^2 - 1}}]}{z^2 - z(2 - k_5(k_3 + k_4)) + 1 - k_3 k_5}$$

(d)

FIG. 88 Obtaining second-order filter functions with the biquad in Fig. 86: *a*, notch filter; *b*, low-pass filter; *c*, high-pass filter; *d*, band-pass filter.

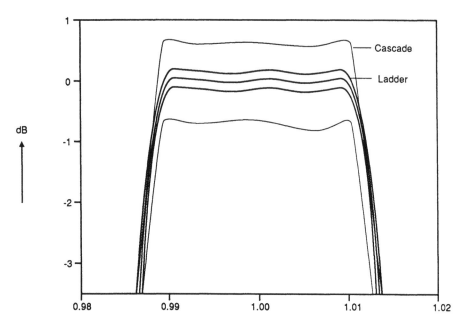

FIG. 89 Sensitivity performance to 1% tolerance of capacitor ratios of sixth-order band-pass filter: a, realized as a simulated inductor-capacitor ladder; b, realized as a cascade of three biquads.

For a high-Q biquad, that is, the poles of $H(z)$ are inside (for stability) but close to the unit circle, r_p^2 is almost (but never equal to) unity. Letting $r_p^2 = 1 - \epsilon = 1 - k_3 k_5$ where $\epsilon \ll 1$, we see that we can obtain high-Q poles without the capacitor spread of capacitors k_3 and k_5 becoming excessively large. Letting $k_3 = k_5 = \sqrt{\epsilon}$ and assuming, for example, that $r_p^2 = 0.99$, we obtain $k_3 k_5 = 0.01$, but $k_3 = k_5 = 0.1$. Thus, this biquad is particularly useful for high-Q pole applications.

The general-purpose biquad of Fig. 86 can provide most of the important second-order filter functions simply by omitting certain capacitors or ensuring certain simple relations. This is shown in Fig. 88. The only second-order function that cannot be realized is that of an all-pass network. This can be seen readily by comparing Eq. (176) with the third expression in Eq. (177). In an all-pass network, it can be shown that in the z-plane $r_z = 1/r_p$. Since r_p must be less than unity (i.e., must lie within the unit circle), it follows that r_z must be larger than unity (i.e., the zeros must lie outside the unit circle). Since $r_z = 1 - k_2 k_5 / k_6$, it is clear that $r_z \leq 1$ and only notch (zeros on the unit circle) but not all-pass biquads can be realized. Fortunately, there are other biquad configurations (that are not as advantageous for high-Q poles) with which all-pass biquads can be obtained readily.

As already pointed out in our discussion on active RC filter networks, the ladder structure is inherently less sensitive to component tolerances than the corresponding biquad cascade. Assuming a 1% tolerance of the capacitor ratios, and comparing a sixth-order band-pass filter designed by cascaded biquads and by an equivalent simulated LC ladder network (see Fig. 85-*d*), we obtain the amplitude tolerances shown in Fig. 89. The superior performance, with respect to component tolerances, of the ladder network is well apparent. Nevertheless, due to the ease with which biquads can be designed — and realized — they commonly are used in systems in which the filter specifications are not excessively severe. For high-precision filters requiring low amplitude tolerance in the pass band, the ladder configuration invariably will provide the better performance. Note that all-pass networks cannot be realized by simulated LC ladder circuits; they must be designed by cascaded biquads. This is hardly a limitation, though, because the pole Qs of commonly used all-pass networks are generally relatively low.

Bibliography

General Filter Theory and Design

Baher, H., *Synthesis of Electrical Networks*, John Wiley and Sons, Chichester, Sussex, England, 1984.

Blinchikoff, H. J., and Zverev, A. I., *Filtering in the Time and Frequency Domains*, John Wiley and Sons, New York, 1976.

Daniels, R. W., *Approximation Methods for Electrical Filter Design*, McGraw-Hill, New York, 1974.

Hasler, M., Neirynck, J., *Electric Filters*, Artech House, Dedham, MA, USA, 1986.

Humpherys, De Verl S., *The Analysis, Design, and Synthesis of Electrical Filters*, Prentice-Hall, Englewood Cliffs, NJ, 1970.

Johnson, D. E., *Introduction to Filter Theory*, Prentice-Hall, Englewood Cliffs, NJ, 1976.

Rhodes, J. D., *Theory of Electrical Filters*, John Wiley and Sons, London, 1976.

Temes, G. C., and Mitra, S. K., *Modern Filter Theory and Design*, John Wiley and Sons, New York, 1973.

Filter Tables

Biey, M., and Premoli, A., *Tables for Active Filter Design (Based on Cauer and MCPER Functions)*, Georgi Publishing, St. Saphorin, Switzerland, 1980.

Christian, E., and Eisenmann, E., *Filter Design Tables and Graphs*, John Wiley and Sons, New York, 1966.

Hansell, G. E., *Filter Design and Evaluation*, Van Nostrand Reinhold, New York, 1969.

Moschytz, G. S., and Horn, P., *Active Filter Design Handbook: For Use with Programmable Pocket Calculators and Minicomputers*, John Wiley and Sons, Chichester, Sussex, England, 1981.

Saal, R., *Handbook of Filter Design*, Allgemeine Elektrizitäts-Gesellschaft AEG-Telefunken, Berlin, 1979.

Skwirzynski, J. K., *Design Theory and Data for Electrical Filters*, Van Nostrand Reinhold, New York, 1965.

Zverev, A. I., *Handbook of Filter Synthesis*, John Wiley and Sons, New York, 1967.

Electromechanical Filters

Johnson, R., *Mechanical Filters in Electronics*, John Wiley and Sons, New York, 1983.

Mitra, S. K., and Kurth, C. F., *Miniaturized and Integrated Filters*, John Wiley and Sons, New York, 1989.

Sheahan, D., and Johnson, R. (eds.), *Modern Crystal and Mechanical Filters*, IEEE Press, New York, 1977.

Temes, G. C., and Mitra, S. K., *Modern Filter Theory and Design*, John Wiley and Sons, New York, 1973.

Active Resistor-Capacitor Filters

Bruton, L. T., *RC Active Circuits, Theory and Design*, Prentice-Hall, Englewood Cliffs, NJ, 1980.

Heinlein, W. E., and Holmes, W. H., *Active Filters for Integrated Circuits*, Springer-Verlag, New York, 1974.

Herpy, M., and Berka, J. C., *Active RC Filter Design*, Elsevier, Amsterdam, 1986.

Huelsman, L. P. (ed.), *Active RC Filters: Theory and Application*, Benchmark Papers in Electrial Engineering and Computer Science, Vol. 15, Dowden, Hutchinson and Ross, Stroudsburg, PA, 1976.

Johnson, D. E., and Hilburn, J. L., *Rapid Practical Designs of Active Filters*, John Wiley and Sons, Long Beach, CA, 1975.

Lam, H. Y.-F., *Analog and Digital Filters: Design and Realization*, Prentice-Hall, Englewood Cliffs, NJ, 1979.

Lindquist, C. S., *Active Network Design with Signal Filtering Applications*, Stewart and Sons, Long Beach, CA, 1977.

Mitra, S. K., *Analysis and Synthesis of Linear Active Networks*, John Wiley and Sons, New York, 1969.

Mitra, S. K., and Kurth, C. F. *Miniaturized and Integrated Filters*, John Wiley and Sons, New York, 1989.

Moschytz, G. S., *Linear Integrated Networks: Design*, Van Nostrand Reinhold, New York, 1975.

Moschytz, G. S., *Linear Integrated Networks: Fundamentals*, Van Nostrand Reinhold, New York, 1974.

Moschytz, G. S., and Horn, P., *Active Filter Design Handbook: For Use with Programmable Pocket Calculators and Minicomputers*, John Wiley and Sons, Chichester, Sussex, England, 1981.

Sedra, A. S., and Brackett, P. O., *Filter Theory and Design: Active and Passive*, Matrix Publishers, IL, 1978.

Williams, A. B., *Active Filter Design*, Artech House, Dedham, MA, 1975.

Discrete-Time and Digital Filters

Antoniou, A., *Digital Filters: Analysis and Design*, McGraw-Hill, New York, 1979.

Hamming, R. W., *Digital Filters*, Prentice-Hall, Englewood Cliffs, NJ, 1988.

Jackson, L. B., *Digital Filters and Signal Processing*, Kluwer Academic Publishers, Hingham, MA, 1986.

Ludeman, L. C., *Fundamentals of Digital Signal Processing*, Harper and Row, New York, 1986.

Oppenheim, A. V., and Schafer, R. W., *Digital Signal Processing*, Prentice-Hall, Englewood Cliffs, NJ, 1975.

Oppenheim, A. W., and Schafer, R. W., *Discrete-Time Signal Processing*, Prentice-Hall, Englewood Cliffs, NJ, 1989.

Parks, T. W., and Burrus, C. S., *Digital Filter Design*, John Wiley and Sons, New York, 1987.

Peled, A., and Liu, B., *Digital Signal Processing: Theory, Design and Implementation*, John Wiley and Sons, New York, 1976.

Proakis, J. G., and Manolakis, D. G., *Introduction to Digital Signal Processing*, Macmillan, New York, 1988.

Rabiner, L. R., and Gold, B., *Theory and Applications of Digital Signal Processing*, Prentice-Hall, Englewood Cliffs, NJ, 1975.

Roberts, R. A., and Mullis, C. T., *Digital Signal Processing*, Addison-Wesley, Reading, MA, 1987.

Strum, R. D., and Kirk, D. E., *First Principles of Discrete Systems and Digital Signal Processing*, Addison-Wesley, Reading, MA, 1988.

Tretter, S. A., *Introduction to Discrete-Time Signal Processing*, John Wiley and Sons, New York, 1976.

Switched-Capacitor Filters

Allen, P. E., and Sánchez-Sinencio, E., *Switched-Capacitor Circuits*, Van Nostrand Reinhold, New York, 1984.

Gregorian, R., and Temes, G. C., *Analog MOS Integrated Circuits for Signal Processing*, John Wiley and Sons, 1986.

Mitra, S. K., and Kurth, C. F., *Miniaturized and Integrated Filters*, John Wiley and Sons, New York, 1989.

Moschytz, G. S., *MOS Switched-Capacitor Filters: Analysis and Design*, IEEE Press, New York, 1984.

Unbehauen, R., *MOS Switched-Capacitor and Continuous-Time Integrated Circuits and Systems*, Springer-Verlag, Berlin, 1989.

References

1. Orchard, H. J., Inductorless Filters, *Electronics Letters*, 2(6):224–225.
2. Muralt, A., Zbinden, P., and Moschytz, G. S., CAD Tools for the Synthesis and Layout of SC Filters and Networks, *Analog Integrated Circuits and Signal Processing*, 3 (May 1993).

GEORGE S. MOSCHYTZ

Milton Keynes UK
Ingram Content Group UK Ltd.
UKHW052026071024
449327UK00027B/2448